American Society for Engineering Education International Forum 2016

New Orleans, Louisiana, USA
25 June 2016

ISBN: 978-1-5108-3579-5

Printed from e-media with permission by:

Curran Associates, Inc.
57 Morehouse Lane
Red Hook, NY 12571

Some format issues inherent in the e-media version may also appear in this print version.

Printed by Curran Associates, Inc. (2017)

For permission requests, please contact American Society for Engineering Education at the address below.

American Society for Engineering Education
1818 N Street, N.W., Suite 600
Washington, DC 20036-2479
USA

Phone: (202) 331-3500
Fax: (202) 265-8504

www.asee.org

Additional copies of this publication are available from:

Curran Associates, Inc.
57 Morehouse Lane
Red Hook, NY 12571 USA
Phone: 845-758-0400
Fax: 845-758-2633
Email: curran@proceedings.com
Web: www.proceedings.com

TABLE OF CONTENTS

S201A – CONCURRETN PAPER TRACKS SESSION I ACCREDIDATION

S201B – CONCURRENT PAPER TRACKS SESSION I STUDY ABROAD PROGRAMS

S201C – CONCURRENT PAPER TRACKS SESSION I CURRICULUM

S501B – CONCURRENT PAPER TRACKS SESSION II SKILLS DEVELOPMENT

S501C – CONCURRENT PAPER TRACKS SESSION II COURSES

ABET Accreditation of International Technology Programs - A Team Chair's Perspective

Dr. Amitabha Bandyopadhyay, State University of New York, Farmingdale

Dr Bandyopadhyay is a SUNY Distinguished Service Professor of Architecture and Construction Management Department at SUNY-Farmingdale State College. He is also the Director of Green Building Institute at the college. He was (2012-13) the Chair of Engineering Technology Accreditation Commission of ABET. He is a Fellow of American Society of Civil Engineers.

ABET Accreditation of International Technology Programs - A Team Chair's Perspective

Introduction

ABET has taken an active role in the global quality assurance process for technical education programs through numerous agreements with organizations worldwide[1]. These include Mutual Recognition Agreements and Memoranda of Understanding. ABET conducts an accreditation review outside the U.S. only with explicit permission from all applicable national education authorities in that program's country or region. In order to seek accreditation by ABET, programs outside of U.S. must have each appropriate education authority, recognition, or accreditation agency complete a request for approval form to be submitted with formal request for evaluation. The ABET accreditation process for programs located outside of the United States is identical to the accreditation process for programs within the U.S. ABET requires a program seeking accreditation in a country where English is not the native language must provide its program name both in English and in the native language. It also requires the English translation of the program name should be based on the technical content and not the literal translation of the program name so that ABET can designate the appropriate accreditation commission and determine the applicable criteria. ABET has started accrediting international technology programs through Engineering Technology Accreditation Commission (ETAC – Technology Accreditation Commission TAC before 2012) about eight years ago.

Background

The author had the privilege to chair teams of evaluators to four international institutions seeking accreditations for their technology programs. All of the institutions, except one, had multiple programs. One institution had only baccalaureate level programs, one institution had only associate level program, and rest of two institutions had both baccalaureate and associate level programs. Medium of instructions in all the programs, except one in South America, were in English.

Geographical and Program Diversity

A great diversity existed in the type of programs, number of programs and geographical location of the institutions. Programs ranged from manufacturing to aeronautics to chemical engineering technology. Geographically, one institution was in South America, two institutions in two different countries in Middle East, and one in South East Asia. One institution had one program and another institution had nineteen programs in five locations.

Typical Institutional Profile

This South American institution was established in 1984 as a non-profit private educational institution. The mission of the institution is to develop people and organizations through education, training, and consulting. It offers professional programs related to the application of technology in design, supervision, operation, and maintenance of industrial sectors. It also offers continuing education programs for engineers and technicians as well as consulting and research services for industry. The institution is recognized by the Ministry of Education of -- as an institution of higher education (Educación Superior), and authorized to offer programs of study leading to the Professional Technologist title.

This Middle Eastern industrial college is a state-owned institution under a Royal Commission. It was established in 1989 and was one of the first colleges of its kind in the region. The current enrollment of the college is around 3600 students. It is located in a planned industrial town. The college's mission is to prepare young citizens for technical and managerial positions for the local industries and for other private and public sector employers throughout the country.

This south east Asian polytechnic is a statutory board under the Ministry of Education, which was set up in 1992 to provide broad-based training for students inclined towards practice-oriented studies at tertiary level. This polytechnic is responsible to its own governing board called the Board of Governors. The Principal & CEO (PCEO) leads the Senior Management team which comprises Deputy Principals (DyPs), Directors, and Deputy Directors of Schools and Corporate Departments. The team is responsible for formulating, reviewing and implementing strategies, policies, systems and processes to drive polytechnic's mission to achieve its vision.

Typical Program Profile

The Diploma in Aeronautical and Aerospace Technology in this south eastern polytechnic was launched academic year 2009/2010 by the School of Engineering in a response for a need for skilled technical personal to support the growing aviation industry in Singapore. The Diploma involves a three year course of study and is conducted during the day on campus. In the first and second years, the students take courses in basic engineering and general studies and in the third year they take more aerospace specific courses. The program is organized in a matrix format in which the Head of the Directorate which has primary responsibility for the program reports to the Director of the School of Engineering. Appropriate faculty members from the Directorate which supports the program and three other related Directorates. There are 14 core faculty and 87 students in the program. The first graduating class was in Spring 2013 and was comprised of 48 graduates.

The Plant Machinery Maintenance Technology program in this South American institute offers a curriculum in the field of mechanical engineering technology. The graduates of the program are prepared to work in a mechanical design, maintenance management, or project management within industry. The program educational objectives are as follows:

- Graduates analyze, design, implement and supervise modern mechanical systems; as well as manage maintenance of industrial plants.
- Graduates identify problems and opportunities for improvement, and they implement solutions applying modern technologies and appropriate procedures.
- Graduates manage resources and work with effectiveness, initiative, creativity, and within teams.
- Graduates are committed with lifelong learning, quality and safety.

- Graduates follow ethical principles and they contribute to the growth of the community.

Typical employers are companies that manufacture steel, mining companies, dairy companies, beverage production, oil and gas service companies, and the airlines. Some of companies hiring graduates are LAN, Gloria S.A., ABB S.A., SN/Power, and Maestranza Diesel SAC.

The baccalaureate program in chemical engineering technology in this middle eastern industrial college is structured for a five-year duration consisting of a preparatory year, two specialization years for the associate degree program followed by another two years for the BS degree program. The number of graduates has increased from 26 to 58 over the last four years. Program graduates have taken up employment as chemical engineering technologists in process industries, including petroleum (ARAMCO, SAMREF, YASREF), petrochemical (SABIC Group, PETRORABIGH, NATPET), environmental protection (RCEPC), desalination units (MARAFIQ), and wastewater treatment plants (MARAFIQ).

The program educational objectives are that graduates will:

- possess the technical knowledge and skills needed for employment in jobs related to laboratory applications, analysis, design, control, technical support and supervision of chemical processes;
- have the understanding and commitment to protect the public interest, health, safety and the environment, and to maintain the highest ethical and professional standards towards employers and community during their professional career;
- recognize the need and engage in life-long learning activities through the pursuit of further studies, on-job training and certification; and
- reach professional success through working and communicating effectively within multidisciplinary team, solving real-world problems, and assuming leadership roles with integrity and high responsibility in their organizations.

Common Theme of Excellence

All the institutions had some common approaches to accreditations. These are in spite of wide differences in educational and cultural backgrounds of the students and faculty. Even between the two institutions in the Middle East local cultural differences were prominent. First, technical core courses, mathematics and sciences were stressed irrespective of countries and degree level. All of the programs exceeded minimum accreditation requirements and stronger than most of the US domestic programs the author had reviewed (the author had the opportunity to review about twenty five institutions over the last eighteen years). All of them had very strong laboratory components. Most of the programs had excellent laboratory and physical plant facility. Overall faculty members were highly qualified and significant portion of the faculty members were trained in western countries. A number of faculty members were hired for technology programs because of their achievement in industry. Most of the programs provided adequate instructions and facility for public speaking and presentation skills. Al most all the programs that offered English only instruction had one year of intensive English language training prior to entry in to the formal technology programs. One exception was where primary and secondary education was given in English. Some of the time technical faculty members were used for such instruction.

All of the programs had prepared excellent self-studies. The self- studies were in great details, well organized, and followed ABET suggested outlines. Often they included charts, tables, graphs, and flow charts. Most of the assessment and evaluation plans were acceptable. All of the programs had excellent industry support. Industrial advisory committee members were actively involved with the faculty members and students. All of the programs had excellent placement records. Employer representatives were very happy with the programs and the graduates were valued by the employers. Many of the programs were already recognized or approved or accredited by various European professional organizations. Many programs had collaborations with one or more Western or United Nations' organizations.

Common Areas of Improvement Needed

Most of the programs offered little flexibility of courses in technical areas. Most of the programs were lacking serious instructions in social sciences and humanities at the college level. Awareness of world current events, except for narrow regional knowledge, and appreciation for diversity were often lacking among the graduating seniors. It appeared that the programs often scarified critical and free thinking to train graduates in deep technical knowledge. Some used course grades for evidence of achievements of student outcomes. Like many US domestic programs, real purpose of assessments were not clear to many of the programs. Most of the programs had collected more than needed data. However, many faculty members were not involved in planning and execution of assessment process and were unaware of significance of assessment. Thus they thought much of the assessments exercise onerous and unnecessary. Because of these, continuous improvement process was often incomplete and neglected

Comparison with U.S. Domestic Programs

As for the technical competency of the graduates, the international programs were either at par or above any comparable U. S. domestic programs. Students appeared to be better prepared when they started technology programs than most U.S. students entering technology programs. They appeared to be more serious and focused students. Very few students transferred from a different major as more students change major in U.S. However, graduates from the international programs seem to know less about other cultures and country. Faculty members overall had more experience in industry, but many lacked terminal degree in technical field.

Conclusion

With overall superior programs and high graduate placement, author and other members of the team wondered why these international programs seeking ABET accreditation. Most common response received was "prestige" to be accredited and affiliated with ABET. International recognition and mobility of the graduates were cited other reasons.

Bibliography

1. Global presence (ABET) www.ABET.org
2. Criteria for Accrediting Engineering Technology Programs www.ABET.org
3. Accreditation Policy and Procedure Manual www.ABET.org

Professional Accreditation of Engineering Programmes and EUR-ACE labels in Central Asia

Prof. Sergey Gerasimov, Siberian Transport University

Professor Sergey Gerasimov (Director of the Accreditation Centre of the Association for Engineering Education of Russia) Date of birth: March 18, 1955 Education: 1972-1977 - Novosibirsk Institute of Electrical Engineering (Faculty of Technical Physics) Degrees: 1977 - Diploma in Technical Physics (M.Sc.) 1985 - Ph.D. (Structural mechanics), Siberian Transport University (STU), Novosibirsk 1998 – Sc. D. (Structural mechanics), STU, Novosibirsk Previous positions: 2005 – present time – professor of the Department of Structural Mechanics of STU 2003 – 2005 - vice-rector for academic affairs 2001 – 2003 - deputy rector for international affairs 1983-2001 associate professor, professor of the Department of Structural Mechanics 1977-1983 - engineer, junior researcher of the Department of Structural Mechanics Teaching experience: general physics, theoretical mechanics, structural mechanics. Research experience: experimental mechanics, optical methods in mechanics of solids, FEM, engineering education; Managing experience: Siberian Transport University managing. Professional development: • Advanced training in English language (Novosibirsk, STU, 1983-1985, aspirantura), • Expert on the Association for Engineering Education of Russia (Moscow) since 2003, • Program evaluation: Ankara, Turkey, 2005; Newcastle, United Kingdom, 2006; Seoul, Korea, 2008; Lausanne, Switzerland, 2011; Porto, Portugal, 2013; Neuchatel, Switzerland, 2013. • International Workshops: ABET Moscow, Russia, 2006, University La-Sapienza, Rome, Italy, 2007, ASIIN, Bonn, Germany, 2009, Swedish National Agency for Higher Education (HSV), Stockholm, Sweden, 2009, • International Engineering Alliance Meetings: Taipei, Taiwan, 2011; Sydney, Australia, 2012. Membership in professional organizations: • Member of the SPIE since 1991 • Member of the SEFI since 2000 • Member of AEER Accreditation Board since 2006 • Member of EUR-ACE Label Committee of ENAEE since 2012

Prof. Yury P Pokholkov, Tomsk polytechnic university, Association for Engineering Education of Russia
Mrs. Kseniya K Tolkacheva, Association for Engineering Education of Russia, Tomsk Polytechnic University

Member of the Association for Engineering Education of Russia responsible for AEER international cooperation activities. PhD in Pedagogical Sciences, graduate degree in "Mathematical Methods in Economics". Also holds a Management degree. Her research focus is in Quality Assurance, active learning and international academic mobility. Has experience in coordinating several TEMPUS projects as well as organizing joint international partnerships in educational and research areas, including national and international projects, organizing conferences, seminars, workshops and training courses.

Prof. Jose Carlos Quadrado Mr P.E., Instituto Superior de Engenharia do Porto

José Carlos Quadrado is a full professor with tenure of the Instituto Superior de Engenharia de Lisboa (ISEL) and is currently under external management assignment in Instituto Superior de Engenharia do Porto (ISEP), Portugal. He is currently the immediate past-President of ISEL and the Vice-President of ISEP. He has a BSc in Energy and Power Systems, a diploma degree in Electrical Engineering, Automation and Industrial Electronics from ISEL, a MSc and a Doctor degree in Electrical Engineering and Computers from Lisbon Technical University. He also holds the Habilitation degree (Aggregation) in Electrical Engineering from Beira Interior University.

Holds the position of immediate past-President of the International Federation of Engineering Education Societies (IFEES) and the position of past president of the Ibero-American Engineering Education Association (ASIBEI). Currently he is a President-Elect of the Latin American and CaribbeanConsortium of Engineering Institutions (LACCEI) and a member of the board of the European Society for Engineering Education (SEFI). He is also a past-Vice President of SEFI. He leads the Portuguese Observatory on European and Latin-American University management strategy best practices (TELESCOPI Portugal) and is the past-President of the Portuguese association of engineering rectors and deans (APDI2E).

He was a member of the general assembly of the European Network for Accreditation of Engineering Education (ENAEE), and has been actively supporting the development of accreditation agencies in Central Asia and Latin America. Being a member and senior member of several engineering societies and engineering education societies in several continents, he is also a visiting professor in many universities around the world and board member of several technological societies.

He holds over 100 international publications (including journals and chapters of books), several patents and some international technical prizes and scholarships, and also held the position of editor and editor-in-chief in some journals. Up to now he has also developed several international engineering projects in the fields of renewable energy, fuel cells, electrical vehicles, intelligent control and engineering education.

Professional Accreditation of Engineering Programmes and EUR-ACE labels in Central Asia

Today the process of accreditation of educational programmes is widely recognized as one the most efficient ways to improve the quality of education and as an entry route to the engineering profession. It is crucial to develop system of continuous programmes improvement at universities through an independent external accreditation of educational programmes by the agencies representing national and international professional community.

External evaluation process leads to several positive impacts. It helps to increase credibility to the programme from different stakeholders. International recognition gained upon successful accreditation of the educational programme facilitates its competitiveness and accountability and leads to real quality improvement of the whole educational programme[1].

The first steps in setting up and implementing a system of Quality Assurance of engineering education in Central Asia were made within the TEMPUS project QUEECA (Quality of Engineering Education in Central Asia [2]) started in 2012. The QUEECA TEMPUS project main aim is to set up and to implement a system of quality assurance (QA) of engineering education in Central Asia (CA), finalised to the accreditation of engineering programmes by the award of the EUR-ACE quality label on the basis of the EUR-ACE Framework Standards and related quality requirements and procedures.

The introduction of easy comparable practices for the accreditation of programmes in the engineering/technology field is hence the main change at national level the QUEECA TEMPUS project is aiming at. The self-sustainability of this strategy is being assured thanks to a massive involvement of relevant actors in all consortium members' countries. Partner countries' Ministries are actively involved in the project in order to comply with legislation obligations as far as Higher Education (HE) system changes are concerned.

The involvement of academics and students at large scale is also being ensured through the active participation of ENAEE and SEFI associations (the main actors in the field of engineering education with a direct involvement in the accreditation issues).

Training a pool of experts in QA issues became another significant outcome of the QUEECA project. A series of training workshops for experts in preparing programs for professional accreditation was hold in 4 CA countries: Kazakhstan, Kyrgyzstan, Tajikistan and Uzbekistan. Training of experts was recognised a crucial step to achieve the main project goal, thus quality of external independent assessment of educational programs vastly depends on the following issues:

- Selection of experts;
- Training of experts;
- Motivation of experts.

Success of external independent assessment substantially depends on correct selection of experts – future participants of educational programs' assessments. Potential experts shall tentatively comply with the following requirements:

- Show interest in engineering education improvement,
- Have higher education management experience and recognition in a certain field of activity,
- Have an academic degree in a corresponding field,
- Be able to use the Internet, e-mail, word processing programs (Microsoft Word) and PDF files.

When talking about a specific field of activity it shall be noted that experts have to deal with various educational programmes in the field of engineering and technology [3].

Experts may be selected from representatives of various Higher Education Institutions (HEIs) and HEIs structures and they may have various statuses, functional abilities and work experience in their respective HEIs. Selection depends on the type of expert information we would like to obtain from future member of expert group during an accreditation visit.

According to HIE status potential experts may form the following groups:

- Administrators – HEI specialists who are well aware of organization of administrative and financial-economic aspects of HEI activities;
- Methodologists – specialists with methodological work experience in the field of creation, implementation and development of educational programs. Such candidates, obviously, shall possess methodological perspective that exceeds the limitations of their own HEI, having knowledge of methodological rules, regulations and materials at the federal level and aware of trends and approaches to methodological work implemented by international academic society;
- Specialists – teacher with a comprehensive experience of teaching and knowledge in the fields relevant to educational programs presented for accreditation.

Any expert group is supposed to be a combination of representatives from three groups mentioned above. Besides, as practice shows, quality of experts group's work increases when it consists of experts of different ages, temperaments, occupational statuses, production and life experience.

Let us consider major principles applied to selection of experts and limitations that shall be noted during selection of potential experts.

First of all, an expert shall not be a person to provide final opinion on the basis of received information even if he/she is a recognized specialist who can contribute a lot to educational program assessment. An expert may only be an equal member of expert group with the right to express his/her opinion (even if this opinion is a dissenting one). Otherwise it may lead to a situation when opinions of other members of experts group will depend on opinion of this specialist and an assessment outcome will be provided on the subjective basis.

Secondly, not every specialist even the most recognized in his field can become a member of an expert group. Some of them, even those with comprehensive knowledge, do not have qualities required for a member of an expert group and are not able to provide adequate opinion which may disrupt assessment especially during meetings with representatives of students and teachers participating in accredited program.

That is why the first criterion for experts' selection is the degree of their competence. There is of course an issue with evaluation of such competence. Usually competence is evaluated on the basis of obvious factors such academic degree and title and length of work in HEI. However, a more thorough selection of potential experts may be carried out with account of their occupational status, number of published works in the subject's field, public recognition of such potential expert as a professional in the field.

Second criterion is the degree of candidate's actual knowledge of the latest scientific and practical achievements in relevant fields, in methodology and management of development and implementation of engineering education programmes. It is also very important to carry out a qualitative analysis of scientific-methodological and practical activity of an expert over the last years.

The third criterion is a reasonable of candidate's niche expertise and his/her general knowledge. Evaluation of candidates in this area required assessment of their methodological and scientific works.

Finally, the fourth criterion is an optimal combination of candidates' individual qualities. This shall include ability to work in a team, tolerance to opinions that differ from their own, ability to provide comprehensive and objective evaluation of problems without being too optimistic or too pessimistic, patience, friendliness and tactfulness.

Expert commission shall consist of specialists in the field of HEI administrative-methodological activities and in specific subject areas that include engineering education programmes. Thus, it is logically assumed that opinions of specialists shall be taken into account mostly in the sphere that is related to evaluated criterion. For example, in order to provide a relevant opinion in human resources part of the programme experts shall possess work experience in a teaching group, knowledge of HEI fundamentals of department's activities financing, etc. At the same time opinion of other experts group members could not be underestimated even if such experts are not highly competent in this sphere.

Any aspect of educational programme implementation could be considered separately but only in close connection with all other elements.
In some cases selection of potential experts may be substituted by their appointment by management. Technically it is a simple way of selection in the form of the head's order of participation in experts' training. However, it is obvious that this practice defies a principle of voluntary participation of experts in public-professional accreditation of educational programmes. Even though cost of experts' selection is minimal in this case, there is a strong possibility that this experts' group will consist of people loyal to management and their future opinions may become similar to a private opinion of management. As a result opinions of experts' group in the process of educational programme assessment may shift to opinions of specialists who are not conflict prone which may also lead to open critics of HEI administration and educational programme management. Another extreme possibility of such appointment is selection of specialists with increased social activity but not always competent which is not beneficial for the process either.
As it was noted earlier it is important that potential experts voluntarily participate in further training and expert visits to HEIs. This voluntary participation and interest of experts are important components of their successful expert activity in future.

Previous paragraph described qualities and characteristics considered during selection of potential experts and evaluation of experts' activity at the stage of their familiarization with educational programme self-assessment materials and during accreditation visit to HEI. Now let us see how experts are trained as it is, obviously, not enough to just be a professional in his/her field and have methodological experience to assess EP quality. Potential experts shall have special knowledge of accreditation procedure and criteria. Besides, it is advisable that they shall have some perspective in this field of activity and understand how these processes are implemented in other countries. A sample agenda for initial training of experts is given below in Table 1.

Table 1

Sample training seminar agenda

Day 1

Activity	**Participants**	**Start time**	**Duration**
Professional accreditation of educational programmes: system development	All	10:00	1:30
Coffee-break	All	11:30	0:15
Criteria and procedure for accreditation of educational programmes	All	11:45	1:30
Lunch	All	13:15	1:00
Self-study procedure (university and educational programmes)	All	14:15	1:30
Coffee-break	All	15:45	0:15
Audit of educational programmes of the university	All	16:00	1:30
Distribution and comments on handout and reference materials	All	17:30	

Day 2

Exercise 1 (Objectives of the educational programme)	By groups	09:00	Step 1 .....0:20 Step 2 .....0:20
Exercise 2 (Learning outcomes)	By groups	09:40	Step 1 .....0:20 Step 2 .....0:20
Coffee-break	All	10:20	0:15
Discussion of the group work results	All	10:35	0:40
Exercise 3 (Correspondence of educational programme objectives and learning outcomes)	By groups	11:15	0:30
Exercise 4 (Compliance with the AEER requirements to the learning outcomes)	By groups	11:45	0:40
Discussion of the group work results	All	12:25	0:40
Lunch	All	13:05	1:00
Exercise 5 (Correlation between learning outcomes and study disciplines)	By groups	14:05	0:40
Exercise 6 (Assessment tools)	By groups	14:45	0:40
Coffee-break	All	15:25	0:15
Discussion of the group work results	All	15:40	0:40

Day 3

Presentation of homework results	All	10:00	1:25
Coffee Break	All	11:25	0:15
Presentation of homework	All	11:40	1:30
Conclusions. Certificates awarding	All	13:10	

As this sample programme shows training of experts start with studies of a general situation in educational programmes accreditation sphere, its influence on engineering education development in the country, promotion of its quality improvement and assurance of international recognition of the system of EPs quality assessment in general and its specific accredited EPs [4]. Further on experts study particular features of public-professional accreditation of engineering education programmes in various countries. Despite the fact that accreditation procedure and criteria applied to activities of all national accrediting bodies are "essentially equivalent", it is important to note specifics and traditions of countries where these programmes are implemented and assessed. Successful work of expert is practically impossible without comprehensive knowledge in this sphere.

Upon consideration of these basic issues experts are engaged in special training which includes detailed study of criteria applied to accreditation of programmes of the first and second educational cycles. An expert shall be familiar with all criteria requirements and methods of EPs qualitative analysis in accordance with criteria basis, as well as with quantitative requirements to EPs of various cycles (content of EPs in general and per subject in ECTS points, number of programme subjects' teachers with PhD and ScD, etc.)
A substantial part of experts' training is allocated to practical classes. During these classes experts learn to analyze EPs goals and outcomes, their compliance with HEI mission and strategy and employers' requirements, efficiency of mechanisms used for adjustment of EP goals and outcomes.
During the final part of their studies experts become familiar with accreditation visit report forms, order and procedure of their preparation. Moreover, experts study the professional ethics code and requirements related to conflict of interest between HIE and experts.

Training workshops were carried out by the Association for Engineering Education of Russia [5] (ENAEE member) in all four member-countries within the frames of the QUEECA project:

- Tashkent (Uzbekistan) - 25-30.05.2014
- Almaty (Kazakhstan) - 30.06.- 2.07.2014
- Dushanbe (Tajikistan) - 20-22.10.2014
- Bishkek (Kyrgyzstan) - 01-03.03.2015

Number of participants varied from 25 to 30 people.
• Representatives of the academic community - 70% of the total number of participants
Faculty members of higher education institutions (professors, associate professors, heads of research and educational units (institutions departments) aware of content and organizational aspects of the educational process;
• Industry representatives - 10% of the total number of participants
Representatives of business and industry, involved in the process of training engineers able to critically assess the competence (learning outcomes) of students and graduates of the educational programme;
• Representatives of the management and administrative staff of higher education institutions / accreditation agencies - 20% of the total number of participants
Proactive and motivated employees who intend to do the work of organizing and supervising the process of accreditation of educational programs in the field of engineering and technology, fluent in English.

Usually after the seminars participants were given a questionnaire to get their feedback on classes' quality. It included the following questions:

1. Seminar duration
2. Seminar content (themes, sections)
3. Teaching level (method, style, a presentation materials
4. Level of the organisation of a seminar (lecture-room, equipment, distributed materials, etc.

Maximum grade was 10 and minimum grade was 1. Below answers' options are given:

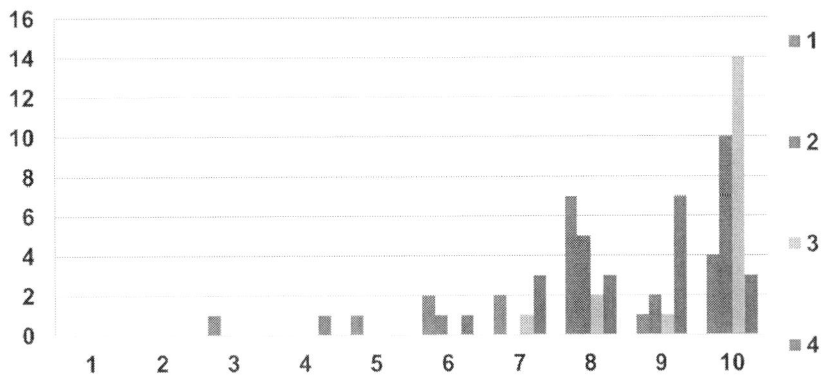

Distribution of experts' grades in Dushanbe

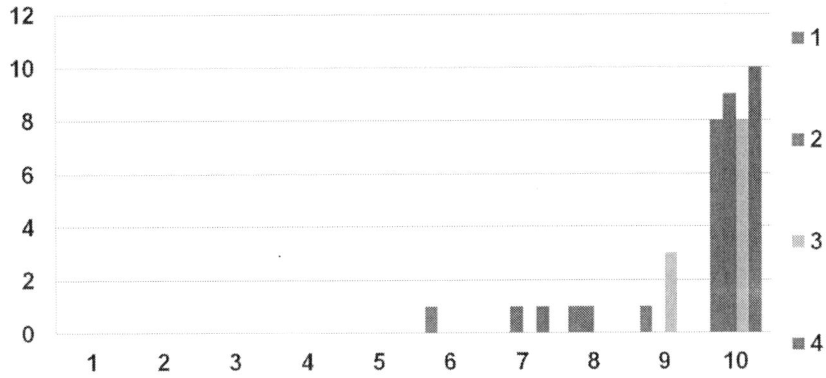

Distribution of experts' grades in Tashkent

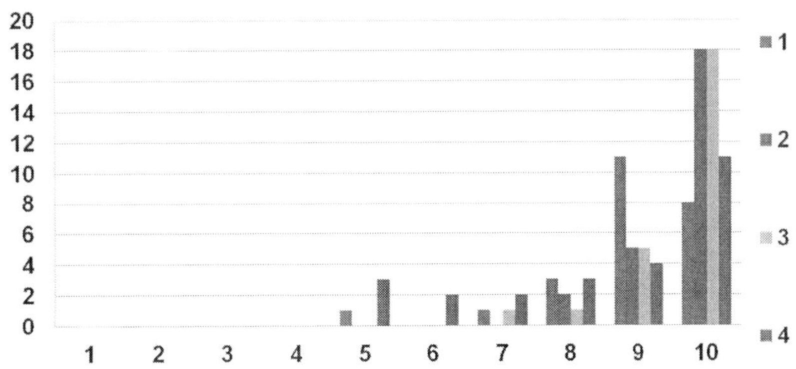

Distribution of experts' grades in Almaty

After training participants become aware of accreditation procedures & criteria and acquire skills needed to prepare educational programmes at their home universities for further international accreditation, including but not limited to EUR-ACE Label. Participants who successfully complete all training tasks are awarded accreditation expert certificate of the Association for Engineering Education of Russia and nominated the title *expert candidate*. Once participated in the on-sight accreditation visit as an audit-team member the candidate is included in the AEER experts' database and becomes *active expert*.

References

[1] Karim J. Nasr, Quality Assurance and Accreditation of Engineering.Programs for a Modern World", Journal IDEAS ("Improving Quality Assurance in Engineering Education"), World Federation of Engineering Organizations, pp. 51-66, Number 17, November 2011.

[2] Borri C., Guberti E., Betti M. (2013). The Quality of Engineering Education in Central Asia. The TEMPUS QUEECA project, Proc. of the 41st SEFI Annual Conference, Engineering Education Fast Forward 1973 > 2013, Leuven, Belgium, 16-20 September 2013.

[3] Gerasimov S.I., Yatkina E.Y. Competence Model for Experts from Accreditation Center of Association of Engineering Education of Russia//Engineering Education. -2011. -№7. -P. 58-67.

[4] Chuchalin A.I., Gasheva Yu., Gerasimov S.I., Pokholkov Yu.P., Shamritskaya P Accreditation of Engineering Programs and certification of Professional Engineers in Russia: a Focus on Life-Long Learning // ASEE Annual Conference and Exposition, Conference Proceedings 122, Making Value for Society. Сер. "ASEE's 122nd Annual Conference and Exposition: Making Value for Society" 2015.

[5] Association for Engineering Education of Russia (AEER), official website. Available at: http://aeer.ru/en/goals.htm (accessed 10 March 2016).

AEER accreditation of educational programs: quality assurance aims and requirements

Mrs. Kseniya K Tolkacheva, Tomsk Polytechnic University, Association for Engineering Education of Russia

Member of the Association for Engineering Education of Russia responsible for AEER international co-operation activities. PhD in Pedagogical Sciences, graduate degree in "Mathematical Methods in Economics". Also holds a Management degree. Her research focus is in Quality Assurance, active learning and international academic mobility. Has experience in coordinating several TEMPUS projects as well as organizing joint international partnerships in educational and research areas, including national and international projects, organizing conferences, seminars, workshops and training courses.

Prof. Yury P Pokholkov, Tomsk polytechnic university, Association for Engineering Education of Russia
Dr. Sergey B. Mogilnickiy, Tomsk Polytechnic University
Ms. Maria Yurievna Chervach, Tomsk Polytechnic University
Prof. Jose Carlos Quadrado Mr P.E., Instituto Superior de Engenharia do Porto

José Carlos Quadrado is a full professor with tenure of the Instituto Superior de Engenharia de Lisboa (ISEL) and is currently under external management assignment in Instituto Superior de Engenharia do Porto (ISEP), Portugal. He is currently the immediate past-President of ISEL and the Vice-President of ISEP. He has a BSc in Energy and Power Systems, a diploma degree in Electrical Engineering, Automation and Industrial Electronics from ISEL, a MSc and a Doctor degree in Electrical Engineering and Computers from Lisbon Technical University. He also holds the Habilitation degree (Aggregation) in Electrical Engineering from Beira Interior University.

Holds the position of immediate past-President of the International Federation of Engineering Education Societies (IFEES) and the position of past president of the Ibero-American Engineering Education Association (ASIBEI). Currently he is a President-Elect of the Latin American and CaribbeanConsortium of Engineering Institutions (LACCEI) and a member of the board of the European Society for Engineering Education (SEFI). He is also a past-Vice President of SEFI. He leads the Portuguese Observatory on European and Latin-American University management strategy best practices (TELESCOPI Portugal) and is the past President of the Portuguese association of engineering rectors and deans (APDI2E).

He was a member of the general assembly of the European Network for Accreditation of Engineering Education (ENAEE), and has been actively supporting the development of accreditation agencies in Central Asia and Latin America. Being a member and senior member of several engineering societies and engineering education societies in several continents, he is also a visiting professor in many universities around the world and board member of several technological societies.

He holds over 100 international publications (including journals and chapters of books), several patents and some international technical prizes and scholarships, and also held the position of editor and editor-in-chief in some journals. Up to now he has also developed several international engineering projects in the fields of renewable energy, fuel cells, electrical vehicles, intelligent control and engineering education.

AEER accreditation of educational programs: quality assurance aims and requirements

Abstract. The article addresses problems of quality assurance in higher engineering education, analyses best international practices in the field, as well as the ways to solve outlined problems. The article indicates that professional accreditation of educational programs is considered as one of the key tools for quality assurance and assessment. Main aims and objectives of professional accreditation, as well as benefits for graduates of accredited programs in building their professional engineering career are described.
Best practices and results of professional accreditation of educational programs in the field of engineering and technology conducted by the Association for Engineering Education of Russia (AEER) are presented in this paper.

Keywords: quality assurance, engineering education, AEER

1 Introduction

The consistent growth of industrial production, research and development that has taken place in many countries over the past decade disclosed lack of highly qualified specialists competent to work with modern technologies, able to design, develop and implement competitive technologies, and focus on innovative professional activity within the real market economy. In the context of rapidly growing globalization and Russian accession to the WTO the trend for drastic technological modernization of Russian economy and transition to the sixth wave of innovation as top priorities for the current decade requires training of future professionals with new competences. Lack of qualified engineers today is the core obstacle for innovative development of prioritized sectors of the country's economy –those that have acquired sufficient investment over the past few years.

One of the key factors of successful economic growth is the sufficient enhancement of human resourcing in organizations and enterprises that develop and utilize breakthrough technologies. Such improvements become possible only with further development of the national system of higher engineering education. In order to meet the modern challenges it is crucial to ensure introduction of new approach to solving key civilization development problems in XXI century. The list of main priorities of the nation's sustainable socio-economic development includes improvement of quality of life, economic growth, science, technology, education, healthcare, culture, ecology and responsible resource management [1]. A new paradigm of engineering education is being developed. Its specific feature is the transition from "knowledge transfer" learning and teaching methods to the practice-oriented continuous education based on fundamental theoretical contents. Therefore, Russian higher education faces a challenge of providing quality assurance and world recognition of Russian educational system, particularly through training of elite specialists and teams of professionals based on the outcome based approach [2].

2 Quality Assurance Tools

The quality of specialists' training in higher education institutions is assured through quality of educational programs and learning and teaching methods, resourcing (including financial resources), level of interaction with strategic partners, efficiency of Quality Management Systems, and quality of pre-university training (newcomers background). Universities all over the world develop their own systems for quality assurance of education based on their programs, information and material resources, methodological framework, human resources, management system and specific requirements from stakeholders (society, individuals and state).

An important role in terms of Education Quality Management Systems is given to the tools to assess the quality of training specialists in HEIs. Different approaches to the assessment of the quality of education are exploited on a global scale: the reputational approach, the student outcome approach, and the total quality (management) approach.

The reputational approach is based on expert (peer review) evaluation method to assess the quality of educational programs and educational institutions as a whole. The student outcome deals with qualitative assessment of outcome indicators of student achievements.

The total quality (management) approach is based on the principles of Total Quality Management (TQM) and the requirements for the Quality Management Systems set by the International Organization for Standardization (ISO), as well as on standards and recommendations of the European Network for Quality Assurance (ENQA) and other international organizations.

In this paper we will focus more on the reputational approach.

The mechanism used within this approach is the accreditation of educational organizations as a whole and educational programs in particular. There are various systems of accreditation, both in governmental (conducted by authorized government bodies) and public (professional, non-governmental) accreditation. It is worth noting that nowadays the modern trend lies in the root of the foreground development of professional accreditation of educational programs. The emphases are put on the development and improvement of such accreditation and its criteria. For instance, one of the main principles of the Bologna declaration is the principle of primary development of professional (non-governmental) accreditation of educational programs [3].

Professional accreditation of educational programs is the process of quality assessment of the program's execution. Accreditation is conducted in line with the rules and criteria developed by professional community (associations, unions) and volunteering experts. Professional accreditation of educational programs is an independent (non-governmental) system for quality assessment of specific educational programs and training activity of educational establishments that train future specialists in the specific fields. Such accreditation is conducted by volunteering experts, professionals of specific fields, representatives of industry (employers) and representatives of academic community. Professional accreditation is a way to ensure accountability of HEI and its educational programs to stakeholders (university entrants and their parents, students, employers, governmental bodies, society as a whole), i.e. existing quality standards meet stakeholders' requirements and expectations.

3 Aims and Objectives of Professional Accreditation

Professional accreditation is aimed at assuring and assessing the quality of educational programs. The system of professional accreditation allows getting considerably quick response to the problems and changes of labor market and business requirements, as well as encourages the development and improvement of education and engineering education in particular, through a system of quality assessment criteria and in line with the requirements from society.

Thus, professional accreditation of educational programs serves the society in large, professional associations, employers, governmental bodies and other stakeholders (including university entrants and their parents) and helps to determine those higher educational institutions and educational programs that meet the accreditation criteria. This implies that the quality of such educational programs is guaranteed through a successful external independent assessment conducted by experts from a professional organization that has not been influenced by any administrative (governmental) bodies managing the system of education. Further on, the certificate verifying successful results of professional accreditation creates added value for an educational program: in a number of countries (USA, Canada, Japan, Great Britain, etc.) a person, who has graduated from an accredited educational program has an opportunity to apply for a status of "professional engineer" that, in its turn, opens new horizons for his/her career development. Students of such educational programs gain specific benefits during their years of study as well: in a number of countries they are able to apply for specific national or regional scholarships. All this underlines "market value" of the accredited educational programs.

For educational institutions and all stakeholders of the educational process professional accreditation provides an opportunity to receive qualified consult and proper methodological basis on terms and conditions of educational programs' development and implementation, introduction of new programs and profiles, improvement of existing programs.

Another important benefit for many educational institutions is the opportunity to state their readiness to compete for the leadership in their field. This can be done by demonstrating their adherence to the quality of educational service and can be confirmed through inclusion of the accredited programs to the existing registers (national and/or international).

During accreditation process HEIs receive valuable recommendations on ways how to improve their educational programs. The recommendations are based on the experience of reviewing experts and represent world's best practices and expertise. This provides extra opportunity to enhance HEI's competitiveness and, therefore, improves competitive advantages of its graduates on both national and international intellectual labor markets.

One more important objective of any national system of professional accreditation is to assure the acknowledgement of high quality of national HEIs and their programs on international level. This problem is addressed through harmonizing and ensuring compliance of existing accreditation criteria and its methodological, juridical and organizational elements with similar documents and guidelines of foreign accreditation bodies and international unions and associations.

The combination of all above mentioned factors allows the system of professional accreditation to sufficiently stimulate the processes of development and improvement of specialists' training.

4 Accreditation by the Association for Engineering Education of Russia

One of the main fields of the Association for Engineering Education of Russia (Association, AEER) activities is the professional accreditation of engineering educational programs. The Association has been working on the development of professional accreditation system of educational programs in Russia since 2002 [4]. In compliance with the Federal Law of the Russian Federation No. 272-FZ of December 29, 2012, "On Education in Russian Federation", AEER is authorized by the employers' community to conduct professional accreditation of educational programs in the field of engineering and technology within a wide range of educational profiles. According to this law such accreditation is given in support of the recognition of quality of education and competence level of those graduates, who have completed an educational program in a specific educational institution that meets the professional standards and labor market requirements.

Compliance with the AEER criteria provides quality assurance and stimulates continuous improvement of HEIs' educational programs. The AEER criteria correspond to the Bologna Declaration and serve for Bachelors', Masters' and Specialists' degree programs (programs of the first and second cycle) [5]. The developed criteria are based on international best practices in the field of engineering education quality assessment. Programs accredited by AEER are included in the AEER register and the ENAEE register; they are presented to the Russian Federation Ministry of Education and Science Federal Service for Supervision in Education and Science; they are published in media and on the official AEER and ENAEE websites.

The accreditation criteria are aimed at assessing learning outcomes that represent a composition of competences, knowledge, skills and attitudes, and methodological culture. They are acquired by students within the study process and are based on the requirements from the professional society. Educational program can be accredited only if it fully corresponds to the criteria mentioned above.

In case of successful accreditation each Bachelor or Master degree program is awarded two certificates (the AEER certificate and the EUR-ACE® label). For the programs of Specialist degree three certificates are awarded: the AEER certificate, the EUR-ACE label certificate, and the certificate recognizing the substantial equivalency with the requirements of Washington Accord.

AEER is the only Russian organization that is authorized to award the EUR-ACE® label to the accredited engineering programs of the first and second cycles and to award certificates recognizing substantial equivalency with the requirements of Washington Accord to programs of the Specialist degree. These certificates are international and are acknowledged in all countries-signatories of Washington Accord and ENAEE members.

During 2003-2014 the AEER Accreditation Center has accredited over 300 educational programs in Russian and foreign universities [6]. Representatives of the Ministry of Education and Science of the Russian Federation, as well as international accrediting and engineering organizations have taken part in the AEER accreditation process as observers. More than 230 programs have been awarded the EUR-ACE® label.

AEER accredited programs are run at leading Russian universities. These programs have high ranking in intra-university ratings. Information about these programs is published in the annual guide "Best educational programs of innovative Russia". Graduates of the accredited programs are in high demand among employers and, most of the time, show positive dynamics in terms of their career development.

AEER is currently working in collaboration with the RUSNANO Fund for Infrastructure and Educational Programs on a joint project on development of a system for accreditation of educational programs in the field of nanotechnology. All the materials for accreditation are developed with regard to the professional standards of this scientific field. 17 educational programs (in the field of metrology and nanoelectronics) of 9 universities (including 1 federal university and 4 national research universities) have already received positive decision within the accreditation process completed at the pilot stage of the project.

In 2014, according to the international requirements, the Association for Engineering Education of Russia has developed and successfully tested accreditation criteria for programs of secondary (vocational) education in the field of engineering and technology. The pilot accreditation has taken place in several colleges of Tomsk.

5 Conclusion

With an aim to improve the quality of training of competitive and in-demand graduates it is important to:

- Continue accrediting best educational programs by national and international organizations, that conduct professional accreditation, such as AEER;
- Set agreements between leading industrial corporations and accreditation agencies in Russia;
- Develop and spread the practice of taking into account the results of professional accreditation when conducting state assessment of HEIs;
- Attract all interested parties into the discussion of engineering education quality improvement.

References

1. Ursul, A.: National Security Strategy of Russia and Education for Sustainable Development. J. Otkrytoe obrazovanie. Iss. 4, 63–73 (2009)
2. Pokholkov, Yu.P., Rozhkova, S.V., Tolkacheva, K.K.: Practice-oriented Educational Technologies for Training Engineers. In: International Conference on Interactive Collaborative Learning, ICL, pp. 691–620 (2013)
3. Birnbaum, R.: The Life Cycle of Academic Management Fads. J. of Higher Education. Vol. 71, 1, 1–16 (2000)
4. Pokholkov, Yu. P., Chuchalin, A.I., Mogilnitsky, S.B.: The national accreditation system for higher education institutions in Russia. J. Higher Education in Europe. Vol. XXVII, 3, 217–230 (2002)
5. Gerasimov, S., Tomilin, A., Tsoi, G., Shamritskaya, P., Yatkina, E., under redaction of Chuchalin, A.: Criteria and Procedure of Professional Accreditation of Educational Programs in Field of Technology: Information Publication. Tomsk Polytechnic University, Tomsk (2014)
6. Accreditation Center of the Association for Engineering Education of Russia, http://aeer.ru/en/acc_center_eng.htm

Triple-Helix and International Collaboration to Design and Implement an Outcomes Based Engineering Curriculum to Better Serve Stakeholders in Valparaíso-Chile

Prof. Uriel Ruben Cukierman, IFEES

Uriel Rubén Cukierman, MS, PE, is President of IFEES, Associate Director of InnovaHiEd and professor/researcher at Universidad Tecnológica Nacional (UTN), the largest Engineering School in Argentina. He has served as the Dean of Engineering at the Universidad de Palermo (UP) and, previously, as the Information & Communications Technologies Secretary (equivalent to a VP position) at UTN for more than 15 years. Recently awarded as "Distinguished Educator" by the IOEM and "Honorary Professor" by the Universidad Ricardo Palma in Lima, Peru. With a BS in Electronic Engineering from UTN and a MS from Universidad Politécnica de Madrid, Uriel has been the Electronic Department founder and former Director at Technical School ORT Argentina, also the former Director of the Learning Technologies Bachelor at UTN. Research interests in Learning Technologies include the areas of mobile devices, learning management systems and the digital divide. Professor Cukierman has contributed to the National Universities Network as a member of its Administrating Committee and has worked as consultant for the National Communication Secretary and National Education Ministry in projects for distance learning and reducing the digital divide. Taught at the university level for over thirty years in electronics, computing, multimedia and learning technologies. Produced three books about Learning Technologies and Engineering Education, one of which was published by Pearson. Authored 5 book chapters and more than 40 technical papers in refereed journals and conferences. (www.cukierman.name)

Mrs. Lueny Morell, InnovaHiEd

Lueny Morell, MS, PE, Ing.Paed.IGIP is President of Lueny Morell & Associates and Founder & Director of InnovaHiEd, a world-class team of experts with extensive academic and industry experience offering services to help higher education leaders in transforming their institutions to better respond to their stakeholders' needs and the socioeconomic development challenges they face. With a BS and MS in Chemical Engineering from the University of Puerto Rico and Stanford University, Lueny is co-founder of NEU, a novel platform to teach engineering in Silicon Valley, California. From 2002 to 2013 she was part of the HP Labs Strategy, Open Innovation and University Relations teams and a full professor of Chemical Engineering at University of Puerto Rico - Mayagüez for 24 years, where she had various academic and administrative positions, including Director of R&D. She is an IEEE Senior Member, an ASEE Fellow and ABET Program Evaluator and has received various awards for her work, including the prestigious US National Academy of Engineering Gordon Prize for innovations in engineering education in 2006. Recognized as one of the Engineering Education Pioneers in the US in 2014, Lueny is passionate about engineering education and innovation as fundamental pillars for economic and social well-being. She is co-founder of IFEES (Past President), GEDC and SPEED. Lueny maintains a blog on topics associated with innovation and engineering education (www.luenymorell.com; www.innovahied.com).

Ms. Rene Alejandro Noel, Universidad de Valparaíso

Full professor at Universidad de Valparaiso. Main research areas are experimental software engineering, software design, and engineering education. 7 years of industrial experience as software engineer, and 4 as faculty member, currently leads curricula innovation in STEM courses at Engineering College of Universidad Valparaíso.

Roberto Munoz, Universidad de Valparaíso

Roberto Muñoz is Computer Engineer (2009) from the Universidad de Valparaíso (Chile) and Master in Computer Science at Pontificia Universidad Católica de Valparaiso (2012). He has recently enrolled in a PhD program at Pontificia Universidad Católica de Valparaíso. Currently, he is full professor of informatics engineering at Universidad de Valparaíso. His research interest are Informatics in Education and Human-Computer Interaction.

Dr. Eduardo Vendrell Vidal, Polytechnic University of Valencia

Eduardo Vendrell Vidal (Carlet, Spain, 1967) holds a Degree in Informatics (1991) from the Faculty of Informatics, Polytechnic University of Valencia (UPV), and a Ph.D. in Informatics from the same university (2001).

Since 1991, he is Associate Professor at the Polytechnic University of Valencia (Universitat Politècnica de València, UPV) in the Department of Systems Engineering and Control, currently attached to the School of Informatics (Escuela Técnica Superior de Ingeniería Informática, ETSINF). His teaching is related to CADCAM in the context of the Degree on Informatics Engineering given at the ETSINF, and the Master on Industrial Informatics and Control at UPV.

He is member of the Robotics research group of the University Institute of Control Systems and Industrial Computing (Instituto Universitario de Automática e Informática Industrial, ai2), at UPV.

His main research interest is in Robotics, CADCAM and 3D modelling and representation. In these fields, he has been participating in several national and international research projects, with published results in conferences and journals.

He has been Associate Dean of Studies at the Faculty of Informatics from 2001 to 2009, and is currently the Dean of the School of Informatics (ETSINF). From 2012, Dr. Vendrell is the President of the Spanish Council of Deans of Informatics (Conferencia de Directores y Decanos de Ingeniería Informática, CODDII) in Spain.

He is also a member of the Committee of Education in Engineering in the World Federation of Engineering Organizations (WFEO), member of the Executive Board of the European Quality Assurance Network for Informatics Education (EQANIE), and advisor to the Spanish Agency for Quality Assessment and Accreditation (ANECA).

From October 2014 he is the President of EQANIE, the European Quality Assurance Network for Informatics Education.

Dr. Carlos Becerra, Escuela de Ingeniería Civil Informática, Universidad de Valparaíso

Carlos Becerra Castro is full professor of informatics engineering at University of Valparaiso and Director of Civil Engineering Computer Career. His background is Engineer, Master and PhD in informatics where his main research interests are software architecture, experimental software engineering and automatic categorization of digital artifacts based on stochastic models.

Prof. Carmen Gloria Prado Mg, Universidad de Valparaiso

Magister en gestión de políticas de educación y cultura Psicopedagoga Licenciada en educación Coordinadora de Docencia facultad de Ingeniería Universidad de Valparaíso

Triple-Helix and International Collaboration to Design and Implement an Outcomes Based Engineering Curriculum to Better Serve Stakeholders in Valparaíso, Chile

Abstract

Following an approach that has been pursued by many engineering programs that are outcomes based accredited, this paper describes the process of designing and implementing the curricula in the College of Engineering of the Universidad de Valparaíso in Chile, using a backwards design approach and also focusing on addressing stakeholders' needs through students' outcomes. Driven by the Chilean government through funding from MINEDUC (its educational development branch), this University has undertaken the task to completely revamp its engineering curricula by engaging faculty, students and industry, together with international collaborators from Argentina, Puerto Rico and Spain. A specific challenge for this redesign is the low performance of Chilean secondary students in international science and mathematics tests, that is tackled by putting an S.T.E.M. emphasis in the overall process. Using the mentioned backwards design approach, the multi-disciplinary, multi-national, multi-stakeholder team will share the process of establishing University-wide learning outcomes, as well as specific outcomes for the engineering program benchmarked through a very novel method and validated by industry and employers' representatives.

The authors will describe their motivations, aspirations and work currently being implemented in the College of Engineering that has also cascaded through the whole University, as well as the lessons learned, challenges, and successes along the road. In addition, the approach for faculty buy-in and capacity building of the new learning/teaching methodologies, outcomes assessment and continuous quality improvement strategies and the internal sharing of the most effective practices will be described.

The authors believe that the process outlined in this paper can help worldwide institutions adopt similar strategies for better serving their constituents.

Keywords

Curriculum innovation, triple helix, STEM, Chile, outcomes assessment

1. Introduction

The University of Valparaíso, is a Chilean state institution with 15.000 students and more than 1.800 faculty. The educational project[1] of the University of Valparaiso, reformulated in 2012, has been defined as "learning outcomes oriented", which is a nuanced interpretation of the learning outcomes educational model. This approach was undertaken to allow the initial introduction of the learning outcomes model in a public and traditional university, where resistance to change and administration formalism are serious threats to a massive curricular transformation. Hence this definition does not necessarily force to completely change all curricula, but rather a conceptual change that involves four key elements[2]:

1. Graduate profiles are formulated in terms of learning outcomes

2. Curricular design must be driven by the graduate profile.

3. Upon graduating, the student must have acquired all the learning outcomes considered in the graduate profile.

4. The achievement of the learning outcomes must be assessed at any time.

The School of Engineering ("Facultad de Ingeniería", or "FACING") of the University of Valparaíso, was established in 2011, gathering engineering careers formerly scattered between Science, Business and Architecture Schools: Construction Engineering, Civil Engineering, Computer Engineering, Oceanic Engineering, Biomedical Engineering, Industrial Engineering and Environmental Engineering. FACING's strategic plan for the period 2012-2016, defines a focus on quality assurance of the science subjects, curriculum innovation, and development of teaching skills. Curriculum innovation has been executed in the context of a state funded project since 2012, which engaged the School in a process of reflection and homogenization of the curriculum of the careers that compose it, leading to the accreditation of their careers by the National Accreditation Commission (CNA), and the certification of their administrative processes (ISO9000).

While this project has achieved the goals set, it was necessary to incorporate specific strategies to address problems of global scope: the development of skills in science, technology, engineering and mathematics (STEM). A new project named MECESUP UVA1409 "Design and implementation of a strategy for evaluation and continued strengthening of STEM skills", funded by the Chilean Government, embraces the challenge of changing teaching practices in order to introduce active learning methodologies for the initial two years of the program, which are intensive in science subjects. Its objectives are to motivate, empower, assess and monitor the development of learning outcomes which are transversal to all the engineering programs, and that are critical to the development of STEM skills. This project was started in March 2015, just after the final implementation's activities of the educational model at the Faculty UV, which led to the launch of the innovated curricula in 2016. In this context, the linkage between both initiatives was relevant and necessary in order to tackle the main problems of the Engineering School: high desertion rates and in time graduation.

	FACING	**UV**
First year retention rate	72%	82%
In time graduation	4,20%	19,90%

Table 1: Retention and in time graduation rates for the school of engineering.

Due to the FACING's recent creation, it also lacks of connection with the relevant society stakeholders and local industry.

The following sections present a triple-helix and international collaboration approach to design and implement an outcomes based engineering curriculum, to better serve stakeholders in Valparaíso, Chile. Section 2 presents the methodological framework of this approach, while section 3 presents main steps of the innovation process. Section 4 presents the conclusions and preliminary results of the curriculum design.

2. Background

2.1 Backwards design

As a field, instructional design is historically and traditionally rooted in cognitive and behavioral psychology, though recently constructivism (learning theory) has influenced thinking in the field. Instructional Systems Design (ISD) or simply instructional design[3] is the practice of creating "instructional experiences which make the acquisition of knowledge and skill more efficient, effective, and appealing"[4]. Thus, ISD focuses on the learning experience. The process consists broadly of determining the needs of the learner, defining the end goal of instruction, and creating some "intervention" (the "catalytic process") to assist in the transition, and thus, resulting in learning. Ideally the process is informed by pedagogically (process of teaching) and andragogically (adult learning) tested theories of learning and may take place in student-only, teacher-led or community-based settings. There are many instructional design models. Many involve five basic phases: analysis, design, development, implementation, and evaluation. This instructional design approach is very much like engineering problem solving! The newly established New Engineering University in Silicon Valley (now GalvanizeU in San Francisco, California) used this design approach for its curricula[5]. We are using a combination of instructional systems and "Backwards Approach (BA)" to re-design UV curricula.

BA design[6,7] is a method of designing curriculum by choosing learning outcomes before instructional methods or assessments. This means one chooses the outcome of the learning experience first, and let's that guide the teaching/learning and the assessment/evaluation. This method challenges "traditional" methods of curriculum planning in which a list of content that will be taught is created and/or selected first and teaching/assessment methodology usually are lectures and laboratories, with written exams as assessment of learning. In backward design, the educator starts with goals, creates or plans out assessments and finally makes lesson plans. Supporters of backward design liken the process to using a "road map"[8]. In this case, the destination is chosen first and then the roadmap is used to plan the trip to the desired destination. In contrast, traditional curriculum planning has no formal destination identified before the journey begins.

2.2 Triple Helix

The concept of the Triple Helix of university-industry-government relationships (Figure 1) initiated in the 1990s by Etzkowitz[9,10] and Etzkowitz and Leydesdorff[11], encompassing elements of precursor works by Lowe[12] and Sábato and Mackenzi[13], interprets the shift from a dominating industry-government dyad in the Industrial Society to a growing triadic relationship between university-industry-government in the Knowledge Society. The Triple Helix thesis is that the potential for innovation and economic development in a Knowledge Society lies in a more prominent role for the university and in the hybridization of elements from university, industry and government to generate new institutional and social formats for the production, transfer and application of knowledge. This vision encompasses not only the creative destruction that appears as a natural innovation dynamics[14], but also the creative renewal that arises within each of the three institutional spheres of university, industry and government, as well as at their intersections.

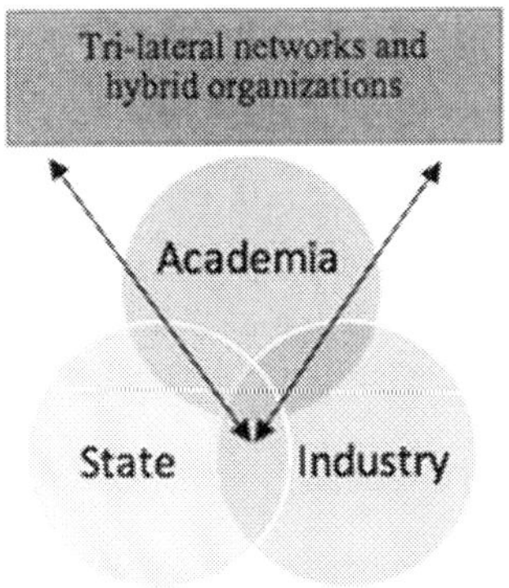

Figure 1. The Triple Helix Model of University–Industry–Government Relations.

In this particular environment, at the Universidad de Valparaiso, this Triple Helix concept has been conscientiously developed and promoted because:

1. There is the University itself,
2. but this is a public, state funded university which follows the state mandates and guidelines and,
3. has effectively involved not only the state, but also the industry in the process, particularly through the External Advice Committee which review the ongoing strategic process (Figure 2)

Figure 2. Review/Advice circle involving the External Advice Committee.

2.3 International Collaboration Team Conformation

As essential as the previously described process is, it is to receive the proper advice and training from external international experts which may add to this process not only their knowledge and experience, but a "non contaminated" vision. And when we say "non contaminated", we are trying to describe in words the fact that these international experts should provide ideas and visions from outside the University and the Country without the bias of local experiences and points of view. As it is usually said, "No One is a Prophet in their Own Land" and, for all the previously explained reasons, an international group of experts has been invited to participate.

This group – formed by members of InnovaHiEd (www.innovahied.com) – includes three engineers who have strong background and experience in universities, both in the academic and management aspects. Lueny Morell, Director and Founder of the consulting firm, has been an Academic for almost 30 years at the Universidad de Puerto Rico and has also have an important experience in HP Labs, leading the University Relations Dept. for almost 10 years. Eduardo Vendrell-Vidal is currently the Dean at the School of Informatics of the Universitat Politècnica de València, Spain, and he is also the President of EQANIE, the European Quality Assurance Network for Informatics Education. Finally, Uriel Cukierman, Associate Director of InnovaHiEd, is currently the President of the IFEES, the International Federation of Engineering Education Societies and also have more than 30 years of academic and management experience at the Universidad Tecnológica Nacional in Argentina. This group of experts has visited the University in several occasions and has offered not only advice, but training workshops for faculties and staff that have helped them to effectively go through this process as it was evidenced in the surveys completed by the attendees.

3. Innovation Implementation

3.1 Innovation Process

In BA the educator is able to focus on addressing what the students need to learn, what data can be collected to show that the students have learned and how to ensure the students will learn. The UV's curriculum design phases are very similar to NEU's process, namely[5]:

Analysis – Understand regional and national economic development needs, technology trends and employer needs. The engineering/technology competencies for UV's Engineering Programs were captured through three ways: 1) conversations with the College of Engineering Advisory Council, which was established in early 2015, considering industry representatives and government representatives, 2) institutional and program accreditation committees that ensure the fulfillment of the accreditation requirements and entanglement with the curricular innovation process, and 3) by benchmarking successful programs worldwide, performing semantic analysis of the competencies and graduate profiles, looking for the most required skills and learning outcomes. This analysis also included the professional engineering profile required by important engineering Chilean and international organizations.

Design – Led by the Engineering Dean and the STEM project team, competencies were distributed across the curriculum ("roadmap") as shown in Appendix 1, and establishing the desired depth of learning the competencies, in terms of three Domain Levels. Each Domain Level covers 2 years of the engineering careers curricula. With the assistance of academic and industry experts, more specific learning objectives and desired outcomes were written for each of the individual courses. The courses' syllabi include specific learning objectives and outcomes, weekly course plan, weekly learning experiences, and outcomes assessment for all the competencies assigned to the course, as well as traditional information like title, description, textbooks/references and other logistics. We used the Felder-Brent definition of learning objectives (or instructional objective): "A statement of something observable and clear that students should be able to do after receiving instruction, plus (optional) conditions under which they would do it and/or what would constitute acceptable performance."[15]

Development – for each learning experience (course), faculty refined learning objectives, defined topics, resources needed, active learning experiences and assessments. Research shows that students learn better when they know why/what they are learning, when they see applications on how the knowledge is used, when having enough time to think and share thoughts with others, and when they engage in active learning. Thus, to design the learning experience we asked course designers to answer the following questions:

- How will students learn?
- What resources are needed?
- When and where will the learning take place?
- Who is responsible?
- What experiences will help students learn the knowledge and develop the skills, attitudes and values?
- How can we address different learning styles?

The teaching/learning strategy should be aligned with course objectives, should also balance facts with concepts, and include a variety of delivery modes. Therefore, all courses have a blended mix of, for example, short lectures (15-20 min) with any activity that keeps students alert, engaged and motivated in the learning process, such as student discussions, in-class demonstrations, laboratory experiences, team projects, and student oral presentations. At the time of preparing this paper, proposals are being sought from experts in industry and academia to develop laboratory/active learning experiences to support the program. All courses will include learning objectives and outcomes assessment (see section below for more details).

3.2 Innovation Results

After being agreed by program managers, the curricular innovation committee and the academic FACING coordination, the proposed STEM competencies were included in the bachelor's degree profile of the FACING careers. The proposed competencies were incorporated into the innovated curriculums of all programs included in the process of curricular innovation. Considering different programs states of progress, the impact on the curriculum is as follows:

- Four programs that have already developed the competencies for curricular innovation, a mapping table is to be drawn between the competencies proposed for each career and STEM competencies.
- The other three programs that were in the process of formulating their competencies, adhere to the Faculty transversal competences (professional ethics and STEM), adding only a limited number of specific competencies, relevant to their discipline.

In both cases, the recommendation is that the specific elements that each career should add to the transversal competences are explicit as learning outcomes indicators, which are ultimately associated with each subject of the new curricula. These activities are currently being deployed among the programs of the School of Engineering, by supporting teacher with three strategies: co-design of learning activities, elaboration of teaching guides including learning activities that are explicitly linked to learning outcomes, and support for teachers in their classroom activities.

4. Conclusions and Preliminary Results

The Triple Helix model has been implemented by means of an External Advice Committee, namely an Advisory Council. While these committees are not so common in certain regions for academic purposes, there are a lot of examples where an Advisory Council is an excellent resource for industry/society – university partnership[16]. By means of a well-defined mission statement and a correct choice of its members, universities can get feedback for programs, goals and strategic planning, taking into account external recognized views. This way, assessment of outcomes, suggestions for new competences or professional profiles, or development of new programs are usual issues that can be discussed in a set of regular meetings held by the Advisory Council. This committee must be compound by external members, including former students that can share a holistic view for the University, but it is important that university leaders will attend all meetings in order to know and assume all comments and advices issued and discussed. Starting from this point, new opportunities for industry partnership could be developed, mainly because members of this external committee could act as a link between the university and society, promoting university capabilities through their commitment. This approach has strengthened the proposed competences, leading to a widespread approval by the faculty. Evidence of a robust and inclusive discussion on the relevance and necessity for STEM competences has established a reliable basis for the next challenge: the intervention of teaching practices to connect learning outcomes with the detailed design of the learning activities.

BA designs requires that competences derive in a learning outcomes definition and assessment. This concept is directly assumed by teachers as they can define low level goals for the students that can be measured and assessed conveniently. These learning outcomes must be well defined, according to a well-known taxonomy, so further analysis can help to verify and re-adjust the syllabus in some iterations. This way, using convenient verbs and nouns when defining learning outcomes can evidence if a course is well oriented or not. For example, while a basic course should be focused on outcomes that lead to learn and assume basic concepts, an elective course, usually organized and taught in last years of the degree, must be oriented to apply and develop concepts. FACING's STEM team is currently training teachers and co-designing learning activities to ensure a properly low level goals definition and evaluation, derived from the competencies roadmap designed.

Finally, it is important to highlight here that several countries are implementing outcome based educations systems and active learning. Some examples to mention in which we were involved are: India (Walchand College of Engineering in Sangli and BVB College in Hubli). Also, there are several institutions that are moving towards this approach in Mexico (several institutions), Colombia (Universidad del Norte), Peru (PUC Lima) and some others. The approach and process described in this paper could be a helpful benchmark for other institutions seeking to re-engineer programs to better serve their stakeholders.

References

1. División Académica. (2012). Proyecto Educativo de la Universidad de Valparaíso. Retrieved from http://divisionacademica.uv.cl/images/pregrado/proyecto_educativo.pdf
2. Centro de Desarrollo Docente. (2012). Extracto del modelo educativo UV sobre el perfil de egreso y la definición de competencia. Complete information at http://cdd.uv.cl/index.php?option=com_content&view=article&id=3:recursosdocentes&catid=3:archivos&Itemid=6

3. Retrieved from http://en.wikipedia.org/wiki/Instructional_design
4. Retrieved from http://en.wikipedia.org/wiki/Backward_design
5. Tamir, Michael, Harichandran, Ronald, Morell, Lueny, "San Francisco's New Start-up is Reengineering Engineering Education", 2015 ASEE National Conference Proceedings, Seattle, WA, June 2015.
6. Merrill, M. D., Drake, L., Lacy, M. J., Pratt, J., ID2_Research_Group. (1996). Reclaiming instructional design. Educational Technology, 36(5), 5-7.
7. Wiggins, G., McTighe, J. (1998). "What is backward design?" In Understanding by Design. (1 ed., pp. 7–19). Upper Saddle River, NJ: Merrill Prentice Hall.
8. McTighe, J., & Thomas, R.S. (2003). Backward design for forward action. Educational Leadership, 60(5), 52–55.
9. Etzkowitz, Henry. The NSF and U.S. Industrial and Science Policy, Science and Technology Policy, Feb. 1993.
10. Etzkowitz, Henry, Enterprises from Science: The Origins of Science-based Regional Economic Development, Minerva 31(3): 326-360, 1993.
11. Etzkowitz, H., Leydesdorff, L., 1995. The triple helix– university–industry–government relations: a laboratory for knowledge-based economic development. EASST Review 14 Ž1, 14–19
12. Lowe, C.U., 1982. The Triple Helix—NIH, industry, and the academic world. The Yale Journal of Biology and Medicine 55, 239-246.
13. Sábato, Jorge, and M. Mackenzi, 1982, La producción de tecnología: Autónoma o transnacional (Nueva Imagen, México).
14. Schumpeter, Joseph A. Capitalism, Socialism, and Democracy. 3d ed. 1942. New York: Harper and Brothers, 1950.
15. Felder, R.M., Brent, R. Designing and teaching courses to satisfy ABET criteria, Journal of Engineering Education, 92 (1), 7-25 (2003).
16. Book, Rebeca G. Maintaining Industry and Academic Partnership Through an Advisory Council. Proceedings of the 2008 Midwest Section Conference of the American Society for Engineering Education. 2008

Appendix 1: STEM Competencies Roadmap

Competence	Domain Level	Learning Outcomes
Apply knowledge of mathematics, science and engineering in solving engineering problems	I.- Solve problems using scientific or mathematical models and formulations that express a simplified engineering problem	Relate the mathematical formulation and the grammatical description of an engineering problem
		Express problems using graphics and mathematical models
		Solve problems using graphics and mathematical models
	II.- Solve problems using scientific or mathematical models and formulations that express a real engineering problem	Implement solutions to problems using graphics and / or mathematical models in real contexts
Design and implement validation activities of engineering proposals for obtaining quantitative and qualitative evidence, for further analysis and interpretation of results	I.- Apply experimental models for empirical validation of solutions to engineering problems	Identify the steps and features of a the design of experimental or empirical validation activities
		Relate the experimental design steps with laboratory experience
	II.- Select the input and output variables to the empirical validation of solutions to engineering problems	Designing an empirical validation activity to answer a research question
Design engineering solutions, such as systems, models and processes in real contexts and under real constraints	I.- Recognize and apply methodologies and tools to design engineering solutions	Identify methods , methodology , tools, techniques , standards applied in the design process engineering solutions .
		Analyze, select and use methods, methodologies, tools, techniques and standards to produce partial results in the engineering design process .
	II.- Build models and systems that can provide solutions to engineering problems in real contexts	Run and evaluate a engineering design process using methods, methodology, tools, techniques and standards
Identify, formulate and solve engineering problems	I.- Develop and implement a model that solves a disciplinary problem	Extract relevant information for understanding and solving the problem
		Understand the objective of the problem
		Establish relationships through information provided by the problem
		Apply different techniques to solve the problem
	II.- Design strategies for solving problems of discipline	Select different techniques to solve the problem according to their suitability to the discipline restrictions
Using techniques, methods and modern engineering tools	I.- Recognize and use the techniques and tools of modern engineering	Identify technical and engineering tools and understand their uses and applications.
		Using techniques and engineering tools in solving problems and / or projects
	II.- Select appropriate techniques and tools to solve problems of modern engineering	Select appropriate techniques and tools to solve problems of modern engineering
	III.- Designing and develop methods, techniques and tools for engineering	Design and develop methods , techniques and tools for engineering
		Build or develop methods , techniques and tools for engineering
Understand the impact of engineering solutions in the global, social, economic context and the environment.	I.- Interpret the impact of engineering solutions in the global, social, economic context and the environment	Infer and explain the impact of engineering solutions in the global, social , economic context and the environment
Knowing political and economic aspects of the context of the engineering performance	I.- Relate the engineering processes in the political, economic and environmental context of the engineering performance	Identify the political, economic and environmental context of the engineering performance

IEET's Mentoring of Myanmar in Engineering Accreditation System

Dr. Mandy Liu, Institute of Engineering Education Taiwan (IEET)

Dr. Liu is currently the Deputy Executive Director of the Accreditation Council and Office Director of Institute of Engineering Education Taiwan (IEET), an accreditation agency for engineering education. Her primary responsibilities are to oversee administration of accreditation and international activities of the Institute. Prior to her current position, Dr. Liu worked as a research associate for the Association of American Medical Colleges (AAMC) in Washington DC, USA, where she conducted research in the area of organization and management studies of medical schools. Dr. Liu received her doctorate in higher education policy and management from Claremont Graduate University in the USA.

Prof. Liang-Jenq Leu, Dept. of Civil Engineering, National Taiwan University

Dr. Liang-Jenq Leu is Professor and Chairman of the Department of Civil Engineering at the National Taiwan University. He is also serving as the President of the Chinese Institute of Civil and Hydraulic Engineering (CICHE), the President of the Chinese Society of Structural Engineering, Standing Director of the Board of the Taiwan Chinese Society for Earthquake Engineering, and Deputy Secretary General and Deputy Executive Director of the Accreditation Council of the Institute of Engineering Education Taiwan (IEET). He is the Subject Editor of the Journal of the Chinese Institute of Engineers (a SCI journal). Dr. Leu earned his Bachelor and Master's degrees from the Department of Civil Engineering of National Taiwan University in 1987 and 1989, respectively. He joined the faculty of National Taiwan University shortly after receiving his PhD from Cornell University in 1994. His area of research includes optimal design of structures, structural health monitoring, earthquake resistant design, and simulation of architectural physics for green buildings.

Prof. Charlie Than, Myanmar Engineering Council

Graduated from Rangoon Institute of Technology in Mechanical Engineering at 1975.Completed the PG Certificate Course in Naval Architecture from University of Newcastle upon Tyne in 1990.Study in Marine Engineering Department of Defense Service Technological Academy for Master and Doctorate Program and completed in 2008.Start the career as Senior Engineer in Inland Water Transport in 1975 and transfer to Myanmar Maritime University in 2001.Start the career as academia in 2001 as Pro-Rector of Myanmar Maritime University.Finally come as President of this University and retired in 2013.Later government appoint in the Myanmar Engineering Council for Quality Assurance of the Technical Universities.At present i am Vice President of Myanmar Engineering Society for contribution to our Society.

IEET's Mentoring of Myanmar in Engineering Accreditation System

Mandy Liu[1], Institute of Engineering Education Taiwan (IEET), Liang-Jenq Leu[2], Department of Civil Engineering, National Taiwan University, and Charlie Than[3], Myanmar Engineering Council

[1]Deputy Executive Director, Accreditation Council, IEET, 7F, No.554, Linsen North Rd., Zhongshan District, Taipei 10453, Taiwan (mandyliu@ieet.org.tw)
[2]Professor and Chairman, Department of Civil Engineering, National Taiwan University; President, Chinese Institute of Civil and Hydraulic Engineering; President, Chinese Society of Structural Engineering; Deputy Secretary General and Deputy Executive Director of the Accreditation Council, IEET (ljleu@ntu.edu.tw)
[3]Chairman, Quality Assurance of the Technical Universities, Myanmar Engineering Council; Vice President, Myanmar Engineering Society

ABSTRACT: Institute of Engineering Education Taiwan (IEET), the accreditation agency for engineering and technological education in Taiwan, is a signatory of the Washington, Sydney, and Seoul accords. It was invited to be a mentor for Myanmar Engineering Society (MES) in its development of engineering education accreditation system in 2013. During the past three years, IEET has helped the Myanmar, eventually the Myanmar Engineering Council (MEngC), develop engineering education accreditation system. In October 2015, MEngC launched its very first accreditation review at eight programs in the Yangon Technological University (YTU), which is the flagship engineering institution in Myanmar. This historical event marked the country's yet another milestone toward internationalization of higher education. This paper is to record the process and outcomes of the mentoring.

IEET became signatory of the Washington Accord in 2007, Seoul Accord in 2009, and Sydney Accord in 2014, respectively. Over the past twelve years, IEET has accredited over 500 programs in 83 universities in Taiwan. It became the country's leading accreditation agency in engineering and technological education. All accredited programs are waived from the Ministry of Education's evaluation. Since the outset, IEET has been promoting outcomes-based education in Taiwan. Specifically, the agency has been paying much attention to the program's training of students' graduate attributes through Capstone Course in recent years. Although Capstone Course is not something new to many programs, only few programs have linked the assessment of the course with attainment of graduate attributes. Now through the requirement of the IEET accreditation criteria, all programs must have Capstone Course and must use the assessment of the Course to show compliance with the IEET's requirements of graduate attributes.

Due to the political situation, higher education in Myanmar has gone through a period of uncertainty and instability. However, as the political and economical situations changed in recent years, the country works hard to catch up with its international peers. MES is one of the most active engines in the country to move and elevate level and quality of engineering education. It joins members in the Federation of Engineering Institutions of Asia and the Pacific (FEIAP) to seek recognition of engineering degree holders from Myanmar. Chinese Institute of Engineers (CIE) is a member of the FEIAP but does not oversee accreditation of engineering education in Taiwan. CIE asked IEET's support to help mentoring MES in the development of engineering education accreditation system in Myanmar. Starting from 2013, IEET delegates have visited Myanmar many times to introduce the engineering education accreditation system to universities and members of MES. Moreover, IEET delegates also provide training workshops to the MES members. When the country finally adopted Professional Engineering Act and recognized accreditation of engineering education as the requirement for engineer licensing, MEngC was thus formed to develop and execute engineering education accreditation. Through several years of working together with IEET, MEngC has published accreditation criteria and policies and launched accreditation visit to YTU. Universities in Myanmar are also showing interest in accreditation now, a departure from years ago when many had doubts in accreditation. IEET hopes that the accreditation of engineering education could serve as a catalyst for the advancement of engineering education in Myanmar.

INTRODUCTION

Institute of Engineering Education Taiwan (IEET), the accreditation agency for engineering and technological education in Taiwan, is a signatory of the Washington, Sydney, and Seoul accords. It was invited to be a mentor for Myanmar Engineering Society (MES) in its development of engineering education accreditation system in 2012. During the past three years, IEET has helped Myanmar, eventually the Myanmar Engineering Council (MEngC), to develop engineering education accreditation system. In October 2015, MEngC launched its very first accreditation review at eight programs in the Yangon Technological University (YTU), which is the flagship engineering institution in Myanmar. This historical event marked the country's yet another milestone toward internationalization of higher education. This paper records the process and outcomes of the mentoring.

ENGINEERING EDUCATION IN MYANMAR

Historical Review on Myanmar Educational Movements

Engineering Education was first introduced with the establishment of Rangoon Institute of Technology (RIT), which was formerly set up as the faculty of Engineering under Rangoon University in 1924, during British Colonial period (Than 2015). At first, the lecturers and instructors were from India, which is also the colonial country under Britain. Thus, at that time, both the education system and engineering education system were mainly based on British system. Then, Myanmar engineers were sent to foreign countries such as Europe and the United States of America to be able to manage programs in engineering education in Myanmar. Therefore, the engineering education system of Myanmar started to change from British system to US System but it was not as strong yet. However, an inconsistency between education systems had arisen since then.

After gaining independence in 1948, most of the scholars were sent to technological universities in the United States of America. When they came back, the engineering education system was changed from the British system to the US system.

Since the caretaker government gained power in 1957, the scholars were sent to East European countries such as Russia, Yugoslavia, Czechoslovakia, East Germany and Hungary. Those scholars tried to change the engineering education of Myanmar according to the education system of Eastern Europe. At present times, the university laws in practice are still based on the East European education laws. Despite the changes of system over the ages, RIT, a leading institute in engineering education in Myanmar, still uses the education system based on the US Education system.

Development of Engineering Education in Myanmar

There were five significant phases in the development of engineering education in the country, namely:

Phase 1: from 1924 to 1940 (beginning of World War II);
Phase 2: from 1946 to 1961;
Phase 3: from 1961 to 1988;
Phase 4: from 1988 to 2010; and
Phase 5: from 2010 to present.

First Phase (1924 - 1940)

In the first phase, there was a Department or Faculty of Engineering under the University of Rangoon, which gave civil engineering, mechanical engineering and electrical engineering courses at College of Engineering which later became RIT.

Second Phase (1946 – 1961)

The system at that time was in such a way that the students had to take common courses in the first 2 years and branched out into different disciplines of choice, starting from the 3rd year. The total contact hours of learning for engineering students were about 30 hours per week and 6 years of undergraduate course after matriculation. The medium of teaching was mainly English. In addition to the lectures, laboratory work and drawing, all the students had to take workshop trainings in carpentry, blacksmith, welding, and in machine shop.

Third Phase (1961 – 1988)

After 1962, the Faculty of Engineering became an independent Institute, called Burma Institute of Technology (BIT), with its own Rector, under the Directorate of Higher Education, Ministry of Education. BIT was changed into RIT after about a year. The reason for the change was that the Ministry of Education planned to open more institutes of technology in other parts of the country at a later date. RIT had its own Administrative

Council and Senate, chaired by the Rector to give policy directions related to respective administrative and academic matters. However, higher Administrative Council and Senate chaired by the Minister of Education were also established which had final say on administrative and academic matters of all the institutes under the Ministry of Education. In order to upgrade the teaching and improve the engineering education, the teaching staff were sent abroad to different countries according to the international aid that was received at that time. During this period, agricultural engineering and automobile engineering disciplines were branched out under mechanical engineering department; petroleum engineering discipline under mining engineering was also formed, and two separate streams called power and electrical communications [later renamed electronics engineering] were established under electrical engineering department. In the late 1960s the total number of students was about 4,000 and the total number of teaching staff was about 200. Therefore, the student/teaching staff ratio was as before about 20:1. The laboratory and workshop facilities were stretched to the limits by this time as additional and up to date equipment could not be acquired due to shortage of funds.

1988 to Present (Fourth Phase and Fifth Phase)

In 1988, throughout the period of the military government, many technological universities and colleges emerged. During this era, Chinese government provided the scholarship programs for higher education institutes in Myanmar. At that time, the developed countries emphasized on outcomes-based education by assessing and accrediting the standard of education system to ensure the further development of the country. Hence, the international agreements, such as Bologna Process and Washington Accords, for accreditation of education systems were developed through six original signing countries. While in Myanmar, due to the lack of access to engineering education accreditation and laws regarding the engineering profession to protect the public property, the quality of the engineering education was a decline resulting in the public criticism. Thus, later in 2009, with the political changes in Myanmar, in order to overcome that problem, and with the help of the Federation of Engineering Institutions of Asia and the Pacific (FEIAP), the professionals attempted to establish accreditation procedures for engineering education through Myanmar Engineering Society (MES). Later, Myanmar higher education system put an effort to change into the outcomes-based education system by approaching the Bologna Process (BP). Engineering education in Myanmar, however, has been benchmarked with the requirements of the Washington Accord.

FEDERATION OF ENGINEERING INSTITUTIONS OF ASIA AND THE PACIFIC (FEIAP)

The Federation of Engineering Institutions of South-East Asia and the Pacific (FEISEAP) was an international non-profit professional organization founded on July 6, 1978 (FEIAP 2016). It was established following an exploratory meeting convened and organized by The Engineering Institute of Thailand under The King's Patronage with the support of the United Nations Educational Scientific and Cultural Organization (UNESCO) on July 03, 1978 in Chiang Mai, Thailand. Being an independent umbrella organization for the engineering institutions in the South-East Asia and the Pacific region, FEISEAP held objectives to encourage the application of technical progress to economic and social advancement throughout the world; to advance engineering as a profession in the interest of all people; and to foster peace throughout the world. The Constitution, as adopted unanimously by the Special General Assembly of the Federation held in Hanoi, Vietnam, on June 2, 2008, incorporated a change of its name to the Federation of Engineering Institutions of Asia and the Pacific (FEIAP).

As of now, there are 18 organizations within FEIAP, representing economies of Australia, China, Chinese Taipei, Hong Kong, India, Indonesia, Japan, Malaysia, Mauritius, Myanmar, Papua New Guinea, the Philippines, Singapore, South Korea, Thailand, Pakistan, Peru, and Timor Leste, respectively. The engineer institution of each of the economies, such as the Chinese Institute of Engineers (CIE) in Chinese Taipei, isis a member of the FEIAP.

FEIAP Engineering Education Guidelines

FEIAP established the Working Group on Engineering Education and the first working group meeting was held in October 2009 in Taipei. The Working Group was charged to draft a set of guidelines to help members that have not yet had engineering education accreditation system to establish one. The objective was to help those economies uplifting quality of engineering education through accreditation system in the respective economies so that they are able to join the international agreement on professional engineers, such as the APEC Engineers. The ultimate goal was of course the mobility of professional engineers among the FEIAP members. The Working Group was chaired by Chinese Taipei, and Dr. John Li of CIE served as the chairman Since CIE is a member of FEIAP but not the engineering education accreditation agency in Taiwan, Prof. Jia-Yush Yen, Secretary General of IEET, the accreditation agency in Taiwan, was invited to serve as the Co-Chair.

Together with Prof. Alan Bradley of Engineers Australia, Prof. Jia-Yush Yen and a team of researchers set out to draft the set of engineering education guidelines for FEIAP. After several revisions and consultations with the FEIAP General Assembly, the *FEIAP Engineering Education Guidelines* was officially presented in December 2010 with the approval of the FEIAP General Assembly in 2011. Since then, the Guidelines have been endorsed by UNESCO as well. Contents of the Guidelines include the following parts: Glossary of terms, Accreditation Criteria Template for the Accreditation of Engineering Education Programs, Accreditation System Model Framework, Mentoring System, Evaluation of Accreditation Agency, and Periodic Monitoring.

As the Guidelines published, FEIAP launched a series of efforts to mentor some of its members toward establishing the engineering education accreditation system. Papua New Guinea and Myanmar expressed interests in being mentored in FEIAP General Assembly in 2012. By 2013, the FEIAP General Assembly approved that Australia (Engineers Australia) and Singapore (Institute of Engineers Singapore) were assigned to mentor

Papua New Guinea, Malaysia (Institute of Engineers Malaysia) and Chinese Taipei (Chinese Institute of Engineers) were for Myanmar.

MENTORING EFFORTS OF IEET

Under the Ministry of Education's support, Institute of Engineering Education Taiwan (IEET) was established in 2003 by a group of senior presidents and deans. It launched accreditation of engineering education system in 2004 and became the signatory of the Washington Accord in 2007, Seoul Accord in 2009, and Sydney Accord in 2014 respectively. Over the past twelve years, IEET has accredited over 500 programs in 83 universities in Taiwan. It is now the country's leading accreditation agency in engineering and technological education. All accredited programs are waived from the Ministry of Education's evaluation. Since the outset, IEET has been promoting outcomes-based education in Taiwan. Specifically, the agency has been paying much attention to the program's training of students' graduate attributes through Capstone Course in recent years.

Since 2009, CIE has secured Ministry of Education's funding to facilitate the project of drafting the FEIAP Guidelines, and IEET became the de facto executing body because CIE is not the accreditation organization in Taiwan. When CIE was nominated as the mentor of Myanmar for accreditation system of engineering education under FEIAP, IEET was inevitably called again to help CIE in this endeavor. Under this project, IEET has invested a significant amount of time and human resources in helping FEIAP setting up the *FEIAP Engineering Education Guidelines*, and it also helped on the mentoring effort of MES in establishing the engineering education accreditation system.

The Ministry of Education played an important role in this endeavor because it provided CIE and IEET with millions-dollar financial support. FEIAP has not provided any financial support so far. Without the Ministry of Education's funding, CIE and IEET were not able to sustain their efforts toward FEIAP goals. The funding of the Ministry allowed for a series of visits by IEET delegates to Myanmar in collecting information about Myanmar's engineering education system, communicating with the local educators and engineers, visiting the universities, training of institutional representatives and potential program evaluators, etc.

From 2012 to 2013, IEET and Myanmar delegates exchanged seven visits. The first visit by delegates from IEM-Malaysia and IEET/CIE took place in July 2012 with the following purposes–

1. to set up an Engineering Accreditation Committee/Body (EAC) in Myanmar;
2. to have an Engineering Accreditation Manual;
3. to train Assessors for program accreditation;
4. to have one or two Universities to volunteer for the accreditation, so that the Manual and the System can be studied and reviewed.

During that first visit, the mentors visited Myanmar Maritime University (MMU), and met up with Rector, Pro-Rector, Head of Dept for Port & Harbor Engineering, and other Senate Members of MMU. Mentors were at the opinion that MMU has the potential to be accredited by EAC should the accreditation system is set up.

The second visit from FEIAP together with the team from CIE/IEET arrived Myanmar in May 2013. During that visit, formal visits were made to Myanmar Maritime University (MMU) and Yangon Technological University (YTU). The team had a fruitful discussion with the team at YTU led by the Chairman of Steering Committee for YTU and MTU (Mandalay Technological University), and Advisor for Ministry of Science and Technology. At the end of the visit, it was agreed that accreditation was now a must for quality assurance of engineering education and for mutual recognition of substantial equivalence throughout the region and the world and also suggested workshops for accreditation to have buy-in from stakeholders: policy makers, professional bodies, industry, academia and even students are needed.

By the suggestion of the second visit in 2013, there were workshops organized by FEIAP that were holding in the following years of 2014 and 2015 which included the following items:

1. Presentation on the need for mobility of engineers, international recognition and benchmarking of engineering degrees, accreditation and quality assurance;
2. Training on outcome-based and competency-based engineering education;
3. Self-assessment and documentation by education providers;
4. Training of program evaluators.

There was a third visit in 2014 for the purpose of conducting the workshop on February 15th, 2014 to facilitate the "Engineer Mobility and FEIAP Engineering Education Guideline". At that time, the Myanmar Engineering Council Law was already established in November 2013. Thus registration of engineers and accreditation of engineering programs were now to be done in Myanmar. Later on March 28th, 2015, IEET organized a workshop for the training of the engineering education evaluators of undergraduate programs to the Accreditation Committee members from Myanmar Engineering Council (MEngC). Not only that IEET delegates visited Myanmar for the purpose of promoting and training, CIE/IEET also provided funding for Myanmar delegates to come to Taiwan and observed IEET's accreditation visit. In October 2014, a group of three observers came and observed IEET's visit to Ming Chi University of Technology. Another observation happened in November of 2015, a group of four observers came and observed IEET's accreditation visit to National Taiwan Ocean University. During those two observations, Myanmar delegates had the first-hand experience on how accreditation review was conducted, and IEET believed that the on-site observation played a key role in the success of Myanmar's move to launch its own accreditation system (Liu 2015).

In October 2015, MEngC launched its very first accreditation review at eight programs in the YTU, which is the flagship engineering institution in Myanmar. This historical event marked the country's further step toward internationalization of higher education. Those programs being reviewed were Chemical Engineering, Civil Engineering, Electrical Power, Electronic & IT, Mechanical Engineering, Metallurgy, Mining, and Textile

Engineering. IEET delegates observed the entire process of the accreditation visit and provided comments and suggestions throughout the entire visit. Since the first batch of graduates will not graduate in three years of time, MEngC granted "provisional accreditation" to these programs after successful reviews and awarded the certificate to the programs in early 2016. More reviews are to take place in the months ahead.

MEngC had a good start on its accreditation system for engineering programs. Although there is a lot of room for improvement, MEngC was able to pull together a set of accreditation system in a mere three-year of time indicated that the country and those who are involved have a high level of motivation and determination of moving the country forward to be benchmarked with the international community. IEET would continue to provide necessary assistance and support to MEngC should there is a need called for. As indicated above, the ultimate goal of MEngC is to achieve signatory status of the Washington Accord. IEET is sure that through a continuous support from the Myanmar government and the local engineering community, this goal will be achieved sooner than it seems.

REFLECTION

As IEET reflects on the past years mentoring Myanmar, the following points and issues are of significance:

1. Development of any accreditation system must consider and remain sensitive of the local education systems and needs, it is not possible and prudent to copy one system into another country without any modification.
2. Internationalization and mobility of engineers are of key purpose for upcoming countries in their development of the accreditation system. In the case of Myanmar, becoming a signatory of the Washington Accord serves as a motivation for Myanmar to develop such accreditation system and to be connected with the outside world.
3. The mentoring experience turns out to be a win-win situation for MEngC and IEET; MEngC was able to establish the accreditation system in a relatively short period and avoid many hurdles that IEET encountered; IEET has gained a lot of confidence on its system as a consequence.
4. Friendship and collegiality are valuable outcomes of the mentoring; both IEET and MEngC became good friends throughout the process, and there are opportunities for further exchanges to take place.
5. From the perspective of the international accords, such as the Washington Accord, more countries develop accreditation system that meets the Washington Accord requirements benefits the betterment of the world-wide engineering education and the engineer community.

CONCLUSION

The mentoring experience for IEET was quite rewarding. Even though FEIAP did not provide any financial support, IEET along with CIE were able to secure Ministry of Education's funding from Taiwan for three years on this project. It is encouraging to see MEngC has launched its own accreditation system in a short time, and IEET hopes that in the years ahead, MEngC will gain sufficient confidence and tunes up its accreditation system to most suitable for the engineering education in Myanmar.

As a result of IEET's mentoring of engineering education accreditation system, IEET was invited to mentor Myanmar Architectural Council (MAC) in setting up accreditation system for architectural education in Myanmar. This development was not expected early on, but it is by all means a very positive outcome between the collaboration between Taiwan and Myanmar.

REFERENCES

FEIAP website. (March 2016). www.feiap.org.

Liu, M. (January 2015). "Delegates from Myanmar, the Philippines and Malaysia observed IEET accreditation visits." Evaluation Bi-monthly, Higher Education Evaluation and Assessment Council of Taiwan. http://epaper.heeact.edu.tw/archive/2015/01/01/6289.aspx

Than, C. (2015). "Engineering education accreditation in Myanmar." Paper presented at the 2015 FEIAP Convention. Taipei, Taiwan.

Comparison of Surveying Engineering Education in USA and Turkey

Dr. Esra Tekdal Yilmaz, Pennsylvania State University, Lehman

Dr. Tekdal Yilmaz is an assistant professor of surveying engineering at Penn State Wilkes Barre Surveying Engineering program. Her research interests include; Terrestrial laser scanning, deformation measurements and building information modelling

Comparison of Surveying Engineering Education in USA and Turkey

Abstract

Engineers with their creative, researcher and knowledgeable identities play an important role for the quality of life and welfare of the society. However the skills listed above are highly correlated with the quality level of engineering education given. Country specific educational habits and opportunities affect the learning pattern of individuals enrolled in engineering programs and therefore have impact on the career and future. In this study, possible improvement suggestions for the engineering education in Turkey will be given while comparing the teaching methods and conditions of two Accreditation Board for Engineering and Technology (ABET) accredited surveying engineering education programs in Turkey and USA.

1. Introduction

Engineering education in Turkey and USA are given with different number of total credits, curriculum, degree and program names which vary depending on the schedule, resources, usability of technology and course content. However there is one common thing in engineering education all around the world, to teach the individual how to reach the required knowledge using or finding correct information. In this way the students can gather vital and sound knowledge during their education life gradually and creatively.

The definition of engineering given by ABET is "Engineering is the profession in which a knowledge of the mathematical and natural sciences gained by study, experience, and practice is applied with judgment to develop ways to utilize economically the materials and forces of nature for the benefit of mankind."

Criteria, compatible with definition of engineering, prepared by ABET for accrediting engineering programs, 2016 – 2017 define the general program evaluation criteria as follows: [1]

1) Students
2) Program Educational Objectives
3) Student Outcomes
    (a) an ability to apply knowledge of mathematics, science, and engineering
    (b) an ability to design and conduct experiments, as well as to analyze and interpret data
    (c) an ability to design a system, component, or process to meet desired needs within realistic constraints such as economic, environmental, social, political, ethical, health and safety, manufacturability, and sustainability
    (d) an ability to function on multidisciplinary teams
    (e) an ability to identify, formulate, and solve engineering problems
    (f) an understanding of professional and ethical responsibility
    (g) an ability to communicate effectively
    (h) the broad education necessary to understand the impact of engineering solutions in a global, economic, environmental, and societal context
    (i) a recognition of the need for, and an ability to engage in life-long learning
    (j) a knowledge of contemporary issues
    (k) an ability to use the techniques, skills, and modern engineering tools necessary for engineering practice.

4) Continuous Improvement
5) Curriculum
6) Faculty
7) Facilities
8) Institutional Support

The goal of engineering education is to give the creativity, research ability and ability to solve problems on their own in addition to the criteria defined under student outcomes of ABET.

2. Widely used accreditation programs in Turkey and USA

There are two widely used accreditation programs MUDEK (Association for Evaluation and Accreditation of Engineering Programs) in Turkey and ABET in USA.

2.1 MUDEK

MUDEK, is a non-legislative organization working for enhancing the quality of engineering education in Turkey by means of the accreditation and providing information services for engineering education programs in different disciplines. It is currently comprised of 73 individual and 2 organizational members.[2]

It was initially established with a name Engineering Evaluation Board as an independent, non-governmental platform in 2002 by the Engineering Deans Council (MDK), which is formed by the deans of faculties administering engineering education programs in Turkey and Turkish Republic of Northern Cyprus (TRNC), to prepare and conduct a comprehensive program for the evaluation of engineering undergraduate programs run by these faculties.

On 26 November 2006 became a member of ENAEE (European Network for Accreditation of Engineering Education) and authorized by ENAEE to award EUR-ACE Label to undergraduate engineering programs starting from January 21, 2009 and authorization is renewed on October 16, 2013 until December 31, 2018.

In 2007, after becoming an association, on November 16, 2007 recognized by the Higher Education Council (YÖK) of Turkey as the National Quality Assurance Agency in accrediting engineering programs for five years and this recognition is renewed for another five years on February 01, 2013.

On June 25, 2010 became a Provisional Signatory of Washington Accord of IEA (International Engineering Alliance) and on June 15, 2011 became a full member signatory.

In Turkey there are 18 public and 2 private university offering normal education (day classes), 7 public university offering second education (evening classes) in surveying and similarly named engineering programs. 8 of them seen in table 1 are accredited by MUDEK and only one program in Turkey seen on table 2 is accredited by ABET.

School Name	Location	Validity Period of MÜDEK Accreditation	*Validity Period of EUR-ACE Label*
Karadeniz Technical University Faculty of Engineering, Normal Education (NE)	Trabzon, Turkey	01.05.2010-30.09.2017	*01.05.2010-30.09.2017*
Karadeniz Technical University Faculty of Engineering, Secondary Education (SE)	Trabzon, Turkey	01.05.2015-30.09.2017	*01.05.2015-30.09.2017*
Bulent Ecevit University Faculty of Engineering	Zonguldak, Turkey	01.05.2015-30.09.2020	*01.05.2015-30.09.2020*
Selcuk University Faculty of Engineering (NE)	Konya, Turkey	01.05.2008-30.09.2018	*01.05.2010-30.09.2018*
Selçuk University Faculty of Engineering (SE)	Konya, Turkey	01.05.2008-30.09.2018	*01.05.2010-30.09.2018*
Yildiz Technical University Faculty of Civil Engineering (NE)	Istanbul, Tukey	01.05.2007-30.09.2016	*01.05.2009-30.09.2016*
Yildiz Technical University Faculty of Civil Engineering (SE)	Istanbul, Tukey	01.05.2007-30.09.2016	*01.05.2009-30.09.2016*
Kocaeli University Faculty of Engineering	Kocaeli, Turkey	01.05.2014-30.09.2016	*01.05.2014-30.09.2016*

Table 1. MUDEK accredited surveying and similarly named engineering programs[3]

2.2 ABET

ABET founded in 1932 as the Engineers' Council for Professional Development (ECPD), committed to education, accreditation, regulation, as well as professional improvement of engineering professionals and students.[4]

ABET is not comprised of individuals but 35 professional and technical organizations specialized in the areas or professions which are eligible (highly qualified) to represent fields ABET accredits in applied computer science, computing, engineering and engineering technology.

ABET encourages alumni of accredited programs and other instructively qualified people who work or practice in the applied science, computing, engineering, and engineering technology professions to make progress toward expert acknowledgment by improving their individual qualifications through licensure and certification.

ABET's accreditation activities are led by four commissions that are in charge of reviewing educational programs and settling on the final accreditation decision for each program. These commissions are Applied Science Accreditation Commission, Computing Accreditation Commission, Engineering Accreditation Commission and Engineering Technology Accreditation Commission.

In USA almost all surveying and similarly named engineering programs are accredited by ABET spread all over the country as seen on table 2.

School Name	Location	Accreditation Dates	Date of Next Comprehensive Review
Alfred State College	Alfred, NY, US	10/01/1992-Present	2018-2019
East Tennessee State University	Johnson City, TN, US	10/01/1992-Present	2019-2020
Ferris State University	Big Rapids, MI, US	10/01/1989-Present	2017-2018
Idaho State University	Pocatello, ID, US	10/01/2004-Present	2017-2018
Michigan Technological University	Houghton, MI, US	10/01/2011-Present	2017-2018
New Jersey Institute of Technology	Newark, NJ, US	10/01/1992-Present	2017-2018
New Mexico State University	Las Cruces, NM, US	10/01/1999-Present	2018-2019
Pennsylvania State University	Lehman, PA, US	10/01/2004-Present	2018-2019
Polytechnic University of Puerto Rico	San Juan, PR, US	10/01/2006-Present	2019-2020
St. Cloud State University	St. Cloud, MN, US	10/01/2004-Present	2017-2018
The University of Akron	Akron, OH, US	10/01/2010-Present	2016-2017
Troy University	Troy, AL, US	10/01/2010-Present	2016-2017
University of Maine	Orono, ME, US	10/01/2005-Present	2020-2021
California State University, Fresno	Fresno, CA, US	10/01/1979-Present	2018-2019
Florida Atlantic University	Boca Raton, FL, US	10/01/2010-Present	2020-2021
Nicholls State University	Thibodaux, LA, US	10/01/2008-Present	2015-2016
Oregon Institute of Technology	Klamath Falls, OR, US	10/01/1985-Present	2018-2019
University of Alaska Anchorage	Anchorage, AK, US	10/01/1995-Present	2016-2017
University of Florida	Gainesville, FL, US	10/01/1986-Present	2018-2019
California State Polytechnic University	Pomona, CA, US	10/01/1992-Present	2017-2018
Kennesaw State University	Kennesaw, GA, US	10/01/2004-Present	2017-2018
Istanbul Technical University	Istanbul, Turkey	10/01/2009-Present	2016-2017

Table 2. ABET accredited surveying and similarly named engineering programs[5]

2.3 Comparison of MUDEK and ABET criteria for surveying and similarly named engineering programs

In both accreditation criteria there are general and program specific measures. General criteria, consisting of 8, defined in ABET are the same with general criteria of MUDEK. But in addition to those MUDEK has one more criteria about organization and decision making stating that all decision making process between the organizations of higher education institution, chancellery, faculty, department and other sub units (if there are any) should be designed to support the achievement of educational objectives.[2]

Discipline specific evaluation criteria for MUDEK states that graduates should prove to be enough for at least one of the following topics: boundary and / or land surveying, geographic and / or land information systems, photogrammetry, cartography, geodesy, remote sensing and other related topics. [2]

Under discipline specific evaluation criteria for ABET for surveying and similarly named engineering programs National Society for Professional Surveyors and American Society of Civil Engineers are listed as lead society and cooperating society respectively. [1]

There are two criteria defined. "The curriculum must prepare graduates to work competently in one or more of the following areas: boundary and/or land surveying, geographic and/or land information systems, photogrammetry, mapping, geodesy, remote sensing, and other related areas." And "Programs must demonstrate that faculty members teaching courses that are primarily design in content are qualified to teach the subject matter by virtue of professional licensure or by educational and design experience."

3. Comparison of surveying engineering education in Turkey and USA

Although surveying engineering education is different in various universities in Turkey most of the problems they are facing are usually common.

In most of the universities grading curves are used for grading. Depending on the average level of the class, students sometimes pass the class by chance without obtaining proper knowledge. Also the idea of competing with each other limits the knowledge and material share between the students.

Surveying engineering education is mostly Turkish however there are some programs with %30 or %100 English as teaching language. Foreign language education in universities are not sufficient enough for engineering education. In addition to that there is a lack of Turkish literature for engineers. Lack of resources and language forces the students to memorize the information given directly by the instructor without judging or considering. This behavior effects the learning pattern badly and therefore affects the future performance of surveying engineers.

The classes are usually crowded compared to most of the classes in USA, this effects the teaching and evaluation performance of the instructor during class and when grading exams. There aren't enough high quality instructors and the wages are low according to the economic conditions of Turkey. These factors affect the motivation of instructors and have a huge impact on students learning abilities.

Research budget and potential are low and cooperation between industry and university is not encouraged and supported. Main authorities in surveying engineering do not share the knowledge and material throughout their web sites as some of the important authorities in USA for surveying engineering like NOAA (National Oceanic and Atmospheric Administration) and NGS (National Geodetic Survey) does.

Second education (night classes), are given by the same academic staff who is in charge of normal education (day classes). This effects the concentration and efficiency of academic staff thus quality of education and graduates.

There is a lack of accreditation in surveying engineering programs only 1 program is accredited by ABET and 8 of 27 are accredited by MUDEK. This also effects the after graduation education of surveying engineers. In most of the USA states graduates of surveying engineering programs take a test to become a licensed surveyor and join classes even after graduation to keep their licenses.

4. Recommendations for improvement

In order to increase the quality of graduates and surveying engineering programs in Turkey some precautions should be taken.

There should be a passing grade criteria for grading system. Students who cannot satisfy the minimum criteria should fail to ensure quality.

Prior to starting engineering education students should increase English, as a foreign language to an adequate level to keep up with the developing technologies all around the world. While doing that number of lecture and research resources should be increased in the libraries to encourage the students for research.

Due to various factors throughout the education life, engineering education has become a know by heart learning system which decrease the ability to think and develop practical solutions for engineering problems. To avoid that, classes should be based on up to date discussion topics about new technologies encouraging the students to attend, gather and share knowledge.

To ensure the affectivity of universities in the country's development, financial resources transferred to the universities should be increased. To keep up with technological improvements, invent new technologies and to accelerate the development of the country, collaboration between universities and industry should be encouraged by developing new cooperation models.

Researches should be supported by raising funds and financial support should be provided to faculty engaged in research.

Majority of the engineering programs are accredited by MUDEK. However discipline specific evaluation criteria of MUDEK is not well defined as in ABET and in addition to that MUDEK mostly consists of individuals instead of organizations. The most important things in engineering education are curriculum, faculty, leading and cooperating supporting organizations. Well defined curriculum together with qualified faculty and supporting organizations specialized in surveying engineering ensures high quality education to students and post-graduation support to graduates.

Final and most important recommendation is to set accreditation for engineering education as mandatory. For Turkey MUDEK, after fixing the weaknesses defined in the previous paragraph, and/or ABET should be acknowledged as accreditation authority.

5. Conclusion

In our developing world rapid technological changes, conditions and needs, forces universities to give education compatible with the rules of changing world. Conditions in engineering education shows that know by heart technical education without forcing students to think and create is not enough for today's competitive circumstances.

In conclusion surveying and similarly named engineering programs in Turkey are in need of accreditation to build the necessary infrastructure for graduating knowledgeable, problem solving, leading, creative and self-confident engineers and to ensure continuous improvement of graduates.

References

1. ABET, Accreditation Board for Engineering and Technology, Criteria for Accrediting Engineering Programs, Baltimore: ABET, Engineering Accreditation Commission, 2015. Available online at: http://www.abet.org/wp-content/uploads/2015/10/E001-16-17-EAC-Criteria-10-20-15.pdf

2. MUDEK, http://www.mudek.org.tr/en/belge/doc.shtm accessed March 11, 2016.

3. MUDEK List of Accredited Programs, http://www.mudek.org.tr/en/akredit/akredite2015.shtm accessed March 11, 2016.

4. ABET, http://www.abet.org/about-abet/, accessed March 11, 2016.

5. ABET accredited program search, http://main.abet.org/aps/Accreditedprogramsearch.aspx, accessed March 11, 2016.

Best Practices of Engineering Education Internationalization in a Russian Top-20 University

Julia Ziyatdinova,

Julia Ziyatdinova graduated from Kazan State Pedagogical University in 1999. Her major areas of study were foreign languages and she finished her University course with honors and qualification of teacher of English and Turkish. She continued her training and obtained PhD in Education degree in 2002. The topic of her PhD study was titled "System of Character Education in the US Schools: Current State and Trends for the Development". She also received additional minor degrees in Management (1998) and Psychology (1999) in Kazan State Technological University.

Julia joined the team of Kazan State Technological University as an instructor at the Department of Foreign Languages and the School of Foreign Languages "Lingua" in 1999 and was rapidly promoted to the position of Associate Professor at the Department of Foreign Languages in 2003. Her teaching career was perfectly balanced by the experience of a translator and an interpreter. She is a well-known person at Kazan international conferences and other events for her high quality consecutive and simultaneous interpreting, such as interpreting for the Academy of Sciences of the Republic of Tatarstan.

The new milestone in Julia's career was the position of the Chair of Department of Foreign Languages for Professional Communication in 2007, when she took over all the responsibilities related to foreign language training at Kazan State Technological University. The teaching and research priorities of her department were then focused on professional and intercultural communication for students in a technical university, professional translation and creation of foreign language environment at a university.

Because of her talents and activities, Julia became one of key figures in university international life. When Kazan State Technological University obtained the new status of a National Research University and joined the list of Top 30 Russian universities, Julia was offered a position of a Head of University International Office. She took over this position in April 2011 and rapidly gathered a strong team of professionals to face the challenges of the new university status and transformed International Office into University International Affairs with two offices covering all the aspects of internationalization.

In addition to her intensive career, Julia is also the Director of Center for Intercultural Communication – a company within the University structure offering excellent language training services for students and adult employees.

Julia is the author and co-author of over 85 publications including monographs, journal articles and study guides.

Dr. Artem Bezrukov

Artem Bezrukov graduated from Kazan National Research Technological University (KNRTU) in 2007. His major program was Chemical Engineering. He has also a minor degree in Translation for Professional Communication. Artem Berukov received his PhD in chemistry at the same university in 2010. His is the head of Protocol Office at International Affairs and an associate professor at the Department of Physical Chemistry at KNRTU. His activity areas include internationalization of higher education, colaboration with universities and research organizations in the USA, Europe, and Asia. Artem Bezrukov is the author and co-author of over 50 publications including papers in peer-reviewed journals, proceeding of international conferences, workbooks and monographs.

Dr. Phillip Albert Sanger, Purdue University, West Lafayette

Dr. Sanger is a professor in the School of Engineering Technology in the College of Technology of Purdue University. His focus and passion is real world, industry based, senior capstone experiences both domestically and internationally. He has successfully developed this area at Purdue and at Western Carolina University. Prior to his career in academia, Dr. Sanger had a successful 30 year career working in and with industry managing and participating in broad range technology development and commercialization.

Dr. Petr Osipov

Best Practices of Engineering Education Internationalization in a Russian Top-20 University

Abstract

Top world universities apply complex internationalization strategies to keep a sustainable leadership confirmed by the global rankings such as QS, THE, ARWU, or to improve their position in these rankings. Russian universities are not an exception and the analysis of the websites of Russian Top-20 universities (according to the National Ranking of Russian Universities, 9 universities of this Top-20 list are engineering universities) demonstrates their strong internationalization infrastructure.

Tracking success stories and best practices of such universities using accurate statistics is a useful tool for sharing their internationalization experiences, finding matching points for cooperation and predicting of their competitiveness potential in the global higher education community.

This paper summarizes 5 years of best internationalization practices and statistics of a Russian Engineering University in the national Top-20 list. The statistical data include dynamics of international students by their total number and country of origin; the statistical analysis of international delegations; the data about foreign professors and guest lecturers; international conferences and other events; the statistics of academic exchange and grants. Best practices and internationalization success stories are correlated with this statistical information and are generally analyzed and characterized using these data as the background to become more formalized and easily shared with other universities.

Apparently good results have been achieved through the following practices: step-by-step collaboration plans with selected top universities in the USA, Europe and Asia; focused Federal Government funding for joint projects (research, guest lecturers and joint conferences) with top world scientists in approved development priority areas; success in Asia through personalized activities carried out by the University's Honorary Professors from China, South Korea and other Asia-Pacific countries; training hundreds of University faculty in the top US and European universities; networking with regional, national and global academic and business partners. Another focus is given to the best practices in the global engineering education environment including collaboration with ASEE and IGIP (such as hosting IGIP conferences and success in IGIP International Engineering Educator Program ING PAED IGIP, plenary session at ASEE International Forum).

These 5 years of activities resulted in achieving the Top-150 position in the QS BRICS list (while the University was not listed in any international rankings before), a 3 fold increase in the number of international students, and the infrastructural changes in the university such as creation of new internationalization infrastructure and a multilingual environment to achieve sustainability in internationalization growth.

Background

Top world universities apply complex internationalization strategies[6,7] to keep a sustainable leadership confirmed by the global rankings such as QS, THE, ARWU, or to improve their position in these rankings. Russian universities are not an exception and the analysis of the websites of Russian Top-20 universities (according to the National Ranking of Russian Universities, 9 universities of this Top-20 list are engineering universities) demonstrates their strong internationalization infrastructure.

Tracking success stories and best practices of such universities based on real statistics is a useful tool for sharing their internationalization experience, finding matching points for cooperation and the prediction of their competitiveness potential in the global higher education community.

This paper summarizes 5 years of best internationalization practices and statistics of a Russian Engineering University in the national Top-20 list. The statistical data include:

- dynamics of international students by their total number and country of origin;
- statistical analysis of international delegations;
- data about foreign professors and guest lecturers;
- international conferences and other events.

Best practices and internationalization success stories are correlated with this statistical information and are generally analyzed and characterized using these data as the background to become more formalized and easily shared with other universities.

The selected best internationalization practices of a Russian engineering university are described below. The statistics is provided for the period of 2007 – 2015 (or 2007-2014 as the 2015 data are still under processing). The university obtained its National Research University status in 2010, so the 2007-2009 period is given for reference. The 2010-2015 block of data is the statistics reflecting implementation of its Development Program and practices intended to boost its internationalization.

Description of best practices

Focusing on internationalization while obtaining a national research university status

Following the experience of the U.S., where the universities are the key national centers of research activities, the Russian Government launched an initiative of creating national research universities in 2009. Such universities were thought to become regional centers of world-class fundamental and applied research. To win this status, the applicants were to submit a Development Program for 2010-2019 with a step-by step strategy of obtaining global competitiveness in research.

The first good decision of the university administration was to add strong internationalization component to its Development Program including the growth of the number of foreign students, attracting funding for international projects, developing international academic programs and training of faculty at top world research centers.

After winning the status of a National Research University in 2010, the university received a USD 300 million grant for the implementation of its development initiatives, including internationalization. It should be noted that internationalization funds are usually provided by academic mobility agencies on a mostly individual basis. For the first time in the university history, it obtained significant funds for internationalization as a part of a huge government grant.

These specific funds provided considerable intensification of the university international activities which can be seen in all the diagrams below.

Step-by-step collaboration plans with selected top universities in the USA, Europe and Asia

The university created a "belt of priority partners" in selected world regions (Fig. 1) and focused on several partners instead of signing dozens of MoUs with all the universities it gets in touch with.

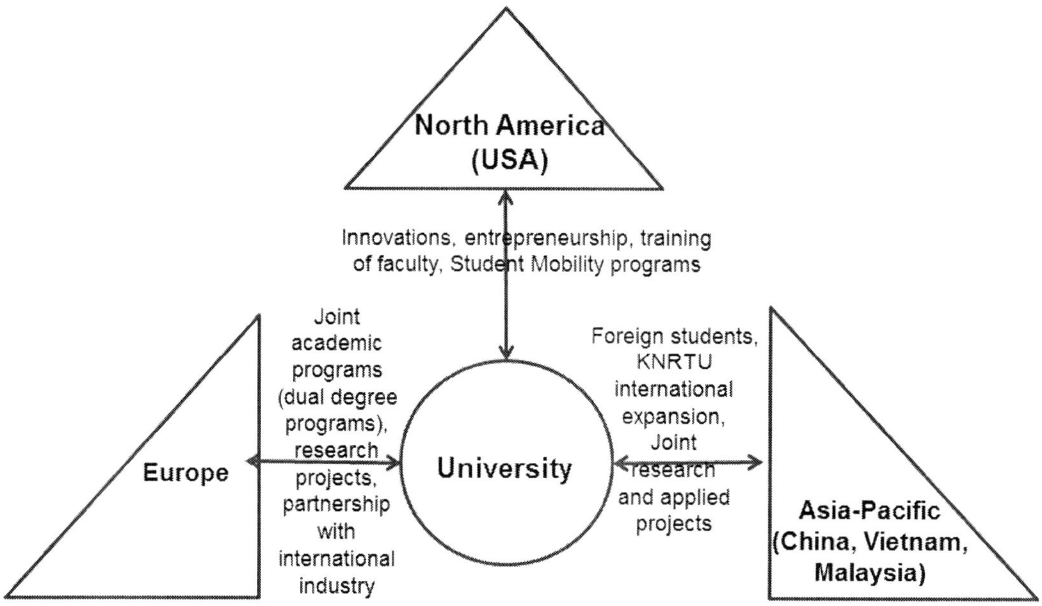

Fig. 1. Priority regions for internationalization.

The USA was selected as the primary country for internationalization with the focus on student mobility programs, training of faculty and innovative research. The university also selected some primary partners in Europe (mostly in Germany and France) and Asia (mostly in China). The total number of active partners is around 15-20, that is more than sufficient.

Success in Asia through personalized activities carried out by the University's honorary professors

The university invited top scientists from China, South Korea and Malaysia to become its honorary professors. Several years of further partnership with these distinguished people resulted in obtaining "agents of influence" in the Asia-Pacific region providing the university with the access to joint projects with Chinese Academy of Sciences, Federation of Asian Chemical Societies etc.

Another success story with honorary professors is the access to organization of local conferences and symposiums. The university organized specialized plenary sessions at conferences in Malaysia and Vietnam hosted by its honorary professors.

Attracting government funding for training hundreds of University faculty in top US and European universities.

A part of the National Research University Development Program was to intensify academic mobility of its faculty. Fig. 1 demonstrates small intensity of faculty visits to foreign universities in 2007 – 2009. After obtaining the national research university status, the university administration decided to send faculty members to the U.S. and European universities for short-term training and making further connections for internationalization. After 2010, there is a 4-fold growth in faculty academic mobility. In 2010-2012, the majority of travels, however, were first-time visits without further continuation. In 2013-2014, the visits became more dedicated and regular: presentations at conferences, lectures, and regular research visits, while the part of 'inefficient' academic mobility decreased. The university found a number of 'stable partners' to collaborate with[2].

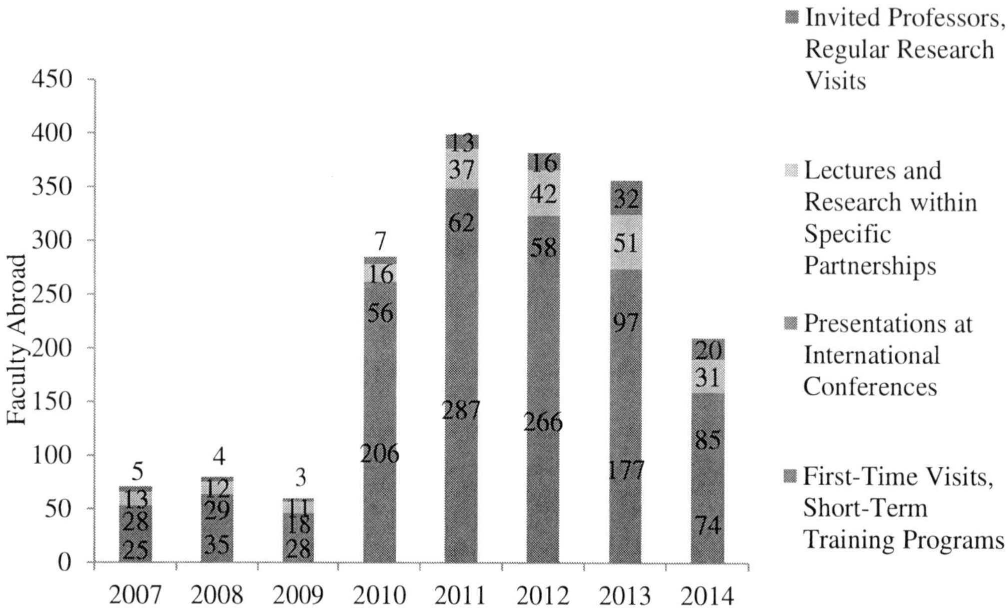

Fig. 2. Academic mobility of the University faculty.

Integration into the global engineering education environment through collaboration with ASEE and IGIP

Another focus was given to the best practices in the global engineering education environment including collaboration with ASEE, AIChE and IGIP (such as organization of IGIP conferences and success in IGIP International Engineering Educator Program ING PAED IGIP. In 2011-2015, the university hosted several events with the contribution of IGIP and participated in ASEE and AIChE annual conferences. While government funds were available for such initiatives in 2011-2013, the University was motivated to attract other funds to continue these activities in 2014-2015, such as international grants etc.

Advertising the National Research University status to attract more international students

Fig. 3 represents the growth in the number of international students. Before 2010, there is definite growth in the number of international students. Growth intensity, however, increased after 2010, when the university obtained its new status. The total number of international students tripled in the 2010-2015 period.

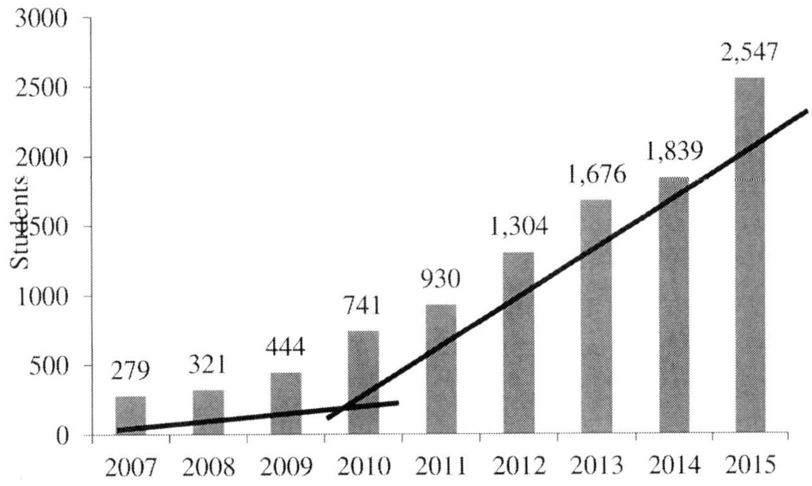

Fig. 3. Dynamics of the number of international students.

The distribution of international students by country (Fig. 4) provides additional clarification. The majority of foreign students is provided by the ex-soviet countries, Africa and Asia. The status of a National Research University resulted in better image of the university in the CIS countries and Asia (for example, Chinese government implements similar initiatives of creating top research universities in the country, so the national research university was considered as a top-level university in Russia). An additional attraction factor is that all the top positions in Russian university rankings are occupied now by national research universities, and prospective students make selection from the top of the list.

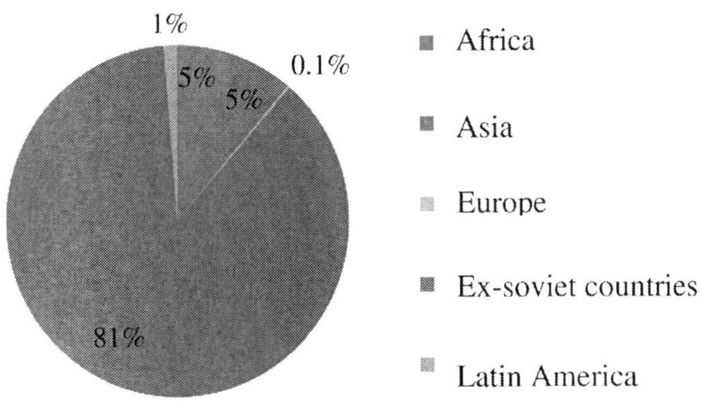

Fig. 4. International students by country of origin in 2015.

Conferences as a tool for massive training of the university faculty

The university succeeded in professional training of its faculty in best U.S. universities in 2010-2012. Another approach was developed in 2012-2015 to share the expertise of top world scientists with more faculty and students. The university attracted federal and local funding intended for professional training of its faculty to invite scientists from the U.S. and European universities. These scientists participated in the conferences hosted by the university. At the same time, they spent several days with students and faculty as lecturers and workshop supervisors, round table moderators and research experts. Such events were officially titled as "Scientific Schools" and were usually organized as conference satellites. Hundreds of students and faculty who participated in these Scientific Schools" were awarded with short-term professional training certificates and got real experience of communication with global experts.

Fig. 5 summarizes the statistics if university conferences (which were accompanied by scientific schools). The university has been hosting about 15-20 conferences on an annual basis since 2010. Each of such conferences provides short-tern training for 100-200 students and faculty.

The university professors also increased their presence at international conferences (Fig. 5, blue columns) due to intensified personalized connections with their foreign colleagues after training in the U.S. and European universities and meeting foreign scientists at conferences hosted by a university.

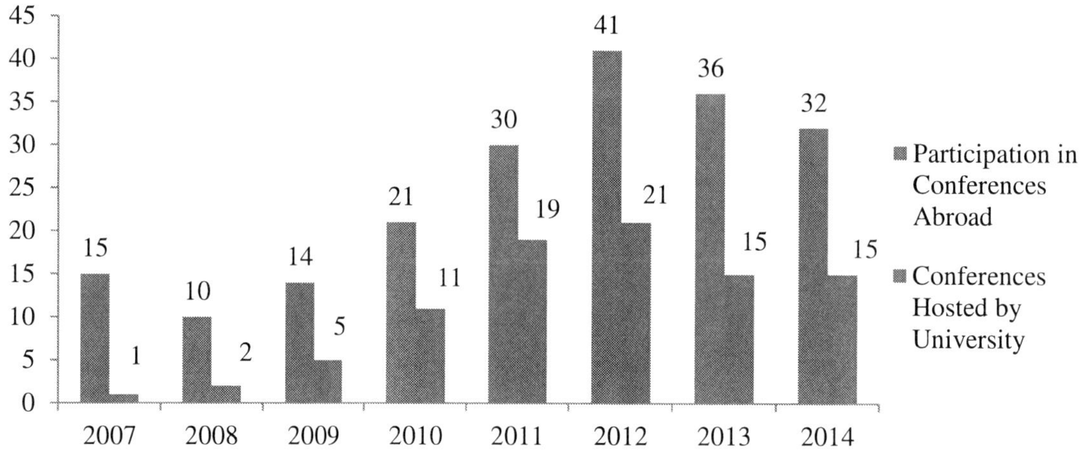

Fig. 5. International conferences at university and abroad.

Results and Conclusions

These 5 years of activities resulted in achieving the Top-150 position of the QS BRICS 2015 list (while the University was only in the Top-200 of the QS BRICS 2014 list and was not listed in any international rankings before), 3 times increase in the number of international students, 2-3 times increase in the intensity of international activities (such as participation in the international conferences, professional training abroad and hosted events with foreign participants).

Another important result is the transformation of these quantitative changes into new quality: infrastructural changes in the university, as the administration faced challenges from the intensification of international activities. New offices have been created, such as Office of Legal Support, which is to overcome administrative bottlenecks in internationalization, Office of Joint International Programs, and Department of Foreign Languages for Professional Communication. This Department implements the Program of Multilingual Environment to achieve sustainability in internationalization growth by overcoming the major barrier: insufficient level of English[1,3,4,5]. This program offers intensive English training for all the university groups: from undergraduate students to deans and administrators to provide professional communication skills.

Thus, addition of the internationalization component to the Development Program of a National Research University allowed achieving new level of internationalization by attracting special funding. Selected practices have been applied to trigger boosting of international activities. It is very important that although the federal funding was over in late 2013-early 2014, these practices continued (Fig. 2-5, years 2014-2015) and were much more intensive than the respective activities before the national research university period (m Fig. 2-5, years 2007-2009). The main reason of this effect is the resulting implementation of a new internationalization infrastructure and changed mindset of faculty and students, who start considering internationalization as a necessary component of their research and academic activities and started seeking additional funding for their overseas partnerships.

These practices may be recommended for other universities in Russia as well as universities in China and India, which apply for similar special government funding programs as well as for the U.S. universities looking for successful approaches to the universities in these countries.

Acknowledgement
The research was funded by Russian Foundation for Humanities grant, project # 15-16-16003.

Bibliography

[1] Bezrukov, A., Ziyatdinova, J. (2014) Internationalizing engineering education: A language learning approach. Proceedings of 2014 International Conference on Interactive Collaborative Learning, ICL 2014, P.229-302.

[2] Kraysman, N.V.,Valeeva, E.E. (2014) Integration of KNRTU into the world community as an example of cooperation with France. Proceedings of 2014 International Conference on Interactive Collaborative Learning, ICL 2014, P.862-863.

[3] Semushina, E. Y. & Galeeva, M. V. (2013). Distant form of mastering translator's competence as a part of engineering education in Kazan Federal University. Proceedings of 2013 International Conference on Interactive Collaborative Learning, ICL 2013, P. 519-521.

[4] Shishova, E.O. (2015) A model of second language learning by means of developing the prognostic component of thinking The Social Sciences, 10: 384-388.

[5] Timirbaeva, G.R. Language learning strategies and their impact on learner autonomy. Proceedings of the 2013 International Conference on Interactive Collaborative Learning, ICL 2013, P.621-622.

[6] Ziyatdinova, J.N, Ivanov, V.G., Bezrukov, A.N., Osipov, P.N., Sanger. P.A. (2015) Going Globally as a Russian Engineering University. 2015 ASEE Annual Conference and Exposition, Conference Proceedings. Paper ID 12035.

[7] Ziyatdinova, J.N., Osipov, P.N. (2012) Integrative approach to intercultural competence development in engineering education. Proceedings of 2012 International Conference on Interactive Collaborative Learning, ICL 2012. INSPEC Number 13248356.

The Case of an Electrical and Computer Engineering (ECE) Department in the Internationalization Process of a Research I Public Institution

Dr. Fabiola P Ehlers-Zavala, Colorado State University

Fabiola P. Ehlers-Zavala was named INTO Colorado State University (CSU)'s Center Director in November 2014, having previously fulfilled the role of INTO CSU Academic Director (March 2013-November 2014). In her CD capacity, she works with Colleges across campus, and has a particular interest in the preparation of international students pursuing engineering degrees at the undergraduate and graduate levels. She earned her B.A. in English Language and Literature together with her teaching certificate from the Pontificia Universidad Católica de Valparaíso in 1992. She then pursued graduate education in the U.S., and she earned both her M.A. (1994) in English and Ph.D. in English Studies (1999) from Illinois State University. Upon her graduation, she worked for Illinois State University (ISU) as Assistant Professor in Bilingual/Bicultural Education until she received her tenure and promotion to Associate Professor. While at ISU, in her last year, she directed the Bilingual/Bicultural Education Program in the College of Education. In 2006, she moved to Colorado to teach in the M.A. in English at Colorado State University where she received her second tenure and promotion to Associate Professor in 2009. Between 2009 and 2013, she directed the M.A. in English (TESL/TEFL). Her areas of expertise include: second language/bilingual reading, second language assessment, and ESL/bilingual teacher preparation. She is the coauthor of Reading Strategies for Spanish Speakers. Her publications in books and journals include "Meeting the reading comprehension challenges of diverse English language learners in K-12: Key contributions from reading research" (2016), "Advocacy in Language Teaching" (2013), "History of Bilingual Special Education" (2011), "Bilingualism and Education: Educating At-Risk Learners" (2010), "How Can Teachers Help Adolescent English Language Learners Attain Academic Literacy?" (2009), "Teaching Adolescent English Language Learners" (2008), "Assessing English Language Learners (ELLs) in Mainstream Classrooms" in The Reading Teacher (2006 & reprinted in 2010); "Bilingual Reading from a Dual Coding Perspective" (2005) in Proceedings of the 4th International Symposium on Bilingualism; "Preparing Quality Bilingual/Bicultural Teachers in the 21st Century: A PDS Model for Educational Change and Success" (2004), "Use of Lexical Borrowings in Sonoran Border Spanish" (2003). She serves on the editorial board of The International Multilingual Research Journal (IMRJ) and TESOL Journal. She is Past President of Illinois TESOL/Bilingual Education, Past Chair TESOL International Bilingual Education Interest Section and Past Chair of TESOL International's Nominating Committee. Most recently, Dr. Ehlers-Zavala served as a member of the International TESOL Diversity and Inclusion Committee (2014-2015), and has been invited to serve in the Editorial Review Board for The Reading Teacher (RT) Volume 70 review year (2015-2016).

Dr. Anthony A. Maciejewski, Colorado State University

Anthony A. Maciejewski received the BSEE, MS, and PhD degrees from the Ohio State University in 1982, 1984, and 1987, respectively.From 1988 to 2001, he was a professor of electrical and computer engineering at Purdue University, West Lafayette. He is currently a professor and the department head of Electrical and Computer Engineering at Colorado State University. He is a fellow of the IEEE. A complete vita is available at: http://www.engr.colostate.edu/ ~aam.

The Case of an Electrical and Computer Engineering (ECE) Department in the Internationalization Process of a Research I Public Institution

Colleges of Engineering at US tier I public institutions have had a long-standing tradition of welcoming international students to their various programs, especially at the graduate level. Recently, the percentage of international students studying at the undergraduate level has also been increasing. Out of all the engineering disciplines, this is perhaps most frequently true for Electrical and Computer Engineering (ECE) programs. Despite the relative decline in US students interested in studying ECE as compared to other engineering disciplines, ECE programs remain highly attractive to international students who wish to come to the US to pursue the best education possible and acquire the knowledge and skills to meet their career and professional goals. Unfortunately, ECE departments, as is the case for the one we will describe in this paper, often find themselves having to turn away many highly qualified candidates because they are confronted with issues of capacity, so admission to highly ranked ECE programs remains extremely competitive in the US.

The changing landscape of higher education in the US is presenting ECE programs with a new challenge, which is the welcoming of a different type of student who is oftentimes perceived as less deserving of access to the highly competitive programs universities offer. We are referring to a group of students known as Pathway (PW) students. PW students are international students who fall short of meeting direct admission requirements when it comes to both GPA and performance in English language proficiency tests for admission to a university. These students enroll in PW programs that offer greater support, and provide them the opportunity to succeed in such a competitive environment. These PW programs are relatively new in the US and are often the direct result of new efforts at some universities to develop comprehensive plans for campus internationalization. Such efforts may be facilitated by the formation of public / private partnerships or Joint Ventures (JVs) between the university and a private company. These new partnerships constitute new creative ways that some universities have found to diversify their campus communities and gain a global perspective. A growing number of universities in the US have been embracing these non-traditional models for rapidly internationalizing their university campuses. It has become evident that the internationalization of university campuses is no longer just a matter of signing memoranda of understanding to facilitate student mobility through the traditional college year abroad experience.[3] In this paper, we describe the specific case of Colorado State University (CSU) and the response of its ECE department to this new development.

The Case of Colorado State University. In the spring of 2012, CSU took an audacious step when it signed a 30-year agreement with INTO University Partnerships (IUP) to form a JV that would become known as INTO CSU. INTO CSU is a limited liability corporation that serves as the new English language center for the university and is equally owned by CSU and INTO IUP. Though run by a Center Director for day-to-day operations, it is ultimately governed by the INTO CSU Board, comprised of three CSU members from Central Administration and three senior officials from INTO. Their shared goals are (1) to increase international student enrollment; (2) to tap into yet another group that can contribute to expand those numbers (i.e., Pathway students); (3) to provide international experiences to the domestic students as well as the university staff, academics, and administrators, who may not be able to go abroad; (4) to serve those coming from abroad with the goal of pursuing an education in the US, given the high value placed on American higher education; (5) to internationalize the campus community at large; and (6) to enrich the off-campus community that will also benefit from welcoming greater linguistic and cultural diversity. Through the formation of this partnership, CSU found a new way to emerge from an era of limited resources and remain competitive at a time when the globalization of higher education programs in the US has become crucial to the preparation of university graduates who need to function in a highly interconnected world and demonstrate they have developed the social and professional skills to function in highly diverse and dynamic environments.

With the advent of INTO CSU, PW programs were developed very quickly at the undergraduate and graduate levels for a Fall 2012 implementation. Initially, among the repertoire of programs, the College of Engineering created multiple PW programs. At the undergraduate level, a two-semester (Standard) PW in Engineering was developed (see Figure 1); and at the graduate level, several PWs were created in Masters of Engineering (ME) degree programs for the following areas: Civil Engineering, Chemical Engineering, Electrical and Computer Engineering (see Figure 2), and Mechanical Engineering. The ME programs were created because these are coursework-only programs that do not overly burden faculty, i.e., they do not have to take on more graduate advisees. It is important to underscore that the development of these PWs occurred very quickly, and it was in the hands of the CSU Academic Working Group that had been formed and tasked with the development of these PWs. The members of this group were varied, representing key areas of the CSU operation: undergraduate and graduate affairs, advising, student affairs, international programs, and so forth. They did not necessarily involve the faculty in the specific departments, which resulted in some considerable backlash as described in the next section using ECE as a case in point.

The case of ECE at CSU. Because the formation of INTO CSU happened so quickly and the development of PWs did not always involve the key administrators and faculty that directly run the degree programs, there was much initial resistance to the new proposition. Unfortunately, many faculty erroneously perceived this effort as being pursued by CSU's Central Administration purely to cope with financial constraints due to rapidly shrinking financial support from the state. The lack of an effective communication plan to articulate the core goals of this new initiative was probably the result of underestimating the level of resistance that would become evident shortly after the inception of INTO CSU. Unfortunately, this initial sentiment continued to exist in varying degrees among many departments at CSU for several years. In the following paragraphs, we describe the journey of ECE in this new context.

The ECE journey relative to INTO CSU. In analyzing and reflecting back on the journey of the ECE department since INTO CSU came to be, it is possible to identify the following four stages:

1) Resistance
2) Awakening
3) Collaboration
4) Innovation

Figure 1. Sample Standard PW for Engineering—Undergraduate

Engineering

Standard Pathway - 2 Semesters

Fall: August 15, 2016
Spring: January 9, 2017

Degree Program Components

125-159 credit hour degree programs
18-22 credit hours apply from Pathway[1,2]
103-141 credits remaining toward degree[1]

Program Information

Entry Requirements

- High school diploma with 2.5 GPA
- Language requirements:
  - TOEFL iBT 60 **OR** IELTS 5.5 (5.5 minimum subscores in reading and writing) **OR** PTEA 44 **OR** Password 6 **OR** completion of the Academic English Level 3 with a minimum 75% in all graded courses

All students should take the online proctored version of the CSU Mathematics Placement Examination (MPE) before arrival.

Semester 1	Course Title	Prerequisites	Credit Hours
EAP 150	**English for International Students I**		6
SPCM 100	Introduction to Popular Culture		3
MATH 160[3,4]	Calculus for Physical Scientists I	See notes 3 and 4	1-5
ENGR 102	Engineering Problem Solving		3
		Total	**13-17**

Semester 2	Course Title	Prerequisites	Credit Hours
EAP 151	**English for International Students II**		3
	AUCC General Education Course[5]		3
MATH 161	Calculus for Physical Scientists II	MATH 160	4
CHEM 111[2]	General Chemistry I **AND**	MPE or MATH 118	4
CHEM 112[2]	General Chemistry Laboratory I **OR**	CHEM 111 or concurrent registration	1
PH 141	Physics for Scientists and Engineers	MATH 160 or concurrent registration	5
		Total	**15**

Progression Requirements

- Cumulative CSU GPA of 3.0
- B or better in MATH 160 or 161
- B or better in PH 141 or CHEM 111
- C or better in CHEM 112
- No grade lower than a C in all Pathway classes, including those in mathematics, science and engineering
- No I or U grades; no unresolved W grade in a required Pathway course; no S grades, except in cases of credit earned by exam
- Submit a Pathway progression application

Notes

1 Total credits remaining toward degree depend on specific engineering program of study.

2 CHEM 111/112 may not apply toward graduation in some engineering programs.

3 Prerequisites are (a) earned transferable Calculus credit, (b) MPE results with MATH 160 placement or (c) MATH 117, 118, 124, 125, 126.

4 Students not starting in MATH 160 may not be able to complete the Pathway program in two semesters. .

5 If students are eligible to complete the second semester of their Pathway during the summer, this course is not required during the Summer semester. Students who begin in the Spring semester and plan to finish during the Summer semester will take fewer than the stated number of credit hours for semester two.

These courses are generally delivered through the INTO CSU Center and are for Pathway students only.

Figure 2. Sample Graduate PW in ECE (old version 2012-2015)

ME in Electrical and Computer Engineering

www.engr.colostate.edu/ece

STANDARD PATHWAY - TWO SEMESTERS

START DATE

- August 17, 2015
- January 11, 2016

DEGREE PROGRAM COMPONENTS

- 30 credit hour degree program
- 6-12 credit hours may apply from Pathway
- 18-24 credit hours remaining toward degree
- In addition to graduate credits earned from the ECE 5xx courses, up to six credits of the ECE 4xx courses will apply toward the degree

PROGRAM INFORMATION

ENTRY REQUIREMENTS

- Undergraduate degree in Electrical Engineering, Computer Engineering, Computer Science, or closely related discipline
- Minimum 2.75 cumulative undergraduate GPA in math, sciences, and technical courses
- Language requirements:
  TOEFL IBT 71 or IELTS 6.0 or completion of the Academic English Program with minimum 80% in all graded courses

SEMESTER 1	COURSE TITLE	CREDIT HOURS
EAP 150	**English for International Students I**	**6**
ECE 4xx or 5xx[1,2,3]	ECE Elective	3
ECE 4xx or 5xx[1,2,3]	ECE Elective	3
	TOTAL	**12**

SEMESTER 2	COURSE TITLE	CREDIT HOURS
EAP 152	**English for International Graduate Students**	**6**
ECE 4xx or 5xx[1,2,3]	ECE Elective	3
ECE 4xx or 5xx[1,2,3]	ECE Elective	3
	TOTAL	**12**

PROGRESSION REQUIREMENTS

- Minimum 3.0 CSU GPA
- GRE score of 146 Verbal, 153 Quantitative, 3.0+ Analytical Writing
- No grade lower than B in all Electrical and Computer Engineering courses
- No grade lower than a B in any EAP class
- No grade lower than a B in any graduate-level class
- No I or U grades; no unresolved W grade in a required Pathway course; no S grades, except in cases of credit earned by exam
- 3 recommendation letters from Electrical and Computer Engineering faculty
- Current résumé and cover letter

NOTES

GRE preparation is provided through INTO CSU as part of the program.

[1] Students are advised to choose ECE courses that maximize their own career development.

[2] No more than 6 credits of ECE 4xx will count toward the 30 credit hours required for the Master of Engineering degree.

[3] Excluded courses: ECE 401, 402, 496.

These courses are generally delivered through the INTO CSU Center and are for Pathway students only.

The stage of resistance. As it should have been predicted, in any organization (big or small) when there is rapid change for which people have not been properly prepared to adopt and / or embrace, there is inevitably resistance to change. The ECE department at CSU was no exception. The introduction of INTO CSU happened too quickly and was very much perceived as an imposition on a department that already had a history of extremely healthy international enrollment, e.g., 80% of ECE's graduate population is not from the US. For this reason, both its head and its faculty failed to understand why they had to be part of the new initiative on which they had not been properly consulted. Given the fact that the value proposition of INTO CSU had not been properly introduced, it is fair to say that the ECE department found itself practically coerced into actively participating in the comprehensive internationalization process of CSU, as it had unfolded very much as a top-down imperative. Initially, ECE reacted negatively to the creation of INTO CSU, and its required participation. Therefore, there was little desire to collaborate with INTO CSU, and the communication between ECE and INTO CSU was difficult, affecting the possibility to engage in meaningful curricular improvements to the PW program. In time, and coinciding with a transition in the INTO CSU leadership, it became evident that there was clearly a lack of understanding as to the ultimate goals of INTO CSU relative to the comprehensive internationalization plan of CSU, and the important role that ECE had in this process. Pathway programs had not been properly explained to the Department, and it had become evident that there were many myths and misconceptions about INTO CSU.

In other words, due to the fact that the ECE participation in the pathway program had been the result of a top-down decision, the leadership, faculty and staff were not prepared for this new development. Consequently, this situation resulted in much push back and resistance during the initial years of INTO CSU. The primary concerns outlined by the ECE leadership that contributed to the push back was their initial perception that this PW program involved: (1) a "back door" admission for less qualified international students; (2) the "bumping" of students that were directly admissible in favor of the admission of students who could potentially bring down the quality of the education being delivered by ECE faculty; (3) the immediate capacity concerns that emerged in light of fewer resources being re-directed to the department by Central administration; and (4) a possible re-direction of university resources to attend to the needs of, what was perceived to be, a "privileged" group of students who could afford this type of very costly education, as is the case of most pathway programs. Thus, in light of all these identified issues, ECE resisted this change, and was reluctant to cooperate for at least the first three years of the INTO CSU partnership. This resistance to change was not surprising, but natural, as the formation of the partnership had taken place very quickly and had not been properly rolled out to the university community. Needless to say, the concerns of ECE were to a significant extent representative of the concerns of the larger university community.

The stage of awakening. As it happens in most cases, resistance to change and to the unknown is a natural human response. Through education and effective leadership, however, challenges of this sort can be overcome, as it also became the case for ECE. This is exactly what ECE has begun to experience in the fourth year of the INTO CSU partnership. Through improvements in communication between ECE and INTO CSU, this fourth year has constituted the renaissance of the ECE pathway. What had been practically impossible to accomplish in the first three years was finally overcome, as ECE began to experience an attitudinal shift towards the whole pathway concept. Through much education and active involvement, the ECE leadership began to understand that this INTO CSU initiative was intended to: (a) assist with a larger institutional goal of CSU, which was the internationalization of its campus at a much faster rate; (b) not just assist a privileged few who could afford this education, but also those who would need greater financial support, as a scholarship program had been established resulting from the matriculation success of pathway students in their corresponding degree programs; (c) critically examine the educational experience of English learners in engineering programs to ensure their success; and (d) promote cross-departmental collaboration between units that contribute different expertise to solve interdisciplinary research problems, e.g., English for specific academic purposes, second language acquisition. In other words, the ECE leadership realized that this could be a way to truly begin to learn on how to best prepare future engineers for a global world.

There were probably two main initiatives that significantly contributed to the shifting of negative attitudes and a move in a positive direction together with the start of building trust across the units. The first one was an agreement reached by the College Dean, the ECE Department Head, and the new INTO CSU Center Director. This agreement was intended to jointly fund an academic advisor position that would be housed in ECE and who would contribute to serve pathway students and all other students in the department. This individual would also liaise with the main academic advisor for pathway students at INTO CSU to ensure the smooth transition of the student into the degree program. The second one was the invitation on the part of INTO CSU to the ECE head to engage in expanding transnational collaborative efforts with outstanding institutions in regions where international collaboration was less developed, as was the case of Latin America. After all, INTO CSU was also intended to contribute to the larger internationalization goals of the university, and through INTO CSU resources could be leveraged to address areas of the world where ECE could engage in new collaborative initiatives. INTO CSU was able to demonstrate the potential it had to contribute to expand the international recruitment capacity of CSU and, in this specific case, that of ECE.

The stage of collaboration. Now that some of the initial skepticism on the part of ECE was overcome, the department head began to productively engage with INTO CSU in the substantial revision of the pathway programs. Such work is certainly evidenced in the example offered in Figure 3, as compared to Figure 2. For a variety of reasons, the initial ECE pathway curriculum had not been made explicit or overtly articulated for the prospective applicant, which resulted in a lower quality experience for the pathway student. The lack of an explicit curriculum posited academic advising challenges and logistical challenges when trying to (a) register pathway students for ECE courses that may or may not relate to the students' direct academic and professional goals and interests, and (b) secure academic support for those students at risk of possible failure, especially at the graduate level.

A step further is the fact that in moving forward, ECE has contributed 9 different plans of study for students interested in the ME program. This action has allowed INTO CSU to be in a better position to provide academic advising to new students and assist them with course registration. It also has allowed us to address administrative pieces that relate to preparing to welcome students who have expressed an interest in a particular plan of study so that capacity issues can be more promptly addressed.

Figure 3. Sample of revised / new graduate PW in ECE.

NEW: ME in Electrical and Computer Engineering - Robotics: Vision

Standard Pathway - 2 Semesters

Fall: August 15, 2016
Spring: January 9, 2017[1]

Degree Program Components

30 credit hour degree program
13 credit hours may apply from Pathway
17 credit hours remaining toward degree
In addition to graduate credits earned from the ECE 5xx courses, up to six credits of the ECE 4xx courses will apply toward the degree.

Program Information

Entry Requirements

- Undergraduate degree in Electrical Engineering, Computer Engineering, Computer Science or closely related engineering discipline
- Minimum 2.75 cumulative undergraduate GPA in math, sciences and technical courses
- Language requirements:
  - TOEFL iBT 71 or IELTS 6.0 or completion of the Academic English Level 3 with minimum 75% in all graded courses

Semester 1	Course Title	Credit Hours
AEEP 8591	**GRE Test Preparation**	0
EAP 150	**English for International Students I**	6
Choose two courses from the following CS or ECE or MATH courses:		
CS 410	Introduction to Computer Graphics	3
ECE 481A2	Introduction to Robot Programming and Simulation	3
ECE 512	Digital Signal Processing	3
ECE 520	Optimization Methods-Control and Communication	3
MATH 560	Linear Algebra	3
	Total	12

Semester 2	Course Title	Credit Hours
EAP 152	**English for International Graduate Students**	6
ECE 555	Robot Motion Planning	3
ECE 513	Digital Image Processing **OR**	3
CS 510	Image Computation	4
	Total	13

Progression Requirements

- Minimum 3.0 CSU GPA
- GRE score of 146 Verbal, 153 Quantitative, 3.0+ Analytical Writing
- No grade lower than a B in all Pathway classes
- No I or U grades; no unresolved W grade in a required Pathway course; no S grades, except in cases of credit earned by exam or the GRE preparation course
- 3 recommendation letters from ECE faculty (2 letters for the Accelerated Pathway)
- Current résumé and cover letter

Notes

1 Students who start the Standard Pathway in the Spring semester will typically complete the Standard Pathway in the following Fall semester.

These courses are generally delivered through the INTO CSU Center and are for Pathway students only.

Accelerated Pathway - 1 Semester

Fall: August 15, 2016
Spring: January 9, 2017

Degree Program Components

30 credit hour degree program
6-7 credit hours may apply from Pathway
23-24 credit hours remaining toward degree
In addition to graduate credits earned from the ECE 5xx courses, up to six credits of the ECE 4xx courses will apply toward the degree.

Entry Requirements

- Undergraduate degree in Electrical Engineering, Computer Engineering, Computer Science or closely related engineering discipline
- Minimum 2.75 cumulative undergraduate GPA in math, sciences and technical courses
- Language requirements:
  - TOEFL iBT 79 **OR** IELTS 6.5 **OR** completion of the Academic English Level 3 with minimum 85% in all graded courses

Fall Start

Semester 1	Course Title	Credit Hours
AEEP 8591	**GRE Test Preparation**	0
EAP 153	**Writing for International Graduate Students**	3
Choose two courses from the following CS or ECE or MATH courses:		
CS 410	Introduction to Computer Graphics	3
ECE 481A2	Introduction to Robot Programming and Simulation	3
ECE 512	Digital Signal Processing	3
ECE 520	Optimization Methods-Control and Communication	3
MATH 560	Linear Algebra	3
	Total	9

Spring Start

Semester 1	Course Title	Credit Hours
AEEP 8591	**GRE Test Preparation**	0
EAP 153	**Writing for International Graduate Students**	3
ECE 555	Robot Motion Planning	3
ECE 513	Digital Image Processing **OR**	3
CS 510	Image Computation	4
	Total	9-10

Finally, and currently unfolding, is the recruitment of other CSU specialists in teaching English learners from the English department who are conducting a needs analysis in ECE courses. The sole purpose is to develop an understanding of the expectations in both undergraduate and graduate engineering education. Results from this work are intended to inform the existing English for Academic Purposes courses in the PW program that are intended to help English learners develop high levels of English proficiency in contexts, such as ECE, where they need to develop and/or acquire English for specific academic purposes.

The stage of innovation. Fortunately, ECE is now beginning to envision the true potential of this partnership. ECE is finding that, by engaging in this cross-departmental effort, ECE has the potential to emerge as an innovator rather than a follower when it comes to the education of future engineers for global citizenship. Future engineers have much to benefit from a campus university internationalization plan. It is the result of these efforts that will allow future engineers to develop the core competencies to effectively function in a global world. These include: the development of an intercultural competence, world mindedness,[4,5] and so forth.

The stage of innovation has also been further nurtured by the collaborative interdisciplinary work that is unfolding between the expertise at INTO CSU and that of Engineering. For example, given that the INTO CSU Center Director is an applied linguist with specialization in the teaching and learning of English to speakers of other languages, she invited the Department Head of ECE to collaborate in a study that intended to better understand the non-verbal processes involved in the process of reading comprehension of engineering texts. Such research was presented at the Annual Meeting of the American Association of Applied Linguistics in April 2016.[2] As these two experts engaged in this type of work, they are closely monitoring student performance to better understand what possible interventions need to be considered in the teaching and learning of PW students relative to the competencies needed for becoming successful ECE students and eventually graduates.

Here are some additional tangible ways in which the partnership has begun to influence the ECE department that can lead to true educational innovation:

- Cross-disciplinary collaboration
- Faculty development
- Curricular re-envisioning
- Resource sharing through collaborative practices
- Research and scholarship
- Departmental competitiveness and global positioning

There is no question that the INTO CSU venture offers a platform for innovation. It represents a new educational challenge, and it has the potential to serve as the springboard that ECE faculty can make use of to have a greater reach of significant impact not only on campus but on the world. "Successful CI [comprehensive internationalization] is not the product of well-meaning but heavy top-down decision-making by presidents and provosts. Neither is it only the result of bottom-up populism, but a product of these two processes in concert to pursue consensus" (Hudzik, 2011, p. 3).

Recommendations / Lessons learned.

The success of INTO CSU, and in particular of its signature program known as "Pathways" will require a long-term shared commitment by departments working in collaboration with the language specialists at INTO CSU. It will require clarity regarding the shared goals relative to what it takes to ensure student success and a true and positive attitude that will lead to collaboration, innovation and continuous education of everyone on campus and beyond.

Based on our shared experiences, we have concluded that the following are essential ingredients for success: (1) clear and effective communication of the value proposition; (2) involvement of key personnel directly affected by the initiative (i.e., administrators, faculty, and staff); (3) strong and effective leadership; (4) clear plans for addressing the issues of concern, such as capacity and student performance; (5) strong willingness to work to plan for growth; (6) high levels of commitment to, and a strong sense of belief in, the joint venture.

Finally, based on our direct experience, we can only agree with the following: "Comprehensive internationalization is a commitment, confirmed through action, to infuse international and comparative perspectives throughout the teaching, research, and service missions of higher education. It shapes institutional ethos and values and touches the entire higher education enterprise. It is essential that it be embraced by institutional leadership, governance, faculty, students, and all academic service and support units. It is an institutional imperative, not just a desirable possibility. Comprehensive internationalization not only impacts all of campus life but the institution's external frames of reference, partnerships, and relations. The global reconfiguration of economies, systems of trade, research, and communication, and the impact of global forces on local life, dramatically expand the need for comprehensive internationalization and the motivations and purposes driving it".[1] True commitment is key for success.

References

[1] Hudzik, J.K. (2011). Comprehensive Internationalization: From concept to action. *NAFSA Executive Summary*, pp. 1-2, doi: http://www.nafsa.org/_/File/_/cizn2011_execsummary.pdf

[2] Ehlers-Zavala, F. & Maciejewski, A. (2016). *Mental imagery experienced by both pathway and non-pathway graduate students in an engineering course at a US Research I institution.* Poster presentation discussed at the annual meeting of the American Association for Applied Linguistics (AAAL), Orlando, FL.

[3] Marmolejo, F. (2011). The future of higher education internationalization. *The Chronicle of Higher Education.* doi: http://chronicle.com/blogs/worldwise/defining-internationalization/28615

[4] Parsons, R. L. (2010). The effects of an internationalized university experience on domestic students in the United States and Australia, *Journal of Studies in International Education, 14*(4), 313-334.

[5] Volet, S. (2003). Challenges of internationalization: Enhancing intercultural competence and skills for critical reflection on the situated and non-neutral nature of knowledge. In *2003 Biennial Language and Academic Skills in Higher Education Conference*, 24 - 25 November 2003, Student learning Centre, Flinders University, Adelaide, 1-10.

Implementing and Integrating International Research into the Engineering Curriculum at Lincoln University, Pennsylvania and the University of the West Indies, Trinidad

Dr. Monica Gray P.E., Lincoln University

Dr. Monica Gray is an Assistant Professor of Civil and Environmental Engineering at The Lincoln University. She simultaneously received her PhD in Civil and Environmental Engineering (Water Resources concentration) and Masters of Public Health (Environmental & Occupational Health concentration) from the University of South Florida, Tampa. She also received a Masters in Biological Engineering from the University of Georgia, Athens and B.S. in Agricultural Engineering from the University of the West Indies, Trinidad. Dr. Gray is a seasoned engineer and educator who has lived, studied and worked in various countries around the world.

Prior to arriving at The Lincoln University, she managed the global curriculum portfolio for over a hundred and twenty programs in sixteen countries at Arcadia University and was instrumental in successfully developing and implementing study abroad opportunities and exchanges for undergraduate engineers from institutions across the country, while internationalizing the engineering curriculum through cooperation, consortia and curriculum integration.

Ms. Constance Loretta Lundy

Implementing and Integrating International Research into the Engineering Curriculum at Lincoln University, Pennsylvania and the University of the West Indies, Trinidad

Introduction

Educating engineers to meet the needs of today's global economy and diverse workforce involves preparing culturally competent individuals as well as providing increased access to members of under-represented, under-prepared and minority groups (UUMs). Minorities and Females constitute about 4% and 15% of the engineering workforce while accounting for over 13% and 50% of the general U.S. population respectively.[1] This status quo is maintained by disproportionately high attrition rates from engineering programs among UUMs, ranging from 49% for Females to 62% among African Americans.[2-4]

Many strategies are currently being employed to reduce this disparity. These efforts include first-year seminars, internships, learning communities, and capstone projects.[2, 5] Study abroad, however, has not been seriously considered as a high impact retention activity. The research shows that UUMs struggle in mathematics and science intensive programs such as engineering, not simply because of academic under-preparation per se, but more importantly, due to feelings of not belonging and lack of confidence in their ability to learn coupled with low self-efficacy and self-direction.[6-8] In fact, the common theme among successful high impact retention strategies is that they provide access to structured opportunities for all students, including UUMs, to develop self-directing competencies in both the cognitive and affective learning domains.[9, 10] Therefore, facilitating skill sets that are *sine qua non* both to successfully learn and complete engineering degree programs.

Studying abroad inherently impact students' personal and academic development.[11] Study abroad exercises students' cognitive analysis ability (*students are able to deal with ambiguity in new, ill-defined situations by formulating models and seeing relationships*), affective organizational skills (*students are able to balance their responsibilities and formulate a cohesive and systematic approach to learning*) as well as self-efficacy, which is a proximal predictor of proficiency in both these domains.[12] Self-efficacy is "*an individual's judgment of his or her capabilities to perform given actions*".[13] Indeed, study abroad can be very challenging, as culture shock requires adjustment to the host culture, and may also elicit "*feelings of not belonging*".[11] However culture shock provides students with sustained opportunities to "*deal with ambiguity in new, ill-defined situations*". Felder[8] describes the cognitive development trajectory of college students as a continuum of knowing everything, "*ignorant certainty*", recognition that context determines answer, "*intelligent confusion*" to a point where students begin to trust their own process of organizing their learning and judgment. A similar process occurs during the study abroad experience. According to self-efficacy framework and intercultural transformation theory, study abroad participants' equilibriums are disturbed when seemingly familiar interactions occur without familiar signs or social cues. This cognitive dissonance leads to "*identity confusion*", an analogue to "*intelligent confusion*".[14] Each such interaction, however, leaves the student better equipped to handle similar encounters because of "*greater cognitive and affective capacity*".[12]

Therefore study, research, internship, and/or service learning abroad experiential opportunities must be added to the hosts of high impact experiential educational strategies currently being employed to increase UUMs' participation in engineering academic programs and subsequently the profession. In addition, spatial mobilization of the engineering sector predicated by a global economy, as well as engineering program accreditation with a student learning outcome requirement to have "*the broad education necessary to understand the impact of engineering solutions in a global, economic, environmental, and societal context*", have led to the market demand for more culturally savvy engineers. It follows therefore that an engineering educational experience abroad not only provides a structured opportunity for UUM students to increase their self-efficacy and self-direction skill sets which will help them persist and do well in their current engineering programs but also prepares them to meet the demands of the global market place. Study abroad has not been previously considered as a high impact activity, there is, however, a movement afoot to change that paradigm. The National Science Foundation in the latest Louis Stokes Alliance for Minority Participation (LSAMP) program solicitation cited researchers Bowman and Sage as follows, "*Preparing a diverse, globally-engaged scientific and technological workforce necessitates strengthening international research opportunities for students under-represented in STEM fields*". [15]

Studying abroad: benefits, challenges and opportunities

According to the Institute of International Education's (IIE) Open Doors 2015 report, over the last two decades, the number of U.S. students studying abroad has tripled to a about 300,000. This indicates that students, faculty and administrators are recognizing the benefits of the study abroad experience and are availing themselves of the opportunities. For example, research shows that students who study abroad have higher GPAs, are more likely to graduate on time, and are more attractive to employers that seek to hire candidates with intercultural competencies.[16]

African American students account for 15% of the overall U.S. undergraduate population but only represent 5.6% of the over 300,000 U.S. students studying abroad. There are several reasons for this disparity: finances, faculty leadership, attrition rates, choice of major, family support, anxiety resulting from travel inexperience, and lack of insight of the correlations between global cultural competence and career goals.[17, 18] It is interesting to note that high attrition from degree programs has been listed here also as a contributing factor for under-participation. Traditionally, study abroad has been marketed as an opportunity to experience other cultures. For minority students, this angle has not been effective as they interact across culture on a daily basis.[19] However, based on the research into the causative factors for under-representation, an opportunity to increase skill sets that can assist in persisting and doing well in their current engineering programs as well as prepare them for continual lifelong professional success might be a better approach to encourage underserved minorities to study abroad.

While STEM is now the top sending group accounting for about 23% of the study abroad cohort, engineering represents only 5% of that total. By and large, STEM fields, have very restrictive, highly sequenced and vertical curriculums, creating an additional barrier that prevents their majors from partaking of the transformative experiences gained from studying abroad. This still continues today especially in engineering. Students are fearful of not being able to find equivalent courses which may result in study abroad credits not transferring back to home institution, falling out of sequence, getting behind, and subsequent delayed graduation and increased educational cost. Faculty are equally concerned with the quality of the academic experience overseas, their lack of control over the curriculum, and possible accreditation implications. Faculty leadership in encouraging study abroad participation under these conditions is, as would be expectedly, low.

As noted above, there are several factors that are driving demand for more engineers to study abroad: ABET accreditation, global economy, and federal funding agencies. In addition, as the number of STEM majors studying abroad increases, this may assuage concerns of both students and faculty about the ability of engineers to study abroad and still graduate on time. Further, an opportunity may exist in the fact that females constitute about 65% of the study abroad participants. This is an inherent strength that must be capitalized upon, given the high female attrition rates from engineering programs.

Strategies for broadening UUM engineering majors' participation in study abroad

The preceding discussion advances study abroad not only as a high impact strategic retention tool to ensure that more UUMs complete engineering programs but also a way to position this cohort of students for successful careers. This paper postulates, however, that to successfully engage African American engineering students to participate in study abroad, the experience must be adopted as part of an institution's overall culture, is not simply an additional item to an already packed curriculum. Internationalizing the curriculum requires strategic direction and system-wide buy-in. Successful strategies include cooperative agreements among U.S.-based and global institutions, development of U.S.-based consortia, curriculum integration, course matching, and scholarship funding, as well as a robust faculty-led program.

Cooperative agreements

Cooperative agreements or memoranda of understanding (MOU) are an effective way for U.S.-based and international institutions with mutual internationalization missions and goals to broker relationships that advance the individual and collective vision of all signatories. MOUs that tend to work and are productive in the long run are characterized by up front, explicitly negotiated needs and strong bilateral support from individual parties. Some items that are typically included in internationalization agreements are scholarly exchanges, of both faculty and students, curriculum and course integration, formal agreements with individual departments, and English as the language of instruction.

For administrators who are concerned about ABET accreditation issues, a conservative option might include finding a partner university in a country that is a signatory of the Washington Accord and ensuring that the perspective institution is accredited by that country's accreditation body. The Washington Accord, signed in 1989, is an international mobility agreement among engineering accrediting bodies in seventeen countries. Each signatory recognizes accredited engineering programs in others' jurisdiction as being substantially equivalent to its own. For example, South Africa is a signatory of the Washington Accord, and its accrediting body is the Engineering Council of South Africa.[20] If Lincoln University has an ABET accredited engineering program and wishes to sign an MOU with the University of Pretoria (UP) in South Africa, which is accredited by the Engineering Council of South Africa, then both programs are recognized as being substantially equivalent to each other. In this case, Lincoln's administrations would be even happier to realize that UP is also ABET accredited which is not a very common occurrence outside the U.S.

Local and regional consortia
Capitalizing on the idea of strength in numbers, local and regional universities sometimes come together to form a consortium. These partnerships may include stakeholders from across each member institution. It is common to include both study abroad officials and faculty. Individual schools may be relatively small and often do not have the enrollment numbers to create a cost effective study abroad contingent. By coming together, a group of schools can create agreements among themselves as well as with the global partner(s). The consortium then sends a large group of students to their partner institution(s) reducing the overall cost per student as well as ensuring the integrity of the educational experience. Member schools sometimes sign off on course credits though a process of triangulation. For example, if Lincoln and University of Delaware (UD) are part of a consortium and UD gave credit for a Statics course taken at the University of the West Indies (UWI), then Lincoln, by virtue of its consortium membership, would be more inclined to also accept those credits for its students studying at UWI.

Curriculum integration and course-to-course matching
Curriculum integration involves the time and offering synchronization of typical, fundamental engineering courses. Both the global and U.S. partner institutions evaluate their individual course sequence to determine common core offerings and work to offer those courses in tandem. Typical core courses that lend themselves to such processes include Statics, Dynamics and Fluids. Inherent in this process is the analysis to determine the equivalency of the courses as well as removal of the guess work about when is the best time for engineering students to study abroad. Courses are therefore pre-approved for engineering credit; thus, students go abroad confident that they can make adequate progression towards completing their degree on time.

Scholarship funding opportunities
Financial constraint is a major concern for all students even though it has become relatively easier to get study abroad scholarships for STEM majors compared to other majors. Many scholarships specifically target minority student populations. The following is a discussion of a few:

- Benjamin A. Gilman International Scholarship – funded by the U.S. Department of State and administered by IIE, this scholarship funds traditionally under-represented cohorts in study abroad and shows a preference for non-traditional destinations. Students can study, research or intern while they are abroad. Funding is typically up to $5000 but can be increased to $8000 if students study a critical needs language. STEM is considered a priority group.
- Vira I. Heinz Women in Global Leadership program – funded by the Vira I. Heinz Endowment and administered by University of Pittsburgh, this program awards scholarships to three women from each participating institution of at least $5000 with an additional $1000 if they study outside of Western Europe. STEM is considered a privileged group.
- Third party study abroad providers – usually offer scholarships to participating students. Universities or consortium can negotiate better rates for their students especially for large cohorts, typically at least 10 students.

Faculty-led programs
Faculty-led programming is an essential component to campus internationalization plans aimed at making study abroad part of the academic culture and not the exception. Programs may include faculty taking students abroad and delivering course(s) in the host country in which the

students are studying. The abroad experience can be a continuation of a course taught on campus and/or a team taught course where U.S.-based and the host foreign country's students are grouped to work on a project during the semester and then come together at the end. Whatever the iteration, faculty-led opportunities ease many traditional fears and significantly impact the internationalization process. First, faculty who take students abroad are more likely to send students abroad. Second, since the faculty member delivers the content, there is no question as to the quality of the experience. Finally, if the abroad experience is a continuation of an on-campus course, then the program fee can be embedded in the course cost and can thus make use of the students' financial aid packages that are already in place.

Case study: partnership between Lincoln University and the University of the West Indies
Lincoln University of Pennsylvania, in collaboration with the University of the West Indies, St. Augustine, Trinidad & Tobago and the Greater Philadelphia Region Louis Stokes Alliance for Minority Participation (Philadelphia AMP), has entered into a series of progressive partnerships in support of its Faculty-led STEM Research Internship Program to the Caribbean. The following details that progression and provides a case study for implementation of the internationalization strategies discussed above.

Lincoln University
Lincoln University, the nation's oldest degree granting Historically Black College and University (HBCU), has an outstanding track record for preparing UMM STEM majors for possible careers and graduate studies in STEM-related fields. Since its inception, Lincoln University has attracted a diverse student population and faculty from the surrounding community, region, and the world. The mission of the university has always been to prepare its students to be leaders of the highest caliber, both at home and abroad. Our alumni include African leaders such as Nnami Azikiwe, first president of the Federal Republic of Nigeria; Kwame Nkrumah, first president of Ghana; and most recently, 1994 graduate Saara Kuugongelwa-Amadhila, Namibia's first female Prime Minister. In the pre and post- apartheid years, Lincoln University served as the orientation center of over a thousand South African students who came to study in U.S. universities. Our strong ties with the African continent continues today such that in the last five years, almost half of our international student population was from African countries. Here in the U.S. our most notable alumni include Thurgood Marshall, the first African American U.S. Supreme Court Justice; Langston Hughes, world-acclaimed poet; Lillian Fishburne, the first African American female U.S. Navy Rear Admiral; and Hildrus A. Poindexter, the first African American to earn both an M.D. *(1929, Harvard University)* and a Ph.D. *(1932, Columbia University)* as well as the first African American internationally-recognized authority on tropical diseases.

Our students study abroad during the traditional semesters with no attendance during the shorter terms such as winter or spring break. Students study in a large contingent of countries with spotty participation in the Caribbean and almost exclusively using third party study abroad program providers. Our study abroad portfolio comprises about 50 graduate and undergraduate students per year, the majority of whom are Business and Language majors, from a total population pool ranging from 1,500 to 1,900. Thus, STEM majors are in the minority of study abroad participants accounting for about 18% of the 240 students who studied abroad in the last 5 years. STEM participants have been self-selecting as our programs have not been intentionally integrated within their curriculum. One of the key barriers for all our students is lack of financial resources. About 96% of our students receive financial aid and are thus economically challenged.

The University of the West Indies – St. Augustine, Trinidad & Tobago
UWI – St. Augustine is one of the four campuses of the premier regional university system serving seventeen countries of the Commonwealth Caribbean and home to the flagship Faculty (College) of Engineering. Its mission is to enhance Caribbean development by collaborating with regional and international partner institutions to establish networks for the next generation of leaders. The university is therefore very well connected throughout the Caribbean and is at the forefront of scientific research, resource management and policy development. As a regional university, U.S. students studying there have the opportunity to interact with students from multiple Caribbean islands simultaneously.

Greater Philadelphia Region Louis Stokes Alliance for Minority Participation (Philadelphia AMP)
Philadelphia AMP is sponsored by the National Science Foundation (NSF) and consists of nine tertiary institutions, including Lincoln University. The mission of the Alliance is to substantially increase the numbers of under-represented minorities earning baccalaureate and advanced degrees by employing recruitment and retention strategies at the college level to promote minority talent in STEM disciplines. NSF has recognized international STEM research as a high impact, successful recruitment and retention strategy. The Philadelphia AMP is now in its 20th year and is therefore classified as a senior level Alliance. With this status comes the additional responsibility of focusing on international STEM research. Therefore the Alliance funds and supports international research opportunities for students under-represented in STEM fields.

STEM faculty-led programing
The Faculty-led STEM Research Internship Program to the Caribbean is currently Lincoln University's only faculty-led STEM opportunity. It first ran in summer 2015 with five STEM students, sponsored by Philadelphia AMP and Lincoln University, and was hosted by UWI-St. Augustine. The general research agenda was natural resource management with a focus on the interrelationship among conservation strategies in individual sectors (e.g., water resource issues in solid waste management) as well as sustainability versus entrepreneurship (i.e., the balance between preserving for future generations versus and the need to make a living now). The program's pedagogy featured a "double-exposure" approach where students learn locally and apply globally. Program objectives included exploration of the unique environmental resources available in the both countries; analysis of how resources are harnessed to foster sustainable development, employment and cultural enrichment; and contextualization of the contradictory balancing of environmental realities, conservation strategies and the need for sustainable livelihood. Upon completion of the program, students were expected to be able to discuss and describe how the unique resources of an environment/county facilitate the emergence of conservation strategies, and cultural and economic activities; evaluate and articulate the challenges of preserving natural resources, cultural traditions as well as ensuring sustainable livelihoods; and perform basic research both in the lab and in the field.

Students were introduced to resource and environmental management concepts in the classroom and through a virtual-hybrid platform with attendant U.S.-based field trips to the communities adjacent to Lincoln University. While all students were expected to become articulate in all presented topics, each student was assigned an area of focus on which to conduct comparative research. The five comparative research projects were water treatment, wastewater treatment, solid waste management, food production and soil management. The international portion of the program was a mirror image of the U.S.-based field studies culminating in PowerPoint presentations in Trinidad. Students continued their research upon return to the U.S., resulting in poster presentations at Lincoln University's fall Annual Science Fair. Notably, three of the five participants placed in the top three of their categories. Participants were also required to attend a series of professional development activities with the Office of Career Services with the aim of helping them update their resumes and articulate the research abroad experience to future employers and graduate schools. All students completed all requirements of the program. Primary feedback were requests for extending the Trinidad leg of the program by about a week and more interactions with Caribbean students. The most challenging aspects to the program revolved around coordinating field trips in both countries.

The summer 2016 instalment will again be hosted by UWI-St. Augustine and funded in part by Philadelphia AMP. Engineering students were the primary target of the STEM majors. The focus this year is energy, featuring peer-pairing of U.S. and Caribbean students on projects to model international, multidisciplinary teams. The summer 2015 program ran in late June when most students at UWI-St. Augustine had already left for the summer. The U.S. contingent will arrive in Trinidad during the first week of May before the end of the spring semester, to facilitate more integration and multinational team work with Caribbean students before their summer break.

Campus internationalization

While Lincoln University has always had a global vision, STEM students have not been directly targeted towards fulfilling the global competency mission. In light of Lincoln's long history serving UUM STEM majors, the imperative to prepare students for the global marketplace, the mandate arising out of membership in the Philadelphia AMP to provide *"international research opportunities for students under-represented in STEM fields"* and the desire to offer this life changing experience to more students, it has been concluded that the STEM fields are areas of synergistic and strategic growth. Consequently, an expert on STEM-related campus internationalization was invited to campus to speak to relevant cohorts across all departments. As an outcome to those meetings, Lincoln's president has formed the Committee on International-Global Initiatives (CIGI) to address the issues of increasing the availability of study abroad opportunities, increasing international student enrollment, creating opportunities for international faculty development and expanding of an internationalized curriculum.

The Faculty-led STEM Research Internship Program to the Caribbean aligns with this strategic direction and is therefore fully supported by the administration. Lincoln University now firmly re-commits its faculty and administrators to the task of exploring and implementing new policies and procedures that will expand the numbers and enhance the education of students, especially STEM majors, who wish to fully embrace research and practical experiences globally. This re-commitment is especially important for the Engineering Science Program (ESP), which has seen tremendous growth in the last year. During the development of the ESP curriculum, spring semester of the sophomore year, was made to mirror spring semester of the first year at UWI-St. Augustine to facilitate future semester-length student exchanges. An aspirational goal of this program is that each engineering student will have at least one study abroad experience before graduation.

References

1. United States Census Bureau. *QuickFacts United States*. U.S. Department of Commerce; 2015.
2. American Society of Engineering Education (ASEE). *Going the distance: best practices and strategies for retaining engineering, engineering technology and computing students.* Washington, D.C.; 2012.
3. National Science Foundation (NSF). *Science and engineering degress by race/ethnicity of recipients: 1997-2006. Mark K. Fienegen, Project Officer*. Divison of Science Resources Statistics, Arlington, VA; 2009. Available at: http://www.nsf.gov/statistics/nsf10300/. Accessed October 15, 2014.
4. National Science Foundation (NSF). *Women, Minorities, and Persons with Disabilities in Science and Engineering.* Arlington, VA; 2015. Available at http://www.nsf.gov/statistics/wmpd/
5. Freeman KE, Sharon TA, Duvon GW. *Do Learning Communities Enhance the Quality of Students' Learning and Motivation in STEM? The Journal of Negro Education.* 2008;77(3):227-240.
6. Gray M, Lundy C. *First Learning, then Lifelong Learning: Engineering Study Abroad to Increase Access and Retention among Minorities and Under-represented Groups. 2015 ASEE International Forum*. Seattle Washington: ASEE PEER; 2015.
7. Matthews M. *Keeping students in engineering: a research-to-practice brief. American Society of Engineering Education, ASEE, Editorial: www asee org.* 2012.
8. Felder RM, Brent R. *Understanding student differences. Journal of engineering education.* 2005;94(1):57-72.
9. Chubin DE, May GS, Babco EL. *Diversifying the engineering workforce. Journal of Engineering Education.* 2005;94(1):73-86.
10. Brainard SG, Carlin L. *A Six-Year Longitudinal Study of Undergraduate Women in Engineering and Science. Journal of Engineering Education.* 1998;87(4):369-375.
11. Yang M, Webster B, Prosser M. *Travelling a thousand miles: Hong Kong Chinese students' study abroad experience. International Journal of Intercultural Relations.* 1// 2011;35(1):69-78.
12. Milstein T. Transformation abroad: *Sojourning and the perceived enhancement of self-efficacy. International Journal of Intercultural Relations.* 3// 2005;29(2):217-238.
13. Schunk DH. *Self-efficacy and academic motivation. Educational psychologist.* 1991;26(3-4):207-231.
14. Zaharna RS. *Self-shock: The double-binding challenge of identity. International Journal of Intercultural Relations.* // 1989;13(4):501-525.
15. Bowman MH, Stage FK. *Personalizing the goals of undergraduate research. Journal of College Science Teaching.* 2002;32(2):120.
16. Hamir HB. *Go Abroad and Graduate On-Time: Study Abroad Participation, Degree Completion, and Time-toDegree*. Lincoln, Nebraska: Educational Leadership and Higher Education, University of Nebraska; 2011.
17. Acquaye LA, Edmonds Crewe S. *International Programs: Advancing Human Rights and Social Justice for African American Students. Journal of Social Work Education.* 2012/12/01 2012;48(4):763-784.
18. Shih K. *Study abroad participation up, except among minority students. Diverse Issues in Higher Education.* 2009;16.
19. Salisbury M, Paulsen M, Pascarella E. *Why do All the Study Abroad Students Look Alike? Applying an Integrated Student Choice Model to Explore Differences in the Factors that Influence White and Minority Students' Intent to Study Abroad. Res High Educ.* 2011/03/01 2011;52(2):123-150.
20. Alliance for International Education. *Washington Accord Signatories*. Available at: http://www.ieagreements.org/Washington-Accord/signatories.cfm. Accessed January 16, 2016.

Toward Success of Collaborative Program In School of Engineering Between the US and China

Dr. Jeongkyu Lee, University of Bridgeport

Dr. Jeongkyu Lee is currently associate professor in Department of Computer Science and Engineering at University of Bridgeport. He received Ph.D. in Computer Science from the University of Texas at Arlington in 2006. Before he pursued his doctorate, he worked as a database administrator for seven years with companies including Hana Bank and IBM Korea. His primary research area is in the multimedia database management system and analysis. Research interests include graph-based multimedia data modeling, indexing structure, big data system/analysis, social media data mining, and user profiling.

Dr. Sarosh H Patel, University of Bridgeport
Brian Lim, University of Bridgeport

Dean of International Program Initiatives and Special Assistant to the President at the University of Bridgeport.

Prof. Roger Dianlei Geng PhD, Wuhan University of Science and Technology

Education: 2008-2015 PhD: Shanghai International Studies University Work: 2010-Present Professor of English and Cultural Studies Director, Office of International Relations/Dean, International School Wuhan University of Science and Technology, China

Prof. Zhigang Jiang, Wuhan University of Science and Technology

Toward Success of Collaborative Program In School of Engineering Between the US and China

Abstract:

In 2013, the University of Bridgeport (UB) in the United States (U.S.) and the Wuhan University of Science and Technology (WUST) in People's Republic of China (China) agreed on an undergraduate collaborative program in electrical engineering. Students admitted by Department of Electrical Engineering (EE) at WUST will study in China for 2 years, then students can be transferred to the same program at UB in the U.S. for the remaining academic requirements at UB. Upon the completion of all graduation requirements at UB, students will receive a Bachelor of Science in EE from UB. Students who complete all four years at WUST or transfer to UB after two years, will be granted a Certificate of Graduation and a Bachelor's degree from WUST after satisfactorily completing of the program requirements. There are several important characteristics of the program between UB and WUST: (1) the program focuses on a specific academic program, so both universities design the program that fits students at WUST: (2) a collaborative program by transferring the credits earned at the other institution: and (3) UB professors visit WUST to teach courses, while WUST faculty members from China visit UB for professional developments as visiting scholars. As a result of the program, eight WUST students have transferred to UB at the junior level in Fall 2015, and about twenty students are expected to transfer to UB in Fall 2016. In this paper, we discuss the details of the program, and characteristics of it. We also present the results of a students and faculty survey conducted to evaluate the expectations and outcomes of the collaborative program.

1. INTRODUCTION

Due to the recent growth of the Chinese economy, it is highly desired for people in China to have global educational opportunities, such as learning English, studying abroad, and participating global programs. On the other hand, the universities in the U.S. have been pursuing diversity in student populations at their institutions with growing number of international students, faculty and programs. Therefore, there have been a lot of collaborative program initiatives between the universities in the U.S. and the universities in China with various types of programs including academic degree programs, exchange programs, summer programs and research programs. According to information released by the Chinese Ministry of Education (MOE) in the International Journal of Comparative Education and Development (Comparative Education Bulletin), by 2013 more than 1,200 Sino-foreign cooperative schools and programs with 30 countries of origin of the foreign partner university were approved by the Chinese authorities [12].

Table 1 shows the summary of global academic program between the U.S. and China selected from ASEE annual conference and ASEE International Forum. For instance,

University of Dayton and Shanghai Normal University have been developing and operating a 3 + 1 joint degree program since 2006 [10]. The first three years of the program are taught and managed by the university in China, then students transfer to the university in the U.S. to get an undergraduate degree in engineering. On the other hand, Worcester Polytechnic Institute sends their students to a university in China during a summer for global project [11]. As shown in the table, there are variety of programs in terms of types, characteristics and participants of the programs.

Table 1. Survey of Global Academic Program between the U.S. and China

University in US	University in China	Type of Program	Major(s)	Year in US	Remarks
AASCU[1]	CCIEE[2]	student's visiting	Electrical Engineering	2Y	1+2+1 program, i.e., return to China after 2 years [4]
Lawrence Technological University	Shanghai University of Engineering Science	Prof. visiting	Engineering	0	LTU professors visits SUES to teach courses within 4~5 weeks [5]
Oakland University	Multiple Universities	dual degree	Engineering	2Y	OU + 3 Chinese Universities
University of San Diego	Southeast University	Training program	Electronics Engineering	0	Developing training program [6]
Georgia Tech	Peking University	Research program	Biomedical Engineering	1Y	China Undergraduate Research Experience [7]
Western Michigan University	Sichuan University	Summer program	Engineering	1M	Summer tour to China [8]
University of Calgary (Canada)	Shantou University	Exchange program	Engineering	10D	Exchange program including activities and lecture [9]
University of Dayton	Shanghai Normal University	Joint program	Electrical Engineering	1Y	3+1 joint degree program [10]
Worcester Polytechnic Institute	Huazhong University of Science and Technology	Project based	Engineering	2M	WPI students visit China for 2 months to perform the global projects [11]

In 2013, the University of Bridgeport (UB) in the United States (U.S.) and the Wuhan University of Science and Technology (WUST) in People's Republic of China (China) agreed on an undergraduate collaborative program in electrical engineering. Students admitted by Department of Electrical Engineering (EE) of WUST will study in China for 2 years, then students can be transferred to the same program of UB in the U.S. for the remaining academic requirements at UB. Upon the completion of all graduation requirements at UB, students will receive a Bachelor of Science in EE from UB. Students who stay all four years at WUST or transferred to UB after two years, will be granted a Certificate of Graduation and a Bachelor's degree from WUST if they are subject to satisfactory completion of the program.

[1] AASCU: American Association of State Colleges and Universities
[2] CCIEE : Chinese universities to the China Center for International Education Exchange

About UB

The University of University was founded in 1927 as the Junior College of Connecticut – the first junior college chartered by any legislature in the northeastern states. In the words of its founders, the college's purpose was to develop in students "a point of view and a habit of mind that promotes clear thinking and sound judgment in later professional and business experience." UB maintains its primary commitments and holds fast to its values. Academic programs are offered through thirteen schools, colleges and institutes. Concern for student development and support predominate. A career-oriented focus in academic programs is complemented at the undergraduate level with a state-of-the-art core curriculum that helps students secure competencies for lifelong learning and knowledge about our world. The School of Engineering of the University of Bridgeport provides comprehensive education and research opportunities to a diverse community in engineering, sciences, and the application and management of technology. The School prepares students for leadership and technology positions in industry, government, and academia and significantly contributes to the profession and community locally, nationally, and globally. The School offers a distinctive education in fundamental and emerging disciplines through its faculty and institutional partners. The education features an application-oriented approach to interdisciplinary issues and opportunities that balances theory with real world state-of-the-art practices. As of Spring 2016, the school offers three undergraduate programs and eight graduate programs including six master's degree and two doctoral degree programs. Specifically, Electrical Engineering program offers both undergraduate and master's degree program [1,3].

About WUST

Located in Wuhan of central China, Wuhan University of Science and Technology (WUST) is a key public higher learning institution claiming traditional strengths in engineering and sciences that are integrated with contemporary popular disciplines such as medicine, economics, management, humanities, art, law, and philosophy. Currently WUST operates under the joint aegis of Hubei Provincial Government, Ministry of Education of PRC and six largest mega steel enterprises in China. For over 32, 000 full-time students hosted by 20 colleges and schools, 32 research centers, WUST now offers 71 undergraduate, 130 plus master's, 36 doctoral and 5 post-doctorate programs, with some claiming to be the strongest in the country. It is also the home of a key national laboratory (in refractory materials and metallurgy), 2 key laboratories accredited by MOE, and 8 accredited by Hubei Provincial Government. In the late 1980s WUST emerged as a pioneer in international education in Hubei province. Its International School (WUST International) is a highly regarded educational institution in the country that provides local students with education of an international standard through partnerships with universities from different parts of the world. At present WUST International co-runs three joint programs at undergraduate level that have been approved by the Ministry of Education, PRC including BE in Electronic Information Engineering jointly run with UB [2]. Graduates of the programs are expected to be equipped with fundamental knowledge, expertise and relearning capacity in related areas as well as a global outlook, science literary and innovative mind, which can prepare them for a fulfilling career in engineering as a researcher, designer, manufacturer, developer or administrator.

There are several important characteristics of the program between UB and WUST: (1) the program focuses on a specific academic program, so both universities co-design the program such that fits students at WUST: (2) a collaborative program by transferring the credits earned at the other institution: and (3) professors at UB visit WUST to teach courses, while WUST faculty members from China visit UB for professional developments as visiting scholars.

In this paper, we present a three-year experience of collaborative program between UB and WUST. This paper is organized in the following way. In Section 2, we describe our design and implementation of the collaborative program between UB and WUST. Next, in order to evaluate the collaborative program, the assessment plan and its results are presented in Section 3 and 4, respectively. Finally, concluding remarks and future directions are discussed.

2. Program Development

In this section, the collaborative program between WUST and UB is summarized based on the agreement of two parties. In addition, the characteristics and history of the program are presented, respectively.

One main objective of the collaborative program between WUST and UB is to build a degree completion program in which students from WUST, who have fulfilled two years of undergraduate course requirements (which will be discussed in the subsection), may transfer to UB and complete a Bachelor's degree at UB within two years.

2.1 Degree Requirements

WUST students may apply to complete a Bachelor of Science in Electrical Engineering program at UB after completing two years of course requirements at WUST. The two years of course requirements for students on the WUST campus are listed in Table 2. As shown in the table, total of sixty (60) credits including combined credits can be transferred to UB when WUST students get an admission and transfer to Junior at UB.

2.2 Admission Requirements

UB will admit qualified students from WUST if they satisfy the admission requirements and conditions indicated below:

- *English Proficiency*: Applicants should have a minimum of 75 (IBT) scores on the TOEFL or a minimum 6.0 score of the IELTS. Applicants who do not have the required TOEFL or IELTS score will be given the ELAB test (i.e., English

Table 2. Two years of Course Requirements of EE at WUST and Transferred Courses to UB

WUST		**UB**	
Course Name	**Credits**	**Course / Category Name**	**Credits**
Moral Cultivation and Basic of Law	3	Humanities I	3
Mao Zedong Thoughts, Den Xioping Theory and "Three Represents"	6	Social Sciences I & II	6
Fundamentals of Marxism	3	Humanities II	3
C Programming Language	3.5	Introduction to Programming	4
C Programming Language LAB	1		
Circuit Analysis Basis (I)	2.5	Network Analysis 1	3
Introductory Practice Experience	0.5		
Circuit Analysis Basis (II)	2.5	Network Analysis 1I	3
Practice Experience on Electronic Tech.	0.5		
Circuit Analysis Basis Lab	1	Network Analysis I Lab	1
Analog Electronics	4	Electronics	3
Digital Electronics	3	Digital Design I	3
Analog Electronics Lab	1	Network Analysis II Lab	1
Higher Mathematics A (I, II)	5, 6.5	Calculus I, II	4, 4
Complex Function and Integral Transf.	3	Calculus III	4
University Physics B (I, II)	3, 1.5	Physics I	4
MATLAB Programming	2.5	EE Elective I	3
Course Project on Software Technology	1		
Probability Theory and Math Statistics	3	Probability & Statistics	3
Single Chip Microcomputer and Interface	2.5	Differential Equations	3
Integrated Design of Electronic Sys (I)	1.5		
Electromagnetic Field and Wave	2.5	EE Elective II	3
Introduction to Electronic Information	1		
Signals and Systems	4	Signals and Systems	3
Total Credits	**68**	**Total Transferred Credits**	**60**

proficiency test administered by the English Language Institute – ELI) upon arrival at UB. If a student achieves the passing ELAB test score, he or she will be able to matriculate into the degree program. Students who did not achieve the minimum required ELAB test score will be required to take an intensive English course at ELI and upon attaining the required English proficiency will be permitted to matriculate into the degree program,

- *Good Academic Standing*: The applicant should have completed, or is completing, two years of undergraduate course requirements as listed in Table 2 at WUST, and be in good academic standing with a cumulative grade point average of at least 2.85 in terms of US grading system, and
- *Transfer Credit*: To ensure that a student has sufficient credits to graduate within two year at UB, WUST should select qualified students who have fulfilled the course requirements as listed in Table 2. The exact number of transfer credits is determined based on the evaluation of the applicant's transcript with the specific program requirements. Only credits with a grade of "C" or better are considered for transfer in terms of US grading system.

2.3 Graduation

To graduate from UB within two years, the enrolled student must be able to fulfill all graduation requirements for a Bachelor's degree in Electrical Engineering major including the course requirements listed in Table 3. A student may need to extend his or her matriculation beyond two full years if satisfactory progress has not been made in the two full academic years. If students are not able to complete all required subjects at WUST while they are in China for the first two year, the students will be required to take any missing courses at UB. The academic advisors will make every effort to make sure the required courses are included within two years of the study. However, in some cases the student will be required to take additional classes to meet the graduation requirements at UB.

Table 3. Proposed Study Plan at UB (BS in EE) during Year 3 and 4

Semester	**Proposed Study Plan at UB (BS in EE)**	**Credits**
Fifth Semester (16 Credits)	PHYS 112 Principles of Physics with Lab	4
	EE 286 Microprocessors	3
	ENGR 111 Intro to Engineering	3
	EE Elective	3
	ENGL 101 Composition & Rhetoric	3
Sixth Semester (16 Credits)	CHEM 103 General Chemistry with Lab	4
	ECON 300 Economics for Engineers	3
	EE 317 Intro to Control Systems	3
	EE 337 Analog Electronics Lab	3
	Fine Arts Elective	3
Seventh Semester (15 Credits)	EE 350/315/316 Communication Lab	3
	EE 364 Programmable Logic Control	3
	EE 346 MEMS (Micro Electro Mechanical Systems)	3
	EE 349 Senior Design Project	3
	EE Elective	3
Eighth Semester (13 Credits)	CPE 315 Digital Design II with Lab	4
	EE Elective	3
	EE Elective	3
	CAPS 390 Capstone	3

2.4 Obligations

The obligations of WUST includes:

- Responsible for obtaining and maintaining any and all necessary approvals and registration to operate the program by the government and all relevant authorities in China,
- Design and prepare publicity and advertising materials for the program and undertake marketing in China to generate applications; notify prospective students of the features and advantages of the program,
- Responsible for teaching organization of the courses offered on the WUST campus including providing appropriate number of qualified and experienced teachers to deliver teaching,

- Provide venues, facilities, materials, equipment and supporting staff that are necessary for teaching,
- Permit and assist UB representatives to make occasional presentations on WUST to students who are interested in completing a degree at UB,
- Recommend qualified students for the program,
- Submit application with support documents,
- Oversee and assist in the admission application and the visa application processes, and
- Provide the applicants with the appropriate departure orientation.

In addition, the obligations of UB includes:

- Responsible for obtaining and maintaining any and all necessary approvals and registration to operate the program by the government and all relevant authorities in the U.S.,
- Participate in the design of the curriculum, introduce courses thereto as required by the program, share curriculum materials and syllabi, and administer course assessment,
- Responsible for teaching organization of courses offered by UB on WUST campus as mutually agreed, including providing qualified and experienced teachers to deliver teaching,
- Provide current information on the program such as academic calendar, costs and fees, and guidelines for application for admission,
- Review the applications and send qualified applicants a Letter of Acceptance and documents for visa application,
- Provide Conditional Acceptance Letter to the student upon enrollment in WUST,
- Arrange on-campus housing and participation in a meal for the students, and
- Provide appropriate orientation for the students upon arrival.

2.5 Program Characteristics

There are several important characteristics of the program between WUST and UB.

- First, the collaborative program focuses on specific academic program, so both institutions mutually design the program that fits to students at WUST. Currently, EE major is operated under the collaborative program,
- Second, the collaborative program allows WUST students to receive degrees from UB and WUST upon the completion of all graduation requirements at each institution by exchanging the credits earned at the other institution, and
- Third, professors at UB visit China to teach courses directly during shorten period time, which is usually four to five weeks. Every semester UB professors visit China to teach selected courses for freshman, sophomore and junior classes. On the other hands, WUST professors also visit UB as a visiting scholar.

2.6 Program History

UB has been recognized one of the most international universities in US with the 15% of international undergraduate student body by the Best Colleges – US News and World Report [11]. With UB's focused international characteristics, the University also has been in collaboration with various higher education institutions in different countries, for a variety of collaborative programs including exchange of faculty for instruction, research, lectures and establishing collaborative academic programs for undergraduate and graduate degrees. Over the course of a few years UB very successfully established and implemented collaborative degree awarding programs with Chinese universities of high academic ranking. One good example of such a collaborative program is the US-WUST 2+2 collaborative program in Electrical Engineering.

In October 2011, after an in-depth study and discussions UB and WUST agreed and signed an agreement to develop a collaborative degree program of Bachelor of Engineering in Electronic Information Engineering which will be offered on the WUST campus. This agreement was subject to the approval by the Chinese MOE. After the institutional agreements were signed the agreement was submitted to the MOE in Beijing, China, so that this collaborative program is fully recognized and accredited by the Chinese government. In Spring 2012, WUST obtained an official approval from the MOE and started recruiting for the first class in Fall 2013. The program has grown very significantly since. The program enrollment since inception has been as follows – 43 students admitted in Fall 2013, 64 students admitted in Fall 2014, and 72 students admitted in Fall 2015.

3. ASSESSMENT PLAN

The evaluation and assessment for the collaborative program between WUST and UB have been conducted under the leadership of Prof. Jeongkyu Lee at UB. In addition, Mr. Brian Lim and Prof. Sarosh Patel at UB are in charge of the evaluation of UB, while Prof. Roger Dianlei Geng and Prof. Zhigang Jiang from WUST conducted the evaluation at WUST.

The objectives of this assessment are two-fold:

1. *Formative evaluation*: to collect the program implementation data to determine the effectiveness of collaborative program, and
2. *Summative evaluation*: to collect the program performance data to support future efforts to assess the program's progress towards its desired outcomes as well as to estimate the impact of the program on its target student populations.

The formative evaluation assess:

- Independent assessment of the collaborative program quality by different parties, such as students, UB professors and non-teaching administrators,
- Students' assessment of the courses that are taught by UB visiting professors at WUST campus, and
- Students' assessment of the program at UB.

Summative evaluation will reflect:

- The students' performance evaluation for the courses which are taught by UB visiting professors at WUST, and
- The evaluation of program outcomes in terms of enrollments at WUST as new students and enrollments at UB as transferred ones.

Prof. Lee is responsible for the development of assessment instruments that can accurately reflect student perception and collect performance data necessary to document project impact and success. Quantitative and qualitative data analysis are combined for evaluation purposes. The proposed evaluation and assessment plan is presented in detail in Table 4.

Table 4. Evaluation and Assessment Plan

Goal	Evaluation Questions (Sample)	Proposed Methods	Frequency
	Formative Evaluation		
Program Quality Assessment	Develop a set of questions to assess effectiveness of program including suggestions for both WUST students and UB professors: *Q1: What was the primary reason when you applied to International College at WUST?* *Q2: Does the collaborative program between WUST and UB meet your expectations?*	Student Survey UB Professor Survey	End of year
Course Quality Assessment	Develop a set of questions to assess effectiveness of courses including suggestions: *Q3: Is UB professor helpful for you to understand the subject?* *Q4: Are you able to follow and clearly understand the course instruction in English?* *Q5: Do the courses (UB professors) meet your expectations?*	Student Survey	End of class
UB program Assessment	Develop a set of questions to assess UB program: *Q6: If you are at WUST, are you going to transfer to UB?* *Q7: How much do you know about University of Bridgeport?*	Student Survey	End of year
	Summative Evaluation		
Students' performance measure	What is the academic performance of WUST students in the classroom taught by UB professors? - *Collecting the average score of each class*	Collecting data at WUST	End of year
Program enrollments	What is the performance of the collaborative program between WUST and UB? - *Collecting the number of enrollments*	Collecting data at WUST	End of year

4. ASSESSMENT RESULTS

In order to assess the collaborative program between WUST and UB, the following surveys were conducted as planned in the previous section:

- *For WUST students*: 160 students responded to the survey and then analyzed, and
- *For UB Professors*: 6 UB professors who visited WUST responded to the survey and then analyzed.

4.1 Program Quality Assessment

To evaluate the collaborative program quality overall, WUST students are asked to response to Q1 and Q2, and their results are shown in Figure 1.

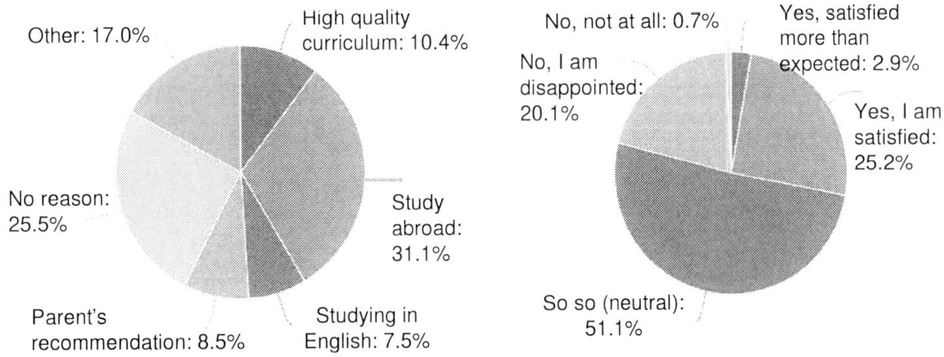

(a) results of Q1 (b) results of Q2

Figure 1. Program Quality Assessment: (a) result of Q1 and (b) result of Q2

As shown in Figure 1 (a), 49% of students selected the international program at WUST because of academic reason including studying abroad (31.1%), curriculum (10.4%), and English (7.5%). On the other hand, 20.1% of students responded negative to the program (see Figure 1 (b)). The results show that overall the collaborative program appeals the students' need and their satisfaction are positive so far.

4.2 Course Quality Assessment

To assess the quality of courses taught by UB professors, Q3 ~ Q5 are asked to WUST students and their results are shown in Figure 2.

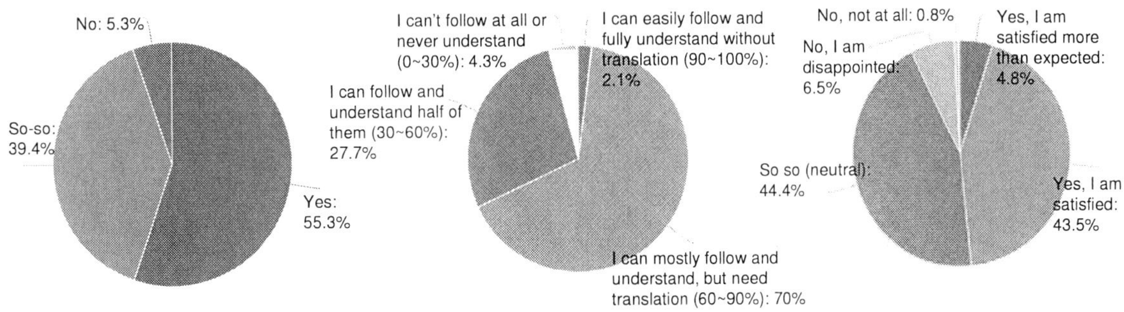

(a) results of Q3 (b) results of Q4 (c) results of Q5

Figure 2. Course Quality Assessment: (a) result of Q3, (b) results of Q4 and (c) result of Q5

As shown in Figure 2 (a), more than 55% of students answered that UB professors are helpful in their classroom, while only 5.3% are negative. Overall, more than 48% of WUST students satisfied with the courses taught by UB professor, while only 7.3% are negative (see Figure 2 (c)). Based on the survey results in Figure 2, the courses taught by UB professors are successful with high students' satisfaction.

4.3 UB program Assessment

Due to the limited number of students who transferred to UB program, we have asked WUST students regarding to UB program. Figure 3 shows the results of UB program from survey.

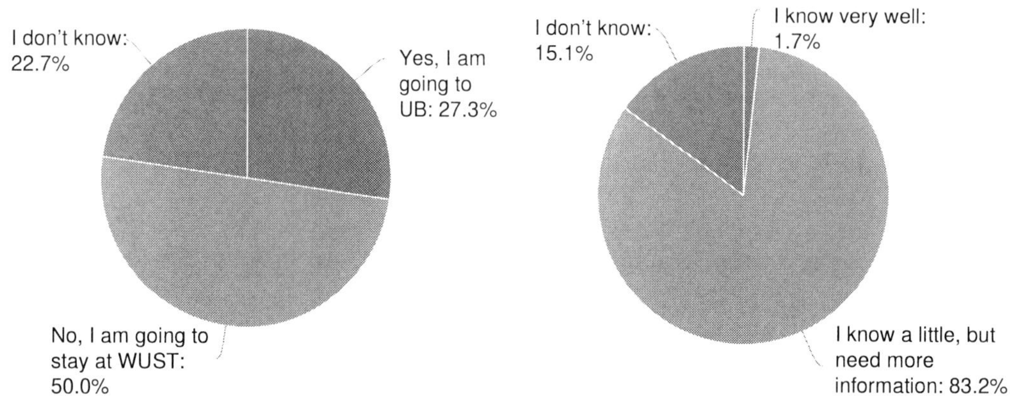

(a) results of Q6 (b) results of Q7

Figure 3. UB Program Assessment: (a) result of Q6, and (b) results of Q7

As shown in Figure3, WUST students do not have enough information about UB, and 27.3% of students decide to transfer to UB after 2 years studying at WUST. The main reasons of not transferring to UB include finance issue and English problem.

4.4 Summative Evaluation

In order to conduct summative evaluation of the collaborative program, two data sets are collected as shown in Figure 4. Figure 4 (a) shows the average score of each class taught by UB professors from 2013 to 2015, while Figure 4 (b) shows the number of enrollments and the average cut-off scores in the national college entrance exam at the program since 2013.

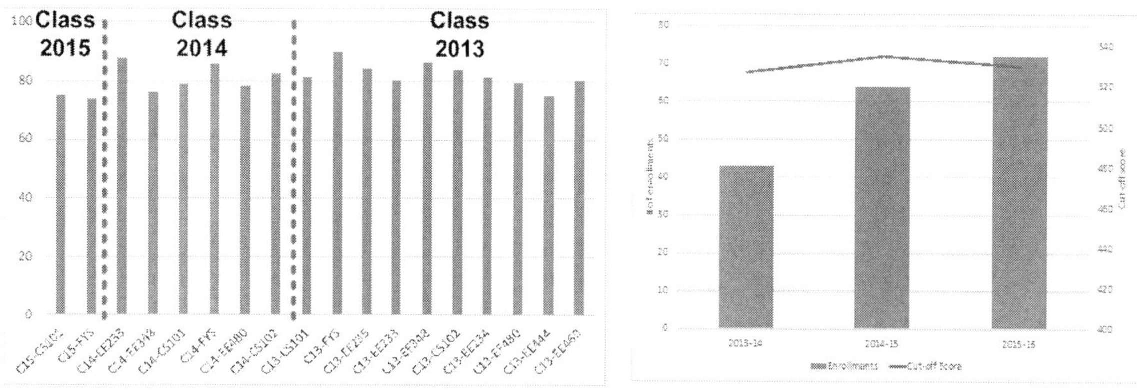

(a) average score of each course (b) the number of enrollments and cut-off score

Figure 4. (a) Average Score of Each Class, (b) Enrollment at the Program

As you can see in the figure, the average score of Class of 2014 and 2015 are over 80 (which are equivalent to B- or B grade in UB grading system), respectively, while Class of 2015 shows 75 on average. The results indicate that the freshman have the difficulty on taking classes with foreign instructors in English. Overall, the results of summative evaluation support that the program impacts on the students at WUST and has been improved.

4.5 Lessons Learned

In order to improve the quality of collaborative, WUST students and UB professors are asked to the following questions:

> *Q8: To improve the collaborative program, which of the following activities should be enhanced first?*
> *Q9: To improve the quality of courses (UB professors), what should be considered?*
> *Q10-1: (Student Only) If you do not transfer to UB, what is the main reason to decide to stay at WUST?*
> *Q10-2:(Professor Only) What is the best number of weeks for one semester long subject at WUST?*

Table 5 shows the top 3 choices for each group. In order to improve the collaborative program, 55% of students suggest the diverse activities, while UB professors (50%) are interested in the better teaching environment, i.e., lab and classroom. However, regarding Q9, both students and professors selected "longer course period" to improve the course quality. Currently, it is 4 ~ 5 weeks for each class. According to the students' response of Q10, it is the financial aspect that prevents WUST students from transferring to UB. This should be addressed for the successful program in terms of quantity as well as quality.

Table 5. Top-3 Choices for Q8 ~ Q10

Q#	Response of WUST Students	Response of UB Professors
Q8	1. Diverse activities (such as short-term visiting, online conference) (55%) 2. Communication between instructors and students (50%) 3. Communication between WUST and UB (49%)	1. Facilities (such as laboratory and classroom) (50%) 2. Communication between instructors and students (50%) 3. Communication between WUST and UB (16%)
Q9	1. Longer course period (currently 4 ~ 5 weeks) (51%) 2. Schedule for future courses (41%) 3. English translation (39%) Communications with UB professors (39%)	1. Longer course period (currently 4 ~ 5 weeks) (50%) 2. Visiting time (currently May and December) (34%) 3. More assisting professors (34%)
Q10	1. Financial issue (70%) 2. English (12%) 3. No interests for studying abroad (9%)	1. 4 ~ 5 weeks (60%) 2. 6 ~ 7 weeks (20%) 3. More than 7 weeks (20%)

5. CONCLUSION

In this paper, we presented development and implementation of an undergraduate collaborative program in electrical engineering between the University of Bridgeport (UB) in the U.S. and Wuhan University of Science and Technology (WUST) in China. Students admitted by Department of Electrical Engineering (EE) of WUST will study in China for 2 years, then students can be transferred to the same program at UB for the remaining academic requirements. Upon the completion of all graduation requirements at UB, students will receive a Bachelor of Science in EE from UB. Students who stay all four years at WUST or transfer to UB after two years, will be granted a Certificate of Graduation and a Bachelor's degree from WUST if they are subject to satisfactory completion of the program requirements. In order to assess the quality of collaborative program, we design and conduct the survey for both WUST students and UB professors. The survey results show that the program is successful, since the enrollments have been increasing at WUST and more students plan to transfer to UB. In addition, the WUST-UB joint program was successfully passed the first audit by China Academic Degree and Graduate Education Development Center (CDGDC) in 2015. However, there are several issues in the program that should be addressed for better quality, such as financial issue, teaching environments and communication.

REFERENCES

[1] The University of Bridgeport Catalog, http://www.bridgeport.edu
[2] The Wuhan University of Science and Technology, http://www.wuhan.edu.cn/
[3] Best Colleges by US News and World Report 2016, http://colleges.usnews.rankingsandreviews.com/best-colleges/rankings/regional-universities-north/most-international
[4] Brzoska, M., & Koh, M., & Rodriguez-Marek, E., & Talarico, C. (2010, June), A Multinational 1+2+1 Electrical Engineering Program Paper presented at 2010 Annual Conference & Exposition, Louisville, KY.
[5] Howell, S. (2005, June), A New Joint Engineering Program Between Shanghai University Of Engineering Science And Lawrence Technological University Paper presented at 2005 Annual Conference, Portland, Oregon.
[6] Tang, Y., & Lord, S. M. (2012, June), Comparison of Practical Training Experiences for Electronics Engineers in China and the U.S.A.: Case Study of Southeast University and the University of San Diego Paper presented at 2012 International Forum, San Antonio, Texas
[7] Fasse, B. B., & Benkeser, P. (2011, June), Developing the Global Biomedical Engineer through a 12-Month International Undergraduate Research Experience in the U.S. and China Paper presented at 2011 Annual Conference & Exposition, Vancouver, BC.
[8] Abubakr, S., & Qi, D. (2007, June), Development Of Global Engineering Education In China For Western Michigan University Engineering Students Paper presented at 2007 Annual Conference & Exposition, Honolulu, Hawaii.
[9] Hugo, R. J., & Brennan, B., & Zhang, J., & Niu, X. (2015, June), Evolution of a Chinese-Canadian Educational Partnership Paper presented at 2015 ASEE International Forum, Seattle, Washington.
[10] Segalewitz, S. I. (2013, June), Seven Years of Success in Implementation of a 3 + 1 Transfer Program in Engineering Technology Between Universities in China and the Unites States Paper presented at 2013 ASEE Annual Conference.
[11] Rong, Y. (2015, June), Ten Year Experience of Global Capstone Design Projects in China Paper presented at 2015 ASEE International Forum, Seattle, Washington.
[12] Iftekhar, S. N., & Kayombo, J. J. (2016), Chinese-Foreign Cooperation in Running Schools (CFCRS): A policy analysis, Int'l Journal of Research Studies in Education, Vol 5, No 4, 73-82.

Strategies for Increasing Student Participation in International Programs

Dr. James N Warnock, Mississippi State University

James Warnock is the Associate Dean for Academic Affairs in the Bagley College of Engineering at Mississippi State University. His background is in biomedical engineering and he has been a big proponent of self-directed learning and active learning in his classes and was the first person to introduce problem-based learning in the department of agricultural and biological engineering at MSU. James is also the Adjunct Director for training and instruction in the professional services department at ABET. In this role, Warnock oversees the development, planning, production and implementation of the ABET Program Assessment Workshops, IDEAL and the assessment webinar series. He also directs activities related to the workshop facilitator training and professional development.

Ms. Galyna Melnychuk, Mississippi State University

Strategies for Increasing Student Participation in International Programs

Abstract

This paper describes the strategies utilized by the Bagley College of Engineering at Mississippi State University, a tier one research university located in the southeastern US. In the institutional strategic plan, the university recognizes the need to apply their academic, scientific, and extension capabilities to global challenges while providing students with an international perspective that expands their future opportunities. To increase international opportunities for students, the College of Engineering has focused on *opportunity, benefit* and *affordability*.

To increase opportunity, courses offered as faculty-led study abroad programs have been selected from those that can be applied to the most number of engineering programs. Starting summer 2015, Technical Writing was offered in Munich in collaboration with Munich University for Applied Sciences. This course is required for all engineering students. Students also have the option to take Engineering Economy in France. This course counts as an elective in all engineering programs. Starting summer 2016, we will offer Engineering Mechanics I in Cardiff, UK. This class is required for seven of the 11 engineering programs. These courses are offered at international universities where we currently have established activities, either through bilateral exchange agreements and/or active research collaborations.

The greatest challenge to increasing participation in international activities is communicating to students the benefits of a global experience, especially in a state that has one of the lowest percentages of passport holders in the country. To increase awareness, the college has initiated a student ambassador program. Ambassadors meet with classes to talk about opportunities, share their personal experiences and testify how the experience has been of benefit, both professionally and personally. The college has also introduced a minor in Global Engineering Leadership as an academic incentive to students.

Finally, to make international opportunities affordable, the university has introduced study abroad scholarships for incoming freshman students meeting certain criteria (in-state residency and minimum ACT scores). In fall 2015, ~240 study abroad scholarships were awarded, 164 of which were offered to engineering students. Additionally, the college has been able to offer substantial scholarships from gifts received from engineering alumni and industry sponsors.

The strategy to increase opportunities for students, communicate benefits of study abroad, and enhance affordability has been effective since first implemented. The college has witnessed a 56% increase in students participating in international programs since 2014. Although this is a modest increase in terms of absolute numbers, the college is on a trajectory to meet its goal of having >200 students per year participate in study abroad by 2020.

Introduction

There is little argument that we live and work in a global economy. As such, it is imperative that engineers have a deeper understanding and appreciation of other cultures. In fact, the

Engineering Accreditation Commission of ABET, states this as an expected student outcome in General Criterion 3 (h) "the broad education necessary to understand the impact of engineering solutions in a global, economic, environmental, and societal context."[1]

Traditionally, engineering programs have sought to achieve this objective through courses alone. Requiring students to take history, foreign language, geography or similar courses was thought to prepare engineers to function effectively in a global economy. Recently this has been found to often be ineffective. As a result, some engineering programs have introduced international engineering programs for undergraduate students, such as those at the University of Rhode Island and the University of Wisconsin-Madison. These programs, especially when combined with appropriate coursework, offer the opportunity for students to learn and experience different cultures at both a cognitive and emotional level. Researchers have quantitatively shown that students who participate in these programs have higher levels of global awareness. [2]

Expanding students' perception of other cultures is critical to their success as engineers. Additionally, if these students are to become leaders in the profession, it is imperative that they understand and appreciate the diversity of cultures in the world. There is an immediate need for this at Mississippi State University, which is located in a state with one of the lowest percentages of passport ownership in the nation at just under twenty-percent while the United States as a whole is just under fifty percent.[3] This is strong evidence that most in-state residents have never travelled outside of the United States. A certain percentage of those have likely not travelled abroad as a result of financial constraints, but there must also be a number who have the financial needs but merely lack the inclination or opportunity to travel. Importantly, while Mississippi ranks at the bottom for percentage of passport holders, this is not an anomaly for addressing issues of limited global awareness. Rather, we are at the center of an entire region with limited global exposure.

Informal discussions with students who have participated in study abroad opportunities have revealed that these students' outlook on the world was changed and their ability to be more accepting of different cultures was enhanced. We believe this also decreases a student's resistance to change and will therefore make them a more effective engineer in today's constantly changing world. In a recent study on the environmental awareness of engineering students and their resistance to change, Weber found that students who lived in or near urban areas had less resistance to change.[4] Given we are a rural state, it stands to reason that engineering students would tend to be more resistant to change than those from other, less rural states.

In addition to producing engineering graduates who can function effectively in today's economy, engineering colleges desire to produce leaders who can thrive in the global economy. This requires a broad understanding of the world, global issues, policy, and leadership fundamentals. Within our college of engineering, which teaches the majority of our in-state student engineers, several programs along these lines exist but they are not unified into a coherent program nor do they have a specific engineering focus. We have built upon our existing strengths and combined the appropriate programs into a unified global engineering program, the minor in Global Engineering Leadership, to debut in Fall 2016.

Opportunity

For many years, the college of engineering has been a part of various consortia and bilateral exchange agreements that have allowed students to spend a semester studying at international partner universities. These programs offer opportunities for students to study in virtually any geographical region from Australia to South East Asia to Europe to South America. However, the number of students that participate in these semester long programs is relatively small and the majority of students are interested in faculty-led summer study abroad programs. These programs offer the security of being taught by faculty from their home institution, providing familiarity, being able to take a class with students from their own department, offering companionship, and often reduce the need to be conversant in a foreign language. Other factors, such as cost, articulation of courses to the home institution and parental anxiety, make faculty-led programs more appealing to students.

In 2013, only one faculty-led engineering program was offered. Students with a desire to study abroad had other options through the university or third-party providers in subjects such as foreign languages or international business, but very few that would count towards their degree program. At that time, the strategy for developing new study abroad programs was to send out a request for proposals to all faculty. This passive approach did not generate many proposals and of those submitted, many were very discipline specific and did not appeal to a wide range of engineering students. Consequently, of the few programs offered, many did not have enough students register for them to be viable and they were cancelled.

In 2014, a more proactive strategy was developed to provide additional opportunities for students to participate in summer study abroad programs. The first requirement was that the courses to be offered should be integrated with the curriculum. The college of engineering offers 11 undergraduate programs and courses were identified that counted towards the most number of degree requirements to maximize the benefit to the most number of students. The first course selected was engineering economy, which was already being offered, and could be included as an engineering elective in all programs. The second program developed was Technical Writing, a course required for all undergraduate students. The third course, introduced in summer 2016, was Engineering Mechanics I (Statics), a requirement for seven of the 11 engineering programs. The study abroad courses and the programs to which they apply are summarized in table 1.

Table 1: Summary of relevance of faculty-led study abroad courses to the different engineering curricula.

	Engineering Mechanics I (Statics)	*Engineering Economy*	*Technical Writing*
Aerospace Engineering	R	E	R
Biological Engineering	R	E	R
Chemical Engineering	R	E	R
Civil Engineering	R	E	R
Computer Engineering		E	R
Computer Science		E	R
Electrical Engineering		E	R*
Industrial Engineering	R	R	R
Mechanical Engineering	R	E	R
Petroleum Engineering	R	E	R
Software Engineering		E	R

R = Required, E = Elective, *Electrical Engineering requires Technical Writing; however, it is taught in conjunction with the senior capstone design course and students must register for the designated section of the course.

In addition to ensuring the programs offered could be integrated in the curricula and were relevant to students, it was important to ensure the offerings were sustainable. The willingness and availability of faculty to teach courses each summer was a main concern. To alleviate this concern we only considered courses that had multiple instructors on campus who would be willing to teach overseas. Our second consideration was location. We wanted to choose destinations that would be appealing to students and their parents and where we would be able to offer the courses for several years. We therefore selected destinations in Europe that were considered relatively safe and affordable. Additionally, we opted to partner with universities where we already had existing relationships, either through bilateral exchange agreements or research collaborations. Table 2 presents data showing the demand for the selected courses based on the number of students enrolling in the on-campus offerings, the international location for the study abroad course, and the type of existing relationship between our institution and the host institution.

Table 2: Data for faculty-led study abroad courses. On-campus data are for the 2015-16 academic year.

Faculty-led Study Abroad Course	*Number of On-campus sections*	*On-campus students enrollment*	*Study Abroad Location*	*Research Collaboration*	*Bilateral Student Exchange Agreement*
Engineering Mechanics I (Statics)	13	639	Cardiff University, UK	✓	✓
Technical Writing	26	485	Munich University of Applied Science, Germany		✓
Engineering Economy	2	437	Albi School of Mines, France		✓

Benefit

Having created various opportunities for students to participate in faculty-led study abroad programs, in courses that were applicable to the majority of students in the college and that were in high demand, it became important to communicate the benefit of education abroad to students and parents and provide incentives to encourage greater participation. Our first tactic has been to increase awareness in students and their parents even before they attend their first class as freshman. Global experience has become a part of our recruitment rhetoric and we focus on the number of engineering companies within the state that are either from overseas or are US companies with extensive international operations. We emphasize the benefit of having an appreciation for the global marketplace, even for students who wish to stay in Mississippi after graduation. During freshman student orientation, as part of the academic session, we tell parents to get their child a passport as part of their high school graduation gift. In doing so, we are creating the expectation among parents and students that an international experience should be part of their engineering education.

To help promote continued awareness among students, we have initiated a student ambassador program. Student ambassadors are easily recognizable by their T-shirts that say, "I studied abroad… ask me where." These act as an ice-breaker and are intended to help start conversations from curious students who want to know more about the study abroad experience. Student ambassadors accompany the director for international programs when she visits classes to discuss the various opportunities and can offer their perspective on study abroad. The Director for IP will visit all engineering introductory courses to continue the message freshman students received before enrolling in courses. As the majority of students don't study abroad until their junior or senior years, ambassadors and the IP director will visit other engineering classes for sophomores and upperclassmen. Ambassadors will also help at the college of engineering booth at the university study abroad fairs.

To increase the incentive to study abroad, the college of engineering has introduced a minor in *Global Engineering Leadership*. The minor is designed to enhance students' knowledge of engineering practice in the global marketplace. Students completing this program develop proficiency in a foreign language, have experience living and studying in a foreign country, and become familiar with engineering leadership and management techniques. A total of 18 hours are required to complete the minor, including a required course in "Challenges in Global Engineering". The curriculum outline is shown in table 3.

The foundation of the minor is the *Challenges in Global Engineering* course. Upon completion of the course students gain a broader perspective of global challenges facing engineers today, including specifics of managing international engineering projects, engineering standards, economic, legal, and environmental issues. Students have an understanding of the role of culture in engineering design and problem solving and global trends in engineering. They are exposed to engineering work in various countries through projects that focus on real-world case presentations and case reports. Students also learn to explore resources, strategies, and options to solve problems in a globalized engineering world.

Students are required to take six hours of a foreign language. Foreign languages not only provide the necessary skills for effective communication in a non-native language, they in fact form the basis for understanding a people's culture. The specific language is selected in consultation with the students' academic advisors and can be used to fulfill the university general education requirements. Ideally, and if possible, the chosen language would be relevant for the student's follow-on study abroad course. However, we believe there is value in completing a foreign language even if a student ultimately completes a study abroad in a country that does not speak the studied language.

Students are allowed to take six hours of approved leadership electives to round out the course requirements for the minor. These courses could include engineering and public policy, law and ethics, cross-cultural management and leadership, and project management.

Table 3: Educational requirements for the minor in Global Engineering Leadership.

Curriculum Outline	***Credit Hours***
Challenges in Global Engineering	*3*
Foreign language [1,2]	*6*
Language I	
Language II	
Relevant Overseas Engineering Experience e.g.	*3*
Faculty-led Study Abroad in an engineering course	
International Engineering Internship	
Semester-long Engineering Exchange Program	
Overseas Service Learning (e.g. Engineers Without Borders)	
Leadership Electives	*6*
International Business Law	
Construction Engineering and Management	
Analysis and Mitigation of Conflicts, Claims and Disputes	
Civil Engineering Comprehensive	
Social and Ethical Issues in Computing	
Managing Software Projects	
Introduction to Engineering and Public Policy	
Engineering Entrepreneurship Seminar	
International Business	
International Economic Development	
International Economics	
Engineering Economy I	
Engineering Administration	
Project Management	
Engineering Laws and Ethics	
Cross-Cultural Leadership	
Socially Responsible Leadership	
Cross-cultural Management	
International Logistics	
International Transportation	
International Supply Chain Management	

Notes: [1.] Must be a modern language. The six credit hours must be of the same language. [2.] The foreign language studied does not have to be the same language spoken in the country of the overseas engineering experience, though this is highly desirable. Students are strongly advised to complete the language component before their overseas experience.

Affordability

The university is located in a rural state where there is a high level of poverty. In 2014, 53% of graduating students had taken out loans to pay for their education and the average level of indebtedness was $29,365. The financial aid statistics for the university are shown in Figure 1. These data indicate that there is a high financial need among students at the university and the cost of studying abroad could be prohibitive for many families. The college of engineering actively sought sponsorship from alumni and industry partners so students could receive scholarships to offset the cost of study abroad. In 2016, the college of engineering awarded $50,000 in scholarships to students at an average of $2,500 per student. In addition to the college study abroad scholarships, the university introduced study abroad scholarships to students entering the university in 2015. In-state students who attained a composite score of 30 or above on the ACT received a study abroad scholarship of $1,000 to $4,000. The amount awarded to each student is determined by their ACT score and high school GPA. The university awarded ~240 study abroad scholarships in fall 2015 to incoming freshmen. Of those, 164 were awarded to engineering students.

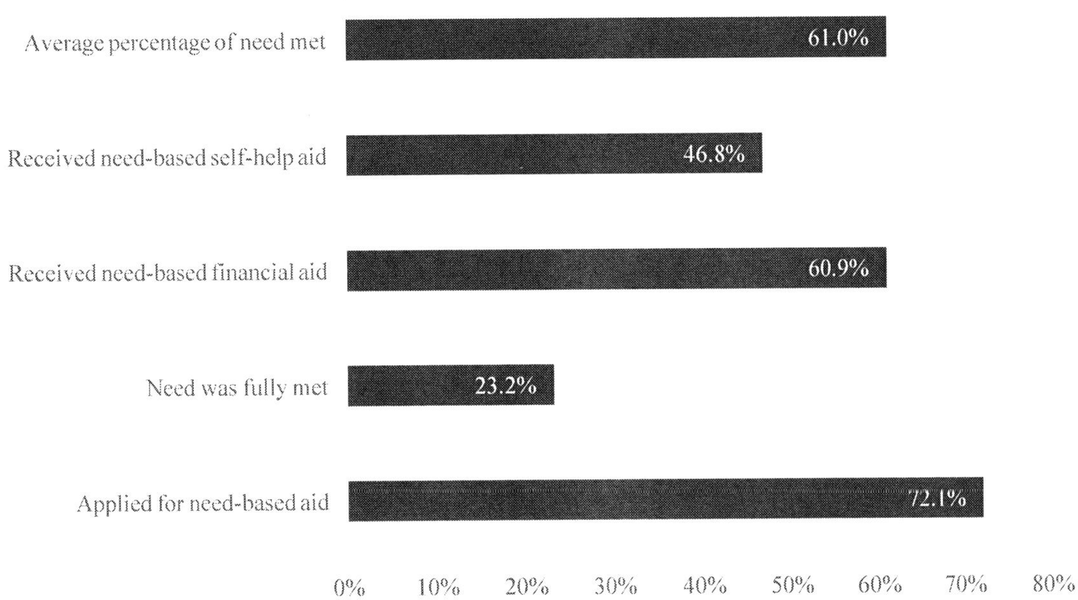

Figure 1: Financial Aid statistics for students attending the home institution in 2014. Data obtained from US News and World Report.

Impact

As seen in Figure 2, there has been a significant increase in the number of students participating in education abroad since the new strategic initiatives were systematically introduced in 2014. The number of students participating in semester or year-long exchange programs has remained consistent with an average of 6.8 ± 1.92 students per year over the past five years. For spring 2016, six students are spending the semester at an international university, five in Italy and one in Australia. The biggest increases in student participation are seen in the faculty-led programs with engineering-led programs increasing by 2.5-fold between 2014 and 2015 and non-engineering-led programs increasing by 3.3-fold between 2013 and 2015. Data for non-engineering-led programs are not available for 2016; however, 46 students are registered for engineering-led programs for summer 2016, showing a further 31% increase over the previous year and a 229% increase compared to two years prior. For the upcoming summer, 22 students will take *Engineering Economy* in France, 16 students will take *Technical Writing* in Germany, and 8 students will take *Engineering Mechanics I* in the UK. Overall, the number of students participating in an international experience increased by 56% between 2014 and 2015, with early indications that this trend will continue into 2016.

These early results are certainly encouraging but it is important to provide context. The number of student participants represents approximately 2% of the undergraduate engineering student body. It should also be noted that different majors are better represented than others in global programs. Biological engineering, industrial engineering and mechanical engineering have the highest representation compared to civil engineering, electrical engineering and computer science that are poorly represented. This indicates the importance of academic advisors in the individual departments promoting awareness of education abroad opportunities with their students and faculty making students aware of the impact of engineering solutions in a global context.

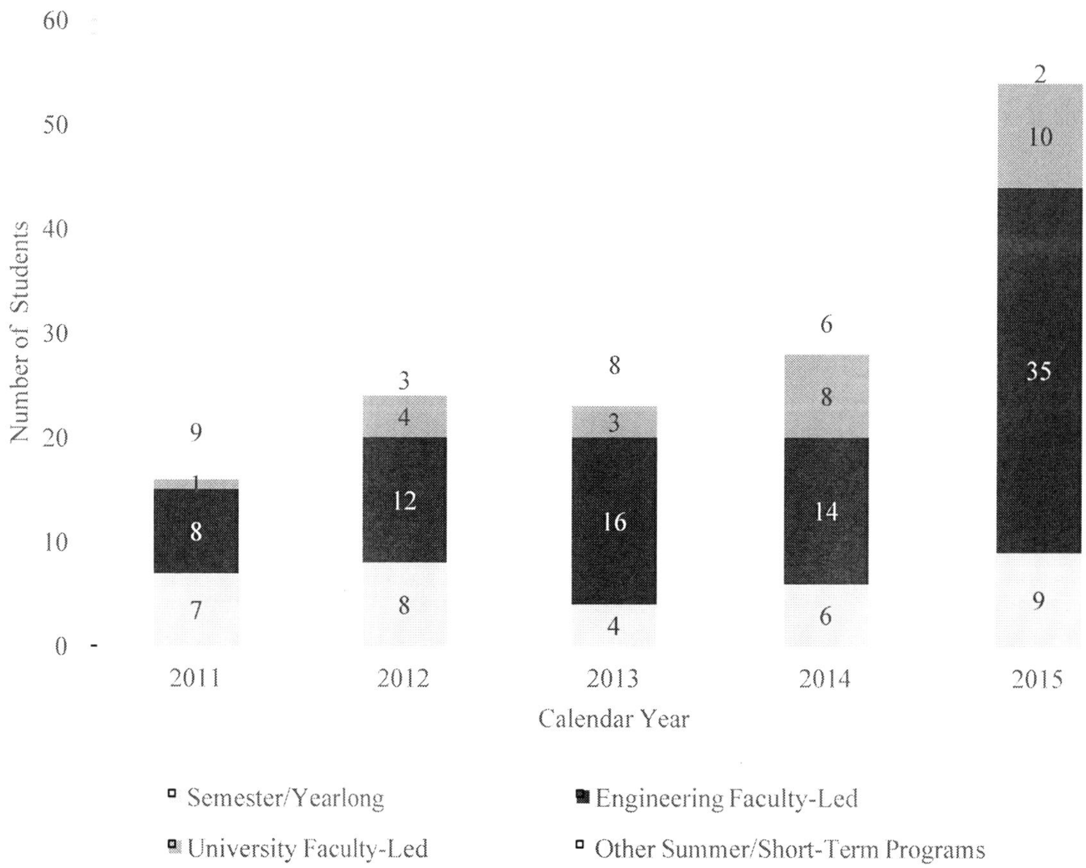

Figure 2: Engineering student participation in international programs between 2011 and 2015.

Summary

The college of engineering has implemented a strategy to increase the number of engineering students participating in international experiences. First, we have introduced faculty-led summer study abroad programs that are integrated with the curriculum and satisfy degree requirements for our students. Second, we have recruited student ambassadors to increase awareness of various opportunities that exist. We have developed a new minor in Global Engineering Leadership as an incentive for students while simultaneously helping to better prepare them to work effectively in a global marketplace. Finally, the college and the university have been able to develop scholarships so students from all financial backgrounds have the opportunity to gain valuable international experience.

References

1 ABET, Inc. (2016) Criteria for Accrediting Engineering Programs, 2015-2016. Retrieved March 12, 2016

2 Shen, Y., Jesiek, B. K., & Chang, Y. (2011). Cultural orientation and global competency: A comparative assessment of engineering students. In *American Society for Engineering Education*. American Society for Engineering Education.

3 Milecards.com (2016). What States Have the Most Passport Holders? Retrieved April 26, 2016, from http://milecards.com/what-states-have-the-most-passport-holders/

4 Weber, N., Dyehouse, M., Miller, C. C., Fang, J., Hua, I., & Strobel, J. (2013). Impact of Household Location on First-Year Engineering Students' Environmental Awareness and Resistance to Change. Journal of Engineering Education, 102(4), 603-625. doi: 10.1002/jee.20028

The Challenges and Lessons Learned in Establishing a Travel Course

Dr. Yanjun Yan, Western Carolina University

Yanjun Yan received her B.S. and M.S. degrees in Electrical Engineering from Harbin Institute of Technology (China), and the M.S. degree in Applied Statistics and the Ph.D. degree in Electrical Engineering from Syracuse University. She is an assistant professor in engineering and technology at Western Carolina University. Her research interests are statistical signal processing, diagnostics, and particle swarm optimization.

Dr. Paul M Yanik, Western Carolina University

Dr. Paul Yanik is an Assistant Professor of Electrical and Computer Engineering Technology at Western Carolina University. His research interests include human-robot interactions, assistive devices, pattern recognition, machine learning, and engineering education.

Prof. Sudhir Kaul, Western Carolina University

Dr. Kaul is an Assistant Professor of Mechanical Engineering at Western Carolina University. His research interests include Fracture Diagnostics, Structural Dynamics and Control, and Motorcycle Dynamics.

Dr. Chip W Ferguson, Western Carolina University

Chip Ferguson is the Associate Dean of the Kimmel School and Associate Professor of Engineering and Technology at Western Carolina University.

Dr. Robert D. Adams, Western Carolina University

Dr. Adams is an Associate Professor of Electrical Engineering at Western Carolina University. His research interests include in digital image processing, biomedical signal processing and engineering education.

Dr. Wes Stone, Western Carolina University

Dr. Wes Stone is an associate professor in the Department of Engineering and Technology at Western Carolina University in Cullowhee, NC. He earned his bachelors degree from the University of Texas at Austin, masters degree from Penn State, and PhD from Georgia Tech, all in Mechanical Engineering. His research interests include manufacturing processes and quality techniques. He also serves as the program director for Engineering Technology at WCU.

Dr. Hugh Jack, Western Carolina University

Dr. Jack is Department Head and Distinguished Professor of the Department of Engineering and Technology at Western Carolina University. His interests include robotics, automation, controls, and project management.

Dr. Jeffrey L. Ray, Western Carolina University

Jeffrey Ray is Dean of the Kimmel School of Construction Management, Engineering, and Technology at Western Carolina University (WCU) and Professor of Mechanical Engineering. Prior to joining WCU, Ray was Dean of the School of Engineering Technology and Management and Professor of Mechanical Engineering Technology at Southern Polytechnic State University (SPSU) in Marietta, Georgia for seven years. Before his tenure at SPSU, he was the Director of the School of Engineering and Professor of Mechanical Engineering at Grand Valley State University (GVSU) for ten years, in addition to leading the multidisciplinary industry-sponsored capstone design courses.

Before joining GVSU he was an Assistant Professor of Mechanical Engineering at Youngstown State University. His degrees include both B.S. and M.S. degrees in Mechanical Engineering from Tennessee

Technological University and a Ph.D. in Mechanical Engineering from Vanderbilt University. While at Vanderbilt, he worked in the Department of Orthopaedics performing skeletal biodynamics research. Before beginning engineering school he completed an apprenticeship and was awarded the title of Journeyman Industrial Electrician. These professional experiences have provided Ray the opportunity to experience the full spectrum of engineering careers.

The Challenges and Lessons Learned in Establishing a Travel Course

Abstract: A travel course takes a significant amount of effort in its planning and execution. The logistics are even more challenging when a travel course is introduced for the first time. In the Engineering and Technology department at Western Carolina University (WCU), a faculty-led travel course has never been taught. The Council on International Educational Exchange (CIEE) offered an inaugural grant opportunity, especially for a travel course in STEM. In response to that, a proposal was submitted after discussions among colleagues and administrators. Although the grant proposal was not funded, the CIEE offered a large discount for the program and the CIEE program manager worked tirelessly to accommodate the requests to revise the program. When the initial abstract of this paper was submitted in January 2016, nine student applications had been received, for travel in May 2016. Unfortunately the class was cancelled after two students withdrew reducing the numbers below the required enrollment and budget limits. However, throughout the development of this course, many successful partnerships were fostered. The collaborating parties included: students, departmental colleagues, administration (the Department Head, Dean, and Associate Dean), the International Service Office, colleagues in other departments, the CIEE Program Manager, WCU's Development Officer, and a private donor. It is no small feat for a travel course to be jumpstarted from scratch, and partnerships are the key for successful implementation. Although our travel course was unsuccessful at meeting the final objective, we gained valuable knowledge from the process. The current paper addresses several factors in establishing a new travel course, such as assessing a fair number of credit hours for a relatively short duration, developing an appropriate budget, and incorporating project-based learning into a short time table. Furthermore, the current paper shares some guidelines that may be helpful in establishing a new travel course, such as using survey tools to understand student needs, making infographics to advertise the course, and encouraging the students to talk to their friends and classmates about the course. These tools have been effective but need to be used carefully to avoid misleading the intended audience. The many lessons that have been learned during the development of this course will also be shared in this paper.

Background

The 2015-2016 academic year is the third year the first author has been employed as a tenure-track assistant professor in the Department of Engineering and Technology at Western Carolina University (WCU). The first author was not born in the United States (US) and has gained a great appreciation for studying abroad from her personal study abroad experiences in Japan and the U.S. This appreciation and passion led the authors to engage in the development of the department's first international travel course for credit. During the information collection stage, the authors investigated the causes for the historical absence of study abroad opportunities in the department, determined if the department would support efforts to develop an international travel course, and understood the process for beginning such an endeavor.

Encouraged by the inaugural Generation Access Grant opportunity offered by the Council on International Educational Exchange (CIEE), a non-profit organization founded in 1947, the authors worked with CIEE to develop a course proposal and grant application. The custom program was shaped to be a faculty-led two-week travel course in Berlin, Germany. The course was scheduled for May 2016, with a focus on Project-Based Learning (PBL) and cultural influences on engineering practices. For the grant application, the authors prepared a proposal as well as a travel itinerary (with the CIEE program manager's help) and a course syllabus, although the last two items were not required for the grant submission. Despite presenting a strong case, the authors were not awarded the grant. The initial response to this decision was to end all efforts and pursue other grant opportunities in the future. However, a week later, the CIEE program manager offered a large per student discount (worth about half of the grant that we applied for), because of the strong efforts and close collaborative work with the CIEE to develop this faculty-led custom program. Although the discount was still less than the grant, it was substantial. Thus we decided to move forward with the travel course proposal. There have been many revisions and refinements of the course plan, involving many people as outlined below in six modules.

The first module in collaboration is with colleagues in the department. Collaboration with colleagues is a fundamental step to identify the pros and cons of such a travel course. When the authors discussed the grant application with a few colleagues, the initial response was that the course might not fit into the already overburdened curriculum in our engineering and technology programs. However, after further discussions regarding the PBL core of our curriculum, it was agreed that exposure to global issues was essential for our students, and this travel course would greatly strengthen our potential for meeting the student learning outcomes of the PBL sequence. Furthermore, some programs in our department incorporate technical electives in their curricula, and this travel course could be used for this purpose. The four program directors in the host department expressed support in welcoming such a travel course into their programs' curricula.

The second module in collaboration is with CIEE, or any third party that one may collaborate. While developing the CIEE grant proposal, the authors discussed with the CIEE program manager extensively via phone-calls and email, and many activities were requested that were well aligned with the course syllabus. The offices of CIEE are spread around the world. Therefore, the CIEE program manager was able to work closely with their Berlin site to provide the required information and connections to enable the creation of the custom program. In working closely with the CIEE, we were able to secure a portion of the initially requested funds.

The third module is collaboration with students. The reason for offering a travel course was not just for the sake of offering course credits, but for our students who are earnest in gaining a study-abroad experience. A Qualtrics© email survey was distributed in early September 2015 to all undergraduate Engineering and Engineering Technology students. The results clearly conveyed a student passion for a travel course. The response served as a validation of the efforts to create such a travel course. Nevertheless, a survey response indicating a student's willingness did not always mean that the student would actually register for such a course. This can be due to various concerns, which are discussed later in this paper. Of the prospective students who indicated a desire to participate in this travel course, one student stood out, and was very eager to go on the trip. This student not only talked to her friends, but also accompanied instructors to visit classes to promote this travel course. This student shared her past experiences from studying abroad, and explained how the students could apply for financial aid, etc. Some of the student's friends also applied for the travel course, and this student's help was instrumental in making progress toward the full development of the travel course.

The fourth module in collaboration is with colleagues on campus who have led summer travel courses. Although there has never been a travel course offering in the department, there have been quite a few travel courses offered at the university. The authors took advantage of the on-campus international fair, an international program workshop, and close contacts to get in touch with several faculty members who have led summer travel courses. Those faculty members were passionate about their efforts to lead travel courses and were open and willing to share their experience. There were several common traits discovered about each of these faculty:

- To keep the costs down, many faculty do not request stipends for course development and delivery, despite that fact that the time and efforts dedicated to preparing such a course is greater than all their other courses combined.
- Due to the timing of the travel course, these faculty members often miss family events that happen at the same time as the travel. In spite of such sacrifices, these faculty members choose to participate for their students.
- These faculty members are very diligent to ensure student safety. Some of them are very thoughtful and even send letters to the parents of student participants explaining the travel process and expectations.
- These faculty members are proactive and flexible to include the most exciting and relevant events into their programs. For instance, the itinerary of some travel courses depends on current or emerging events during the travel, which cannot be pre-planned. Hence, those faculty members need to always be alert for such uncertainties, and the students trust them to have their best interests at heart.

Common to all faculty members interviewed was the sentiment that a travel course is far from a vacation for either faculty or students, and it involves a huge amount of preparation and execution. The 10 days or 2 weeks of travel, although seemingly short, is a convergence of all the hard work, and proves to be very intense.

The fifth module in collaboration is with administration, especially when a new travel course is offered. The collaboration with administration was occurring throughout the entire planning phase, to keep all the parties involved and updated. The first thing that the authors were grateful for is the WCU International Office Director distributing the CIEE grant opportunity to all faculty on campus. This grant especially encouraged STEM majors to apply, so the authors were motivated to explore this possibility. More than providing the grant information, the director also provided detailed information on what resources were available on campus, introduced the authors to a faculty member who had led a travel course and was able to explain the procedure in proposing a travel course, and included the first author in a meeting with the CIEE manager when she visited the campus. The second item the authors were very grateful for is the consistent support of their Department Head, Associate Dean and Dean. Everyone was engaged in discussions about whether to continue the efforts or not, the number of credits for the course, recruitment strategies to attract more students, and other logistical items. All the administrators were extremely understanding and provided many good suggestions. For instance, when we ran into an under-budget situation where insufficient insurance was included in the original budget, our Department Head offered to pay for the difference beyond what we had mentioned to our students earlier.

The sixth module in collaboration is with donors. An experienced travel course leader might have ample experience to accurately estimate the travel costs, but a new instructor of a travel course may be not so good at it. That's what happened to us. When the authors conducted the student survey, we proposed a cost figure based on our knowledge then, but when all the required items were added into the budget, the final budget left no room for project materials. One way to make the budget work was to recruit more students so that the amount per student was reduced. However, the price of such a practice is that the students may lose the small-group environment and the workload of the faculty may exceed the level that makes the teaching effective. Therefore, we solicited donor sponsorship on project materials and T-shirts. In our college, there is a development officer who works closely with donors and she knows which board member would be likely to help our travel course. Through the development officer's connections, a donor gladly accepted our request for course materials and T-shirts. On the T-shirts, we would help advertise the donor's company, making their help more rewarding for them.

Given all the modules in collaboration, a successful launching of a travel course comes from not just a single module, but an integrated efforts of them all. All the modules in collaboration are intertwined, and they are fostered throughout the entire duration of the travel course's development and implementation.

To foster an effective collaboration, the key is to find out the trustworthy collaborators, trust them, learn from them, and appreciate them.

Through the collaborations with a wide-range of colleagues and friends, we discovered a colleague had previously drafted a travel course proposal, but the proposal was declined by the administration. That is why there had never been a travel course offered in our department before. The common concerns include:

1. Why is the travel course, of 10 to 15 days, worth 3 credit hours while an on-campus course of 3 credit hours takes 16 weeks to finish? (incidentally, a 3 week travel course is often worth 6 credit hours.)

2. How should the travel course be incorporated into the program curriculum?

3. In our case, the travel course is part of a Project Based Learning (PBL) course sequence. How feasible is it for the students to finish their projects in a relatively short time?

All of these concerns were valid and required answers before a travel course was offered.

To answer the first question, there are many lessons to be learned from the colleagues in other departments of the university who had led travel courses [1]. The common practices to include travel in a course are mainly in two categories: a travel component course and an independent travel course. An example of a travel component course is a course that can be offered in the spring semester with clear indication that it involves a travel component in the summer. The instructors will assign student grades as incomplete at the end of the spring semester and assign the grades after travel. The advantage of such a course is that the students will have taken a course on campus, and they get to know each other as a cohort before traveling together. So the extra component of travel is not regarded as unworthy of credit hours, as there is no added credits due to travel. However, at WCU, this practice is not allowed, because it is not allowed to assign an incomplete to an entire class section. However when a travel component is scheduled within a semester, such as in the fall break or spring break, then it is allowed. Some courses are designed in this manner. In the department, there is no flexibility to quickly create a dedicated semester long course with a travel component, and this option is not presently available. On the other hand, an independent travel course is a stand-alone course that lasts for 10 days, two weeks or three weeks, with possibly different numbers of credit hours. A common number of credit hours is 3 but has been as high as 6. Sometimes two courses have been offered on the same trip so that the entire trip is worth 6 credit hours, with each course accounting for 3 credit hours. The justification behind these hours is the amount of hard work that the students undertake, the extent of learning that they gain through travel but not by other means, and the course design that helps the students to achieve the learning outcomes in the course syllabus effectively. Typically, a 3 credit course is offered in a 16-week semester, but some universities may offer it in 10-week quarters. A travel course of 10 days or 2 weeks is another point on the spectrum to compress the learning time, by intensifying the learning through direct experiences, with the same or even more contact hours. In reflection when we have experiences that if we were to learn about something we have never seen or experienced, it takes lengthy descriptions, illustrations, metaphors, and analogies, and yet we may still not gain a thorough understanding. However, by seeing or experiencing the objects or events, we immediately discern their characteristics and they form a long-lasting impression in our memory, often with associated emotional responses, as well. Thus, the goal of taking a travel course is to facilitate learning through first-hand experiences.

To answer the second question, we needed to talk to our program directors. Better yet, this included all of the colleagues in the discussion. The department offers two engineering programs and two technology programs, each with mechanical or electrical concentrations.

PBL is implemented across all four undergraduate programs. PBL has been a hallmark of our curriculum, exposing students to different disciplines and fostering inter-disciplinary collaboration. In formal terms an ABET EAC accreditation criteria [2] for an engineering program is

"*h. The broad education necessary to understand the impact of engineering solutions in a global, economic, environmental, and societal context*".

An ABET ETAC accreditation criterion for an engineering technology program is

"*j. A knowledge of the impact of engineering technology solutions in a societal and global context*".

Although there is no plan to include the first travel course for assessment in the self-study for ABET accreditation, eventually the assessment plan will need to consider the learning outcomes that may be used in the future. Three out of the four programs include technical electives in their curricula, and this travel course can be a technical elective. The other program does not include a technical elective in its existing curriculum, so this travel course does not count as a major course, but can still count towards the total credits for graduation.

To answer the third question, it was agreed that the travel course should be an elective in the PBL course sequence, and a project would be an essential component. The PBL course sequence ranges from freshman seminar to senior capstone projects, and gradually deepens the students' knowledge to carry out practical projects, and prepares our students to succeed in their capstone projects as well as engineering jobs in industry after graduation. This travel course aims at incorporating the global perspective while allowing each student to work on a team project that is appropriate for their expertise and level. Students are welcomed from all four programs, and their friends in STEM majors or in entrepreneurship majors, at all undergraduate levels to enroll in this travel course. Hence, we have a diversity of expertise and experience levels, which can be both challenging and effective in the teaming.

Team-work and collaboration are key to ensure success in such a course. If this travel course were to happen, students would be encouraged to propose a project they are passionate about. The original plan was to have students find and buy the components they may need for the project in Berlin during our stay so that they would immediately learn about different standards, units, and cultural influences on engineering designs. A classroom would have been reserved with a workshop with tools in a nearby university, so that the students would have the resources to construct prototypes of their projects. At the end of the class, they would present their prototypes to their peers and submit final reports to summarize what they had learned through this process. The prototyping cycle would be intense, but not uncommon in a real world situation. The course would also teach the students how to compose their resumes and how to present themselves in job interviews, especially about their travel experiences.

Another big hurdle to offer such a travel course is the expense. At WCU, there is a scholarship dedicated to study abroad programs, but to be eligible for the scholarship, the students need to take another 3 credit hours during the same summer term. Financial aid in summer also requires the students to take 6 credit hours. Therefore, for the students who plan to apply for financial aid, they are strongly recommended to apply for the scholarship, and apply for it early to increase their chance of success. Apparently some students in other programs did their internships in Europe right after their travel course, so the students were encouraged to apply for internships in Europe that would start immediately after this travel course, enriching their travel experiences.

Another common concern of the students on the travel course is the fear of unknowns. We had planned to meet regularly before the travel for students and faculty leaders to meet and get acquainted, explain the activities they would participate in during the travel, and answer any questions the students may have. Working with CIEE and having experienced on-site support staff would bring us peace of mind, too.

As the PBL capstone projects are typically funded by industry partners through a center liaison between our college and industries, the travel course should help foster any possible sponsorship of such a capstone project from the site visit hosts. It would be especially beneficial if the projects in this travel course could be a starting point for much more extensive capstone projects later on.

Several useful practices were identified that were effective gauging student interests. However, these should not be the only measure to estimate interest.

1. **Email survey by Qualtrics**. Qualtrics is a survey service sponsored by our university and offers many features to streamline data collection through surveys. We used it twice to reach out to all of our students. The first time was in September 2015 when we were originally planning a travel course and asking ourselves whether we should submit the course proposal to the upper administration or not. We asked the students two questions: (1) if there were no constraints, whether they'd like to join, (2) considering the price range and timing, whether they expect themselves to be able to join. Quite a few students responded yes to the first question but no to the second question. Thus we knew that we should find ways to support our students more. There were also 30+ affirmative answers to the second question, reflecting a strong interest in this travel course from our students, which motivated us to move the idea forward. So we submitted the course proposal. Despite the 30+ affirmative answers to the second question, when our course was officially approved and we could accept course applications in November, many of the students who said yes earlier didn't follow up with an application. Later when the idea of a spring/fall break course was considered, we worked closely with CIEE to develop an adapted program. We sent out the second email survey with a comparison of the two options (a summer course versus a spring break course). We asked the students to tell us their choices as soon as possible. In spite of our impression and expectation that the spring break course would fill up in no time, there wasn't much interest in the spring break option. Instead, the students who had committed to the summer course still preferred the summer option. However, we were very glad that we had sent out the second email to all the students, because some students who gave affirmative responses to the second email had not responded to the first email. If we were to only contact the small pool of students who responded to the first

email positively, we would have missed out on those students who responded affirmatively to the second email survey but did not respond to the first email.

2. **Talk to classes in person**. After our first email Qualtrics survey revealed that the majority of the interests, most were from junior and sophomore students, we identified several major core courses at those levels and visited their classes. Many classes took great interest in it, although not all students were responsive due to concerns of costs and timing. In the several classes we visited earlier, we always talked in a mode to promote this course, and some students responded with some resistance. Then in a later class, a colleague helped us to ask his class why the students still had not registered for this travel course yet. This turned out to be a salient question, as most students immediately opened up and shared their concerns on the costs and timing. A few students asked for spring/fall break travel courses as they wanted to do internships during summer more than taking a travel course. The real-time interactions helped us understand our students' needs. However, later when we proposed a spring break course option and compared it with the summer course that had been approved earlier, no students actually signed up for the spring break option. This made us wonder if some students didn't want to sign up, they would give some reasons or excuses. Even if we did what they asked for, they might still have other reasons to show that we couldn't satisfy their needs. However, this exercise had really benefited our decision process as the students who were committed to the summer course still preferred the summer course, and we knew that we made the right choice for our students in the first place. Also, the exercise we did for a spring/fall break travel course could be useful for later planning, too.
3. **Create infographics and post flyers in the hallways.** An infographic is a graphical representation of information that is more enticing for students to read than pure text. We can embed hyperlinks easily in the infographic. We used piktochart.com to create infographics about the course, and the full version contains multiple pages (see Appendix 1). Then we created a one-page flyer for each infographic to include a QR code link to the full version of the infographic and provide a tinyurl link that the students can rip away (see Appendix 2). The infographic web link can be represented as slideshows on computers, tablets and cellphones. The first infographic was about the details of the summer course, and the second infographic was about the comparison of the spring break option and the summer option, as well as FAQ on questions that we had received by then. Besides the one-page flyers that we posted at almost every turn of the stairways in our building, we had also printed the full version of infographic pages and taped them at two entrances of our building, so our students would definitely see them. Maybe not all students would read them carefully, but a student told us that he did go through all the infographic pages and he had found all the answers he would like to know.
4. **Let the students talk to their friends.** We are lucky to have a student who is an avid traveler. She has accompanied us to promote this course to some classes, and

she has also talked to her friends and classmates to encourage them to sign up this course. The persuasion from a friend is very effective.

The biggest lesson we have learned in student recruitment is that we should recruit early. We started the discussions of a travel course in September 2015, but since we were not experienced in budgeting, we overestimated the student enrollment number and underestimated the budget, so we had to request departmental and donor support to make the budget feasible. Also, because the support came from our department or alumni, we hoped our own students would benefit from them, so we focused our recruitment only within our department. When we realized our recruitment goal was too high in December 2015, we reached out to other STEM and business majors. One of the student applicants was a friend of our own student and from another department. However, since our recruitment goal was too high, we needed to work with CIEE to reduce some activities to make the budget to work, the student from another department did not feel like the trip was worth the money any more, despite the fact that our program had a $1000 discount per student and we had donor support for course materials. In comparison, our own students were very understanding and flexible, as they understood the efforts we put into creating this travel course for the first time. After the student from the other department withdrew, we could not make our budget work, so we had to cancel the class. If we had started recruiting earlier, we could have established better relationship with all the student applicants, and we could have had the time to recruit more students.

Summary

This paper presented a case study for setting up collaborations in multiple modules in order to create a travel course. These include collaborations within the department (with both faculty and students), across campus, with a third-party provider and their connections abroad, and with local industry donors. The authors would also like to acknowledge the influence of the governmental incentive, because the CIEE grant was in response to the President's strategic plan initiative on study abroad programs [3] and the Generation Study Abroad Initiative [4,5].

We received nine student applications in total but two had to withdraw. Therefore, we could not get this course off the ground this year. However, if we did not overestimate the student enrollment number and underestimate the budget, a travel course with seven students could have been feasible.

The experiences and lessons learned during this process were very helpful for us and will allow us to do a better job in the future. We hope that this paper will inspire other institutions that have never offered a travel course before to give it a try, to benefit from the lessons that we have learned, and to provide the students a safe, educational, and exceptional faculty-led study-abroad experience.

References

1. LaVelle H. Mills, David Deviney, and Betsy Ball, "Short-Term Study Abroad Programs: A Diversity of Options", The Journal of Human Resource and Adult Learning, Vol. 6, Num. 2, December 2010
2. ABET EAC criterion: http://www.abet.org/accreditation/accreditation-criteria/
3. President Obama's 100,000 Strong Educational Exchange Initiatives, including 100,000 Strong China launched in 2009, and 100,000 Strong in the Americas launched in 2011: http://www.state.gov/100k/index.htm
4. The Generation Study Abroad Initiative: http://www.iie.org/en/Programs/Generation-Study-Abroad/About
5. CIEE's pledge to the Generation Study Abroad Initiative: https://www.ciee.org/generation-study-abroad/downloads/CIEE_Pledge.pdf

Appendix 1: A one-page flyer used in Dec 2015

On the next page.

Appendix 2: An infographic website with introduction of this travel course and our answers to students' questions.

The link is below:

https://magic.piktochart.com/output/8591453-engr493_berlin_2016

The print-out is attached after the next page in the last part of this paper.

Engineering Abroad @ Berlin, Germany 3 credits

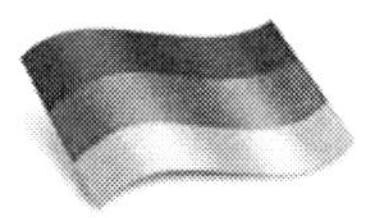

May, 14-29, 2016

ENGR493 - Special Topics: Engineering Principles and Practices in a Global Setting

Instructors

Yanjun Yan
yyan@wcu.edu

Paul Yanik
pyanik@wcu.edu

Do you like building things & traveling? Then join us!

All ET, CM, Business/Entrepreneurship, CS, STEM majors are welcome!

- 3 Credit Hours
- 10 students on the trip (8 seats filled so far). Act fast!
- 14 nights in Germany (WWII Bunker tour, BMW plant tour, VW plant tour, etc), 4 site visits, 4 guest speakers, 1 one-day trip to Dresden, 1 overnight excursion to Hamburg, cultural events, 1 free day, and 8 days of classroom/workshop usage
- $3583 paid to OneStop (covering programs, lodging, breakfasts, two dinners, a transportation card with unlimited usage during your stay, insurance)
- Flight tickets (est. ~$1200) and meals (lunches and other dinners, est.~$300) are on your own. In other words, you can use your frequent flier miles, if any, and you won't get stuck on a meal plan that you don't like.

Course info: http://tinyurl.com/nfgrype
Application Form: http://tinyurl.com/ofk4nbo
Email us for more info, and submit your application by Jan 15, 2016!

ENGR493, Berlin, 2016
Slide Show, PRESENT:
http://tinyurl.com/nfgrype
Application form:
http://tinyurl.com/ofk4nbo

ENGR493, Berlin, 2016
Slide Show, PRESENT:
http://tinyurl.com/nfgrype
Application form:
http://tinyurl.com/ofk4nbo

ENGR493, Berlin, 2016
Slide Show, PRESENT:
http://tinyurl.com/nfgrype
Application form:
http://tinyurl.com/ofk4nbo

ENGR493, Berlin, 2016
Slide Show, PRESENT:
http://tinyurl.com/nfgrype
Application form:
http://tinyurl.com/ofk4nbo

ENGR493, Berlin, 2016
Slide Show, PRESENT:
http://tinyurl.com/nfgrype
Application form:
http://tinyurl.com/ofk4nbo

ENGR493, Berlin, 2016
Slide Show, PRESENT:
http://tinyurl.com/nfgrype
Application form:
http://tinyurl.com/ofk4nbo

ENGR493, Berlin, 2016
Slide Show, PRESENT:
http://tinyurl.com/nfgrype
Application form:
http://tinyurl.com/ofk4nbo

Engineering Abroad @ Berlin, Germany 3 credits

May, 14-29, 2016

ENGR493 - Special Topics: Engineering Principles and Practices in a Global Setting

Instructors

Yanjun Yan yyan@wcu.edu

Paul Yanik pyanik@wcu.edu

Link to course application form
- already prefilled with course info
- download it to fill in pdf and then print it out
- sign it with a pen (electronic signatures are not accepted)
- submit it to Belk 334 or 338 in the first week of Spring 2016

Act fast if you want to join!

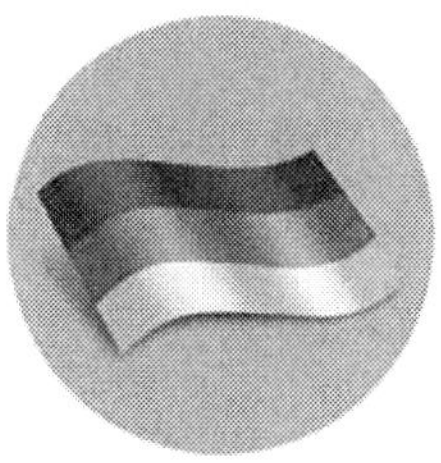

COURSE DESCRIPTION

This is an international experiential course undertaking the study of engineering practices and principles, teaming, project planning, and conceptual design processes.

The 3 credits earned in this course may be applied to E&T technical elective degree requirements. Please talk to your advisor regarding your specific degree evaluation.

Construction management, CS, business/entrepreneurship, and STEM majors are all very welcome.

OBJECTIVE

TOPICS

1. Project management
2. Project requirements and constraints
3. Project planning
4. Project implementation
5. Process metrics and quality
6. Effective communication
7. Project demonstration and presentations
8. Cultural impact on engineering practices

WHY JOIN?

- See and experience
- Boost your resume
- Broaden your horizons
- This is the 1st travel course in our E & T Dept
- Exposure to cutting-edge technologies
- Opportunities yet to be revealed

This trip needs 10 students to happen. Act fast if you want to join.

Well, we understand that this travel course requires a big commitment financially and in other aspects. We will try our best to offer travel courses in the future to different destinations. If you can not make it this time, you are welcome to join later.

ESTIMATED COSTS

Pay to OneStop (include program, lodging)	**$3583**
Flight ticket	**$1200**
Meals	**$300**
	$5083

Blessing 1: The price above has already included a $1000 per person discount offered by CIEE, which is rare. A travel course is expensive, but given our extensive program activities, our course really is a great deal.

Blessing 2: The E & T Department Head has kindly offered to pay for the OneStop payment difference beyond $3500, up to $83, as reimbursement.

Blessing 3: The fee for project materials is NOT charged to the students. Instead, we asked for donor support, and a generous donor has kindly agreed.

Other expenses depend on you:

- The flight ticket is estimated to be $1200 (check StudentUniverse.com). We will share info in the travel group and/or buy together to save.
- The breakfast, as well as the welcome dinner and the farewell dinner, are included. You need to buy/cook your own lunches and other dinners.
- The Berlin airport pick-up/drop-off is included as a group (not individually). You will be given a Metro card to have unlimited usage of public transportation.
- E-passport waives the visa.

In collaboration with CIEE:

Lodging
Transportation
Event arrangement

Link to CIEE

DAILY ITINERARY IN BERLIN ★

5/14/16 Sat	Leave USA	In flight
5/15/16 Sun	Check in. Welcome dinner	Berlin
5/16/16 Mon	City tour. Orientation	Berlin
5/17/16 Tue	Site visit 1 (TU of Berlin). Guest lecture 1 (ABO Wind)	Berlin
5/18/16 Wed	Guest lecture 2 (Railway engineering)	Berlin
5/19/16 Thu	Site visit 2 (BMW plant tour)	Berlin
5/20/16 Fri	Field trip to Dresden (VW plant, Baroque Saxon tour)	Berlin
5/21/16 Sat	Site visit 3 (WWII Bunker tour). Community engagement	Berlin
5/22/16 Sun	Depart for Hamburg for an over-night excursion	Hamburg
5/23/16 Mon	Urban development visit. HafenCity REAP Group	Berlin
5/24/16 Tue	Site visit 4 (Berlin Adlershop tour)	Berlin
5/25/16 Wed	Guest lecture 3 (Telekom Innovation Labs)	Berlin
5/26/16 Thu	Guest lecture 4 (SCAITI)	Berlin
5/27/16 Fri	Class	Berlin
5/28/16 Sat	Free day. Farewell dinner	Berlin
5/29/16 Sun	Leave Berlin for USA. Arrive on the same day	Home

★ Subject to change, depending on actual reservations in spring 2016. The guest lectures might be different, but similar to above.

The schedule will include 8 days of classroom and workshop usage at a nearby university. The classroom time will primarily focus on project design and prototyping.

SIGN UP FOR THE TRAVEL COURSE

Jan 15, 2016 — Interview with the instructors, submit application.
Link to course application form

Jan 15, 2016 — If you wish, submit WCU Away Scholarship application (up to $1500, for GPA>=2.75), which requires you to take another summer course from WCU, either on campus or online. Financial aid also requires you to take 6 credits in summer.
Link to scholarship application form

★Feb 15, 2016 — Pay $500 nonrefundable deposit at OneStop. The travel group will regularly meet after Feb 15 (roughly once every two weeks).

★ Apr 1, 2016 — Pay $3083 at OneStop.

Apr 1, 2016 — Buy your flight tickets. It is preferred, but not required, that the class travel together.

★ We have to follow the hard deadlines at OneStop. Because the $500 deposit is nonrefundable, please first let us know that you are joining. We will confirm with all the prospective participants to make sure that all are sure to go, and then you go to pay for it.

COURSE EXPECTATIONS

PREPARED
Be prepared, focused and ready to learn.

RESPONSIBLE
Be punctual.
Submit all assignments on time.
Actively participate in all activities.

PROACTIVE
Be open-minded, flexible, and adventurous. Share your thoughts and work as a team!

RESPECTFUL
Be respectful towards teachers and classmates. Encourage one another and stay united as a team!

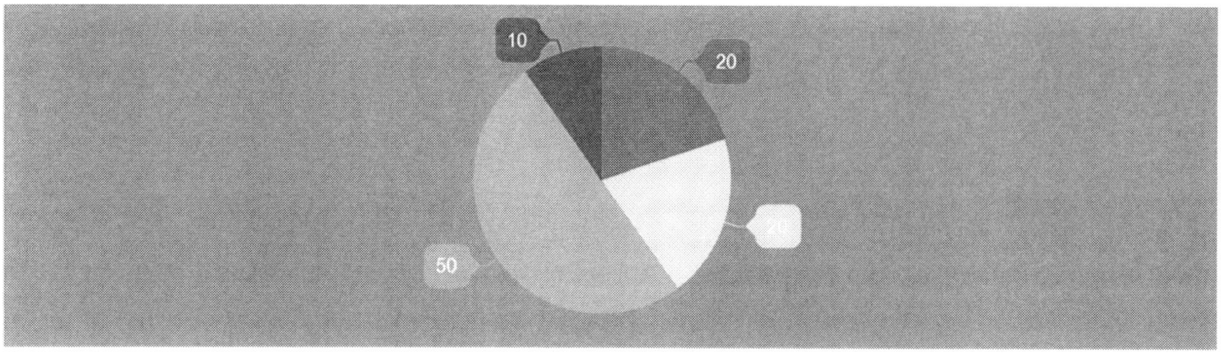

GRADING

Pre-trip Participation	**20%**
During-trip Participation	**20%**
During-trip Projects	**50%**
Final Report	**10%**

100%

KEEP IN CONTACT

Belk 334, 338 @ WCU
CIEE G27 site,
Gneisenaustraße 27,
10961, Berlin @ Berlin

Check your
campus email

Go to all events
on time

TEACHING PHILOSOPHY

"As engineers, we value critical thinking, clarity, and intellectual honesty. We envision ourselves as facilitators who share the educational experience by maintaining an interactive learning environment to which everyone can contribute. We, working together, will make sure that this trip is safe, educational, and exceptional."

Detailed course syllabus
More info from next page

Contact: Yanjun Yan
yyan@wcu.edu

Paul Yanik
pyanik@wcu.edu

FINANCIAL HELP

Financial Aid	?	Require 6 credit hours in summer
WCU Away Scholarship (up to $1500, GPA>=2.75)	?	Require another 3 credit hours course from WCU in summer
Department Head Promised Help	✓	Reimburse up to $83 difference (above $3500)
Donor Support	✓	Project materials

FOR YOUR DEGREE

Credit Hours	✓	3 Credit Hours
No Tuition	✓	For current WCU Students
In-Course Project	✓	Designed to suit your expertise and level

This trip needs 10 students to happen, and we have got 10! We'd love to have more students to join. E&T students, construction management students, business/entrepreneurship, computer science, and other STEM major students, are all welcome.

This course is blessed by so many people's willingness to help financially and in other ways, and we are looking forward to your joining us!

PROGRAMS

Program	Included	Notes
Double or Triple Hostel Stay	✓	Meininger Hotel in Mitte, or alike, easy access to bus
Airport Transport	✓	Group, not individually
Prepaid Transit Card to Use Buses	✓	
On-site Orientation	✓	
Welcome Dinner	✓	
Breakfast	✓	
Farewell Dinner	✓	
City Tour	✓	
BMW Plant Tour	✓	
WWII Bunker Tour	✓	
Berlin Adlershof Tour	✓	

PROGRAMS, Continued

- Day field trip to Dresden (Volkswagen Plant, Baroque Saxon Walking Tour)
- Neighborhood University Classroom and Workshop
- Project Manager (PM) Talk by ABO Wind
- PM Talk by Bombardier Transportation
- Quality & Usability Talk by Telekom Innovation Labs
- Quality Assurance Talk by DCAITI
- TU-Berlin Tour
- Two-day Overnight Excursion to Hamburg
- Program Assistant from CIEE
- iNext Travel Insurance, Health Insurance, Trip Cancellation Insurance

FAQ

Q1: I want to do internship in summer. What should I do?

A1: Employers often value studying abroad experiences, and if they like you, they would work with you to allow you to start your internship from June, for example. Or, find an internship in Europe and go to that internship right after this travel course! If nothing above works, I am sorry that you can not join this Maymester trip. Instead, let us know what time in the year you prefer to take a travel course, and we will consider it in the future.

Q2: How much financial aid can I get?

A2: Every student's situation is different. Please make an appointment at OneStop (Killian Annex) to talk to Ms. Jennifer Thomas about it. The financial aid won't be approved until mid-spring for the Maymester trip, but at least you will know how much you can expect to take out.

Q3: Am I qualified for the WCU Away Scholarship?

A3: Below is the answer from the lady who's in charge of it.

- GPA is 2.75 or above.
- There are a limited number of scholarships so applying early is important.
- The scholarship is actually a reimbursement process. Students have to make their purchase of plane tickets or hotel accommodations in advance and can receive up to $1500 (based on their expenses) after they've completed their summer courses.
- All receipts turned in by the students must have their name on the receipt. The funds do not cover items such as food, clothing, class materials.
- The additional course does not need to be linked to the travel course in some way but the student does need to enroll in a summer WCU course. WCU Away funds are only available to undergraduate students.

Q4: Can I get back the full amount of $1500 Scholarship?

A4: The scholarship applies to lodging and plane tickets, but not food. The official response we got on 12/14/15 is that "Students who apply for WCU Away funding can only be reimbursed if they pay for their travel expenses on their own. That is to say not via One Stop or a third party booking agency. Each student would need to submit receipts with their name on it in order for us to reimburse them. I know this makes use of WCU Away funds difficult for some trips but we have to operate under state regulations."

We have made a request to CIEE to see whether we can separate lodging from the program fee so that you pay for the lodging directly yourself, but not through OneStop. However, it is not determined yet.

Q5: Can my friend from another department or another university take this course?

A5: Your friend is very welcome, as the friendship shared in a cohort could help you enjoy your studying abroad experience immensely. Please be aware of the rules below:

If your friend is a student at WCU, s/he follows exactly the same financial rule as you (no tuition and possibly s/he can apply for financial aid). This course may not help with his/her major credits, but should count towards the total credits for graduation.

If your friend is from another university, s/he has to apply to be a visiting student at WCU ($75 application fee with filled forms, etc), and s/he also has to pay for the tuition (depending on the credit hours) besides the travel costs.

We have got a special discount in the Maymester program from CIEE, and our course is blessed with the generous support from our department head and a donor, so we hope that you will take advantage of all these.

Q6: Where do I use the credit earned from this course?

A6: The Maymester course's 3 credits can be applied as technical elective for BSE, EE and ET, but not yet ECET. Still, whatever major you are in (CM, CS, BE, or STEM), this course' 3 credits count towards the total credits for graduation.

Q7: Can I choose to take this course with or without credit?

A7: Sorry, it is not a personal choice. A travel course is either with no credit for the entire class, or is with credit for the entire class. This course is with 3 credit hours. If you prefer a travel course without credit for a shorter time, let us know, and we will consider it in the future.

Q8: Can an international student take this course?

A8: Yes, absolutely. Please talk to WCU IPS to make sure that you can obtain the Schengen visa to visit Germany and the US visa to come back to campus. You should have your I-20 form signed by the IPS.

Q9: How many students will be on this trip?

A9: The CIEE proposal is based on 10-14 students or 15 students, and the price for 15 students is less.

Originally when we talked with CIEE, we aimed at recruiting 15 students in order to get a better rate ($3585 to OneStop, otherwise it would be $400 more per person). Our proposal was based on this rate, and we can not charge any more than our proposed price.

Later we got to know that many travel courses at WCU had about 4 students, 5 students, 7 students, or 10 students, so 15 students can be a bit of stretch. So we contacted CIEE again on the possibility to give us the same rate for 10 students, and they suggested two options:

(1) If we were to cut some of the programs, we would need to cut the Dresden day field trip and one or two talks, which were not desirable.

(2) If we would revise our lodging option from CIEE site to a hostel, we could manage it to work without altering any of our events. The hostel they had in mind is Meininger Hotel in Mitte, whose online photos actually look pretty neat. You can google them. If CIEE can't reserve this hostel, they will book something similar for us.

So we opted for the second option, and our recruitment goal is to have 10 students on this trip. Act fast if you want to join!

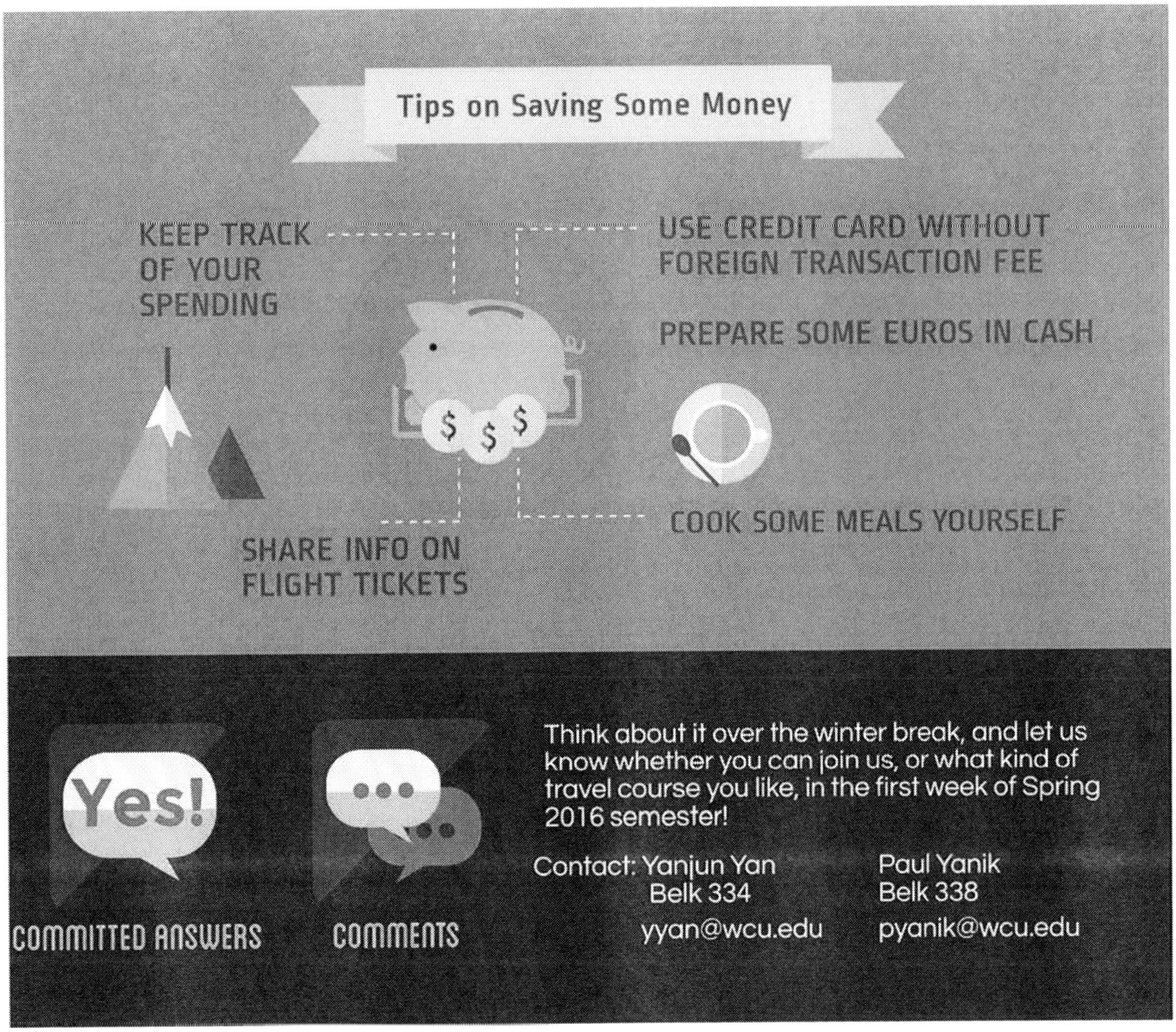
Tips on Saving Some Money
KEEP TRACK OF YOUR SPENDING
USE CREDIT CARD WITHOUT FOREIGN TRANSACTION FEE
PREPARE SOME EUROS IN CASH
SHARE INFO ON FLIGHT TICKETS
COOK SOME MEALS YOURSELF
Yes!
COMMITTED ANSWERS
COMMENTS
Think about it over the winter break, and let us know whether you can join us, or what kind of travel course you like, in the first week of Spring 2016 semester!
Contact: Yanjun Yan
Belk 334
yyan@wcu.edu
Paul Yanik
Belk 338
pyanik@wcu.edu

Best Practices in Globalizing Engineering Students

Ms. Allison Wright, Texas Tech University

Allison Wright is an SIT Graduate Institute Alum focusing on Education Abroad and is currently the Director of Texas Tech University's International Engineering Programs overseeing the implementation of the International Experience Initiative. Previously she has worked for an international provider in London and for North Dakota State University. Allison is a two-time alum of international programs in Caen, France and Istanbul, Turkey. She is an active member of the International Education field, attending workshops in non-traditional majors, internships, student advising, and best practices.

Ms. Ashley Nicole Haseley, Texas Tech University

Ashley Haseley has a master's degree from Concordia University, Irvine in International Studies with a focus on Education. She currently works for the International Engineering Programs Office at Texas Tech University. She spent a year completing her master's degree in Shanghai, China and did research on job satisfaction with Chinese teachers. She has previous experience with service abroad in Sabah, Malaysia and studied abroad throughout different countries in Europe. Since January 2015, she has been with the Edward E. Whitacre Jr. College of Engineering where she continues to grow programs and provide guidance for the International Experience Initiative. She is also an active member in the international field, presenting, attending workshops and focusing on best practices.

"Best Practices in Globalizing Engineering Students"

Allison Wright, M.A.,Texas Tech University

Allison Wright is an SIT Graduate Institute Alum focusing on Education Abroad and is currently the Dircctor of Texas Tech University's International Engineering Programs overseeing the implementation of the International Experience Initiative. Previously she has worked for an international provider in London and for North Dakota State University. Allison is a two-time alum of international programs in Caen, France and Istanbul, Turkey. She is an active member of the International Education field, attending workshops in non-traditional majors, internships, student advising, and best practices.

Ashley Haseley, M.A., Texas Tech University

Ashley Haseley has a master's degree from Concordia University, Irvine in International Studies with a focus on Education. She currently works for the International Engineering Programs Office at Texas Tech University, she oversees programs in Asia, Africa, the Pacific/Australia, and the Middle East. She spent a year completing her master's degree in Shanghai, China and did research on job satisfaction with Chinese teachers. She has previous experience with service abroad in Sabah, Malaysia and studied abroad throughout different countries in Europe. Since January 2015, she has been with the Edward E. Whitacre Jr. College of Engineering where she continues to grow programs and provide guidance for the International Experience Initiative. She is also an active member in the international field, presenting, attending workshops and focusing on best practices.

Abstract
In the fall of 2013, Texas Tech University (TTU) and the Edward E. Whitacre Jr. College of Engineering (WCOE) implemented a requirement for all undergraduate engineering students to participate in an international experience as part of their degree plan. The intent behind this initiative is to facilitate the preparation of future engineers at the college level to become part of an ever-increasing international profession.

In the implementation of the International Experience Initiative (IEI), the members of the International Engineering Programs (IEP) Office discovered a series of best practices to help globalize engineering students. These best practices include leveraging advisors and peer advisors called the Engineering Diplomats to educate and recruit students on their international opportunities, the marketing of programs, and maximizing partnerships and creating customized programs.

Marketing is an essential tool needed to promote programs. Since the IEI's inception, IEP updated its website, created flyers for all programs, and established email marketing for engineering students. Throughout this paper, we will provide examples of creative material, data on the increase in student participation and of web traffic as a direct result of marketing.

Through partnerships with providers, IEP collaborates to create diverse programs that meet student interests, personal needs, and future career goals. This is done by examining past, present, and future programming with specific partners. In the past, IEP created an ongoing summer research exchange with Universidad del Norte (UniNorte) using the support services of International Studies Abroad (ISA). Today, IEP and Butler University Institute for Study Abroad (IFSA-Butler) collaborated to create a program that has curriculum available for multiple universities that focuses on engineering and culture. Looking toward the future, IEP and Cultural Experience Abroad (CEA) are creating three geographically diverse programs while integrating engineering curriculum and cultural immersion.

Background

Texas Tech University (TTU) was founded in 1925 and located in Lubbock, Texas. The 2015 U.S. World and News Report ranked the Engineering program 94th as one of the "Best Engineering Schools" in the United States. Since 2010, engineering student enrollment has increased more than 32.3%. Within the Whitacre College of Engineering (WCOE) there is an office dedicated to the International Experience Initiative (IEI) called International Engineering Programs Office (IEP). The WCOE is committed to providing our students with quality international engineering experiences that prepare them to work in a global environment. In addition to increasing students' global perspectives, an international experience will be one of the most memorable times of their lives. The chart below is an overview of the organizational structure within the IEP office.

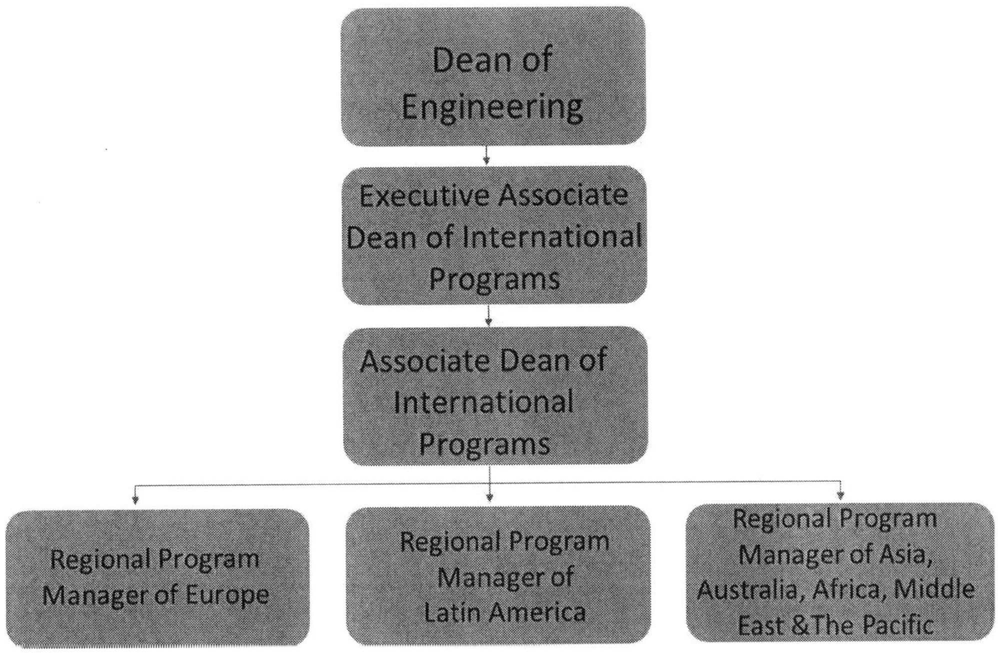

Figure 1.1: International Engineering Programs Office Organizational Chart

Going abroad is still an area of improvement in the United States. Currently, less than 1 in 10 American students have an international experience during their academic career (Bidwell, 2014).

Statistics have shown:

- Study abroad experiences in the early stages of a student's career statistically improve 4 year, 5 year and 6 year graduation rates, with the greatest improvement in the 4 year graduation rates.
- Student with a study abroad experience have higher GPAs at graduation than a student that did not study abroad.
- Study abroad experiences improve retention of underrepresented groups, such as engineering (Rhodes and Sutton, 2012).

Although there is room for improvement to get American students abroad, a positive trend is the increase in student participation in the STEM field. In the last two years, more STEM programs have been developed and more STEM students have gone abroad than in previous years. It's been reported that STEM students have outpaced all other fields in U.S. Higher Education in going abroad (Bidwell, 2014). The chart below shows the increase in STEM participation abroad. This report is from the Open Doors Report of 2014.

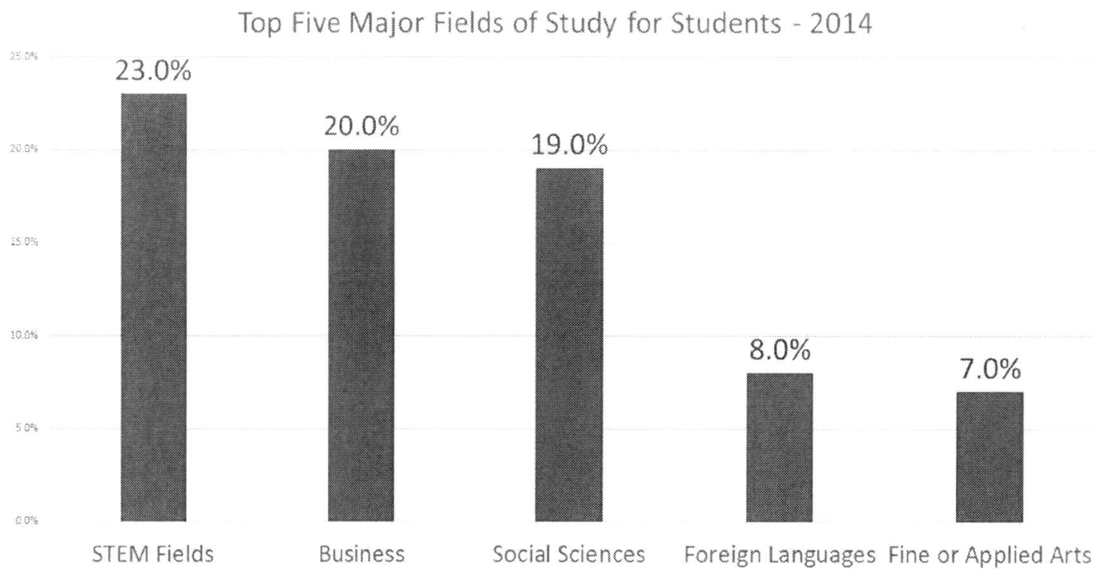

Figure 1.2: Students Going Abroad 2014

Looking at these trends, the IEP Office is dedicated to preparing its graduates for the ever-changing global landscape, therefore starting in fall 2013 the WCOE at TTU implemented an international requirement for engineering students. This was created to prepare engineering students to work in a global environment where they will need to learn new communication styles, work with others from different backgrounds (Hurt, 2015), and will set them apart from other engineering graduates (WCOE, 2012). Students can complete the requirement in four different ways: study, intern, research or service abroad. Study continues to be the most popular with 97.3% of students choosing this method. An additional 1.4% of students choose internships, followed by 0.3% for research, and 0.9% for service. Since the requirement was implemented (fall 2013) there has been a 25% increase in engineering students going abroad.

The requirement guidelines vary based on the student's' catalog. Students who entered between fall 2013 and summer 2016 must have the equivalent of three credit hours abroad. Beginning in fall 2016 students must be abroad for a minimum of six weeks. This change came from the desire for TTU engineering students to gain further cultural immersion.

The international requirement compliments the Office of International Affairs' strategic goals to

- "increase the number of students studying abroad to 4,000 per year"
- "identify two appropriate study abroad programs for each major offered on campus"

- "encourage semester-long programs through establishing course equivalencies in all degree plans" (Office of International Affairs, 2014).

In order to develop a successful international department, best practices need to be in place. This paper will focus on a couple of the best practices in the implementation of the international requirement. These best practices are from the IEP advisors' experiences over the past two years working in the WCOE. These topics are leveraging advisors to educate and recruit students on their international opportunities, marketing of programs, and maximizing affiliated partner relationships through the creation of customized programs.

Best Practices: Advisors and Peer Advisors

The IEP has found one of the best methods of informing students on the international requirement is to have a unified message from key allies within the WCOE. Two of these allies are academic advisors and peer advisors also known as Engineering Diplomats.

Best Practices: Advisors

In the WCOE, students must meet with their academic advisors in order to register for the next semester. The IEP found that working with academic advisors is one of the most consistent ways to reach students and promote programs. Academic advisors know all the idiosyncrasies of the engineering and computer science degree plans including course prerequisites, course sequence, and if courses on campus are full. Previously if courses are full on campus and being offered abroad, IEP and academic advisors have worked together to encourage students to take the course abroad. The IEP and the Lead Academic Advisor work closely together and attend the same monthly meetings, where IEP disburses program flyers that are specific to each advisor's' major. These flyers are hung up in academic advisors' offices where students can see them. Attending monthly meetings allows for the exchange of information, building personal relationships with the advisors, and keeping the lines of communication open. These open lines of communication are especially important when working on student's applications and making sure they remain on track in their degree plans. It is also important when a difficult situation arises, such as a student does not meet the program or course requirements to communicate with the advisor. The IEP has experienced students trying to go around either set of advisors to participate in a program they are not qualified for. This is mitigated by communicating when issues appear.

Academic advisors are invited to the breakfast every fall semester before the Study Abroad Fair. The IEP personally invites each advisor to attend and sends email reminders leading up to the event. During this breakfast affiliated providers and faculty-led program professors can discuss their programs that work for different disciplines. After the breakfast, there is time to socialize and IEP advisors introduce academic advisors to affiliated providers best suited for their students' needs. The IEP is the only office on campus that brings academic advisors and affiliated providers together and has received positive feedback from both groups. Both parties have the opportunity to learn from each other, ask questions, and contribute ideas for new programs.

Best Practices: Peer Advisors

As discussed in NAFSA's Guide to Education Abroad (2005), peer advisors can be the best ambassadors for programs in which "they participated and for education abroad in general." With this in mind, the Engineering Diplomats were created in summer 2014 as the ambassadors of the IEI. They are a recognized student organization using students who have already completed their international experience requirement. One of the main purposes of the organization is to help the WCOE achieve its goals regarding its international initiatives. The organization is overseen by an IEP advisor.. Currently, the Engineering Diplomats are comprised of 12 active undergraduate members which include a five person officer group. Members receive a scholarship each semester and gain experience in organizing and planning events, recruiting and hosting information sessions. The Diplomats advise their peers to achieve the recruiting expectations of the WCOE (Student Organization Proposal, 2014).

Diplomats undergo a rigorous selection process that includes a GPA of 3.0 or higher, having spent a minimum of one long semester on TTU's campus, and a panel interview with the current officers (Student Organization Proposal, 2014). After they have been selected to the organization, Diplomats undergo thorough training (NAFSA, 2005) by the organization's advisor every semester. Training includes a two hour session with the advisor discussing the advising process, programs, the IEP website, and expectations. Senior Diplomats are paired up with new members to help them learn about IEP policies and best practices to promote programs.

Duties of the Diplomats include:

- Weekly office hours
- ENGR 1315 classroom visits each semester
- Two to three major or location specific information sessions a semester
- Two general information sessions a semester
- Attendance at WCOE Events, such as the Study Abroad Fair, Job Fair, and Kick-Off
- Assistance in IEP events, such as program meetings

Between 10:00am to 4:00pm from Monday to Friday, the Diplomats have office hours. These hours have been strategically placed in the front office of the Engineering Opportunities Center (EOC). The EOC is a location within the WCOE that offers resources on tutoring, career opportunities, community, and retention (Engineering Opportunities Center, 2016). During these hours, students can walk-in and ask general questions to the Diplomats. Popular program marketing materials are available for Diplomats to hand out and they utilize computers to help students navigate the IEP website. These three things are some of the largest uses of time for IEP advisors and by utilizing the Diplomats, IEP advisors are able to work on projects and other duties.

The IEP has worked hard since the organization's creation to grow membership, expand duties, and increase their presence within the WCOE. The Diplomats have a distinct logo and polo worn during office hours. Diplomats are also featured on the IEP website, utilize contacts in residence life to present in the residence halls, and attend other events. Due to these efforts to increase their presence, Diplomats are identified outside of their office hours as a resource to students. Several have reported students asking them questions on their walk to class and off campus.

Best Practices: Marketing

Marketing is an essential tool needed to promote programs. "For education abroad to become a more viable, visible, and prominent part of undergraduate education, the field needs to learn to use appropriate promotion and publicity tools to build support in every way possible, on and off campus" (NAFSA, 2005).

The IEP focuses on different types of marketing strategies. "There are three simple rules to successful promotion: variety, repetition, and appropriateness" (NAFSA, 2005). By offering a wide range of strategies this allows the IEP to create variety in the message and tone, have consistent repetition and establish the appropriate message based on the targeted audience that is being reached. These strategies are:

- Revised Website
- Promotional Materials
- Email Marketing and Newsletter
- Social Media
- Attending Events/Info Sessions

The WCOE is consistently working on collecting student feedback and accurate data for programs. "Collecting and reporting accurate study abroad data is one of the most compelling ways for study abroad offices to demonstrate how they support the mission of their campus and contribute to the educational experience of their students" (NAFSA, 2016). The TTU engineering students are constantly asking the IEP, Diplomats and the academic advisors on where they can go to not only fulfill their international requirement, but also not fall behind in their degree plan. Therefore, the IEP generates updated data based on the student information pulled from their online application. This data is used to educate the Diplomats and academic advisors. An example of the IEP data is located in figure 1.3.

These figures show the second and third top locations by major where students have fulfilled their international requirement. Spain is by far the most popular option for students because it has been the main faculty-led program location. Although, moving forward the IEP wants to grow student participation in other programs such as Latin America, Asia and Australia. This is why the data is primarily focusing on the top two and three destination for engineering students (keep in mind foundational students are mostly freshmen students who have not met the requirements to declare an engineering major).

From this data, it shows Europe is the most desirable and course efficient regions for TTU engineering students

Chemical ENG	**9.6%**

Civil ENG	**7.7%**

Location	Percentage
Italy	1.4%
Czech Republic	0.2%

Location	Percentage
United Kingdom	0.7%
Italy	0.5%

Electrical/ Computer ENG	**3.6%**
Location	**Percentage**
Ireland	0.4%
Italy	0.4%

Computer Science	**1.6%**
Location	**Percentage**
Ireland	0.4%
Greece	0.2%

Construction/ Environmental ENG	**2.5%**
Location	**Percentage**
Germany	0.3%
Italy	0.2%

Foundational ENG	**35.6%**
Location	**Percentage**
Italy	4.6%
Ireland	1.5%

Industrial ENG	**3.1%**
Location	**Percentage**
Germany	0.4%
Italy	0.4%

Mechanical ENG	**27.6%**
Location	**Percentage**
Germany	2.9%
Colombia	1.5%

Petroleum ENG	**8.8%**
Location	**Percentage**
Germany	0.7%
Italy	0.4%

Figure 1.3: The data was pulled from spring 2014 to fall 2016 programs.

Website

The feedback that was given from faculty, staff, students and parents was one of the main reasons as to why the IEP decided to revise its website. The goal for updating the website was to make it more user friendly. One of the initial major changes was creating a world map on the front page that shows all the different areas students could go to fulfill their international requirement. The map is also color coordinated by area of interest such as, study, internship, research and service abroad. The tool bar was also updated to help provide users an easy way to explore important resources such as dates and deadlines, contact information, scholarships and frequently ask questions. During the redesigning stage, the IEP, the WCOE marketing team, academic advisors, and the Diplomats beta tested the website from an advisor, a recruiter, and a student's point of views. Since the majority of the traffic is from TTU engineering students, using the Diplomats was an essential tool needed in providing the appropriate feedback.When IEP first introduced the new website, an email was sent out to 4,989 undergraduate and graduate engineering students. From that number, 2,683 students actually opened the email and clicked the link to view the new website. The website went live in January 2016 and since then IEP has received positive feedback from faculty, staff, students, parents and partner institutions.

Promotional Materials

A new initiative that IEP started this past year was developing a "Guide for Going Abroad" one-sheet for its students. On this one-sheet, there is detailed information on the appropriate steps a student should follow when going abroad. The one-sheet is laid out in a color coded chart form and checklist. The TTU engineering students respond well to charts due to the fact that most of their degree plans are set up as a chart system. The color coded information is to help the student understand what their responsibilities are and who oversees that portion of the application; and the checklist is helpful for advising. During Diplomat peer advising or IEP advising the advisor will go over the checklist with the prospective student to make sure the student understands the appropriate steps to take and what still needs to be done. Incorporating both styles helps with reaching different learning styles for TTU engineering students.

Our main promotional material that we use are highlighted program focused flyers that have a distinct look. The IEP flyers are either made by the IEP or sent from affiliated providers. The flyers include a photo, location, dates, deadlines, pricing, contact information and academic information. It has been found that the key to making a successful flyer for TTU engineering students is to include course information for each program. If the course information is not included it will be the first question that is asked from the student. The IEP uses the flyers at all recruiting events, fairs, orientation, information sessions and general engineering events, as well as, during Diplomat peer advising, academic advising and IEP advising. Even though IEP is moving toward a more digitally focused industry it's still found that there is a demand from students for paper materials. With the flyers, the IEP has created a business card with a QR code that takes you to the website.

Email Marketing & E - Newsletters

The IEP office obtains a bi-weekly report that shows all TTU undergraduate engineering students who have not fulfilled the international requirement. The IEP uses this report to tailor messages targeted to those students who still need to go abroad. The emails consist of informing students of future events, fairs, information sessions, and deadlines. Announcements of new programs with the attachment of the program specific flyer is "one of the most useful functions of email" marketing (NAFSA, 2005). Another initiative that has been in the works is establishing an e-newsletter for the IEP office. The focus of the e-newsletter would include things like, featured stories, featured programs, student testimonials, important dates, international student involvement, financial aid and scholarship information, and general abroad facts.

Social Media

The IEP has a dedicated Facebook page for its TTU engineering students. The page was created in summer 2013 and since then the page has received 330 likes. This is a continuing project of the IEP and after the fall 2015 IEP photo contest the facebook account increased in its numbers. The IEP office posts daily on program information, information sessions, study abroad memes, study abroad articles, etc. During the fall 2015 semester, the IEP office found that providing images with all articles and/or links has resulted in an increase in clicks from the TTU engineering students. For example, the image below of Spongebob Squarepants was related to an extension date for our Chile programs. The headline was, "There's still time to apply to Chile programs! The deadline has been extended to this Friday". Within 15 hours it had reached 90 people.

Figure 1.4: Facebook Meme Example

Another marketing tactic that the IEP found successful was using the program flyer as the image to inform students about their opportunities. The IEP did this with three different programs and all three different post received over 160 clicks. Areas of growth that has been started or plan to start in the future are, having an active blog site, utilizing twitter and the creation of an instagram account.

Attending Events

The IEP has a goal of sending 720 students abroad each year. Only relying on social media and email marketing is not enough promotion for this size of group. "Good and effective promotion involves knowing the target market" (NAFSA, 2005). Therefore, IEP attends every event that is held for TTU engineering students. These events included, but are not limited to; Freshmen and Transfer Orientations, Resource Fairs, Welcome Events, Study Abroad Fairs, Career Fairs, Engineering Week, International Week and others. The Engineering Diplomats also provide weekly information sessions and give presentations to all introductory engineering courses at the beginning of each semester.

Best Practices: Utilizing Provider Relationships

Since there is a goal to send out 720 students a year, one way to reach this goal is to leverage TTU's partnerships with its affiliated partners. Affiliated partners offer a wide array of programs for a large amount of students and usually at a lower cost than faculty-led programs (Moseley, p 3). In addition, it is important to use providers that already have contacts and logistical support in a region when designing programs (Spencer and Tuma 2007). Currently the WCOE is working extensively with three separate partners, International Studies Abroad (ISA), Institute for Study Abroad-Butler University (IFSA Butler), and Cultural Experiences Abroad (CEA) to create customized programs in the summer to fit a variety of engineering majors. The following are some examples of programs created with providers.

Past

The IEP's original customized program is with ISA and the Universidad del Norte (UniNorte) in Barranquilla, Colombia. The program stems from the 100,000 Strong in the Americas initiative. A joint research exchange of five engineering students in the areas of mechanical engineering (ME) and electrical and computer engineering (ECE) with a focus on renewable energy. This eight week program features five students from both universities. Starting in the first week, students from Texas Tech traveled to Barranquilla and spend one week at a research symposium with UniNorte students and researchers to discuss potential topics. At the start of the second week, the UniNorte students traveled to Texas Tech to begin their research, while Texas Tech students remained in Colombia to begin their research. Over the course of the subsequent six weeks, students worked in their research areas and took a course to supplement their research and to fit in with the theme of the exchange. At the end of the seventh week, TTU students traveled back to Texas Tech to join the UniNorte students for a research poster presentation (Gracia and Wright, 2015). This presentation was open to the TTU community and included attendance from students, faculty, and staff in the WCOE.

In Colombia, Texas Tech students were housed in homestays as a way to become culturally immersed in society. The ISA provider acted as the on-site logistics coordinator of the customized program. In Texas, UniNorte students were housed in TTU dormitories as a way to experience university life in the United States. Both groups had additional cultural activities to facilitate further cultural understanding. In Colombia, ISA was further utilized for these activities. In Texas, a member of the WCOE acted as the coordinator of cultural activities.In the planning of this research exchange program, representatives from both institutions and the provider travelled to both locations to explore facilities and community (Gracia and Wright, 2015).

While five students were planned for each group, TTU was able to send three students and UniNorte two students. Given that this was the first year of the program, both institutions expect the numbers to reach the full five students and perhaps expand program numbers in the future.

Present

Currently, IEP is partnering with IFSA-Butler for a program located in Derry, Northern Ireland and open to non-TTU students. This program began with representatives from IFSA-Butler and IEP discussing the issues of sending engineering students abroad and potential semester long options. One of the main issues discussed is how difficult it is for engineering students to go abroad for a semester due to their "semester-by-semester academic planning and internship expectations" (Hurt, 2015). Considering capacity and costs, the University of Ulster for a six-week summer session was decided upon. The IEP staff looked at courses needed for a wide variety of engineering students and selected Dynamics and Microcontrollers as the engineering courses to be offered. Dynamics is needed for mechanical, civil, industrial, and petroleum engineers. Microcontrollers is needed for electrical and computer engineers and computer science. Students select either Dynamics or Microcontrollers and pair it with an Irish Studies course. This course focuses on the history and culture of Ireland and culminates in a one-week cultural stay in a Gaelic speaking community.

All three groups work together to provide program components. The University of Ulster provides the delivery of the academic content, use of academic facilities, and on-campus housing. IFSA-Butler provides the logistics of the program including orientation, cultural excursions, visits to engineering firms in Northern Ireland, and 24-hour emergency support. TTU provides the syllabi the courses are based off.

To set it apart of other customized programs in the IEP portfolio, this program is listed on IFSA-Butler's website and is open to non-TTU Engineering students. There are several benefits to having the program opened to non-TTU students, including not needing to rely on TTU to meet the minimum program participation numbers, keeping the program cost low, and student diversity. As of March 1, 2016, 22 students had applied to the program.

Future

Looking toward the future, IEP and CEA are planning on creating three geographically diverse programs while integrating engineering curriculum and cultural immersion.

> "Producing globally competent engineers for the 21st century is increasingly important in a world that is 'flat' and more globally connected. Engineers are now facing the distinct possibility that they will be working for a multinational company, or NGO that requires multicultural awareness, some degree of foreign language proficiency, and the ability to communicate effectively across cultures and time zones" (Kennedy & Jones, 2015).

These geographical areas include; Paris, France, Buenos Aires, Argentina and Shanghai, China.

At the CEA Center in Paris, students must take oral communication and calculus III. These courses are chosen because it fulfills the degree requirements for all TTU engineering and computer science students. In Paris, students will stay in shared apartments for six weeks with CEA coordinating all on-site program logistics. This is the first year the IEP office has ran this program and there are 11 TTU engineering students participating in this program. In the future, the CEA program will be modified so that TTU can collaborate with other U.S. Higher Education Institutions and make this program open to students outside of TTU. This will help generate more student participation, as well as, become more cost effective. TTU is also expanding into a semester-long program at this location.

In Buenos Aires, Argentina at the University of Belgrano, students will take one engineering and one culture course. The program will be six to eight weeks in length, students will live in homestays and participate in cultural excursions. These program logistics have yet to be finalized but is on the agenda to be finalized later in 2016 and to run in summer 2017. Similar to other programs, CEA will coordinate on-site logistics.

At East China Normal University in Shanghai, China students will study over the course of ten weeks taking an East Asian Studies course and an engineering course. The engineering options are currently being researched based on the desire to collaborate with other U.S. Higher Education Institutions; this will help with student participation and the cost of the program. Students in this program will be housed on campus. On-site logistics such as, cultural excursions including a trip to Beijing, and communication with other institutions will be coordinated by CEA. The IEP is working closely with CEA staff to create, promote, and recruit for the program and hopes to have a minimum of 10 TTU student participants.

Conclusions

According to a World and U.S. News report, "For the first time, American students studying abroad who are pursuing STEM fields outnumber those in every other field of study" (Birdwell, 2014). The WCOE at TTU continues to strive to create great engineers, and with the establishment of the IEP office, this not only helps the students with the requirement, but also makes them stand out from other U.S. Higher Education Institution graduates.

At TTU there is a high turnover of academic advisors and to keep a consistent message, IEP will continue to maintain relationships with all advisors. The IEP hopes to grow the member numbers of the Diplomats and remain updated on IEP policies and programs. The IEP will utilize current marketing strategies, grow social media, and remain current on marketing trends.

With the establishment of WCOE international requirement and the growth of STEM abroad, the IEP office is consistently working on best practices and program growth for its engineering students. One area IEP would like to expand further is internships, service, research, and semester long programs. Each of these areas will take long-term planning in order to accomplish good results.

References

Bidwell, Allie. (2014). U.S. Falls Short in Studying Abroad. *U.S. News and World Report.* Retrieved from: http://www.usnews.com/news/blogs/data-mine/2014/11/17/how-studying-abroad-has-changed-in-the-last-decade.

Engineering Opportunities Center. (2016). Engineering Opportunities Center. Retrieved from: http://www.depts.ttu.edu/coe/undergraduate/.

Gracia, Zaida and Allison Wright. (June 14, 2015). Development of a Novel Interdisciplinary Engineering Student Exchange Program. *ASEE.* Retrieved from: https://peer.asee.org/development-of-a-novel-interdisciplinary-engineering-student-exchange-program.

Hurt, Emily. (2015). Why Study ABroad is Crucial for Engineering Students. *GoAbroad.com.* Retrieved from: http://www.goabroad.com/articles/study-abroad/why-study-abroad-is-crucial-for-engineering-student.

Moseley, William G. (December 11, 2012). Don't Go Soft on Study Abroad: a Call for Academic Rigor. *The Chronicle of Higher Education.* Retrieved from: http://chronicle.com/blogs/worldwise/dont-go-soft-on-study-abroad-a-call-for-academic-rigor/31082.

NAFSA. (2016). Marketing Education Abroad: Best Practices for Education Abroad Offices and Functions. Retrieved from: http://www.nafsa.org/findresources/Default.aspx?id=8378

NAFSA. NAFSA's guide to education abroad. NAFSA: Association of International Educators, Washington DC: 2005.

Office of International Affairs. (2014). 2015-2020 Strategic Plan. Retrieved from: http://www.depts.ttu.edu/international/attachments/oia/OIAStrategicPlan.pdf.

Rhodes, Gary and Richard Sutton, Glosssari Project, University of Georgia. Presented at the NAFSA Conference, Houston, TX, May 2012.

Spencer, S. E. & Tuma, K. (2007). The guide to successful short-term programs abroad (2 ed.).

Comparison of Two Project-Based Learning Experiences in Panama City, Panama

Dr. Aaron Richard Sakulich, Worcester Polytechnic Institute

Aaron Sakulich is an assistant professor in the Department of Civil and Environmental Engineering at Worcester Polytechnic Institute. His research interests include the development of sustainable cementitious materials for infrastructure applications and international project-based learning. The recipient of a 2007 Fulbright award to Morocco, he is now the co-director of an off-campus project site in Panama City, Panama.

Comparison of Two Project-Based Learning Experiences in Panama City, Panama

Abstract

In 1970, Worcester Polytechnic Institute adopted The WPI Plan, a project-based educational philosophy that requires students to complete three research projects during their undergraduate careers, each the equivalent of a full term's worth of work. Almost immediately, this curriculum evolved to incorporate off-campus project opportunities, with the first off-campus Project Center established in 1974. This paper compares two groups of undergraduates that recently carried out NSF-funded research at the Panama City, Panama, Project Center in order to fulfill one of their project requirements. The first group was composed of five seniors, selected based on academic standing, each working under the guidance of personnel from the Panama Canal Authority on a specific technical project as part of their capstone design project. The second group, selected based on financial need, investigated the effects of the ongoing Panama Canal Expansion Project on the Food-Energy-Water (FEW) nexus in the canal watershed to fulfill their junior year 'Interactive Qualifying Project' requirement. The findings of each team, logistical considerations for advisors, the results of student evaluations of the experience, and unique difficulties faced by the students are discussed, as well as lessons learned for the improvement of the program in future years. These projects represent two distinctly unique examples of project-based learning.

Background

In 1970, Worcester Polytechnic Institute adopted The WPI Plan, a project-based educational philosophy that requires students to complete three term-length research projects, advised directly by faculty, during their undergraduate careers [1]. During their junior year, students generally complete their "Interactive Qualifying Project" (IQP). This is a critical time in a student's professional development, as junior-year students have the technical skills necessary for a productive research project, but likely have little research experience and are close enough to graduation to be seriously thinking about the future and focusing more on academic work [2, 3]. The final requirement is the "Major Qualifying Project" (MQP), generally completed during a student's fourth year of study. The MQP is similar to senior design projects implemented at other universities, focusing specifically on developing engineering solutions to technical problems; as such, student teams are generally major-specific.

The IQP is in many ways the cornerstone of the WPI Plan; whereas in 1970 there were other schools that had capstone design and humanities projects as a part of their STEM curriculum, the faculty observed a need for a project in which engineering skills would be built in a societal context. This was due primarily to the failure of post-war technological progress to deliver on the utopian promises of the 1950s: despite incredible scientific breakthroughs, domestic and international social, economic, and political goals were not being met. The authors of the WPI plan reasoned that this was because technological innovations were being implemented without careful consideration of the social context – that a technology cannot exist in a void, but must be developed, implemented, and refined in a space as close to the stakeholders as possible [4].

Project-based learning has been shown to improve student outcomes in various professional competencies, including communication and teamwork skills; to prepare students for self-directed research projects in industry or academia; to demonstrate to students the value of interdisciplinary inquiry; to encourage students in seeking graduate degrees; and to provide students with a better understanding of core concepts [5, 6]. In recent decades, many in the engineering education community have concluded that professional competencies alone will not be sufficient for success in the increasingly transnational world of industry – students must develop global competencies that will enable personal and professional success outside of their home countries. These skills include the ability to work in diverse teams with individuals that hold differing perspectives; communication skills, including foreign languages and an understanding of etiquette, that enable the transfer of knowledge and inclusion in international networks; and the ability to recognize and adapt to the social factors that may determine how project partners approach research problems and tasks [7]. While it is generally agreed that curricula aiming to instruct students in global competency require coursework in international studies, a study abroad experience, and the building of foreign language skills, Lomann *et al.* argue further that these three factors must be connected, and that they must be relevant to a student's major, to be effective [8].

While the number of students studying abroad has risen dramatically over the past few decades, the length of the average student experience has steadily declined, with programs of one to five weeks duration becoming increasingly popular. In a study of over 3,700 alumni of study abroad programs, Dwyer concluded that longer programs were not necessarily better – students participating in study abroad opportunities as short as six weeks in length reported statistically significant gains in a variety of personal and professional outcomes [9]. Almost immediately after its inception, the project-based system at WPI evolved to incorporate off-campus opportunities for global competency development, in which students perform project work for one quarter (seven weeks). Study abroad itself is not necessarily the only means by which to develop global competencies, however – Parsons concluded from a study of 1,302 students that experiences such as attending cultural events at home, having friends with diverse national backgrounds, and taking conventional courses with international content were also strong predictors of higher development of key competencies such as cognitive and behavioral adaptability and respect for other cultures [10]. For this reason, all WPI students travelling to an off-campus project center are required to take a pre-departure orientation course to supplement their off-campus experiences and prepare them for travel. In academic year 2015/2016, 712 IQP students and 84 MQP students applied to pursue a project off-campus. This paper reviews recent experience with both IQP and MQP teams at the project center in Panama City, Panama.

MQP Research at the Panama City Project Center

An off-campus Project Center was established in Panama City in 2009 through networking with WPI alumni, who are particularly active in Panama. The Panama Canal is the most important single civil engineering structure on earth, reducing the distance of maritime travel between the east and west coasts of the United States by over 12,500 km (7,800 miles) and accommodating nearly 10% of global trade [11, 12]. In 2007 the people of the Republic of Panama voted in favor of expanding the Panama Canal by a roughly three-to-one margin at a total cost of over $5 billion, and ports worldwide have invested billions of dollars in upgrading their facilities to accommodate the "New Panamax" ships that will soon be able to transit the Canal [13].

The Project Center initially hosted only MQP teams working with the Panama Canal Authority (ACP) on technical projects such as the energy efficiency of water pumping stations;

investigating water quality in Lake Gatún; and helping to design grouting systems for the Borinquen 1E dam [14, 15]. MQP students are currently supported by Grant #1357667, titled *IRES: Environmental Impact of the Panama Canal Expansion Project*, funded by the National Science Foundation's Office of International Science and Engineering. This grant aimed to provide students with projects related specifically to the sustainability of the Panama Canal Expansion Project, as well as to improve the MQP experience by doubling project length to roughly 15 weeks; providing additional cultural programming; supporting students in presenting their work at a professional conference; and encouraging students to submit manuscripts based on their work to journals.

The first cohort of grant-supported students carried out five distinct projects in 2015, which the students summarized in a report titled *The Panama Canal Expansion Impacts: Connecting Us to the Future* [16]. The first project involved investigating the practicality of updating lighting at the El Prado and Corozal complexes and the Gamboa Dredging Division, from high-pressure sodium high-intensity discharge systems to more energy efficient LED-based systems. It was concluded that replacing a total of 211 light poles hosting 343 individual fixtures would reduce ACP energy usage by two thirds (over $60,000 each year) with a payback period of 7.1 years. The second project investigated land usage on the banks of Lakes Gatún and Alajuela. The raising of the operational level of Lake Gatún by 0.4 m (1.5 ft.) as part of the expansion project has led to the ACP working to identify unauthorized habitations on shore and either evict the individuals, or establish legal certifications and appropriate taxes. The student prepared a framework for a carrying capacity study as well as a shoreline management plan for Lake Gatún. The third student compared several options for visual navigational aids along the path of the expanded Canal. Such aids, which include buoys, lighthouses, and range towers, help a navigator align a ship to the center channel of a waterway, and are often preferred by mariners over more advanced, computerized systems. This project recommended the implementation of glass-reinforced plastic towers as navigational aids in cases where existing aids were being removed or replaced. The fourth project investigated new designs for the replacement of the Gatún spillway, which is no longer sufficient for regulation of the water level of the manmade lake at the center of the Panama Canal system, and recommended that the ACP emplace large steel pipelines through Lake Gatún and pump spillway construction debris off-site in the form of a slurry. The final project analyzed drinking water from several tugboat landings and the storage tanks of two in-service ACP tugboats, the *Veraguas II* and *Guía*. Water quality was checked against health targets based on World Health Organization benchmarks for the presence of total coliforms, residual chlorine, and turbidity.

IQP Research at the Panama City Project Center

In the spring of 2015, the NSF released a dear colleague letter requesting proposals for research on the Food/Energy/Water (FEW) Nexus. Introduced to the environment and development literature in 2008, the FEW Nexus is one of the more recent methods used to understand the varied ways in which livelihood dimensions affect, and are affected by, each other and are dependent on 'chains' of environmental goods and services. These relationships are often reciprocal. Agriculture, for example, requires energy inputs but may also provide energy outputs in the forms of food or biofuels. Although it has been widely considered a logical and efficient way to think about resource and livelihood interconnections, the FEW Nexus approach has been criticized for not being explicit about the politics and inequality surrounding outcomes and policies, and in particular, for avoiding questions of environmental justice [17, 18]. The economic impacts of the Canal Expansion Project are generally agreed upon, with projections showing Canal revenue increasing by roughly 12% and roughly 40,000 jobs being created. The environmental impacts, and the impacts on local ecosystems and livelihoods, are much less clear. The government's optimism is dogged by the criticisms of environmental experts regarding the wider impacts of the Canal expansion on local livelihoods, rural communities, and regional water supply [19]. The benefits and damages of the Expansion Project are not likely to impact all stakeholders equally – despite rapid economic growth, Panama is routinely ranked as one of the most unequal countries in the world, with nearly 40% of the population living in poverty [20]. Supplemental funding was awarded to the above-mentioned IRES grant, which was used to support a five-person IQP team that investigated the effects of the Expansion Project on the FEW Nexus in the Panama Canal watershed.

The students divided their work into four phases: acquiring contacts, conducting interviews, analyzing findings and identifying stakeholders, and building a framework describing the FEW Nexus in Panama. The team carried out 18 interviews with the personnel from government, non-government, educational, and private organizations. In a final report titled *Food, Energy, Water (FEW) Nexus Analysis in the Panama Canal Watershed* the students identified eight interconnected areas that may be present in any population center, but that have unique aspects in Panama, and made recommendations regarding potential future areas of research. The individuals interviewed by the students expressed a particular concern regarding water security. On a per person basis, Panama consumes twice the average amount of water than any other Latin American nation – and 50 liters (13 gallons) per person per day more than the average inhabitant of the United States [21]. Further, Panama City is experiencing the same stresses of any growing city – water treatment plants are rapidly approaching full capacity, infrastructure is corroding and deteriorating, and newly constructed neighborhoods are experiencing flooding and water management difficulties as they become integrated into the urban fabric. Unlike other cities, the expansion of the Canal will lead to an increase in water usage, even with the implementation of advanced water-saving technologies. Currently, roughly 50 million gallons (190 million liters) of freshwater are expelled into the ocean during each of the roughly 40 daily ship transits through the Canal [22]. Multiple interviewees stated that although there are emergency measures aimed at reducing water consumption in force, and existing laws aiming to prevent the contamination of the drinking water supply, neither carry strong enough penalties to be taken seriously by the local population [23]. Waste management was a concern closely related to water usage. All of the waste produced by Panama City was discharged into the Gulf of Panama without any treatment. In 2010 a wastewater treatment plant was constructed at Tocumen, serving roughly a third of the city; since reaching full capacity in 2013 the plant has been operating at reduced efficiency [24]. Finally, the food supply of Panama is being affected by the expansion of the Canal. The migration of individuals from rural areas to Panama City results in a reduced agricultural workforce – decreasing the amount food produced in Panama, while encouraging remaining farmers to adopt less labor intensive methods such as slash-and-burn land clearing.

Project Evaluation and Comparison

During a recent assessment of WPI's project-based educational model, over 2,500 WPI alumni answered questions regarding 39 areas of long-term impact [25]. The results were unambiguous: In off-campus projects, students consistently achieved learning outcomes at levels significantly higher than on-campus projects. Students working with partner communities outside of WPI not only showed improvement in terms of professional preparation, but in terms of their outlook and personal growth. In the words of an alumnus interviewed as part of this assessment project: "*My entire experience after the [project] was different than before. I felt like a different person when I came back… I saw the reason why I went to college, because I saw something taken to completion in the real world.*"

Both student teams selected for work in Panama participated in university-standard evaluation activities, which included both quantitative data (Likert Scale ratings) and qualitative data (responses to open-ended questions) focusing on every aspect of the program. Although the sample set was extremely small (five MQP and five IQP students), that roughly 90% of questions were answered with a 'four' or 'five', on a scale of zero to five, indicates broad satisfaction with the experience. (The other 10%, mostly ratings of 'three' received from MQP students, were scattered and did not indicate a clear trend with regards to one particular aspect of the program.)

Reaction to the actual project work was mixed – two MQP students rated their projects as 'five' with the others rating theirs 'two' or 'three'. While the topics were "*interesting*", students expressed some concern that some of the projects were not directly related to the undergraduate classwork they had completed and/or did not include elements considered necessary by the department (i.e. experience with professional licensure and a design element). This may have been due to ineffective communication between the ACP human resources department and the actual supervisors, who did not always know the date on which students would be starting their project. Although the projects sometimes lacked focus or direction, the students ended up viewing this positively in terms of professional development– producing something of value required self-motivation and the development of skills that increased their independence as a researcher. One student wrote "*I think that I personally learned how to research better and formulate my thoughts better from doing this project because it was more independent as it was only me working on it whereas IQP it's a group. So it really puts the responsibility on you to learn how do these things on your own.*" The IQP team, on the other hand, embraced the unstructured nature of their project, writing, "*I loved the independence and flexibility of our project and the ability to collaborate and figure new things out with our group*" and "*The unstructured nature of the project gave us a great opportunity to be heavily involved in every stage, from formation to completion*". One student did note, though, that the unstructured nature of the project "*was sometimes challenging, as five voices clashed and caused a lot of tension - especially in the last few weeks*".

In general terms, these evaluations reflect the experience of Chieffo and Griffiths in a study of over 2,300 students – 93% of students that studied abroad generated unique, qualitative comments about their program experience, while only 57% of their on-campus did so. In general, the comments of the former group were more introspective and related to personal gains, while the latter group commented primarily on classroom learning and logistics [26]. Of the 15 students participating in on-campus projects advised by the author during this time, fewer than half provided qualitative comments, and of those, most were related to logistical issues – "*Our project did not deserve a C*", "*[I disliked] finding out WPI does not have a proper ball mill*", and "*[I disliked the] lack of help from grad students at the end of project*" being representative.

While certainly not conclusive, this implies that some aspect of the off-campus experience may lead students to think more specifically and introspectively about their personal competencies.

Students were positive about the social and cultural aspects of their work, particularly the opportunity to improve their Spanish. The MQP students participated in a one-week orientation course at a Spanish Language school in Panama City, of which one student wrote "*[I] learned more conversation in that week than I had in 8 years of classes*". This course was not possible for the IQP team, as their work was carried out in the middle of the academic year. A recent paper by Briggs suggests that study abroad experiences alone are not sufficient for building vocabulary in foreign languages, as there is no way to ensure that students will not revert to English outside of classes [27]. This study concluded that students would benefit if the institutions hosting them made concerted efforts to ensure that they had a variety of opportunities to communicate in their non-native language outside of the classroom. Although both groups became involved in cultural experiences based on relationships established as part of their project work, particularly with the personnel of the housing facility, further activities were encouraged. Each team took part in an 'ecotourism' trip to the country's western provinces; attended at least one soccer match (which Panama lost); were invited into the homes of their Panamanian colleagues; and so on. Potentially more important than gains in vocabulary or linguistic ability, studies suggest that even short-term language experience abroad has a significant positive impact on student attitudes towards developing language skills [28].

The final, obvious area of improvement regards addressing homesickness. One of the five MQP students withdrew from the program early for this reason. Although the student's project was still completed, both the project and the student would have benefitted from completing the 15 week program. IQP students reported experiencing varying degrees of homesickness, although none took the step of returning home possibly because logistical arrangements (finding a replacement project, arranging classes, etc.) would have been significantly more difficult during the school year (as compared to the MQP projects, which begin, and could be potentially abandoned, during the summer). One student wrote, "*I had a difficult adjustment to studying abroad in my first week, but there was never any dislike for the project itself*". Although the effects of study abroad on the mental health of students has not been as thoroughly investigated as the effects of study abroad on developing language skills or global competencies a recent summary of the literature concludes that stress and homesickness can significantly reduce the likelihood of students attaining learning goals, and that creating genuine social relationships with fellow students, native individuals, or even program administrators is the most important mechanism by which such difficulties can be addressed [29]. This is supported by student responses – one student wrote "*we went from 5 to a gang of friends*" and that becoming socially involved with coworkers "*helped keep us busy ... gave us something to look forward to ... having friends keeps your mind from home*." One way in which the issues of homesickness and loneliness could be better addressed in future years is by adjustment of the pre-departure orientation course. Goldstein and Keller reported that students are more likely to blame culture shock on external factors (communication difficulties, unfamiliar food, inability to find one's way in an unfamiliar city, etc.) when, in fact, the educational literature attributes culture shock to internal factors (ineffective coping strategies, identity confusion, prejudice, etc.) [30]. This student attitude was reflected in the pre-departure orientation course, taught by the author, in which students questions focused on matters of sociocultural adaptation. A future emphasis on psychological adaptation may be more appropriate.

Conclusions

The IQP and MQP projects supported by NSF IRES funding in Panama each had the goals of educating students in engineering sustainability, but were significantly different experiences. The student feedback and advisor experiences have led to two parallel goals for future years of this program: First, WPI faculty will press for better communication with the ACP, including direct contact between faculty and project supervisors at an early stage, rather than all contact being carried out through the ACP's human resources department. Second, preparatory activities will include more of a focus on independent research – if a project is unsatisfying or not appropriate for program goals, students should arrive in-country knowing what steps they can proactively take in order to improve their situations. Ensuring that students work on teams of at least two individuals, rather than independently, may help to meet this goal, by providing students with someone that can help them adapt. Finally, the experience has prompted a review of the methods by which students are recruited to the program – students selected on the basis of financial need rather than academic standing were, generally speaking, more adaptable, more independent, more satisfied with their experience – and may not otherwise have been able to participate in off-campus project work.

Acknowledgments

This work was supported by grant #1357667, titled IRES: Environmental Impact of the Panama Canal Expansion Project, awarded by the National Science Foundation's Office of International Science and Engineering.

Bibliography

1. *2013-2014 Undergraduate Catalog, Section 1: The WPI Plan*, 2013, Worcester Polytechnic Institute: Worcester, MA.
2. Brainard, S.G. and L. Carlin. *A longitudinal study of undergraduate women in engineering and science*. in *Frontiers in Education Conference, 1997. 27th Annual Conference. Teaching and Learning in an Era of Change. Proceedings*. 1997.
3. Black, K.M., *An Industry View of Engineering Education*. Journal of Engineering Education, 1994. **83**(1): p. 26-28.
4. Bothelo, A., et al., *Creation of the IQP: Bridging the socio-technical divide in the postwar era*, 2014, Worcester Polytechnic Institute: Worcester, MA. p. 21:17.
5. Tryggvason, G. and D. Apelian, *Shaping our world: Engineering Education for the 21st Century*. 2011, Hoboken, NJ: John Wiley and Sons. 320.
6. Sunthonkanokpong, W., *Future Global Visions of Engineering Education*. Procedia Engineering, 2011. **8**(0): p. 160-164.
7. Shen, Y., B.K. Jesiek, and Y. Chang. *Cultural orientation and global competency: A comparative assessment of engineering students*. in *American Society for Engineering Education*. 2011. American Society for Engineering Education.
8. Lohmann, J.R., H.A. Rollins, and J. Joseph Hoey, *Defining, developing and assessing global competence in engineers*. European Journal of Engineering Education, 2006. **31**(1): p. 119-131.

9. Dwyer, M.M., *More Is Better: The Impact of Study Abroad Program Duration.* Frontiers: The interdisciplinary journal of study abroad, 2004. **10**: p. 151-163.
10. Lindsey Parsons, R., *The Effects of an Internationalized University Experience on Domestic Students in the United States and Australia.* Journal of Studies in International Education, 2010. **14**(4): p. 313-334.
11. Avery, R.E., *America's Triumph at Panama.* 1913, Chicago, IL: The L.W. Walter Company.
12. Haskin, F.J., *The Panama Canal.* 1914: William Heinemann.
13. Rodrigue, J.-P., *Factors impacting North American freight distribution in view of the Panama Canal expansion.* 2010: Van Horne Institute.
14. Lee, B., Z. Lorch, and T. Melong, *Evaluation of the erosion control methods implemented by the Panama Canal expansion program*, 2012, Worcester Polytechnic Institute: Worcester, MA. p. 96.
15. Amevor, D.F., et al., *Maintenance and Expansion Evaluation of the Panama Canal*, 2015, Worcester Polytechnic Institute: Worcester, MA.
16. Brakenwagen, A., et al., *The Panama Canal Expansion Impacts: Connecting Us to the Future*, 2015, Worcester Polytechnic Institute: Worcester, MA.
17. Middleton, C., et al., *The Rise and Implications of the Water-Energy-Food Nexus in Southeast Asia through an Environmental Justice Lens.* Water Alternatives, 2015. **8**(1): p. 627-654.
18. Middleton, C. and S. Allen, *The (Re) discovery of "the Nexus": Political economies and dynamic sustainabilities of water, energy and food security in Southeast Asia.*
19. Gonzalez, C.G., *Environmental Impact Assessment in Post-Colonial Societies: Reflections on the Proposed Expansion of the Panama Canal.* Tenn. JL & Pol'y, 2008. **4**: p. 303.
20. Berrey, C.A., *Inequality, demography, and variability among early complex societies in Central Pacific Panama.* Journal of Anthropological Archaeology, 2015. **40**: p. 196-212.
21. Wedgwood, A. and K. Sansom, *Willingness-to-Pay Surveys: A Streamlined Approach: Guidance Notes for Small Town Water Services.* 2003: WEDC, Loughborough University.
22. Carse, A., *Nature as infrastructure: Making and managing the Panama Canal watershed.* Social Studies of Science, 2012: p. 0306312712440166.
23. Fotos, M., Q. Newcomer, and R. Kuppalli, *Policy Alternatives to Improve Demand for Water-Related Ecosystem Services in the Panama Canal Watershed.* Journal of Sustainable Forestry, 2007. **25**(1-2): p. 195-216.
24. Foster, V., A. Gomez-Lobo, and J. Halpern, *Designing Direct Subsidies for Water and Sanitation Services: Panama—A Case Study.* 2000.
25. Vaz, R. and P. Quinn. *Long term impacts of off-campus project work on student learning and development.* in *Frontiers in Education Conference (FIE), 2014 IEEE.* 2014.
26. Chieffo, L. and L. Griffiths, *Large-Scale Assessment of Student Attitudes after a Short-Term Study Abroad Program.* Frontiers: The interdisciplinary journal of study abroad, 2004. **10**: p. 165-177.
27. Briggs, J.G., *Out-of-class language contact and vocabulary gain in a study abroad context.* System, 2015. **53**: p. 129-140.
28. Khoroshilova, S., et al., *Academic Mobility: The Impact of Short-Term Language Courses Abroad on the Development of Language Competences.* Procedia - Social and Behavioral Sciences, 2015. **214**: p. 992-999.
29. Hunley, H.A., *Students' functioning while studying abroad: The impact of psychological distress and loneliness.* International Journal of Intercultural Relations, 2010. **34**(4): p. 386-392.
30. Goldstein, S.B. and S.R. Keller, *U.S. college students' lay theories of culture shock.* International Journal of Intercultural Relations, 2015. **47**: p. 187-194.

Administering A U.S. Based M.S. Degree in Kilimanjaro, Africa – A Global Benchmarking in Regulatory Science

Dr. Mitchell L Springer PMP, SPHR, Purdue University, West Lafayette

Dr. Mitchell L. Springer, PMP, SPHR

Dr. Springer currently serves as the Executive Director for Purdue University's Polytechnic Institute located in West Lafayette, Indiana. He has over 35 years of theoretical and industry-based practical experience from four disciplines: Software Engineering, Systems Engineering, Program Management and Human Resources. Dr. Springer possesses a significant strength in pattern recognition, analyzing and improving organizational systems. He is internationally recognized, has contributed to scholarship more than 150 articles, presentations, books and reviews on software development methodologies, management, organizational change, and program management. Dr. Springer sits on many university and community boards and advisory committees. He is the recipient of numerous awards and recognitions; most recently he was awarded the Purdue University, College of Technology, Equity, Inclusion and Advocacy Award.

Dr. Springer received his Bachelor of Science in Computer Science from Purdue University, his MBA and Doctorate in Adult and Community Education with a Cognate in Executive Development from Ball State University. He is certified as a Project Management Professional (PMP), Senior Professional in Human Resources (SPHR & SHRM-SCP), in Alternate Dispute Resolution (ADR), and, in civil and domestic mediation. He is a State of Indiana Registered domestic mediator.

Dr. Kari L. Clase, Purdue University, West Lafayette

Kari Clase is an Associate Professor in the Department of Technology Leadership and Innovation in the Polytechnic Institute and the Department of Agricultural and Biological Engineering in the College of Agriculture at Purdue University. Dr. Clase is also the Director of the Biotechnology Innovation and Regulatory Science (BIRS) Center. The mission of the BIRS Center is to develop global programs to ensure sustainable access to medicines for Africa and developing nations, and to advance discovery in manufacturing technology, quality of medicines, and rare disease research. This mission is accomplished through innovative knowledge-based programs in STEM areas with an emphasis on interdisciplinary collaboration. Dr. Clase teaches multiple courses covering topics in biotechnology, bioinformatics, biological design and drug discovery to engineers, scientists and technologists. Her currently funded projects include collaborators from multiple disciplines and an impact that spans K-12 to graduate education.

Mrs. Lauren Ann Terruso, Purdue University

Lauren Terruso is the Operations Manager of the Biotechnology Innovation and Regulatory Science Center in Discovery Park at Purdue University.

Administering A U.S. Based M.S. Degree in Kilimanjaro, Africa – A Global Benchmarking in Regulatory Science

Dr. Mitchell Springer (*Purdue University*), Lauren Terruso (*Purdue University*), Mary Speer (*Purdue University*), Zita Ekeocha (*Kilimanjaro School of Pharmacy*), Dr. Stephen Byrn (*Purdue University*), and Dr. Kari Clase (*Purdue University*)

Abstract

In the world of pharmaceuticals, regulatory science is an emerging field that has goals of developing tools, drugs, devices, and practices to increase benefit and lower risk concerning safety, quality control, and effectiveness. The dangers of having a subpar, or non-existent, regulatory science practice can be costly at best, or lethal at worst. Formal education at the graduate level to train professionals is a relatively new field. Research has been done on the practice of regulatory science, but not on the institutions educating these scientists. Purdue's investment, in particular, has the ultimate goal of getting sustainable medicine to Africa. The collaboration of Purdue's Biotechnology Innovation and Regulatory Science program with the Kilimanjaro School of Pharmacy is helping to provide good regulatory practices in Africa. This prompted the need to perform benchmarking activities comparing and contrasting the Purdue program to both domestic and international certifications and degrees.

The methodology employed to assess the potential list of resources was to make an exhaustive list (as possible, since higher education is constantly changing) of graduate and post-baccalaureate institutions that provide certification in regulatory science fields. First, the categories of cost, availability, and curriculum needed streamlining for comparison purposes. Then, best practices needed to be identified, for both domestic and international programs. Common themes of coursework in the life sciences, biotechnology, and policy/ethics were identified. Often, programs would require or prefer a bachelors in health science, life sciences, or engineering with common requirements of the advanced degree mandating courses in tissue engineering, stem cell engineering and related disciplines. The international programs focused more on global processes than domestic ones. In addition, there was a common trend, both domestically and internationally, that pharmaceutical sciences background/employment experience was preferred and the programs were designed to be part-time so the degree/certification could be earned while remaining employed (some required a project be completed at the student's place of employment). The most typical degree earned in this field was a Master's degree and while ranging in cost from $23,000 - $100,000 in the domestic programs, the cost was typically much less expensive in the international programs.

Overview

The Biotechnology Innovation and Regulatory Science (BIRS) program transitioned to the Purdue Polytechnic Institute in the fall semester 2014, after 12 successful years as the Regulatory and Quality Compliance program in the Purdue College of Pharmacy. This transition allowed the program to expand its pharmaceutical regulatory science and quality offerings to include medical devices & diagnostics, combination drugs and biotechnology innovation.

This science-based program is a 33 credit hour MS that includes 21 required hours and 12 elective credits. The curriculum aligns with identified key competencies for regulatory science programs and also prepares the student for further study if desired. Included in the required hours is a directed project. The directed project is an independent study course that allows the student to apply the program learnings to a project that proves to be value added within their company.

A Biotechnology Quality and Regulatory Compliance graduate certificate program is also available. It is comprised of 4 courses (1 course per semester) and compliments bachelor, MS or Ph.D. degrees.

These programs are designed to offer an education that fits the busy working professional. Students have flexibility in registering for courses and can enroll in 1 to 3 courses per semester. Courses are offered in a blended format. Basic information is presented online and the advanced material, including difficult concepts and group projects, are presented on weekends (3 weekends per semester) on the Purdue University campus located in West Lafayette, IN. The program welcomes students to enter fall, spring or summer semesters.

Program Administration

In the final analysis, it was determined the best approach to administering an MS program in Kilimanjaro, Africa, would be a distance-hybrid delivery model; with the face-to-face component being on-site in Kilimanjaro. The administrative organization for the program was/is the Purdue Polytechnic Institute's Center for Professional Studies in Technology and Applied Research (ProSTAR).

On June 11, 1998, the Purdue University College of Technology initiated the process for University, and subsequently, Indiana Commission for Higher Education, approval of a non-traditional delivery medium, fee-based weekend alternative to Purdue's traditional campus tuition-based Master of Science with a major in Technology degree[1].

On October 13, 2000, the Indiana Commission on Higher Education (ICHE) approved the University request for delivering a hybrid distance-based alternative to traditional classroom-only programs. The entire process from conceptualization to full final approval took two years and four months. Noteworthy, from the proposal excerpt, the concept of evaluation was integral to the program concept from the onset.

In the fall of 1998, the COT's Department of Industrial Technology took a lead role in implementing, pursuant to authorization, the first weekend master's program (WMP) in Technology on the campus of Purdue University in West Lafayette, Indiana. The original offering was cohort-based and it employed a weekend format; meeting from Friday through Sunday. The cohort met three times a semester, twice in the summer semester, for a total of five semesters (Fall, Spring, Summer, Fall and Spring). Because of its non-traditional approach, the state's authorization included the establishing of a different fee structure than normal classes which resulted in a cost that was higher than conventional campus-based instruction.

The Center for Professional Studies in Technology and Applied Research (ProSTAR) was approved by Purdue University under the College of Technology as an academic Center in February 2009. At that time, the underlying foundation for ProSTAR's professional education activities was a Master of Science degree with a primary focus in technology leadership and innovation skills including tools for process improvement and quality management.

As well, the program incorporated other innovations beyond its delivery system, schedule and fee structure. To be consistent with its goal of developing practical skills and knowledge immediately, or at least quickly, applicable to business and industry, its plan of study[2] incorporated a base of essential core studies, flexible and easily tailored courses to insure relevance to emerging technologies, and a guided, industry focused applied research and development project called simply the Directed Project (DP). The latter DP was deliberately designed to require work commensurate to what is typically expected of a master's degree thesis[3].

ProSTAR is entirely self-funded from fee-based programs. It offered its first 100% distance program in 2010.

The initial on-campus distance-hybrid offering in the fall of 1998 spawned a comparable off-site instantiation of this sole program in 2005. The 2005 instantiation was delivered in an on-site format at the location of the target corporate partner. This industry- and corporate-specific instantiation provided for the first time, significantly increased enrollments outside of the main campus and the prior and on-going on-campus distance-hybrid baseline program. Figure 1.0 below depicts the suggested first wave of increased enrollments.

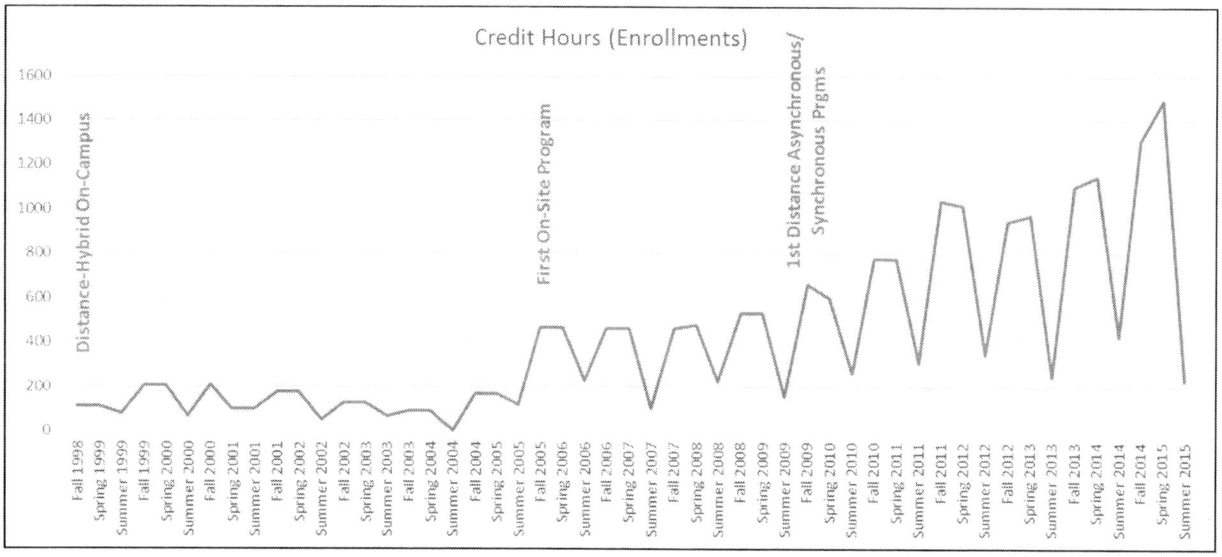

Figure 1.0 – Delivery Induced Enrollment Growth

The geographical limits of on-campus programs, even taking into consideration the distance-hybrid aspects of the program, created a self-imposed constraint. Moving the programs to a customer's location, geographically distant from the main campus program, spurred enrollment growth, but it too created enrollment limitations based on corporate sponsorship and geographical specificity.

In 2010, ProSTAR, in collaboration with the academic departments, entered the distance education market with three 100% distance programs. Distance offerings created a significantly greater market, one not bounded by geography. Most growth to-date has been the result of distance offerings. It is further anticipated that most future growth providing the greatest opportunity for sustainment of an on-going administrative organization will materialize through distance delivery models.

The 2010 offering of distance programs came in two noticeably different delivery methodologies: namely, synchronous and asynchronous. Synchronous delivery of a given distance program implies the recipient of the instruction is receiving the instruction in real-time as the instruction is being provided. While this methodology supported the distance element thereby not requiring the student to attend class on the university campus or, in a previously discussed format, at an employer's location, it still limited student participation by requiring a set time by which the instruction was to be given and therefore a set time by which the student must be available and prepared to receive said instruction. This concept created yet another limitation to full enrollment possibility or potential for maximum enrollment opportunities.

Distance programs offered through the latter delivery mode of asynchronous delivery, fixed, or removed, the barriers other predecessor program delivery models created.

While the on-campus distance hybrid, on-site distance-hybrid, and 100% distance asynchronous delivery models still exist in multitude of program offerings, the 100% distance asynchronous delivery model creates the greatest promise for maximum enrollments; only limited by the desire to obtain the offered degree program.

Figure 2.0 below depicts the delivery modes employed and their respective limitations.

Delivery Mode	Limitations		
	Geographical	Employment	Real-Time
1998 Distance-Hybrid; on campus			
2005 Distance-Hybrid; on-site, off-campus			
2010 Distance Synchronous Programs			
2010 Distance Asynchronous Programs			

Figure 2.0 – Enrollment Limitations by Delivery Mode

Methodology

There was a need to perform a benchmarking study to compare the existing Purdue program and the programs offered by others throughout the world in order to identify strengths and potential gaps in the Purdue BIRS Program, specifically associated with global regulatory science education. A comprehensive literature review yielded an exhaustive list of institutions with graduate degrees and certifications in fields related to Regulatory Science (some less descriptive – like Biotechnology or Biomedical Engineering; some more descriptive – like Management of Drug Development) and the specific results were compiled. Further investigation was performed on each institution, looking at cost, availability, and curriculum.

The methodology employed encompassed a time-phased set of inter-related activities as described below.

- A compare and contrast was performed by identifying and normalizing categories of cost, availability, and curriculum
- Program offerings, nationally and internationally, were mapped to the normalized data for cost, availability and curriculum
- A compare and contrast was performed by characterizing and normalizing best practices, nationally and internationally, across identified programs
- Comparison of normalized data (other identified programs) to the Purdue University data was performed to determine criteria of preeminence; considering such factors as cost, mode of delivery and curriculum

Table 1 graphically depicts the time-phased activities performed over a sixteen week period.

Activities Performed	Weeks							
	1-2	3-4	5-6	7-8	9-10	11-12	13-14	15-16
Criteria Identification and Normalization								
Program Identification and Mapping to Criteria								
Identification of Best Practices								
Comparison of Normalized Data								

Table 1.0 – Time-Phased Research Activities

Findings[1]

Through the above research, it was determined institutions offering elements of programs in regulatory science were distributed throughout the United States. There were western schools in Arizona, California, and Washington; mid-western schools in Minnesota, Illinois, and Wisconsin; eastern schools in Pennsylvania, Maryland, Massachusetts, and Washington, D.C.; southern schools in Georgia. Of the twenty two domestic institutions researched, fourteen included terms in their degree title (or, at least their degree's concentration) that were clearly describing the regulatory science field. Of the ones that did not use the term "regulatory" or "regulatory science" explicitly, the degrees were in more peripheral disciplines, such as biotechnology, however, their curriculum indicated that the program goals aligned with regulatory science. University B, for example, offers a Master's in Biotechnology, and specific course offerings include: Development of Vaccines to Infectious Diseases, Pharmacokinetics and Drug Design, and Molecular Targets of Drug Discovery. The course titles suggest that the content may include some topics peripherally related to regulatory science. While these courses are important, courses found within specific regulatory science degree programs were more focused in regulatory science content and its applications. An example of this category and representative course included, University C Clinical Research Management degree with a Regulatory Science concentration: Fundamentals of Regulatory Affairs course that is an overview of the role of ethical clinical research in new product development; or, Medical Device Development and Regulation course, that explores the regulatory framework for the design, development, approval, and marketing of medical devices.

It should also be noted that certain programs, such as University D and University E, that have more general degrees, such as Biotechnology/Biomedical Engineering, in addition to a separate Masters in Regulatory Science, also had a wide range in price. For example, at University E, a full-time, in-person graduate degree in Biotechnology or Biomedical Engineering costs a student over $70,000. A Regulatory Compliance graduate degree provided in the evening costs $37,521. It also appeared that most regulatory science programs across the country cater to working

[1] To maintain anonymity, the names of researched universities have been changed to University A, B, C... The authors maintain the mapping and bibliography for future research purposes and reference.

adults. Very few regulatory science programs are full-time and in-person, with the only institution being University E and two of its master's degrees: Biotechnology and Biomedical Engineering.

Many of the other programs offer in-person and online hybrids (or solely online), evening and weekend classes, and the option of part-time or full-time. Depending on the institution, cost (tuition only, no extra fees, books, or living expenses were included in these amounts) could vary significantly. Some were as low as approximately $20,000 for the whole degree (University C's fully online program) and some were as high as $84,000 (University F). It was difficult to compare course requirements in regards to units, since some institutions only had "8 units" (University B) and others were required "168 units." The time to complete the degree was comparable and suggests that overall, the course expectations were similar. Most degrees can be completed (if enrolled full-time) in a span of two years.

Many programs were similar to Purdue's 33-credit hour requirements. Those which were closest to the approximate $33,000 Purdue University program cost are: University G ($30,420), which is offered entirely online; University D's Master's in Regulatory Affairs, which offers six concentrations ($28,440) and is offered part-time and online; University E's Regulatory Compliance Master's degree ($37,521) which is offered full-time evening courses; University H's Master's in Regulatory Affairs ($29,250), which is offered entirely online; University I's Master's in Advanced Studies in Clinical Research ($30,000), which is offered in-person late afternoons and early evenings; and University J's Master's in Regulatory Science ($28,000) which is offered online and part-time. University K's Master's degree in Biomedical Regulatory Affairs is the only institution that included a single cohort of students progressing together throughout the program for two years, going full-time, and meeting in-person in the evenings. The cost is also comparable to Purdue ($32,400), lasting two years.

In addition to master's programs, some institutions offer certificates as well. These are all post-baccalaureate, but some are prior to a master's degree, while others are obtained after a master's degree. The certificates are more similar than the graduate degrees, as far as cost, length of program, and availability. The nine institutions are distributed throughout the country, east coast, west coast, mid-west, and south. The cost range for these certificates ranges from $6500 (for a 3-course certificate at University K) to $19,500 certificate at University L. The courses are relatively comparable with titles like "Clinical Research for Regulatory Affairs" (University G), "Practical Quality Management" (University E), "Current Good Manufacturing Practices" (University J), and "Managing the Guidelines for Quality" (University M).

As for international programs that specified Regulatory Science, rather than a general Biotechnology or Pharmacy degree, their coursework lists were similar to those of the domestic institutions. However, graduate certificates were more common than the master's degree and the requirements of the international graduate certificates were comparable to a master's degree. TOPRA in conjunction with the University N offers a master's degree with courses like

"Regulatory Control of Clinical Operations" and "Strategic Planning in Regulatory Affairs" that costs the equivalent of $19,500. University O, Toronto, Ontario, offers a Regulatory Affairs Postgraduate Certificate, which costs approximately $7000 for domestic students and $21,000 for international students. This appears higher than the certificates offered in the states, but is more rigorous, covering three semesters and 14 courses.

Conclusions

When comparing the other domestic institutions to Purdue's BIRS master's and certificate programs, there are some best practices that surface.

- Worker-friendly administration of the programs. Due to the applied nature of this field, many individuals are working while earning the degree or the certificate. Most of the institutions offer some kind of flexibility (including Purdue), whether it's online, in-person on the evenings or weekend, or some hybrid combination.
- Program Cost - Purdue is very cost competitive with most of these degrees. There are a few that are less expensive, but the ones that provide a similarly styled program, curriculum and/or delivery, are comparably priced.
- Program delivery – the benefits (contact and cost) of a distance-hybrid style of program delivery appears to be a best practice when delivering a program outside of the U.S. The face-to-face element of the Purdue program delivery model provides the dual benefit of personal contact and efficiency of delivery; yet, never compromising on effectiveness.

In the final analysis, the Purdue University global regulatory science program, delivered internationally in Kilimanjaro, Africa, benefited significantly from the collective understanding gained from the market research performed. The findings support that the program cost is economical and the flexible program delivery and program administration that aligns with the schedule of working professionals helps contribute to long term success and sustainability, both nationally and internationally.

References

[1] Springer, M. L., Schuver, M. T., & Dyrenfurth, M. J. (2011). Long term alumnus performance and evaluation after graduation from a distance learning hybrid weekend master's degree program in technology. *ASEE 2011 Annual Conference Proceedings*. Vancouver, B.C.

[2] Springer, M., Bertoline, G., & Schuver, M. (2011). Building an Academic Center Infrastructure for Professional Development. In *Proceedings of the 2011 American Society for Engineering Education Annual Conference*. Washington, D.C.: American Society for Engineering Education.

[3] Springer, M. L., Schuver, M. T., & Dyrenfurth, M. J. (2011). Long term alumnus performance and evaluation after graduation from a distance learning hybrid weekend master's degree program in technology. In *Proceedings of the 2011 American Society for Engineering Education Annual Conference*. Washington, D.C.: American Society for Engineering Education.

The Pathway to Systems Education for the Global Engineer

Dr. Alice F Squires, Washington State University

Dr. Alice F. Squires is an Associate Professor at Washington State University (WSU) with over 30 years of technical leadership experience. Prior to joining WSU, Dr. Squires served as Manager of Systems Engineering at Aurora Flight Sciences, Senior Researcher for the nationwide University Affiliated Research Center in Systems Engineering and Online Technical Director for SSE at Stevens Institute of Technology, Senior Systems Engineer consultant to LM, IBM, and EDO Ceramics, Senior Engineering Manager at GD, Senior Engineering Manager at LM, and Advisory Engineer/Scientist at IBM. Dr. Squires is a contributing author and editor to the Systems Engineering Body of Knowledge (sebokwiki.org) and the Graduate Reference Curriculum for Systems Engineering (bkcase.org/grcse). She is certified by PMI as a Project Management Professional, and by INCOSE as a Certified Systems Engineering Practitioner, including in Acquisition. Dr. Squires is a lifetime member of the Beta Gamma Sigma, Tau Beta Pi, and Eta Kappa Nu Honor Societies. She is a Senior Member of the IEEE, a member of and Director on the Systems Engineering Division board of ASEE, and a member of the ASEM, NDIA, INCOSE, and PMI. Degrees earned include a BSEE from the University of Maryland; a MBA from George Mason University; and a Ph.D. in Systems Engineering from the Stevens Institute of Technology. Her areas of expertise include Systems Engineering, Engineering and Technical Management, Project Management, Systems Thinking, Online Education, Engineering Education, and Competency Development.

Prof. Jon Patrick Wade, Stevens Institute of Technology (School of Systems & Enterprises)

Jon Wade is a Distinguished Research Professor in the School of Systems and Enterprises at the Stevens Institute of Technology and currently serves as the Director of the Systems and Software Division and Chief Technology Officer for the Systems Engineering Research Center (SERC) where he is leading research in the use of technology in systems engineering education and complex systems. Previously, Dr. Wade was the Executive Vice President of Engineering at International Game Technology where he managed corporate wide research and development. Dr. Wade spent ten years at Sun Microsystems during which time he managed the development of Enterprise Servers. Prior to this, he led advanced development of supercomputer systems at Thinking Machines Corporation. Dr. Wade received his SB, SM, EE and PhD degrees in Electrical Engineering and Computer Science from the Massachusetts Institute of Technology.

Dr. Nicole A.C. Hutchison, Stevens Institute of Technology

Dr. Nicole Hutchison is a Research Engineer at the SERC. Her primary work through the SERC includes the Helix project. Previously she was a member of the BKCASE research team. Before joining Stevens, she spent 5 years working for Analytic Services, supporting the US Departments of Defense and Homeland Security. She holds a PhD in systems engineering from Stevens and her INCOSE CSEP.

Building a Pathway to Systems Education for the Global Engineer

Abstract

As systems become increasingly global in nature, those involved in the conception, design, development, support, sustainment, and end of life of such systems must have a global perspective on both the operational context in which those systems will be used, and the challenges faced by the development, production, deployment, and support teams involved in their creation and sustainment. To build a pathway to systems focused education, the foundations of systems knowledge and skills should be instilled in engineering students early in their education, starting with systems thinking in K-12 and building on more advanced system concepts through undergraduate education and beyond. This paper describes how societal needs combined with global trends produce systems challenges that must be addressed by engineers who are systems thinkers. After analyzing three separate undergraduate engineering education research efforts in process, the authors propose a pathway to support engineering education to mature systems engineering knowledge and skills in the engineering profession.

Introduction

Solutions to meet fundamental human needs often lead to large and complex engineered systems that can only be successful if they are socially acceptable and provide value to our global society. Yet, as humanity attempts, through engineering and technology, to make the world a better place, our ever-evolving society creates new and ever greater challenges. With these challenges in mind, the International Council on Systems Engineering (INCOSE) formed an international task force of experts in 2012 to develop the *Systems Engineering Vision 2025*[1] report on the future of systems engineering. This vision is shaped by the global environment and includes human and societal needs, policy and business challenges, and rapidly evolving technologies that serve as the foundations for developing systems. The experts found that while the systems engineering field needs to grow and thrive, this was not sufficient. The evolving work environment, following global trends, both constrains and enables the manner in which systems engineering is practiced. The multi-disciplinary perspective, critical to system product innovation, defect reduction and customer satisfaction, could no longer be the purview of only the systems engineer. The infusion of systems thinking across a broad range of the engineering and management workforce was needed to meet the demands for a growing number of systems-capable engineers with the necessary technical and leadership competencies. The experts concluded: *Leaders need to be systems thinkers; all engineers need some education and training in systems engineering knowledge and skills; and all systems engineers need to be well versed in a broad set of socio-technical and leadership skills, serving as a central, multi-disciplinary focal point for systems development with stakeholders from all walks of life.* With this as the backdrop, this paper addresses three separate research efforts focused on advancing undergraduate engineering education and integrates the findings into a proposed pathway to systems education for the global engineer.

Characteristics of Global Engineers

In 2008, the American Society of Engineering Education (ASEE) in collaboration with the International Federation of Engineering Education Societies (IFEES) formed an international team to research the attributes of the global engineer. Hundley[2] reports on the outcomes of both an "*Attributes of the Global Engineer*" survey delivered in 12 languages between October 2010 and September 2012, and a series of workshop events held from 2012 to 2015. Based on the 1,027 usable survey responses and feedback from focus groups and workshops held in the United States, Finland, Belgium, Colombia, Argentina, Japan, U.K., Spain, U.A.E., and India, twenty top attributes for global engineering effectiveness were identified and sorted into the following five categories[2]:

- ***Technical Attributes:*** Engineering-related knowledge and skills needed for disciplinary success (4 attributes)
- ***Professional Attributes:*** Workplace-related competencies needed for effective global performance (5 attributes)
- ***Personal Attributes:*** Individual characteristics needed for global flexibility (4 attributes)
- ***Interpersonal Attributes:*** Skills and perspectives needed to work on interdependent global teams (2 attributes)
- ***Cross-cultural Attributes***: Societal and cultural understanding needed to embrace diverse viewpoints (5 attributes)

Of these, the top eight attributes (assumed order of perceived importance) reported by Hundley[2] are shown in Table 1.

Table 1. Top Eight Attributes of the Global Engineer (adapted[2])

Category	Attributes
Professional	Communicates effectively in a variety of different ways, methods, and media (written, verbal/oral, graphic, listening, electronically, etc.)
Personal	Possesses the ability to think both critically and creatively
Personal	Shows initiative and demonstrates a willingness to learn
Interpersonal	Functions effectively on a team
Personal	Possesses the ability to think both individually and cooperatively
Technical	Demonstrates an understanding of engineering, science, and mathematics fundamentals
Technical	Demonstrates an understanding of information technology, digital competency, and information literacy
Personal	Maintains a positive self-image and possesses positive self-confidence

Of note, the collective stakeholders from the focus groups felt that the following needed to be emphasized or added to the list for engineers to be successful in the global context[3]:

- Cultural sensitivity / tolerance to other people and perspectives
- Open-mindedness and ability to adapt
- Ability to behave ethically across cultures / social responsibility
- Research, analytical thinking, problem-solving, and improvement capabilities
- Innovation and entrepreneurship

Given Cross-Cultural attributes missing from the top eight list, and the observations of the focus groups, the missing Cross-Cultural attributes, of importance to this research, are listed below[3]:

- Demonstrates an understanding of political, social, and economic perspectives
- Demonstrates an understanding of the ethical and business norms and applies norms effectively in a given context (organization, industry, country, etc.)

- Possesses an international/global perspective
- Possesses fluency in at least two languages
- Embraces an interdisciplinary/multidisciplinary perspective

Systems Engineering Expert Skills

Since 2012, the Helix team in the Systems Engineering Research Center (SERC) has been investigating what makes systems engineers effective. The team has conducted in-depth interviews with 288 individuals from 20 organizations to support this work: 263 systems engineers from a mix of the defense industry and commercial industry (healthcare, transportation, telecommunications, etc.). Validation exercises were conducted with 25 individuals who work with systems engineers (program managers, classic engineers, etc.) across a variety of organizations. Each organization participating in Helix was asked to confirm that the individuals who participated in the study were considered effective; organizational representatives assessed all individuals from "average" to "very high" effectiveness.

This data is being used to build a theory of what makes systems engineers effective, called *Atlas*, preliminary versions of which were published in 2014 and 2015.[4,5] *Atlas* includes consideration for proficiency – an individual's level of knowledge, skills, abilities, demonstration of behaviors, and cognitions – which are critical to the effectiveness of systems engineers. From the data, the Helix team has built a model of six proficiency areas, as shown in Figure 1.

The colors in Figure 1 delineate the types of proficiencies required: purple (top right pair) highlights the skills for engineering; blue (bottom pair) for engineering systems; and orange (top left pair) for working with people to engineer systems. Within each of these pairs, working clockwise the first item is an enabler for the second. For example, the fundamentals of math/science, and general engineering enable an individual to understand and build skills within the context of a specific domain and set of operational parameters. Each of the proficiencies are described briefly below[5]:

- ***Math/Science/General Engineering***, which includes the foundational concepts from mathematics, physical sciences, and general engineering required to perform basic engineering and to understand the work of other engineers.
- ***System's Domain & Operational Context***, which includes the relevant domains, disciplines, and technologies for a *given* system and its operation. This will change depending on the types of systems being developed.
- ***Systems Engineering Discipline***, which includes the foundation of systems science and systems engineering knowledge; these are the elements thought of as "traditional" systems engineering and include methods, processes, and tools.
- ***Systems Engineering Mindset***, which includes the skills, behaviors, and cognition associated with being a systems engineer, in particular the ability to view things holistically or understand multiple perspectives.
- ***Interpersonal Skills***, which are skills and behaviors associated with the ability to work effectively in a team environment and to communicate and coordinate across the problem domain and solution domain.
- ***Technical Leadership***, which includes skills and behaviors associated with the ability to create and guide a diverse team of experts toward a specific technical goal.

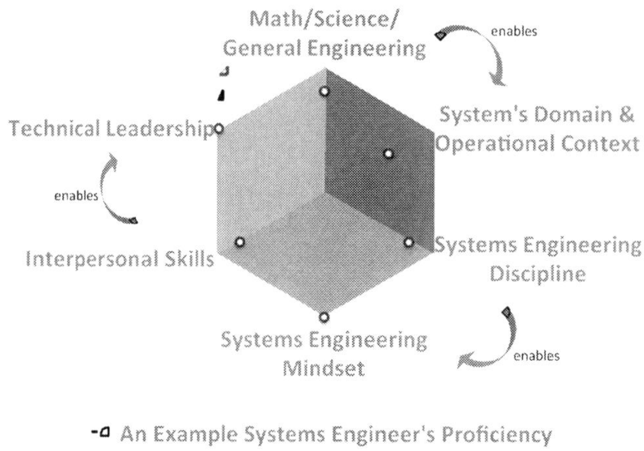

Figure 1. Helix proficiency model.

The proficiency areas that focus on working with technology – Math/Science/General Engineering, System's Domain and Operational Context, and Systems Engineering Discipline – were seen as important foundations for systems engineers to understand the systems being created and the work done by other engineers across diverse teams. However, chief systems engineers and non-systems engineers consistently cited the proficiency areas that focus on working with people – Interpersonal Skills and Technical Leadership – as areas where individuals *must* excel to be effective. Those who were skilled in the technical aspects of systems engineering only were seen as less effective and as less likely to become chief systems engineers. In terms of SE Mindset, *big-picture thinking and paradoxical mindset – the ability to simultaneously understand and hold in opposition differing views – were seen not only as critical skills, but also provided value and critical insights directly to projects*, according to the all of the non-systems engineers interviewed.

Integrating Systems Education into Undergraduate Engineering Education Workshops

In 2015, engineering and systems engineering faculty met at an International Council on Systems Engineering (INCOSE) Academic Forum in May and at the American Society of Engineering Education (ASEE) annual conference in June to discuss the need for integrating systems engineering knowledge into the education of all engineers. As part of these workshops, twenty-nine systems engineering knowledge areas of importance were identified and placed into four main categories, defined in the ASEE workshop[6] as:

1. ***Systems Science and Fundamentals:*** Formal science dealing with the nature of systems; the science "behind" systems engineering. (5 areas)
2. ***Systems Thinking:*** Holistic thinking: critical thinking guided by systems theory. (7 areas)
3. ***Design and Analysis:*** The examination of a problem from all aspects and the development of an effective solution. (10 areas)
4. ***Technical and Project Management:*** Applying knowledge, skills, tools, and techniques to manage the cost, schedule, and technical aspects of a project to meet the project requirements. (7 areas)

Between the two workshops, thirty-six faculty members participated through a voting scheme to determine the top system knowledge differentiators in undergraduate engineering education listed in Table 2. Each differentiator listed received from 3 to 30 votes. Using the final vote counts for the top 13 differentiators, relative importance is shown in the % column.

Table 2. Results of Voting – Top 13 System Knowledge Differentiators (adapted[6])

Categories	System Knowledge Differentiators	%
Systems Thinking	Interdependencies (interactions, interfaces, relationships) among multi-disciplines	28%
Systems Thinking	Problem Analysis (goals and objectives, needs statement, requirements elicitation)	12%
Design and Analysis	System Modeling	8%
Design and Analysis	Trade off and Decision Analysis	8%
Design and Analysis	Requirements Analysis	8%
Systems Science	Understanding Complexity	8%
Design and Analysis	Systems Architecture and Analysis	6%
Design and Analysis	Systems Integration	5%
Systems Thinking	Total Life-cycle View	5%
Systems Science	Systems Theory	4%
Technical and Project Management	Dealing with Uncertainties and Change	4%
Systems Science	Emergence	3%
Systems Thinking	Multiple and holistic perspectives	3%

Proposed Pathway to Systems Education for the Global Engineer

To build a pathway, first the previous research outcomes, shown in Table 3 by research area, are integrated into a set of categories, which capture the essence of the findings.

Table 3. Consolidation of Findings from Mutually Exclusive Research Teams

Thinking/Mindset Approaches	
Global Engineer	Possesses the ability to think both critically and creatively; and individually and cooperatively
SE Expert	Systems Engineering mindset; Ability to view things holistically
Integrating SE	Interdependencies (interactions, interfaces, relationships) among multi-disciplines; Problem Analysis (goals and objectives, needs statement, requirements elicitation)
Multiple Perspectives	
Global Engineer	Possesses an international/global perspective; Demonstrates an understanding of political, social, and economic perspectives; Embraces an interdisciplinary/multidisciplinary perspective
SE Expert	Systems Engineering mindset; Ability to understand multiple perspectives
Integrating SE	Total Life-cycle View; Multiple and holistic perspectives
Understanding Context	
Global Engineer	Demonstrates an understanding of the ethical and business norms and applies norms effectively in a given context (organization, industry, country, etc.)
SE Expert	Understands system's domain and operational context including relevant domains, disciplines, and technologies for a given system and its operation.
Integrating SE	Understanding Complexity; Systems Theory; Emergence

Foundational Technical Skills	
Global Engineer	Demonstrates an understanding of engineering, science, and mathematics fundamentals; Demonstrates an understanding of information technology, digital competency, and information literacy
SE Expert	Understands foundational concepts from mathematics, physical sciences, and general engineering required to perform basic engineering; Understands the work of other engineers; Understands the foundation of systems science and systems engineering knowledge, including methods, processes, and tools.
Integrating SE	System Modeling; Trade off and Decision Analysis; Requirements Analysis; Systems Architecture and Analysis; Systems Integration
Personal/Interpersonal Skills	
Global Engineer	Communicates effectively; different ways, methods, and media (written, verbal/oral, graphic, listening, electronically, etc.); Shows initiative and demonstrates a willingness to learn; Functions effectively on a team; Maintains a positive self-image and possesses positive self-confidence; Possesses fluency in at least two languages
SE Expert	Has ability to work effectively in a team environment and to communicate and coordinate across the problem domain and solution domain; Includes skills and behaviors associated with the ability to create and guide a diverse team of experts toward a specific technical goal.
Integrating SE	Dealing with Uncertainties and Change

Next, a proposed pathway to support the maturation of the global engineer of 2025 and beyond was developed using the same categories from Table 3 and additional background research on building system concepts starting in Kindergarten. The approach is applied over the lifelong education pathway of the global engineer and is presented in Table 4.

Table 4. Proposed Pathway to Systems Education for the Global Engineer (Thinking/Perspectives/Context adapted[7])

Early Years: 5 - 10 years	*Thinking/Mindset Approaches*	Understands linear cause and effect; recognizes simple patterns.
	Multiple Perspectives	Understands simple concepts (e.g. sharing) from another person's viewpoint; identifies internal feedback related to simple continuous processes in one's own daily events.
	Understanding Context	Recognizes relationships of parts with a context (e.g. member of a family); distinguishes internal from external feedback.
	Foundational Technical Skills	*Follow STEM guidelines.*
	Personal/Interpersonal Skills	Exposed to group work; begins second language; uses more than one media for written and verbal communication.

Middle Years: 11 - 13 years	*Thinking/Mindset Approaches*	Recognizes simple non-linear causal relationships with real-world variables; graphs behavior over time; understands and uses causal loop diagrams and system archetypes; identifies system goals and develops more complex pattern recognition.
	Multiple Perspectives	Runs and analyzes case studies; identifies states and state changes.
	Understanding Context	Recognizes and defines context; relates impact of changes in environment to change in familiar behaviors.
	Foundational Technical Skills	*Follow STEM guidelines.*
	Personal/Interpersonal Skills	Continues group work; achieves elementary level of second language; uses multiple media for written and verbal communication.
Teenage Years: 14 – 17 years	*Thinking/Mindset Approaches*	Recognizes system archetypes; generates system models and deduces system behaviors.
	Multiple Perspectives	Creates system models with internal and external feedback and with multiple states and state changes.
	Understanding Context	Rigorously analyzes and tests system models for robustness (level of insensitivity to changes in environment/context).
	Foundational Technical Skills	*Follow STEM guidelines.*
	Personal/Interpersonal Skills	Understands importance of communicating effectively across multiple media; participates in positive team experiences; achieves middle school level of second language; achieves success in most subjects due to developed intrinsic motivation to learn.
Young Adult: 18 – 21 years	*Thinking/Mindset Approaches*	Understands more than one discipline in sufficient depth to apply a multidisciplinary approach to system model development with accurate modeling of interactions, interfaces, and relationships; reliably predicts system behaviors when changes are made.
	Multiple Perspectives	Integrates a total life cycle view and multiple perspectives into design approaches; able to optimize solutions to address the right problem, factoring in both development and operational considerations.
	Understanding Context	Applies systems theory concepts including complexity and emergence into system modeling and analysis.
	Foundational Technical Skills	Achieves application level of: system modeling; trade off and decision analysis; requirements analysis; systems architecture and analysis; and systems integration.

	Personal/Interpersonal Skills	Achieves application level for communicating effectively across multiple media and team effectiveness and in two or more languages; values lifelong learning; possesses self-confidence and positive self-image; effectively deals with uncertainty and change.
Adult: > 21 years	*Achieves Knowledge and Skills listed in Table 3.*	

Conclusion

As the community comes together to mature the building of the pathway to systems education for the global engineer, the principles that we believe are critical for success in the undergraduate systems education of global engineers focus on individualized, agile, experiential, and learner centric processes where multiple disciplines are integrated in a setting that recreates the practicing environment. Our pathway to developing a global engineer starts with systems thinking concepts being introduced alongside the traditional reductionist and critical thinking approaches found in our standards today, and continues through the pursuit of lifelong learning.

Acknowledgements

The authors wish to thank several teams for the foundational contributions to this work. These teams include the ASEE Attributes of the Global Engineer team, the INCOSE Vision 2025 team, the Systems Engineering Research Center (SERC) Helix team, and the INCOSE Academic Forum team.

References

1. Friedenthal, S., Beihoff, B., Nichols, D., Oster, C., Paredis, C., Stoewer, H, Wade, J. (2014, October), A World in Motion: Systems Engineering Vision 2025, International Council on Systems Engineering, http://www.incose.org/docs/default-source/aboutse/se-vision-2025.pdf?sfvrsn=4
2. Hundley, S. (2015a, June), The Attributes of a Global Engineer: Results and Recommendations from a Multi-Year Project Paper presented at 2015 ASEE International Forum, Seattle, Washington. https://peer.asee.org/17160
3. Hundley, S. (2015b, June), The Attributes of a Global Engineer Summary presented to the Corporate Member Council at the ASEE 2015 Annual Conference and Exposition, Seattle, Washington.
4. Pyster, A., P. Dominick, D. Henry, N. Hutchison, C. Lipizzi, M. Kamil, S. Manchanda. 2014. Atlas*: The Theory of Effective Systems Engineers, version 0.25*. Hoboken, NJ: Stevens Institute of Technology. SERC-2014-TR-038-4. November 21, 2014.
5. Pyster, A., D. Henry, N. Hutchison, C. Jauregui, M. Clifford. 2015. Atlas: *The Theory of Effective Systems Engineers, version 0.5*. Hoboken, NJ: Stevens Institute of Technology. SERC-2015-TR-108. December 1, 2015, http://www.sercuarc.org/wp-content/uploads/2014/05/Helix-Report-Atlas-0.5-December-2015.pdf
6. Squires, A., Looft, F., Virani, S.S. (2015), "Integrating Systems Education into Undergraduate Engineering Education plus INCOSE Spring Academic Forum Summary Workshop Report", Report of Workshop delivered at the 2015 American Society for Engineering Education (ASEE) Annual Conference and Exposition, Seattle, Washington, June 14, 2015.
7. Squires, A., 2007, *Systems Thinking and K12 Education*, Doctoral Qualifying Examination, Stevens Institute of Technology, School of Systems and Enterprises, November 21, 2007.

A Localized National Engineering Education and Research Outreach Model for Engineering Workforce Pipeline

Dr. Otsebele E Nare, Hampton University

Otsebele Nare is an Associate Professor of Electrical Engineering at Hampton University, VA. He received his electrical engineering doctorate from Morgan State University, Baltimore, MD, in 2005. His research interests include System-Level Synthesis Techniques and K-16 Integrative STEM education.

Vitaly Khaykin, Hampton University
Dr. Hoshang Chegini, Hampton University

Dr. Chegini received his doctorate, masters, and bachelors' degrees in Mechanical Engineering from Old Dominion University, Norfolk, Virginia. His academic background is in thermal science of mechanical Engineering and his areas of expertise is in multi-phase computational fluid dynamics (CFD). Dr. Chegini joined Hampton University School of Engineering & Technology as an associate professor in September 2002. Before joining Hampton University, he served as a faculty and chair of Mechanical Engineering department at K.N.T. University of Technology in Tehran, Iran for over 12 years. Currently he is working as a CO-PI on a research project funded by Department of Energy.

Mrs. Chandra T Oaks-Garcia, Time Out 4U, Inc.

Professional Biography

Chandra T. Oaks-Garcia is a secondary technology teacher in the Career and Technical Education Program. She earned her B.S in Mechanical Drafting and Design Technology from Alabama Agriculture and Mechanical University and her M.Ed. from Averett University. She is currently studying postgraduate STEM Education at Old Dominion University. Mrs. Oaks-Garcia's expertise is in the field of Computer-Aided-Design. She has over twenty-five years of professional experience as a Senior Designer with extensive knowledge in 3D Modeling Software and Computer-Integrated-Manufacturing collectively in the areas of Civil, Electrical, and Mechanical Design.

Mrs. Oaks-Garcia has supports her local school districts within the community working with the National Institute of Aeronautics (NIA), Hampton, Virginia as a "Teacher In Residence" in 2009. She served on conference planning committees for VTEEA and TSA locally over a period of time. She also worked with the Division's Career Academy Team for Hampton City Schools in 2010.

Mrs. Oaks-Garcia has proudly served a community of learners in the Hampton City School District in various capacities for the past fifteen (15) years. For her teaching strategy, Mrs. Oaks-Garcia incorporates real world learning into the classroom to promote and enhance student's critical thinking and problem solving skills. In September 2001, her Future Engineering Team consisting of 11th and 12th grade students of Hampton High School tackled a real-life engineering challenge following the terrorist attacks in three U.S. locations by designing, manufacturing and defending solutions to increase airline safety after 911. Mrs. Oaks-Garcia and her engineering students were featured in the Hampton Daily Press and on WAVY TV in 2001, and also in the NASA Newsletter in March 2002 for their innovative design solutions.

Mrs. Oaks-Garcia currently teaches STEM and Technology Education at the secondary level. She serves as a mentor for Hunter B. Andrews's team of STEM students in the Virginia Ship Repair Foundation Annual Competitions. In 2010, the STEM students won 1st place with their futuristic Ship Design. Currently, Mrs. Oaks-Garcia is the facilitator of the after school CAM program at Hunter B. Andrews Pk-8 School. She is committed to empower the next generation of 21st century learners at an early age through STEM.

Mrs. Oaks-Garcia is the founder and CEO of Time Out 4U Incorporated, Youth Engineering Program, established in 2013. She serves as the Administrator for Operations and oversees activities. The organization is involved in grant-supported projects focused on Computer Additive Manufacturing and STEM Outreach for after school programs. The organization sponsors an annual STEM Symposium.

Mrs. Oaks-Garcia is from Mobile Alabama and currently resides with her husband in Hampton, Virginia and their four sons.

Dr. Vadivel Jagasivamani, Hampton University

A Localized National Engineering Education and Research Outreach Model for Engineering Workforce Pipeline

Abstract

A consortium of eight universities and three US Department of Energy (DOE) national laboratories was established in 2014. This consortium is implementing a model for industry, academia, and government partnership focused on pipeline development of skilled workforce through advanced manufacturing. This paper reports on the expansion of STEM outreach activities with a focus on advanced (additive) manufacturing for K-20 as well as the enhancement of education and training of individuals as implemented in one of the consortium member institutions. The outreach components involved the training of undergraduate students through summer exchange at universities as well as through summer internship placement at national laboratories. A local outreach to middle / high schools was established through the implementation of an advanced manufacturing skills development after-school program for middle school students. The paper also presents the model curriculum for the outreach program and shows the integration of undergraduate students and K-12 teachers working together.

Introduction to MSIPP Program

In 2012, the National Nuclear Security Agency (NNSA) established the Minority Serving Institutions Partnership Program (MSIPP). This program is a consortium-based program with the principal goal of establishing a sustainable career pipeline for minority students pursuing Science, Technology, Engineering, and Mathematics (STEM) degrees. The collaboration between Historically Black Colleges and Universities (HBCUs) and DOE/NNSA sites and labs in STEM disciplines in the area of advanced manufacturing brings a heightened awareness of nationally important research at NNSA plants to HBCUs with a common interest in STEM research fields. To attain the principal goal of establishing a sustainable HBCU/MSI workforce pipeline of STEM professionals, the MSIPP is focusing on the following four areas:

- Building and maintaining a skilled STEM pipeline of students that will be capable of meeting future DOE/NNSA needs;
- Establishing collaborative efforts between the HBCUs and the DOE plants and labs to foster advanced research;
- Establishing collaborative relationships between the HBCUs and the DOE plants and labs to enhance the education and training of individuals in advanced manufacturing and technology development;
- Establishing collaborative relationships between the HBCUs and the DOE plants and labs to expand the STEM outreach opportunities with a focus on advanced manufacturing for K-16 students.

Several consortiums were established by this program and Hampton University is currently a member of the Consortium for Advanced Manufacturing (CAM). This paper focuses specifically on the outreach component of the consortium with a purpose to educate and introduce students into the STEM field by providing innovative and interactive program focused on the advanced manufacturing topics.

Collaboration Structure

NNSA developed the initial structure by creating eight different consortiums, however later, in 2014, the decision was made to reduce the number of consortiums from eight to four to avoid duplicate effort. The two Advanced Manufacturing consortiums were merged into one consortium. A lead university and several other universities along with a design agency and/or production site were assigned within each consortium. The CAM is comprised of the following HBCUs and industry partners:

- North Carolina A&T (NC A&T) – Lead University
- Alabama A&M (AAMU)
- Clark Atlanta University (CAU)
- Hampton University (HaU)
- Honeywell FM&T Kansas City Plant (KCP)
- Howard University (HoU)
- Lincoln University (LU)
- Oak Ridge National Laboratory (ORNL)
- Southern University of New Orleans (SUNO)
- University of the District of Columbia (UDC)
- Y-12 National Security Complex (Y-12)

Outreach Component Goals

The intent of the CAM STEM pipeline outreach component is to improve STEM educational experiences of students in K-16 so that participants increase their employability and enter the job market with experiences in STEM gained from a young age. The CAM provides a host of professional expertise and resources that allow middle, high school, undergraduate and graduate students to experience a myriad of STEM related activities at the various campuses. The outreach component of the CAM assists individual universities to plug gaps that currently exist in their individual STEM programs and activities targeted at their respective communities, stakeholders, and future student populations. The inclusion of KCP, Y-12, and ORNL strengthens the CAM partnership by providing students with opportunities to expand their educational experiences via lab visits, career exposure, and faculty-student internships.

Outreach Project Goals: The pipeline focus of CAM is the enhancement of the workforce pool. All projects supported by the CAM outreach component meet one or some of the project goals as outlined below:

- To increase the number of students, internships, programs, and collaborative research between CAM partners;
- To provide and preserve educational opportunities for underrepresented students in STEM in grades K- 16;
- To maintain/raise academic standards by enabling learners to acquire competencies needed for success in college;
- To provide opportunities for students in grades K-16 to make connections between in-class curricula and its significance in a real world setting;

- To provide opportunities for students in grades K-16 to develop skills and attitudes necessary for the attainment of academic, career, and life goals;
- To provide opportunities for students in grades K-16 to understand and utilize the scientific and build/design methods of inquiry;
- To provide opportunities for students in grades K-16 to use appropriate technology related to Advanced Manufacturing and other STEM career fields;
- To ensure proper placement by assessing each learner's level of preparedness for secondary, post-secondary, and graduate school coursework.

Outreach Component Efforts at Hampton University

Developing a Regional High School Expansion Program Model: The goal of this program is to enhance relations between consortium members and regional schools by developing a comprehensive High School Expansion Model with an Advanced Manufacturing focus. As part of the Regional High School Expansion Program effort, Hampton University is currently working on the pilot Advanced Manufacturing STEM after school program at Hunter B. Andrews PK-8 school in the city of Hampton, VA. This pilot program is serving as one of the components for implementing the CAM outreach's regional pipeline schools expansion model. The program targets 15 to 18 students, who meet on twice a week (currently Mondays and Wednesdays from 4pm to 5pm) and is being coordinated by a designated STEM teacher at the school with the help of undergraduate Hampton University engineering students. The initial project effort was mainly local, but the project is expanding the collaborative reach by engaging two CAM industry partners. The program is projected to grow into the community of teachers and students outside Hunter B. Andrews through the mobility of TimeOut4U Inc. partnership.

Hampton University teamed up with TimeOut 4U, Inc. (*www.timeout4u.org*) - a non-profit focused on engineering programs for youth that strives to diversify the 21st century workforce by broadening STEM exposure and career readiness to minorities and other underrepresented youth during early academic years to explore other funding opportunities. The National Society of Black Engineers (NSBE) student chapter at Hampton University has also volunteered to help with the program. During the first year, a program curriculum was developed in collaboration with TimeOut 4U, Inc. The curriculum breakdown included the following general modules: overview of advanced manufacturing; career exploration; computer-aided design; creative problem solving; manufacturing career track; rapid prototyping & 3D printing; and workplace readiness skills. During the first year all the modules were delivered by the program coordinator and the undergraduate engineering students. The program is being enhanced to include module delivery by industry partners, and contributors from other collaborating university partners. Hampton University leads the project and coordinates the partnerships as well as seeks resources for sustainability. Collaborating partner universities provide material support based on past experiences in other outreach programs (e.g. relevant modules/projects). In the second year, Hunter B. Andrews will continue hosting the after-school program and TimeOut 4U, Inc. will disseminate the program through holding mobile workshops to the community and technology teachers. The industry partners will give online based workshops to the students via skype type of delivery. Each partner commits to delivering at least one module. The industry presentations to the students are aligned with the modules focused on careers.

The program activities require the participants to engage in STEM related modules demonstrating the infusion of engineering concepts and applied mathematics highlighting manufacturing advancements. Activities are planned for the academic year period for a minimum of 50 hours. Advanced manufacturing activities that have been developed at other consortium universities for the summer programs are incorporated into the program as well as the advanced manufacturing program modules developed specifically for the program. The program coordinators focus on:

- Coordinating program activities and projects;
- Developing and ensuring quality project activities;
- Recruiting program participants;
- Maintaining and tracking data on student participants;
- Planning, implementing, and managing regular activities;
- Enhancing the program curriculum materials;
- Training and supervising the undergraduate students and volunteers.

As an example, the program modules connect the engineering design concepts with practical applications in real-world engineering, where student perform problem solving, design, manufacturing, and testing of products, and the program coordinators train the students in CAD tools and 3D printing technology.

Students begin by learning manufacturing and rapid prototype processing using the 3D printer. To capture the imagination of middle school students and build interest in manufacturing, students sketch and design a useful object that will be made using a 3D printer. Opportunities are introduced to allow students to create, refine, and/or assemble a number of items, giving their work an end result they can hold in their hands.

Program Curriculum:

Module #1:	Introduction Week - What is manufacturing – Reverse Engineering/Systems - Explore current and emerging careers in manufacturing
Module #2:	What Makes A Great Print? - Design Concepts - Problem Solving Skills
Module #3:	Explore and Declare Your Idea! - Engineering Design Process - Eco-Green - Newton Laws of Motions – Energy Transfer
Module #4:	Creating a Design: - Computer Software/Applications (Inventor/AutoCAD) - Technical Drawing
Module #5:	What is 3D Printer/Additive Manufacture? - Identify what 3D printers currently can and cannot print.
Module #6:	How to Print I: Explain how 3D printer technology works - Limitations of 3D Printing

- Working within Constraints

Module #7: How to Print II: Making a Connection
- Computer-Integrated Manufacture
- Input-Process-Output

Module #8: Rapid Prototype and Real World Testing
- Quality and Efficiently
- Functionality

Module #9: Modify and Market
- Communicate
- Feedback

Module #10: Explore Types of Manufacturing, and Related Careers
- Colleges and Universities
- Companies and Opportunities

Some of the results obtained from implementing the curriculum above included the students brainstorming and selecting a popular toy design to demonstrate what a 3D printer can do. The process of the activity included utilizing 3D modeling software such as AutoDesk Inventor. They were able to implement the engineering design process to generate rapid prototypes and concluded with real world testing. Once all components were manufactured and assembled, they performed real world testing for functionality, ergonomics, and efficiency. Students were effective in problem solving, modifying to improve upon their 3D printed product. Excitement and curiosity grew among the students when they saw their very own drawing becoming a reality within hours. Students have gained a clear concept of STEM related educational pathway to careers in advanced manufacturing engineering from their very own detail research and oral presentations. Images of their toy product are shown below in figures 1, 2, and 3.

Figure 1: Example toy product designed and printed by students.

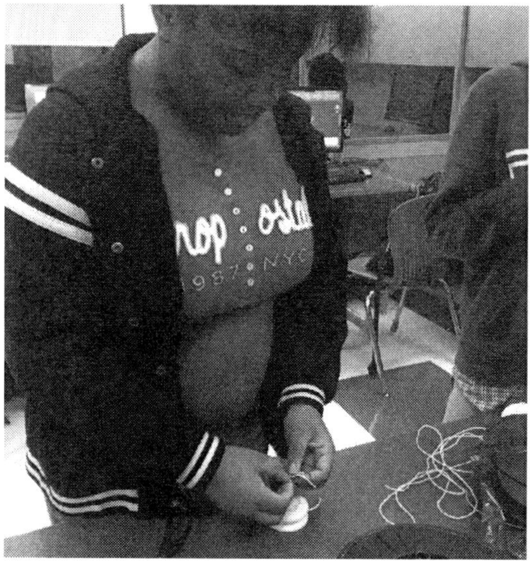

Figure 2: Assembling of toy product components.

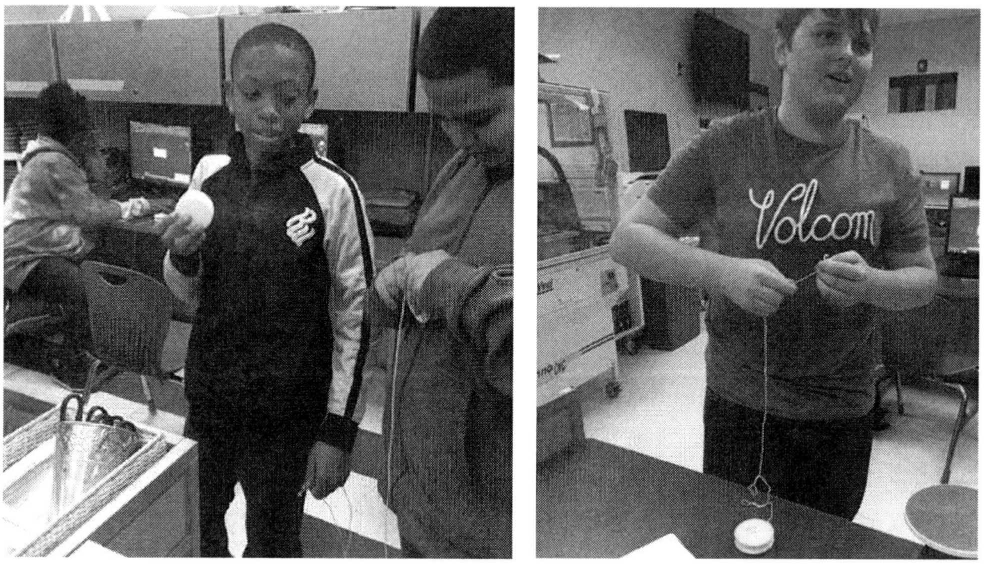

Figure 3: Toy product testing.

Summer Exchange Program: The consortium sponsors students from partner universities to enhance their skills by exposing them to workforce skills that may be unavailable at their home institutions. The program also assists partner institutions in enhancing their outreach program offering through training undergraduate students to support programs at their home institutions as well as co-advising. Hampton University initiated an undergraduate summer exchange program with Lincoln University. The students assist in a two-week summer STEM program for a period of up to four weeks. During the summer assignment, students primarily serve as one of the program counselors for the Business of Engineering summer program at Hampton University along with engineering and business undergraduate students. As part of the responsibilities, students:

- Assist in preparing project activities for the Business of Engineering Summer program at Hampton University;
- Assist in project enhancements and implementation;
- Are available each program day from 8:45am- 4:15pm;
- Attend each field trip and assist with logistics;
- Accompany students to cafeteria for lunch;
- Assist and oversee completion of student projects and activities;
- Serve as a role model by limiting use of electronics during instructional periods;
- Organize and oversee HU scavenger hunt;
- Encourage positive team dynamics;
- Organize/create program newsletter and assist with closing ceremony program (i.e. superlatives, assist students with PowerPoint, overall program organization);
- As a team, complete program summary in paper format (paper to include literature review, data collection from surveys, program synopsis);
- Learn about the CAM after-school outreach program on advanced manufacturing;
- Learn the operation of Ultimaker2 3D printer.

Industry Day: The consortium outreach model also offers an Industry Day initiative where workshops and lectures in advanced manufacturing and/or other engineering areas of interest are presented to students and faculty of university partner outreach programs. Industry personnel from industrial partners are invited to university partners for a day to connect with students and partners as well as learn about collaboration opportunities. At Hampton University, the industry day plan involves having workshop presentations on modeling and simulation to students and faculty as well as presentations on advanced manufacturing to middle school students.

Tutoring: The outreach program is extended to most of the undergraduate students through tutoring programs offered by each university partner. The tutoring program at Hampton focuses on engineering courses mainly first year and second year courses. Third and fourth year courses are also considered for tutoring based on the availability of tutors.

K-12 STEM Enrichment Program: The teachers in the K-12 schools of partner universities are also offered opportunities for professional development. The university partners recruit teachers from their region to attend programs such as the SECME Summer Institute. Hampton University has supported a local school teacher to attend SECME and the teacher supports the regional high school expansion program in turn. SECME has a 40 year legacy of being a professional development Institute dedicated to focusing on equity in STEM initiatives by bringing together K-12 educators, university faculty, and industry and government experts promoting inclusion in learning and doing STEM. The selected K-12 teachers receive training on the current pedagogies of STEM and follow up with strategies to promote STEM in their schools and region through formal and informal learning environments.

Outcomes and Assessment

Several outreach initiatives have been started at Hampton University as part of the CAM project supporting the broad goal of establishing collaborative relationships between the

HBCUs and the plants and labs to expand STEM outreach opportunities with a focus on advanced manufacturing for K-20 students. These included a Regional High School Expansion program, Summer Exchange program, Industry Day, and Tutoring program. The general outcome for all these projects is to inspire the students to become engineers or to at least to delve into the STEM field and these programs help brace students for their career path in the STEM field by challenging them with obstacles that require an engineering design process.

The outreach component committee within the CAM holds an annual Program Assessment conference with the purpose of presenting the oral and poster program progress reports, discussing the effectiveness of the projects, discussing plans of sustaining the program, and providing an overall guidance for the outreach component. The metrics for the assessment include the number of students targeted by the outreach program, assessment of the knowledge gained by the students, their feedback regarding the program, and the student retention rate within the STEM field.

Conclusions

A collaboration program has been established between eight HBCUs and three DOE/NNSA plants and national labs with the goal of building a successful model for sustainable workforce pipeline within the STEM field with applications for advanced manufacturing topics. The outreach part of the consortium has been tasked with developing programs and projects that would involve K-20 students and attract and retain them in the STEM area. Several such projects have been developed and initiated at Hampton University, including the Regional High School Expansion, Summer Exchange, Industry Day, and Tutoring programs. It is possible to use this model approach in local, national, and international engineering/STEM collaborations.

Bachelor Degree Program "Software Engineering" in the Higher School of Economics: Background and Perspectives

Prof. Sergey M Avdoshin, National Research University Higher School of Economics
Dr. Valery V. Shilov, National Research University Higher School of Economics, Moscow, Russian Federation

Valery V. SHILOV

Born:

August 18, 1954 in Moscow

Education:

Secondary school 13, Moscow Region, Gold medal, 1971 Moscow State University, Diploma in applied mathematics, 1976 Higher Institute of management (Moscow), Diploma with distinction in law, 2001

Candidate of technical sciences degree, 1989

Affiliations:

1976-1989: Researcher, Central Scientific Research Institute 45 of Ministry of Defense 1989-1997: Senior Researcher, Scientific Computer Centre Russian Academy of Sciences 1997-2004: Full professor, Computer Systems Design Department, MATI – Russian State Technological University 2004-2015: Head of Computer Systems Design Department, MATI – Russian State Technological University 2015-: Head of Computer Systems Design Department, Moscow Aviation Institute (National Research University) (part time) 2015-: Full professor, Academic supervisor of BS program in Software Engineering, National Research University Higher School of Economics (full time)

Member of the Editorial board, "Informacionnye Technologii" magazine Member of the Editorial board, "Kibertonia" magazine (Kiev, Ukraine) Scientific Advisor, Great Russian Cyclopaedia Member of Advisory board, Russian Virtual Computer Museum

Member IT History Society

Invited lector, Moscow Institute of Physics and Technics ("Fiz-Tech"), 1999-2011 Invited lector, Hanoi University of Technology (Hanoi, Vietnam), 2005-2008 Invited lector, Hanoi Open University (Hanoi, Vietnam), 2009-2012

Awards:

IEEE Computer History Competition CHC'60 Award, 2006 IEEE 2010 Student Competition Best Architecture Prize, 2010 Yuri Gagarin medal from Federation of Cosmonautics of Russian, 2010

Books (in Russian):

Informatics (Moscow, 2003) (with coauthors) The Career in IT (Moscow, 2003) (with coauthors) The Chronicle of computational and information technologies. People. Events. Ideas. Parts 1-4 (Moscow, 2004-2007) Charles Stanhope. Scientific biography (Saint-Petersburg, 2011) (with Yu. Polunov) This wonderful history of informatics (Moscow, 2011) The History of Logical machines (Moscow, 2014)

Scientific papers:

More than 300 scientific papers in parallel architectures, compilers, history of computing etc. in Soviet, Russian and International editions.

Conferences:

Member of Organizing Committee, "SoRuCom 2011" IFIP Conference (Novgorod, Russia) Member of Organizing Committee, International Gagarin Conference (Moscow, Russia, 2005-2015) Member of Organizing Committee, IV International Anatoly Kitov Conference (Moscow, Russia, 2014) Member of Program Committee, "SoRuCom 2014" IFIP Conference (Kazan, Russia) Member of Program Committee, V International Anatoly Kitov Conference (Moscow, Russia, 2015)

Dr. Sergey A. Silantiev

BACHELOR DEGREE PROGRAM "SOFTWARE ENGINEERING" IN THE HIGHER SCHOOL OF ECONOMICS: BACKGROUND AND PERSPECTIVES

Sergey M. Avdoshin, Valery V. Shilov, Sergey A. Silantiev

National Research University Higher School of Economics
Moscow, Russian Federation

savdoshin; vshilov; ssilantiev; @hse.ru

The Higher School of Economics (HSE), Moscow, Russia, was established in 1992 with the aim to promote economic and social reforms in Russia through education of a new generation of researchers and practitioners. Currently, HSE has a prestigious status of National Research University and is considered as a multi-discipline Centre of study and research. It established itself as a leading university in Russia in the field of economics, social and political sciences, informational technologies and mathematics. It is ranked among the top-three most popular universities in Russia. Our University enjoys effective partnership with leading foreign universities, international programs and organizations, research consortia and scientific periodicals/editions, foreign companies and transnational corporations (over 130). The HSE is rapidly developing university which actively perceives the most advanced ideas. So it was with software engineering.

Software engineering is a relatively young scientific discipline. For the first time, the term "Software Engineering" was proposed in 1968 at the NATO conference at Garmisch-Partenkirchen devoted to the so-called "software crisis" that has arisen with the development of computers of the third generation, allowing the realization not previously implemented software projects. There was a need for new technologies and methods of management for the development of complex large software systems. In Russia, the HSE in 2006 began to develop first domestic standard on software engineering. In 2010 it was adopted by Ministry of Education and Science of Russian Federation and our University started to train specialists in this field. Other Russian universities began to enroll students on this specialty only next year.
At the present time, the bachelor degree program "Software Engineering" is realized on the Faculty of Computer Science at the Department of Software Engineering established at HSE in 2014. The aim of this program is training of technical experts, skilled developers, software architects and software quality managers. The need for such specialists is dictated by the ever-growing demands of the information and communication industry: the organizing of IT-technology parks, the rapid development of offshore market and custom programming, computerization of government agencies, the private commercial business needs, etc. It also worth to mention that HSE is the only Registered Education Provider of the IEEE Computer Society in Russia.

The educational BS program "Software Engineering" is fully consistent with international guidelines on the teaching of software engineering in higher education: Computing Curricula 2005, Computer Science Curricula 2001/2008/2013 and Software Engineering 2004/2014. It covers all aspects of the world's best knowledge and practice in software production: mathematical, methodological, economical, legislative, marketing, etc. In 2011 our program was honored the prestigious IBM Faculty Award.

Our students study core disciplines such as mathematical analysis, algebra, discrete mathematics, probability theory and mathematical statistics, statistical and empirical methods of computing, programming, data structures and algorithms, distributed computing, architecture of computer systems, operating systems and databases. Professional cycle includes software architecture, design of software systems, quality control and testing, software project management, psychology in IT, group dynamics and communication in professional practice of software engineering, economics of the firm, economics of software engineering, intellectual property law and some others. Above this, every student chooses elective disciplines, research seminars, computer practicums, and university-wide minors. The study of English language is of great importance. Future specialists are expected to work in international environment and be acquainted with the most modern world achievements in computer science. Some disciplines of professional cycle are taught in English as well.

The benefits of the program are the following:

1. *Combination of theoretical training with practical work experience in companies.*
We have partner relations with more than 60 leading domestic (Yandex, Kaspersky Lab, etc.) and international (IBM, Microsoft, etc.) companies. It gives our students the opportunity to get knowledge in a wide range of real IT-projects during educational and technological internship, professionally oriented and graduation practices. Among the lecturers there are representatives of business sphere which promotes the integration of training and practice, the application of business ideas in the content of the SE courses. During the Sophomore Year students are encouraged to fulfill the software development projects in teams in framework of the discipline "Group Dynamics and Communications in Software Engineering Professional Practice" where real situations of team development and interaction of team members are considered. During the Senior Year students learn the discipline "Group Project on Software Engineering" where students' teams develop a software product in cooperation with the customer of one of the partner companies in the real conditions. Thus, the educational and scientific process is focused on practical activities. Students throughout the training are acquainted with the real conditions of practical activity and many of them are invited to work in companies upon the results of team projects.

2. *High level of training*
Highly qualified staff from the Software Engineering Department and some other Higher School of Economics departments and faculties takes part in educational process. In addition, the staff of high-tech companies participates in the teaching as well. The representatives of leading Russian research organizations in the field of IT including the Institute for System Programming, Institute for Systems Analysis and Institute for Information Transmission Problems of the Russian Academy of Science are also involved in the training.

3. *Individual educational tracks.*
It is possible for students to choose various educational tracks by selecting elective disciplines, research seminars, computer practicums and university-wide minors.

4. *Active project and research work.*
Projects are the significant part of the entire educational program. Students can use all the resources of the Higher School of Economics as a research University, as well as partner companies. During the training students obtain the individual and teamwork skills in typical software development. Above this, they participate in real projects of leading Russian and foreign companies specializing in the software development, consulting and software project

management. Students from the first year of study involve in the research seminars of various directions, perform interdisciplinary course work. For the implementation of course projects, students use the knowledge and expertise gained during the study of the program disciplines. At the same time the themes of course works would suggest that the students have to find independently their own methods to apply acquired knowledge and technologies. The mandatory part of a course work is the development of software technical documentation. Results of some course works are introduced in training process. At research seminars students make presentations on actual scientific topics, conduct research, develop team projects, and learn how to prepare scientific publications. All these factors allow our students to participate successfully in various national and international competitions.

5. *Work perspectives.*
Graduates of the SE program work as the developers and software architects, software quality managers in industrial companies, research centers and governmental agencies. Philosophically, the aim of the program leading to the Bachelor of Software Engineering degree is to provide students with a strong theoretical and practical background in computer software along with the engineering analysis, design and implementation skills.

In 2014, Higher School of Economics modified its educational standards on the bachelor degree level. The main purpose of the reform was to improve the quality of education by decreasing the number of courses students take simultaneously (4 to 5 + foreign language) and strengthening emphasis on project and research activity, to give a chance to students to follow university-wide non-core courses (so-called "minors"), etc. As a result of changes made in the program, it has acquired the unique character for the Russian educational system. At present time Freshman and Sophomore Year students follow modified syllabi, but Junior and Senior Year students still follow the previous syllabi, which will cease in 2017. But, it is worth to note that most of the core required disciplines, all research/project seminars, some elective disciplines remained in modified curriculum. Thus, BSSE program continuity is preserved.

BSSE program is administered by Program Academic Supervisor and Program Academic Board.

Program Academic Board:

- develops educational program, submits it for approval to Faculty Academic Board and further to the HSE Directorate of Basic Education Programs and Vice-Rector for Academic Affairs;
- clarifies the curriculum for each academic year and suggests teaching staff for the program courses;
- submits a petition to the Department Chair for replacing lecturers in case of unsatisfactory results of monitoring the quality of teaching of specific courses;
- approves the course syllabi;
- generates suggestions for changes in the program educational standard and curriculum, based on international experience in realization of similar educational programs abroad;
- considers reports monitoring students' opinions about the quality of teaching.

Program Academic Supervisor:

- organizes a work of Program Academic Office (PAO) and teaching staff;
- determines the tasks for Program Manager and PAO;

- organizes interaction between the specialists from Tutor's Center and students;
- cooperates with managers and staff of HSE structural subdivisions on matters relating to the realization of the program;
- monitors the quality of teaching.

Graduation requirements for the program are the following. The name of the degree awarded in this program is **Bachelor of Software Engineering** regardless of track. Students are expected to complete their education in a timely manner. Their progress is monitored through an online ASPA (Applicant, Student, Postgraduate Student, Alumni) System. The ASPA's reports or audit are available to the students. Graduating students are carefully monitored to ensure that the curricula has been followed and successfully completed. Mentor of Bachelor degree Graduate Qualifying Work (GQW) and PAO control constantly from December to April of the fourth academic year the process of GQW preparation to be sure all requirements for graduation have been met. They also provide on request of the student the necessary consultations, which may also include the GQW public defending procedures. All students are required to present a final GQW variant not later than 10 days before public defending date.

The SE degree requirements consist of:

- ✓ 64 credit hours of mathematics and science;
- ✓ 112 credit hours of engineering topics;
- ✓ 49 credit hours of general education;
- ✓ 15 credit hours of internships and practices.

Three to five years after graduating with this degree graduates will be able to:

1. Excel in a career utilizing their education in Software Engineering;
2. Continue to enhance their knowledge;
3. Be effective in multidisciplinary and diverse professional environments;
4. Provide leadership and demonstrate professional integrity.

These are the main program educational objectives which are related with program student outcomes. The general outcomes of the Software Engineering Program are as follows. Graduates must be:

1. Able to learn and acquire new knowledge, skills, including areas different from a professional one;
2. Able to identify the essence of scientific problems in professional area;
3. Able to solve problems in a professional activity using the methods of analysis and synthesis;
4. Able to assess the need for resources and use it in solving the problems in professional activity;
5. Able to work with information: locate, evaluate and use the information from various sources necessary for solving of scientific and professional problems (including a systematic approach);
6. Able to carry out research, analyze problem, formulate goals and objectives, select research methods and assess its quality;
7. Able to work in a team;
8. Able to communicate competently considering the objectives and situation;
9. Able to critically assess and reassess the experience (own and others);
10. Able to work in an international environment.

The following table maps the program educational objectives to the program outcomes. Every program outcome is related to one or more program educational objectives, and every program educational objective is related to at least one program outcome.

Table 1.

Mapping of Program Outcomes to Program Educational Objectives.

	Excel in career	**Continue to enhance knowledge**	**Effective in diverse environment**	**Leadership and integrity**
...learn and acquire new knowledge, skills, including an area different from a professional one	x	x		
...identify the essence of scientific problems in professional area		x		
...solve problems in a professional activity using the methods of analysis and synthesis	x		x	
...assess the need for resources and use it in solving the problems in professional activity	x	x	x	
...work with information: locate, evaluate and use the information from various sources necessary for scientific and professional problems	x			
...carry out research, analyze problem, formulate goals and objectives, select research methods and assess its quality	x	x		x
...work in team	x		x	x
...communicate competently considering the objectives and situation	x		x	x
...critically assess and reassess the experience	x		x	x
...work in an international environment	x		x	x

The process for direct assessment of outcomes has three components (items *a*, *b*, *c* below).

a) Quantitative measurement of achievement of each outcome assessed in a subset of the required courses in the program.

This analysis is performed and reported by the PAO and instructor of each course. They establish the instruments to be used to assess each outcome. These are typically questions embedded in student assignments, exams, interviews, questionnaires` or other evaluative mechanisms. PAO also supplies the relevant statistics for the course. These include the number of students, the grading scale and the average score for the embedded question, the percentage of students who achieved the outcome. Finally, the course instructor makes any relevant comments regarding the achievement of the outcome. In addition, he prepares a set of course materials, which includes the course syllabus, copies of the instruments used to assess the outcomes and sample graded student work. PAO stores this information. These materials are the primary source of information for the next level of the assessment process – the SE Program Academic Board.

b) Qualitative evaluation of the achievement of all outcomes assessed in each course.
This evaluation is performed and reported by SE Program Academic Board. This board makes recommendations and suggestions for improvements in the course and its relation to other courses in the curriculum, improvements in the achievement of the outcomes, and improvements in the process itself. Twice per year the board is convened by SE Academic Supervisor and outcomes are assessed.

SE Program Academic Board is tasked with the following:

- To evaluate the course in terms of its contents and its place within the curriculum.
- To perform a qualitative analysis of the quantitative data in the course and course materials supplied by the PAO.
- To examine, evaluate, and ratify the quantitative criteria used, the instruments chosen, the statistics provided, and the sample student graded work.
- To generate suggestions/recommendations in three categories:
    1) Recommendations to future lecturers.
    2) Recommendations to curriculum governance.
    3) Recommendations on improvement of the process.

The results of SE Program Academic Board work are collected by the SE Academic Supervisor for the third and final component of the program outcomes assessment process.

c) Overall analysis of the achievement of each outcome across all courses in which it is assessed.
This analysis is performed by the SE Academic Supervisor who analyzes the reports produced by each individual course instructor and PAO, collects (and generates further) recommendations for improvements at all levels, directs those recommendations to the proper governance bodies, and follows up on actions triggered by those recommendations. He also gathers any feedback from other program-level indirect assessment mechanisms. The SE Academic Supervisor refers suggestions and recommendations to the proper governance bodies, particularly the SE Program Academic Board for consideration and/or action. The supervisor is also charged with following up in subsequent terms on such actions and determining whether recommendations initiated earlier to address any shortcomings have engendered program improvements.

Quantitative data are obtained in direct form from assessments administered by course instructors and PAO. These are validated by the SE Program Academic Board and performance is evaluated by SAO (Percentage of Students Achieving the Outcome that is a percentage of students who received course final grade not less than 6 according to 10-point scale). Course lecturers apply the results of their assessments and SE Program Academic Board evaluations directly within their classes to remediate any shortcomings.

The SE Program Academic Board evaluates courses and outcomes measured within each course qualitatively and their reports are sent to the SE program academic supervisor. He considers these issues at the program level. Each outcome is considered across all courses, and areas where improvement can be made in specific courses in the curriculum as a whole, in the facilities, or in the assessment/improvement process itself, are considered. Appropriate recommendations are forwarded to the appropriate governance or executive body. Issues from previous terms are tracked and effects of changes monitored. The SE Academic Supervisor may also consider quantitative data from national benchmarks, course grades and evaluations as well as qualitative data from alumni surveys, e-mails, HSE Supervisory Board discussions, governance bodies, etc.
Indirect assessments are carried out via students' and alumni interviews and surveys.

The following is the BSSE Program Assessment Methods Table that shows the type of assessment method, frequency of assessment, media, responsible person and evaluation.

Table 2.

BSSE Program Assessment Methods.

Assessment Method	Frequency of Assessment	Data Collected Media	From Whom Collected	Responsible for Collection	Who Evaluates Data
Faculty course assessment reports	Course cycle	Online assessment reports	All faculty	Program Academic Supervisor and Program Academic Office	Program Academic Board
Student surveys	Course cycle	Online questionnaires	All students	Program Academic Office	Faculty and Program Academic Board
Meetings of Program Academic Board	Every year at the beginning of the I semester	Consideration and analysis of course programs	Course instructors	Program Academic Supervisor and Program Academic Office	Program Academic Board
Alumni surveys	Every 3 years	Questionnaires, interviews	Alumni within past 3 years	HSE Center of Internal Monitoring	Program Academic Board
Employer surveys	Every 3 years	Questionnaires, interviews	Employers identified by alumni	HSE Center of Internal Monitoring	Program Academic Board

In 2013 it was made self-examination of BSSE program and self-study report was issued upon the results. The following items have been examined:

- Legislative regulations of educational process;
- Organization of the educational process;
- Organization of scientific and methodical work;
- Organization of research work;
- Cooperation with other scientific, educational institutions, companies and enterprises, international cooperation;
- Facilities;

- Directions and methods to improve the implementation of SE program.

Upon the results of self-examination the following inferences have been made:

1. Structure and content of the curriculum, educational programs of the curriculum disciplines, textbooks and other teaching and methodical materials, faculty professional qualification, their teaching and research potential, requirements for the graduate qualifying works, the quality of students' knowledge, facilities fully meet all the requirements of the HSE educational standard and provide a high level of training quality;
2. Faculty and students of SE program carry out active research work in framework of scientific seminars and laboratories, including the international ones. Teachers demonstrate a high publication activity;
3. Dynamics of the basic quality indicators of SE educational program, level of applicants allow suggesting that this program will be actively developed.

The whole training process is organized on the basis of the curriculum which is developed according to the HSE and international educational standards and clarified for each academic year. Curriculum includes a list of all disciplines, their credit hours and sequence. The academic year is divided into two semesters, consisting of two modules (quarters) each. Duration of each module is 8-12 weeks. The exact start and end dates for each module depends on the weekends and holidays of the current year. There is an exam session after each of a module. Time to prepare for the exams is not provided. Students can be appointed 7 exams within seven days of the session, but not more than one exam per day.

Students graduating with the Bachelor degree in Software Engineering are prepared for engineering practice. Curriculum requires mathematics, social sciences, computer practicums in combination with courses in Software Engineering, and research seminars. Computer laboratories and classrooms ensure practical computing experience for every student in the program. The program education component of humanities, social sciences, and economics assures the students have achieved a well-rounded education. Exposure to aspects of engineering professionalism assures the students have been introduced to contemporary engineering issues. All these factors contribute to the achievement of Program Objective 1, to excel in a career utilizing their education in Software Engineering, and to Program Objective 2, to continue to enhance their knowledge.

In the general education courses, students gain fresh perspectives, methods and tools for understanding the traditional and the newly discovered knowledge. The Mathematical Sciences courses help students acquire concepts and skills in logic, inductive and deductive reasoning, and abstract and quantitative thinking. The Humanities courses enable students to think critically about what thinkers (past and present) have to teach us about the non-material qualities of human beings and human values. The Social Sciences courses enable students to investigate human behavior in its social and psychological context. The English language courses enable students to communicate in international environment and be acquainted with the most recent world achievements in the field of computing. The Philosophy course introduces students to the basic concepts of science and the scientific method and enhances awareness of scientific developments and their impact on society and the environment. All these factors contribute to the achievement of Program Objective 3, to be effective in multidisciplinary and diverse professional environments, and to Objective 4, to provide leadership and demonstrate professional integrity.

Besides the basic disciplines, which form the basis of future profession (*major*), students study elective disciplines. Large part of the time is allotted for the research seminars, project, course and practice work. Students are expected to solve real problems by using knowledge and skills obtained during the training. These problems can be both of academic and applied character. Project work includes student activities aimed at creating of a new not previously existing and unknown product (including analytical research). Various categories of employees such as academics and research staff from the HSE departments and institutions even external to the HSE can provide a project and course work. This work develops not only the professional but also universal and social student outcomes.

Curriculum provides also such important educational element as internships and practices. The practice of students is part of the basic educational program and is carried out in accordance with the educational plan and schedule of training process. Its main goal is to acquire the professional skills, to deep and consolidate the knowledge and outcomes obtained by students during a theoretical training.

BSSE program has the following types of practice: educational and technological internship, professionally oriented and graduation practices. Their distribution is as follows:

1. Educational Internship (Freshman Year) – 3 credit hours;
2. Technological Internship (Sophomore Year) – 3 credit hours;
3. Professionally Oriented Practice (Junior Year) – 3 credit hours;
4. Graduation Practice (Senior Year) – 6 credit hours.

Educational Internship is carried out in order to consolidate, extend and deep theoretical knowledge, get initial practical skills in solving of specific problems. Technological Internship and Professionally Oriented Practice are carried out in order to study the teaching, guidance and regulatory materials, specialized literature, to consolidate practical training skills, as well as to collect, systemize and generalize the materials for the graduate qualifying work. Graduation Practice is carried out in order to collect materials for the implementation of the GQW, to acquire professional experience, to improve the outcomes and to test readiness of the future specialists for individual work in professional sphere.

The internships and practices are carried out on the basis of agreements between HSE and enterprises, institutions, organizations irrespective of their organizational-legal forms and forms of ownership. In accordance with agreements these enterprises, institutions and organizations are obliged to provide a places for practical training of students.

The programs (tasks) of internships and practices are developed at SE Department according to the HSE educational standards. After completing the internship or practice students must present to the special commission reviews of practice supervisors and their own reports. The status of the practice bases (enterprises, companies, institutions, etc.) corresponds to appropriate state requirements and the list of companies where our students are trained is constantly reviewed and extended. The most contemporary information technologies are used during the practice training.

The important aim of a learning process is to obtain the individual and teamwork skills in typical software development by participation in real projects of the leading Russian and foreign companies. Our partners are the companies such as:

- *Group-IB* is one of the leading international companies for the prevention and investigation of cybercrime and fraud with the use of high-tech.
- *International company "Prognoz"* develops data visualization systems and advanced analytics for clients focused on improving the efficiency of management and implementation of innovative technologies.
- *Group PMSOFT* is Oracle Platinum level partner.
- *Company "DialogNauka"* is a system integrator, consultant and supplier of integrated solutions in the field of information security.
- *Company "ROSA"* is a Russian center for the development of system software, including the requirements of the Russian legislation in the field of information security.
- *International Center for Informatics and Electronics* is one of the leaders in consulting, project management, development and implementation of large complex projects on information technology and telecommunications.
- *Company "WebGames"* is one of the largest developers of free-to-play games in Russia.
- *NetCracker Technology* is the proven, strategic partner delivering mission-critical solutions to communications service providers around the world
- *Yandex* is a Russian Internet company which operates the largest search engine in Russia with about 60% market share in this country. It also develops a number of Internet-based services and products. Yandex ranked as the 4th largest search engine worldwide, with more than 150 million searches per day, and more than 50 million visitors (all company's services) daily.
- *R-Style Softlab* is a leading developer and integrator of banking software, belongs to the international group of companies Asseco.
- *FORS* – group of companies, whose activities cover the full range of tasks for the supply of software and hardware systems, designing informational systems, developing infrastructure solutions, rendering technical support and training.
- *Kaspersky Laboratory* is an international company working in the field of informational security.
- *InterProgma* is one of the leading Russian manufacturers of hi-tech and software.
- *Empatika Consulting Company* develops system solutions for business.
- *AT Consulting* is one of the strongest players in the market of information technology and IT services to telecommunication, financial, state, fuel, energy and other companies.
- *JetBrains* provides developers with smart professional tools that help them write clean, quality code and deliver better software faster.
- *Group HeadHunter* provides the searching of a job in various fields of human activity.

We want to confirm the high international educational level of our program. Taking into consideration the world recognized reputation of ABET HSE authorities adopted the decision to accredit BSSE program at ABET Engineering Accreditation Commission. In 2015 the process of accreditation was started and preliminary self-study report has been sent. We hope that program's level corresponds to the world recognized educational standards and accreditation will confirm this fact.

In our report we would like to share our ten-year experience in developing and implementing BS program in the continuously changing field of software engineering.

ASEE 2016
INTERNATIONAL FORUM
NEW ORLEANS, LOUISIANA / JUNE 25, 2016

Paper ID #17523

Developing Innovative Interdisciplinary Biomedical Engineering Programs in Nigeria: Lessons Learned

Prof. David W. Gatchell PhD, Northwestern University

David Gatchell is Director of the Manufacturing and Design Engineering (MaDE) program at Northwestern University. He is a clinical associate professor within the Segal Design Institute, Biomedical Engineering Department, and the Mechanical Engineering Department. He holds a PhD in Biomedical Engineering from Boston University and an AB in physics from Bowdoin College.

Dr. Robert Linsenmeier, Northwestern University

Robert Linsenmeier is a Professor of Biomedical Engineering, Neurobiology, and Ophthalmology. His interests are in the microenvironment of the mammalian retina and engineering education. His teaching is primarily in physiology for both biology and BME majors. He is a fellow if the Biomedical Engineering Society, American Institute of Medical and Biological Engineering, and Association for Research in Vision and Ophthalmology.

Prof. Robert L. Murphy, Northwestern University

Dr. Murphy is Professor of Medicine and Biomedical Engineering at Northwestern University where he holds the position of Director, Center for Global Health.

Prof. Akinwale Oladotun Coker P.E., university of ibadan
Dr. Akinniyi A. Osuntoki, University of Lagos, Lagos, Nigeria

Developing Innovative Interdisciplinary Biomedical Engineering Programs in Nigeria: Lessons Learned

abstract

An interdisciplinary team comprising faculty from the medical school, business school and school of engineering at Northwestern University is in its third year of cultivating "locally grown" healthcare solutions in Nigeria through a grant provided by NIH's Fogarty International Center. Through a series of training sessions at multiple sites in multiple countries, faculty at the University of Lagos (UINLAG) and the University of Ibadan (UI) are gaining the expertise and developing the infrastructure necessary to identify, design, and commercialize solutions to their own healthcare issues.

The specific aims of the grant are three-fold: 1) Establish new and strengthen existing training programs in biomedical engineering in sub-Saharan Africa; 2) Train biomedical engineers and medical doctors to evaluate newly developed HIV-related and other testing and therapeutic devices; 3) Train post graduates from business schools to scale-up development and launch new healthcare practices. In support of these specific aims we have created four distinct annual training sessions: 1) week-long annual planning meetings hosted by either UNILAG or UI and facilitated by the NU team; 2) ten-week long biomedical engineering training programs at the University of Cape Town (UCT) facilitated by UCT and NU faculty ; 3) participation in two-week long Kellogg Field Studies in Africa; 4) four-week long mentoring visits to NU and the greater Chicago area, including participation in the annual BMES conference. An integral component of the training at NU involves Northwestern faculty members working with faculty from UI and UNILAG to develop BME degree granting programs at their respective universities.

This paper describes these training programs and assesses their ability to meet the specific aims listed above. In addition, we share insights gained through direct and in-direct assessments on how well these programs are addressing the specific aims laid out in grant, identifying what requires modification going forward. We also update the progress made in developing BME degree programs at both universities.

introduction

Medical care and population health in low and middle income countries (LMIC) lag behind wealthier nations because of limited infrastructure and financial and human resources[1,2]. The disease burden is radically different in the developing world with higher mortality rates from human immunodeficiency virus/acquired immunodeficiency syndrome (HIV/AIDS), tuberculosis (TB), malaria and other treatable infectious diseases. Infant and maternal mortality surrounding childbirth is unacceptably high. Transplanting Western healthcare systems and technologies to resource-limited settings is a slow and expensive process, and in many instances inappropriate or impossible. Developing new and creative healthcare delivery methods and devices tailored to these environments and conditions will be more effective and less expensive path to improved health. Local research-focused universities can (and should) play a critical role in this scenario[3,4].

Academic institutions have always played an important role in healthcare. Historically the Research University is a site of discovery and innovation as well as a place where future healthcare workers are trained. In the United States and other developed countries great advances have come from a decades-long integration of engineering and medicine, however, in the developing world engineering faculty and physicians collaborate with much less frequency. The objectives of this program address the challenging development issues facing Global Health in relation to the creation of new, innovative, life-saving technologies. Specifically, the objectives are: 1) expand the base of expertise needed and available to address the most pressing problems in global health by developing trans-institutional, problem solving-based research training programs in biomedical engineering that bring together widely diverse experts who work together on research problems in Global Health; 2) Stimulate new knowledge, approaches and solutions in Global Health by putting "innovation" in the Global Health context; and 3) Integrate Global Health research communities within and among institutions by raising awareness and building interdisciplinary biomedical engineering capacity where it is needed the most, in this case, sub-Saharan Africa.

The theme of our program involves the development of effective, affordable, and easy to use innovative biomedical devices that can advance diagnostic and treatment approaches that result in improved survival. We are doing this by establishing comprehensive interdisciplinary training programs in biomedical engineering that include identification of critical healthcare needs, product design, delivery, clinical evaluation, scalability and product launch. Our cross-disciplinary approach involves three key areas of research which only together can lead to improved health and survival: 1) biomedical engineering, 2) medical evaluation, and 3) commercialization. This interdisciplinary training program will provide LMIC faculty and students with a unique comprehensive hands-on learning experience focused on developing innovative global health technologies. The LMIC partners include the University of Cape Town in South Africa, the University of Ibadan, University of Lagos, and the Nigerian Institute for Medical Research in Nigeria.

This paper describes these training programs and assesses their ability to meet the specific aims listed above. In addition, we share insights gained through direct and in-direct assessments on how well these programs are addressing the specific aims laid out in grant, identifying what requires modification going forward. We also update the progress made in developing BME degree programs at both universities.

training at Northwestern University

For each of the past three fall quarters, the biomedical engineering department at Northwestern University has hosted 2-4 engineering faculty from the University of Lagos (UNILAG) and the University of Ibadan (UNI) – See Table I. For 3-4 weeks the faculty remain in the United States where they experience the following:

- Participation in the annual meeting of the Biomedical Engineering Society (BMES). This includes submitting an abstract to the conference and giving a 20 minute presentation on that topic (examples of submitted titles for the 2014 conference are shown in Table II).
- Participation in upper-level BME courses at Northwestern University, including the capstone biomedical engineering design course.
- Visits to Chicago-area biomedical engineering departments (e.g., Marquette University, Illinois Tech).
- Meetings with faculty in the McCormick School of Engineering and Applied Sciences, Weinberg School of Arts and Sciences, Feinberg School of Medicine.
- Tours of engineering labs, medical labs, design facilities, prototyping shops including a rapid prototyping shop comprising over $2M worth of equipment.

Table I. Nigerian faculty participating in Chicago training (2013-2015)

Year	Name	Discipline	University
2013			
	Coker, Akinwale*	Civil Engineering	UI
	Lawal, Taiwo	Pediatric Surgery	UI
	Osuntoki, Akinniyi*	Biomedical Engineering	UNILAG
2014			
	Anyaeche, Osita	Industrial and Production Eng	UI
	Ajibola, Olawale	Systems Engineering	UNILAG
2015			
	Dare, Ademola	Mechanical Engineering	UI
	Adeleye, Rotimi	Systems Engineering	UNILAG

*Professors Coker and Osuntoki have participated in the Chicago training all three years of the grant.

Table II. Abstracts and manuscripts submitted by Frameworks investigators from Northwestern University, and the Universities of Cape Town, Ibadan, and Lagos to the Biomedical Engineering Society (BMES) 2014 meeting, San Antonio, TX.

Titles	Authors
"Development of Design-Oriented BME Degree Programs in Nigeria"	M. Glucksberg[1], A. Coker[2], A. Osuntoki[3], T. Douglas[4], and R. Murphy[5]
"Numerical Modeling of Magnetic Micropump for Biogenic Bulk Transport in a Biomimetic Microchannel"	E. Ig, A. Dare, and A. Coker[2]
"Design and Construction of a Blood Glucose Meter for Use in Nigeria"	A. Zubair, O. Ibe, and A. Coker[2]
"Design and Construction of a Portable Low Cost Electrical Safety Analyzer for Biomedical	O. Ajibola, F. Ogunwolu, O. Ibidapo-Obe, V. Olunloyo, and A. Osuntoki[3]

Devices"	
"Modelling Gait Syndrome in Huntington prime or minutes disease: the Genetic Algorithm Approach"	A. Osuntoki[3], O. Olawale[2], E. Ajibola[3], C. Esezobor[3], and S. Nwaneri[3]
"Establishment of an Interdisciplinary Biomedical Engineering Programme In Nigeria: Preliminary Observations From The University Of Lagos"	A. Osuntoki[3], O. Olawale[1], E. Ajibola[3], C. Esezobor[3], and S. Nwaneri[3]

[1]Northwestern University, Evanston, IL; [2]University of Ibadan, Ibadan, Nigeria; [3]University of Lagos, Lagos, Nigeria; [4]University of Cape Town, Cape Town, South Africa, [5]Northwestern University, Chicago, IL

Kellogg field studies

Many of the innovative products in Northwestern's medical technology portfolio were identified and refined through market research performed by Kellogg School of Management (KSM) students participating in medical field studies in Africa. Led by Professor Kara Palamountain, co-PI on the Frameworks grant, the Kellogg Field Studies program has engaged more than 300 MBA students over the past ~10 years. These students have comprised over 50 teams and have conducted market entry analysis in over a dozen developing countries, including ten African nations. In addition to product design research, these students have identified the formal and informal processes for launching products in these countries. By participating in these field studies, the Kellogg students learn how to apply theoretical market research methodologies toward the development of specific medical products.

One of the primary goals of the Frameworks grant is to train post graduates from Nigerian business schools to scale-up development of, and launch, new healthcare practices. By participating in the *Kellogg Field Studies* program these graduates will learn how to inform the design and development of innovative medical technologies for developing countries as well as how to train students to collect and analyze information by interviewing and/or observing key individual and organizations. The *Kellogg Field Studies* program is innovative in that it not only considers potential end-users of medical technologies, but also those stakeholders that approve, finance, procure, distribute and maintain these technologies. These stakeholders include:

- Healthcare providers;
- National and regional governments;
- Distribution companies;
- NGOs and international agencies.

Each year two post graduates join Professor Palamountain and her students for two week long field studies in Africa. The objectives of the studies include:

- Identifying the unmet medical needs of the country of focus;

- Defining medical technologies that can address the unmet need;
- Understanding the business, institutional, political and financial framework for implementing new medical technologies in that country;
- Obtain feedback on specific medical technologies underdevelopment at Northwestern University's Center for Innovation in Global Health Technologies (CIGHT).

During the first two years of the Frameworks grant, the *Kellogg Field Studies* program has taken faculty from UI and UNILAG (see Table III) to Tanzania (2014, three participants) and Zambia (2015, two participants).

Table III. Nigerian faculty participating in Kellogg field studies (2014-2016)

Year	Location	Name	Discipline	University
2014	Tanzania			
		Diji, Chuks	Mechanical Engineering	UI
		Orekoya, Samuel	Economics	UI
		Sulaimon, Abdul-Hameed	Business Administration	UNILAG
2015	Zambia			
		Olaniyan, Olanrewaju	Economics	UI
		Kuye, Owolabi	Business Administration	UNILAG
2016	TBD	TBD		

training at the University of Cape Town (UCT)

Biomedical engineering at the University of Cape Town dates back to the 1950s with work done in nuclear medicine by Alan Cormack that led to computer-assisted tomography and the *Nobel Prize in Physiology or Medicine* in 1979. Since that time, the Department of Human Biology has been created which merges biomedical engineering with anatomy, cell biology, physiology and sports science. In 2001 the Healthcare Technologies Management (HTM) Program was added to the department. HTM recognizes that proper management of medical technologies and physical infrastructure is essential for the attainment of affordable and sustainable quality healthcare. In addition, HTM aims for optimal acquisition and utilization of healthcare technologies, including proper planning, evaluation and assessment of health facilities, as part of efficient, cost-effective, multidisciplinary healthcare services. Northwestern University's formal association with UCT began in 2005 with the establishment of the Global Health Technologies (GHT) Program where Northwestern engineering students participate in design and development of appropriate healthcare technologies.

For the past three years faculty from UI and UNILAG (see Table IV) have participated in a ten-week Biomedical Engineering Training Program at the University of Cape Town. The visiting faculty participated in the GHT program, attending and participating in activities related to the following courses:

- BME 388-SA: Healthcare in Resource Poor Environments;
- BME 389-SA: Healthcare Technology Assessment and Planning;
- BME 391-SA: Healthcare Technology Innovation and Design;
- GBL HLTH 314-SA: Health and Community Development in South Africa

In addition, the faculty attended the following UCT courses:

- Biomechanics of the Musculoskeletal System – this course provides background for the design of rehabilitative prosthetic and orthotic devices;
- Introduction to Medical Imaging and Image Processing – this course provides background on the use of imaging as a tool in biomedical engineering, broadly.

Faculty also visited biomedical research facilities at UCT as well as the Heart Museum at Groote Schuur Hospital. They had discussions with UCT faculty involved in medical device innovation, were introduced to the current World Health Organization (WHO) initiative on local production of medical devices as a means of improving access, and attended lectures and project-based discussions offered by NU's GHT program.

Visitors were given the opportunity to attend and participate in the following HTM courses:

- Healthcare Technology Planning and Acquisition;
- Project Management;
- Clinical Engineering Practice;
- Medical Devices and Instrumentation Overview;
- Asset Management of Healthcare Technology and Infrastructure;

These courses expose visiting faculty to the broader context of medical device and healthcare technology innovation, i.e., needs assessment, planning, evaluation, operation and life-cycle management. In addition, faculty are given the opportunity to interact with professional students in the HTM program – all traditional HTM students are employed in the healthcare sector across the African region.

Table IV. Nigerian faculty participating in UCT training (2014-2016)

Year	**Name**	**Discipline**	**University**
2014			
	Zubair, A.R.	Electrical Engineering	UI
	Nwaneri, Solomon	Biomedical Engineering	UNILAG
	Esezobor, Christopher	Pediatrics	UNILAG
2015			
	Akintayo, Folake	Civil Engineering	UI
	Lawal, Oluwatoyin	Oral Pathology	UI
	Adetona, Sunday	Electrical Engineering	UNILAG
	Habeebu, Muhammad	Radiology and Radiotherapy	UNILAG

2016			
	Coker, Morenike	Pharmaceutical Microbiology	UI
	Obed, Rachel	Physics	UI
	Balogun, Adeola	Electrical and Electronics Eng	UNILAG
	Umesi, Donna	Restorative Dentistry	UNILAG

development of biomedical engineering educational programs in Nigeria

Our two educational partners in Nigeria, the University of Ibadan (UI), and the University of Lagos (UNILAG) have been working for the past few years to establish graduate-level biomedical engineering degree granting programs. To put these efforts in context, current materials from UI refer to just two bachelor's level programs at Nigerian universities. One of these programs is at the Federal University of Nigeria at Owerri (FUTO), and was established in 2009; the other program is at the School of Biomedical Engineering Technology, Ahmadu Bello University, in Zaria, and was established in 1993[3]. These programs are designed primarily "to train qualified technical experts to manage and maintain hospital equipment and systems," according to materials from UI, and accordingly the program at FUTO is a Bachelor of Technology in Biomedical Engineering (http://www.nigerianbme.org/nibefutolinkage.html). FUTO also has MS and PhD programs in Biomedical Engineering Technology[3].

While educational programs in BME are rare, Nigeria does have the Nigerian Institute for Biomedical Engineering (NIBE), which was established in 1999, and has held annual BME conferences since that time. In 2009, NIBE created an arm called the College of Biomedical Engineering and Technology (http://www.nigerianbme.org/cbet/), which is responsible for accreditation, development of curricula, setting standards for professional practice, and promotion of biomedical engineering within the country. NIBE is part of the International Federation of Medical and Biological Engineering, so not only is there a professional infrastructure in the country for this field, but links to international organizations as well.

Once in operation, graduate training at UI and UNILAG will work somewhat differently than it does in the US. At UI, students are admitted after a bachelor's level degree in engineering, science or medicine, which is designed to take 10 semesters. Enrolled graduate students are expected to first obtain a Certificate in BME through one or two semesters of full time study (12 units), a Diploma is issued after two to four semesters (24 units total), and finally a Master's degree after four to six semesters (48 units total). Students may leave the program at any of these stages with what is anticipated to be a useful credential for work in the emerging biomedical industry. Typical courses are 2 or 3 units, where 1 unit represents at least 15 hours of contact time, and up to 25 hours if the course includes a laboratory or design component. The Master's degree does not require a thesis, but it does require a 6 unit project course. At each level, there are 6 tracks, each having some required and some elective courses. Five of these tracks are similar to those that would be found at many US institutions: Biomechanics, Imaging,

Systems Modelling, Biomedical Materials, and Biotechnology, and the choices are dictated partially by the perceived needs in the country, and partly by the expertise of available faculty. The other track is Clinical Engineering, which would not typically be found in a US biomedical engineering program, but addresses the needs in Nigeria.

Because the university does not offer a bachelor's of science undergraduate program in biomedical engineering, some of the graduate courses at UI are ones that would be found in most undergraduate BME programs in the US. These include a Survey of Engineering in Medicine and Biology, Biomechanics, Biomaterials, Medical Imaging, Instrumentation, and Biochemistry. Other courses are similar to those that would be expected for upper level undergraduates or graduate students in the US. Some topics, notably imaging and fluid mechanics and transport, are each covered by a sequence of several courses. There is no comprehensive physiology course, which is a mainstay of biomedical engineering programs in the US, but there are specific courses in Cardiovascular Dynamics, Mechanics of the Human Body, and Membrane Biophysics. There is also a lack of courses in cellular and tissue engineering, which even in the US are primarily research fields, so this is appropriate for a postgraduate program preparing students for industry. There are currently no courses specifically listed as design courses, but there is a course in regulatory requirements, which will compensate for the lack of material specific to the medical field in undergraduate engineering programs. The project course may involve design, but its description indicates that enrolled students are to, "undertake a project that will utilize theories to solve problems related to biomedical engineering issues," so it is not necessarily directed toward design.

At UNILAG the Postgraduate Diploma (PGD) program started in 2013-2014. As at Ibadan, the PGD is designed to take two semesters (26 units) and the MS about twice as long (an additional 32 units). The mission statement of the department reflects the dual missions of providing engineers for clinical and technical support for hospitals and manufacturers on the one hand, and research and design of new equipment and software. It is capped by the overarching goal of producing engineers "who will assist the medical team by providing the technological know-how to stabilize the healthcare delivery system and make for self-reliance in health matters in our environment."

The distribution of topics taught in biomedical engineering at UNILAG is similar to that at Ibadan, but at present there are fewer choices, and students at UNILAG will obtain more breadth and somewhat less specialization. All students take a range of courses including numerical methods, biomaterials, biomechanics, instrumentation, and imaging, as well as a course called Basic Biomedical Sciences, which is largely physiology, but also includes cell biology. A somewhat stronger focus on design is evident in course descriptions at UNILAG. The description for the Research Seminar states that "Student teams conceive, design, specify, implement, evaluate, and report on a software and hardware project in the domain of biomedicine." Then the Research Project course, required for the PGD, also emphasizes design: "The student will apply the principles of engineering or bioinformatics design to produce products which meet functional needs of people with Medical disabilities [and] design and construct either as a team or individually, a working prototype of a new device which solves a problem without current solution." This is challenging because it is to be done in a semester that also includes four other courses. A second project is required for the MS.

Administratively, biomedical engineering at UI is housed in the Department of Mechanical Engineering. At UNILAG the degrees are housed in the Medical School as one of the basic science departments, which was established in 1974. For a time it awarded Diplomas, but then stopped until the recent resurgence of effort[3]. The degree programs are administered through the School of Postgraduate Studies. These programs are almost invisible for the visitor to the websites of the two universities. There is very little information available about the programs, courses, expected career paths, or faculty at either institution. The Mechanical Engineering Department link at UI has nothing about this emerging program. Considering that these are new programs designed to help address serious problems in Nigeria and Africa more broadly, it is interesting that there is so little information available. This is in contrast to US programs, which all have a wealth of information at the department and school websites, and marketing is an emphasis of almost all universities. It is difficult for us to understand how students would obtain enough information about the programs at UI or UNILAG to even apply. The NIBE website is easy to find, but there are no links to these new graduate programs.

results

Participants in each of the training sessions (with the exception of the workshops in Nigeria) are asked to provide feedback on their experiences and how they map onto the specific aims of the Frameworks grant. Below we capture the key insights shared by the program's participants Positive observations are denoted (+); opportunities for improvement are denotes (-).

feedback – training at Northwestern University

- (+) "Gives one first-hand exposure to established BME programs. This helps to see what can be 'borrowed' or adapted for our nascent BME program (in Nigeria)."
- (+) "Critical exposure for evolving interdisciplinary thinking in individuals with the traditional discipline based training."
- (+) "Helps in identifying potential collaborators for research between Nigerian faculties and U.S. faculties."

feedback - Kellogg field studies

- (+) " The involvement of business faculty in the development of business plans for technology/medical products is important and will be very useful in helping to identify priority projects and develop business models for products developed by BME teams."
- (+) "Informed the structure and content of our new MBA program which was being reviewed for accreditation."

- (+) "Provided first-hand experience for the current director of the MBA program (Dr. Sulaimon)."

feedback – training at the University of Cape Town (UCT)

- (+) "Provide unique opportunities for inter-professional learning."
- (+) "Helping to educate faculty with a diversity of backgrounds to gain a clear view of biomedical engineering within the healthcare system."
- (+) "Helping to strengthen interdisciplinary collaboration and develop faculty for an interdisciplinary BME program."
- (+) "The capacity training will further position the University of Ibadan in its quest to train manpower that will provide creative and inexpensive health care solutions for use in Nigeria and Africa in general. This will promote the efforts of the Northwestern University to develop trans-institutional, cross disciplinary and innovative training programs in biomedical engineering in Africa."
- (-) "The training module should include more hands-on training where participants from diverse backgrounds in engineering and medicine will actually be involved in medical product development."

feedback – overall

Frameworks training …

- (+) erodes barriers to interdisciplinary training;
- (+) strengthens intra- and inter-institutional collaboration;
- (+) helps to demystify the process of device development;
- (+) promotes innovative thinking and innovation;
- (+) helps us to strengthen each other in our areas of weakness.
- (-) inter-university collaboration has been slow to develop and hinders progress on the development of medical devices. The use of information and communication technology is still limited in Nigeria.

discussion and future work

Seven Nigerian faculty have visited Northwestern University to participate in the training program, and another 11 have participated in the program based in Cape Town. Five faculty have participated in fieldwork in Tanzania and Zambia with Northwestern's Kellogg School of Management. Training is proceeding through a combination of traditional classroom-based learning and structured in-country field research in partnership with students and faculty from Northwestern University and the University of Cape Town. Through these activities faculty and clinicians are participating in the identification and development of healthcare technologies appropriate for their local and regional settings. Healthcare technologies can include anything that significantly affects outcomes and cost-effectiveness of healthcare delivery including medical devices, therapeutic interventions, information systems and mobile communication technologies. Selected trainees with biomedical and behavioral science backgrounds in adult and pediatric medicine are directed towards evaluating the newly developed technologies. Faculty trainees in business are instructed and mentored in defining the most appropriate medical technologies needed in their home country along with the tools needed to approve, finance, procure, distribute, use and maintain these medical technologies. These trainings provide opportunities for inter-professional learning, enhance interdisciplinary thinking and project focused cross-disciplinary team building skills in faculties who were trained in the traditional discipline based mode. The ultimate result of these efforts will be to support the expansion in size and scope of the recently established departments in biomedical engineering at the Nigerian institutions.

The faculty training programs supported by the Frameworks grant continue to build upon the experiences of the Center for Global Health in the Feinberg School of Medicine, the Center for Innovation in Global Health Technologies in the McCormick School of Engineering and the Global Health Initiative at the Kellogg School of Management at Northwestern University, as well as the long record of academic excellence at the University of Ibadan, the University of Lagos, and the University of Cape Town. The close collaboration of basic, applied, and clinical research involving medical, engineering, and business schools may serve as one model for cross-disciplinary innovation in health and healthcare technologies; more likely, however, our efforts will result in a model more suited to particular situations in Nigeria, a country rich in resources and with significant technical capacity within their universities that has not yet developed medical products focused on addressing local and regional needs. Like most medical systems in LMIC, available technologies are imported from the US, Europe, and Japan. Much of the population of these countries and others in sub-Saharan Africa suffers from medical, economic and social conditions that differ in both degree and kind from the environment where these technologies were developed. Local BME programs focused on appropriate technologies that understand the challenges of these environments are urgently needed to facilitate development and commercialization of the devices and processes needed to improve healthcare in Africa.

Moving forward there is a definite need to better communicate the learning objectives for each of the four training experiences to the grant's trainees. While effort has been made to map these experiences onto the specific aims proposed in the Frameworks grant, detailed learning outcomes have not been uniformly identified (e.g., training at NU), and, as a result, have not been integrated into an assessment tool for measuring the learning occurring through these experiences. One potential solution to this problem is to use the reflections that we have collected to date to identify what the participants have learned through each of their experiences and what they would have liked to have learned but have not. These skills and concepts can be integrated into a self-efficacy pre- and post-test survey instrument (see the work of McKenna and Hirsch[10] for an example of a similar effort to measure learning of the design process by undergraduate engineering students). In addition to the summative reflections that we have collected (i.e., reflections collected after the training has been completed), we should complement these with more frequent journaling of the experiences using prompts provided by the trainers/faculty leading the experience. These reflections would provide feedback about specific aspects of the training as opposed to the training as a whole. The feedback could then be used to inform, and potentially, upgrade the training experiences being offered.

References

1) Mattson, CA, Wood, AE. Nine Principles for Design for the Developing World as Derived from the Engineering Literature. *Journal of Mechanical Design*. Vol. 136 (2014).
2) Richards-Kortum, R, Oden, M. Devices for Low-Resource Health Care. *Science* 342, 1055 (2013).
3) Nkuma-Udah, KI, Okoye, GC, Ndubuka, GI. "Biomedical Engineering Education and Training in Nigeria." *World Congress on Medical Physics and Biomedical Engineering,* September 7 – 12, 2009, Munich, Germany: Vol. 25/12 of the series *IFBME Proceedings*, pp 71-74.
4) Nkuma-Udah, K. I., & Mazi, E. A. (2007). Developing Biomedical Engineering in Africa: A case for Nigeria. World Congress on Medical Physics and Biomedical Engineering (pp. 3828-3831). Springer Berlin Heidelberg.
5) Hirsch, P., & McKenna, A. (2005, June), *Evaluating Student Confidence In Engineering Design, Teamwork And Communication* Paper presented at 2005 Annual Conference, Portland, Oregon.

Individual peculiarities of engineering students attending minor degree programs. Research on psychology and education.

Prof. Farida Tagirovna Shageeva, Kazan National Research Technological University

Doctor of Education, Professor of the department of Engineering Education and Psychology, Dean of the Faculty of Additional Education at the Kazan National Research Technological University. Scientific interests: educational technologies, innovations in educational practice.

Dr. Inna Mikhailovna Gorodetskaya, Kazan National Research Technological University

Inna M. Gorodetskaya is associate professor at the Department of Engineering Education and Psychology of the Kazan National Research Technological University (Russia). She has Ph.D. in Social Psychology and also works as head of the minor degree program in Psychology at the university. Scientific interests: motivation, value system of a person, self-development, diversity issues.

Dr. Liliya Vasilievna Prikhodko

Individual peculiarities of engineering students attending minor degree programs. Research on psychology and education.

Professional activity of a contemporary global engineer requires constant improvement of engineering training[1]. One of the ways to make the educational system more flexible and correspondent to the real labor market environment in Russia is the so-called "Additional professional education" (minor courses). At the Kazan National Research Technological University there is an option for the top students to acquire an additional degree in Education, Management, Law, Psychology, Foreign Languages, etc. at the Faculty of Additional Education (FAE)[2]. Our reseach showed that students who attend the FAE courses have some psychological peculiarities. The study was aimed at creating a psychological profile of the minor degree students majoring in engineering. A survey was given to the students and to the faculty who studied the following minors: "Corporate management", "Education", "Professional translation", and "Legal fundamentals for business processes". The sample group consisted of 50 minor degree students (21 male and 29 female) aged 19-22. The survey results allowed concluding that minor degree gives more opportunities for individualization.

As a rule the most gifted and target-driven students are enrolled at the Faculty of Additional Education. A student's decision to apply for a minor degree is determined by strong intrinsic motivation (academic interest, professional ambitions, personal development) that cannot be fully satisfied in the major.

Empiric research of professionally important qualities business efficiency[3] showed that FAE students are characterized by a better psychological and business potential. The survey revealed the following parameters:

- *personal orientation* – values, attitudes, motivational trends, moral and ethical patterns;
- *business efficiency* – business qualities, necessary for efficient management;
- *dominance need* - desire to have force and to be significant for others, strive for success, strong achievement motivation;
- *assertiveness* – independence, self-confidence, communicative competence;
- *conservatism* – persistence, resistance to changes, ultimatism;
- *negativism* – vindictiveness, suspiciousness, petulance, unforgivingness, criticism, distrustfulness, resistance to any influence;
- *compliant behavior* – social maladjustment, low self-esteem, lack of initiative, shyness;
- *dependence* – infantilism, limpness, need for being guided, helplessness;
- *conformism* - commitment to the others' opinions, public views, strive for social approval, suggestibility, sociability;
- *responsiveness* – sympathy, empathy, sensibility, devotion, unselfishness, helpfulness.

The average indexes of the above-listed parameters of the survey sample ("Corporate management", "Education", "Professional translation", and "Legal fundamentals for business processes" minor degree students) are presented in Table 1. The quantitative-qualitative correspondence is the following:

1-4 – nominal zone;
5-8 – potential zone;
9-12 – advanced zone;
14-15 – super-zone.

Table 1. Average indexes of personal qualities of FAE students

Personal quality	Minor degree program			
	Corporate management	Education	Professional translation	Legal fundamentals for business processes
Personal orientation	9	12	9	10
Business efficiency	12	8	8	11
Dominance need	8	8	6	9
Assertiveness	10	10	7	7
Conservatism	7	8	4	8
Negativism	2	2	2	5
Compliant behavior	4	4	4	4
Dependence	3	3	3	3
Conformism	5	6	10	4
Responsiveness	7	12	8	7

Typical personal profiles of students who attend various minor degree programs were made basing on the survey data.

As it is seen from Table 1, the students have high indexes of personal orientation and business efficiency, therefore FAE students have leadership potential.

There were some peculiarities revealed between the minors. Future managers are characterized by high assertiveness and business efficiency, and future educators are very responsive. Students who study professional translation have a more "smooth" profile, but are characterized by high conformism and the least conservatism. Future lawyers have comparatively higher negativism and are the least conforming. Thus the personality profiles of FAE students reflect the professional specifics of the chosen minors.

Professors who teach at the FAE were also given a questionnaire that consisted of two parts: assessment of the students' cognitive skills (analytical, systemic, logical thinking) and of their personal traits (independence, responsibility, reliability, etc). In general faculty gave big points to the FAE students as intellectual, creative persons with broad outlook, original way of thinking and leadership potential[4]. The 5-point scale was used. The survey sample comprised 43 professors with the teaching experience from 5 to 27 years.

According to the professors' opinion FAE students have better cognitive skills and personal qualities than those who study only their major. The generalized portrait of a FAE student given by the teaching staff revealed the following qualities: developed intellect, creativity, wide general outlook, original thinking, initiative, target-orientation, enthusiasm, curiosity, independence, responsibility, persistence, respectfulness, sociability and leadership. They are able to work with information, and have a more analytical and critical way of thinking than other students of the University.

Within engineering major a student has not enough flexibitity to choose courses according to their personal peculiarities and motivation, meanwhile in minor degree courses their individuality is reflected[5]. In general students who prefer to study additional courses simultaneously with their engineering major have integrated personality, are fully aware of their values and motives and have businesslike character. At the same time every particular program attracts students of specific type of personality. Self-confidence, assertion, and leadership potential characterize those who choose minor courses in management. Empathy and outgoingness make them choose pedagogy and psychology. Students who choose foreign languages and professional translation are communicative, conforming and the least conservative. Students with critical thinking and low conformism study law. Thus their individual traits correspond their minor.

Additional professional education of engineering students is per se an exclusive possibility for the top students to increase their marketabitity and to realize their personal potential in the future engineering career.

References:

1. S. Dyakonov, V. Ivanov, L Ovsienko (2003). Additional aducation as a new possibitily / Higher education in Russia. № 2. Pg. 3-13.
2. V.Ivanov, P.Osipov, F.Shageeva (2010). Additional education of students as a priority for engineering university / Education and self-development. № 2 (18). – Pg.37-42.
3. M.G.Rogov, N.Sh.Valeeva (2000). Socio-psychological fundamentals of management., Nab.Chelny – 240 pg.
4. T.A.Starshinova (2012). Psychological and pedagogical mechanisms of professional competence development within the framework of pedagogical integration. /KSTU newsletter № 3. – Pg. 210-212.
5. R.V. Kupriyanov, I.M.Gorodetskaya (2015). Global trends in higher education and thier impact on engineering training in Russia. / International Conference on Interactive Colaborative Learning (ICL), 2015. Pages 244-250, DOI: 9510.1109/ICL.2015.7318033

A Global Framework for Understanding Cross Cultural Teaching Experiences Gained in Japan

Dr. Douglas Moore Schutz, Tokyo University of Science

Douglas M. Schutz is an Associate Professor of Information Systems, International Business, and Management at the Tokyo University of Science (TUS) in Japan. He received a Ph.D. in Business Administration focusing on Management Information Systems (MIS) from the Fox School of Business of Temple University at Philadelphia, an MBA in Information Management from the McCombs School of Business of the University of Texas at Austin, and a B.S. in Electrical Engineering from the United States Naval Academy at Annapolis focusing on digital electronics design. Doug was one of 40 Ph.D. students selected globally to present his research at the International Conference on Information Systems (ICIS) Doctoral Consortium in Shanghai, China. His research has been nominated Best Paper at the Hawaii International Conference on Systems Science (HICSS), and his work has been published three times in publications of the Japan Society of Information and Management (JSIM). He was recently selected to teach a one week intensive course to graduate students at the Regensburg University of Applied Sciences in Germany. Prior to academia, Doug worked as an IT manager and consultant in the electric utility industry out of New Orleans for two Fortune 500 companies, where his responsibilities included IT disaster recovery from Hurricane Katrina. Previously, he served as an unrestricted line officer in the U.S. Navy onboard a guided missile destroyer and the second Aegis enabled cruiser.

Prof. Dante Dionne, Korean Air

Dante Dionne is an Organizational Leadership PhD candidate and a Senior Innovation Technology Manager at Korean Air. The past 20+ years of his career has centered on management and professional services consulting. Where, he has specialized in leading multi-national project teams in the design and implementation of digital marketing, mobility and innovative technology solutions.

Dante holds an MA in Industrial/Organizational Psychology from the Chicago School of Professional Psychology and a BS in Business Management with a minor in Computer Science. Dante is an adjunct faculty member at the Chicago School of Professional Psychology with Graduate level teaching credits in Psychometrics, Data Management and as a Visiting Scholar at the University of Science in Tokyo. Dante's dissertation research spans several specializations in psychology including: Organizational, Social, Cultural, Developmental, Cognitive, Performance, Sports, and Positive Psychology. Dante is also an active member of American Psychological Association (APA) Division 46 (Society for Media Psychology and Technology), Division 14 (Society for Industrial and Organizational Psychology), Division 13 (Society of Consulting Psychology) and Division 47 (Society for Sport, Exercise and Performance Psychology).

Prof. Yong-Young Kim, Konkuk University

Yong-Young Kim is assistant professor of Division of Business Administration and Economics at Konkuk University in Korea. He earned his Master's and Doctoral Degree in Management Information Systems from Seoul National University in Korea. His research interests include Smart Work, online games, IT experiential learning processes, IT convergence & platform, and ubiquitous computing. His papers have appeared in Information Resources Management Journal, Cluster Computing, International Journal of Advanced Media and Communication, Asia Pacific Journal of Information Systems, and Journal of Korean OR/MS Society and also have been presented at many leading international conferences (ICIS, HICSS, PACIS).

A Global Framework for Understanding Cross Cultural Teaching Experiences Gained in Japan

Abstract

The two of the three largest economies in the world today, the U.S. and Japan, share common globalization challenges and opportunities, particularly in the area of engineering education. The globalization of engineering education calls for international educators who are able to gain a cross-cultural knowledge of their students, institutions, and countries. This is critical, because both Japan and the U.S. need to attract and develop larger numbers of engineering students with the technical skills for successfully innovating into the future as their populations increasingly age.

This paper provides best practices experienced from teaching in Japan in order to help address and learn from gaps between foreign cultures, so that teaching can be more motivational and effective for learning. This leads to the following research question:

How can engineering education in Japan benefit from cross cultural understanding?

This paper draws from the literature streams of knowledge sharing, social influence, and social psychology. From these streams, a theoretical framework is developed and applied to teaching and learning in Japan. Through this framework, best practices are developed that can be applied for greater cross cultural understanding for engineering education across borders. The framework can be applied to not only Japan, but to other counties for greater understanding for teaching and learning globally.

Introduction

Trends in globalization and demographics are leading to the call for international engineering educators who are able to gain a cross-cultural knowledge of their students, institutions, and host countries. This is important for not only developing countries, but also developed countries such as Japan and the U.S. whose demographic trends are yielding increasingly older populations. Globalization and innovation play key roles for both the U.S. and Japan. In addition to trade, both the U.S. and Japan have the largest foreign direct investment (FDI) outflows of all developed countries; they also now both use FDI income to help counter trade deficits [1].

Japan and the U.S. more than ever need to attract, motivate, and retain larger numbers of engineering students for developing innovations for global organizations. The organizational use of innovation leads to productivity gains which, in turn, leads to greater national economic prosperity [2]. This interpretive study presents and applies a framework for understanding best practices experienced in Japan so that engineering education across borders can be more successful [3].

We next present a literature review of knowledge sharing and social influence. Both of these streams of knowledge serve as the foundations of the dimensions of our theoretical framework. We then present our framework applied to teaching and learning in Japan compared to the U.S. The paper concludes with implications for engineering education in foreign cultures as well as recommendations for future research.

Knowledge Sharing for Learning

Knowledge is created and learned through socialization and internalization for active learning by the individual [4]. Through the lens of practice, the understanding of knowledge extends beyond development or delivery, and includes the consumption and shared participation between employees and managers as well as between students and professors [5].

Knowledge is information that has been personalized through meaningful relations which can be useful [6, 7]. Knowledge is created by combining elements that are previously unconnected, or by creating new ways for combining previously connected elements [8]. Hence, intellectual capital is cultivated by combining knowledge in new ways from different stakeholders of varying expertise and experiences [9]. Teachers and students exchange and recombine elements of knowledge with one another [7].

Organizations need employees to share knowledge with one another. Similarly, universities need professors and students alike to share and absorb each others knowledge in the classroom. Employees as well as students who do not frequently draw from others in similar situations run the risk of path dependency from the reinforcement of their own learning-by-doing [10-12].

In sum, knowledge sharing needs not only students listing to teachers, but also teachers listening to students, and students listening to other students. Incentives such as personal recognition can be used to encourage knowledge sharing [11]. However, while a larger number of knowledge contributors can be beneficial, the quality of the knowledge that is shared is more beneficial than the quantity [7, 13]. There can be diminishing returns with teachers or students who just talk without listening. How much a student shares their knowledge in the classroom, is influenced by the balance between order and interaction based on the social psychology of the culture. This, in turn, is also impacted by the social influence of others in that culture which we next discuss.

Social Influence for Learning

In addition to knowledge sharing, social influence also serves as a key factor for learning and is also influenced by the cultures of countries. Students learn by the diffusion of knowledge through the observation of others [14]. Through observation, a student's beliefs and evaluations, based on the relative advantage and ability to learn new concepts, can have a positive influence on attitude for learning [15]. Informational influence also draws from the theory of social learning in that the individual learns how to use new systems from the experience of others, thus avoiding the opportunity costs of trying to learn on his or her own [16].

From social learning theory, human behavior is described as an ongoing interaction among cognitive, behavioral, and environmental factors [16]. One learns how to execute new behaviors by observing others, which enables individuals to model this information for their own subsequent actions [16]. At universities, students learn by observing and mimicking their professors as well as their fellow students. Similarly, employees in organizational teams mimic the successful, observed behaviors of others for finding expertise and coordinating knowledge with others [17]. Developing and sharing innovative ideas is mainly a reciprocal, social activity relying on the interactive linking of ideas between individuals and collective knowledge [18]. Learning would take much more effort if it was limited to one's own autonomous actions [16].

Individuals that both absorb and share content with others in the classroom are the most active learners. This is in contrast to students that just listen to lectures without sharing (i.e. lurkers) and those that just share information without listening (i.e. spammers). Those that neither listen nor share are "missing in action" and do not learn or contribute in the classroom [19]. In sum, there needs t o be a balance of both listening and participating in classroom discussions.

Deutsch and Gerard [20] break down the social influence of group norm between the 1) normative influence and the 2) informational influence. Normative social influence is, "the influence to conform to the positive expectations of another", [20, p. 629]. Normative influence is the "pull" of the individual to conform to his or her perceptions of the wishes of others.

Informational influence, on the other hand, is "the influence to accept information obtained from another as evidence about reality" [20, p. 629]. In this situation, the user is influenced to use a new system by directly observing the benefits of information from the adoption and use of that system by others [20]. Informational influence is the self-driven "push" by the user to attain the same perceived informational benefits. From social information processing models, an actor's perceptions of the benefits of learning new information are shaped by the actor's social networks [21]. Positive communication about the use of a new information can positively influence an individual in the same social networks to learn [22]. Informational influence also draws from the theory of social learning in that the individual learns new concepts from the experience of others, thus avoiding the opportunity costs of trying to learn on his or her own [16].

The potential added value from the social influence of network externalities of a user base of active participants from the business unit (or classroom) and overall organization (or engineering school) should further motivate users beyond normative and informational social influence. The social influence of the network externalities of others using smartphone platforms for the reciprocal seeking and sharing of knowledge through interpersonal interactions should, therefore, extend beyond unidirectional normative or informational influences. Hence, the social influence from network externalities and critical mass need to be considered in order to fully understand and take into account how and why individuals are motivated from the benefits of using smartphone platforms for exchanging knowledge.

In sum, the social influence of network externalities and critical mass, therefore, provide a mechanism for taking into account the increased potential benefits that "spill over" to new users due to the increased size of the user base using the smartphone platform [22-24]. University students who are active learning are not just unidirectional consumers, they are also producers of their own content [19, 25]. Network externalities and critical mass extend beyond normative and informational social influence that is sequential, focusing on unidimensional use [20, 26, 27]. Network externalities and critical mass enable the understanding of social influence that is also reciprocal (i.e. in both directions), between the multiple dimensions of seeking and sharing and vice-versa, which, in turn, can quickly lead to greater benefits among the participants.

We next combine the concepts of knowledge sharing and social influence, taking into account social psychology from different cultures in developing our theoretical framework. By culture, we mean the unwritten norms and values that are mentally passed on to new group members [28].

Global Learning Framework Dimensions: Coupling and Hierarchy

We now extend the concepts of knowledge sharing and social influence with respect to two dimensions in our Global Learning Framework (Fig. 1): coupling and hierarchy. Coupling measures flexibility and adaptability, while hierarchy measures the the number of organizational levels and centralization [29]. The X-axis of the framework is coupling with respect to knowledge sharing. The Y-axis of the framework is hierarchy with respect to social influence. This framework enables the classifying of learning in countries based on organizational as well as cultural influences from social psychology.

Organizational structure can be mechanical or organic [30]. These variations in topology have been used for contingency predictions with the understanding that there is no one optimal way for firms to organize since internal and external environments vary and change [31]. The speed of learning within the organization has been traditionally viewed as spatial with respect to the size of the population and distance from its centers [14]. From social cognitive theory, individual behavior is influenced by personal cognitions impacted by social influences and structures of the environment [32]. Social structures are groups based on differentiated parameters of social relations, through interaction and communication, and hierarchical level [33].

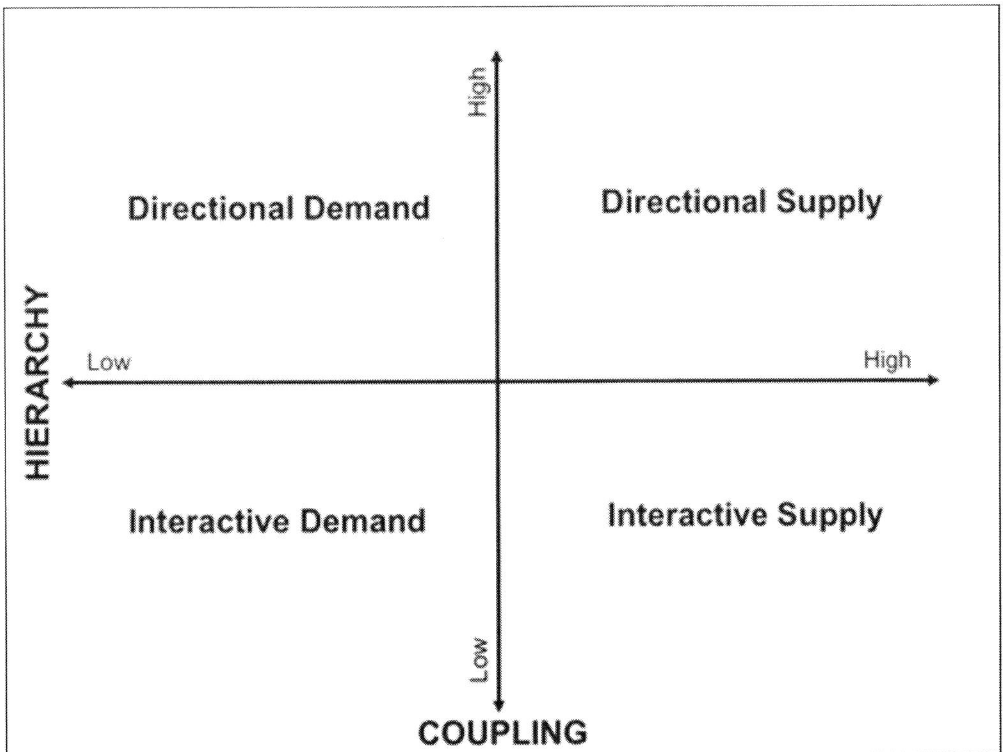

Figure 1. Global Learning Framework

Tightly coupled organizing and communicating occurs in groups with stronger norms, while loosely coupled groups are more flexible [29]. From social psychology of different cultures, we use masculinity, how achievement is viewed, for measuring the level of coupling in our framework [28]. In masculine societies with respect to education, the norm is the best student where failing is a disaster. In feminine societies, the norm is the average student where failing is not a major incident. In masculine societies, brilliant teachers are admired, while in feminine societies, friendly teachers are appreciated [28]. Hence, with with respect to education, Japan has a higher level of masculinity compared to the U.S. Hence, Japan has tighter coupling for sharing a learning knowledge, while the U.S. has looser coupling.

Organizational topology can be centralized or decentralized [34]. The parameter centralization ranges from centralized to decentralized and describes how vertical in structure decision-making authority is spread in an organizational group [35]. In centralized groups, decision-making authority is concentrated vertically in higher hierarchical levels. Here, decision-making is top-down, following a "command and control" flow [36]. At universities, the traditional means for sharing knowledge has been the lecture style from professors down to students. This illustrates a top-down, one-way, hierarchical approach for delivering knowledge typically used at many Japanese universities.

In decentralized groups, decision authority is distributed horizontally at lower hierarchical levels across more autonomous employees having similar hierarchical levels in flatter topologies [35]. In student-centered pedagogy learning theory at many universities in the U.S., learning shifts from top-down to bottom-up. This can be particularly helpful at engineering schools where classes are being turned "upside-down" with peer-led team learning workshops. This active-learning approach has led to increased confidence, intellectual curiosity, and interest in teaching among students [37]. Through student-centered learning, the individual student is responsible for setting learning goals, working towards them, monitoring feedback, and making appropriate adjustments [38]. We use power distance as our measure for hierarchy. Power distance is the emotional distance between employees and managers, or in our study students and professors [28]. Research show that power distance in Japan is typically higher than that in the U.S.

Centralized and decentralized topologies, have their own strengths and weaknesses, with the former being more focused, and the latter being more broad [39]. Hierarchical levels have less vertical influence on each other in decentralized groups since there may be multiple and even conflicting goals between them [40]. Decentralized structures enable greater lateral communication and interaction conducive for adapting to innovations as a continuous process [41]. While decentralization can mean less integration and control within local processes, a stable equilibrium can be reached over time leading to greater overall integration and use of new IT tools at the organization-wide process level [42].

Since the authority for making decisions in centralized structures is concentrated towards the top, leaders are positioned "higher" in hierarchy in these structures [41]. Communication and interaction among employees in centralized topologies are typically vertical with tasks assigned via command and control from the top-down to functional specialists in lower hierarchical levels [30]. These hierarchical structures are able to achieve efficient processing through policies and goals that standardize information [43]. Decision-making contingency theories in the past have proposed that centralized decision making by a leader is more efficient, compared to decision making in decentralized structures among different team members by consensus [44].

As shown in Figure 1, the horizontal axis of the Hierarchy-Coupling (HC) Global Learning Framework is divided into lose (low) and tight (high) coupling. The vertical access represents hierarchy divided into bottom-up (low) and top-down (high). Hence, the lower left quadrant has loose coupling with bottom-up hierarchy.

Through this framework and using the measures of power distance and masculinity for different countries, we introduce four different classification quadrants of global learning between professors and students with examples. These classifications are: 1) interactive demand for the lower left quadrant, 2) directional demand for the upper left quadrant, 3) directional supply for the upper right quadrant, and 4) interactive supply for the lower right quadrant.

Applying the Global Learning Framework for Classification

As an example illustrated in Figure 2, we classify the teaching and learning experience in the U.S. in the lower left quadrant based on measures of power distance and masculinity [28]. The upper left quadrant has loose coupling with top-down hierarchy. This is the quadrant where China is classified. The upper right quadrant has tight coupling with top-down hierarchy. Japan fits in this quadrant. Finally, the lower right quadrant has tight coupling with bottom-down hierarchy. This is where Italy is classified.

The U.S. tends to have an interactive demand learning environment where students are more likely to actively participate and friendly professors are more appreciated. China should tend to have a directional demand learning environment where knowledge mainly flows from professor to student and friendly professors are more appreciated. Japan, on the other hand, is more likely to have a directional supply learning environment where knowledge tends to flow from professor to student and brilliant professors are more likely to be admired. Finally, Italy tends to have a directional demand learning environment where knowledge flows are likely to be more interactive and brilliant professors are admired.

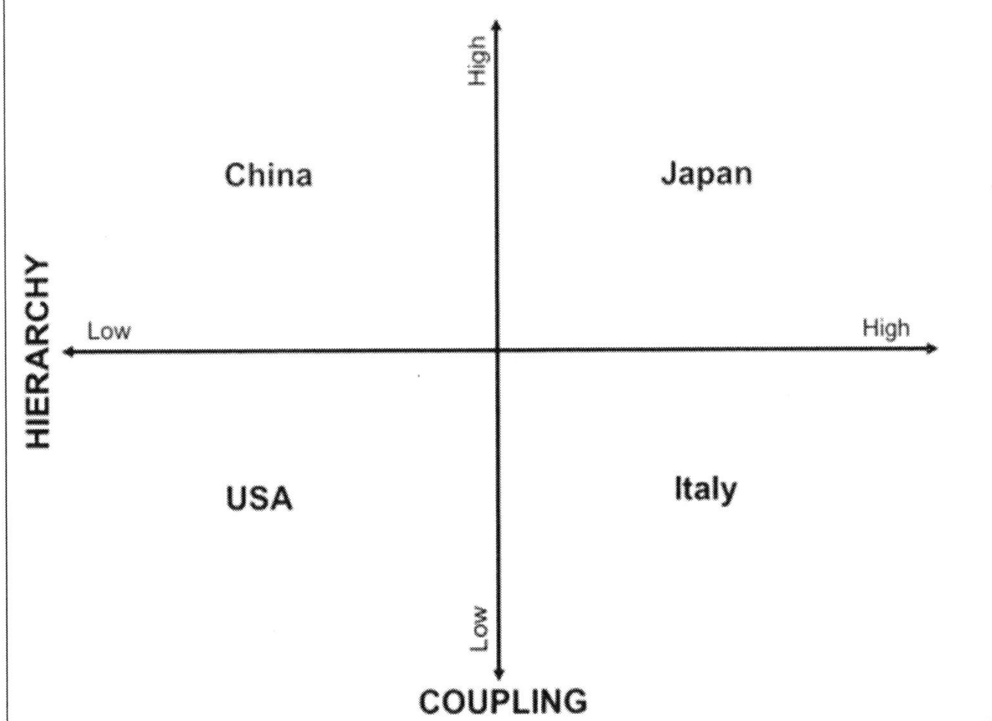

Figure 2. Global Learning Framework Classification Example

Cross-cultural Engineering Education Implications and Future Research

This paper shows that the learning experience in Japan is very different than that in the U.S. We classify Japan as being directional supply (Figures 1 and 2), with centralized social influence hierarchy as measured by power distance and tight knowledge sharing coupling as measured by masculinity [28]. This is opposite to the learning experience we find in the U.S. which is observed to be interactive demand with decentralized hierarchy and lose coupling.

Engineering education in Japan can benefit from these cross cultural understandings by recognizing opportunities and challenges in the classroom. For example, American professors should look for opportunities for exposing students to interactive and bottom-up approaches. This is especially true if the classroom discussions and assignments are in English for improving English language speaking, listening, and writing opportunities for Japanese students.

The following grading criteria recommendations can help facilitate interactive participation in the classroom in Japan: 1) attendance and class participation, 2) written weekly homework, 3) midterm and final exams. By requiring students to attend classes and by assigning written homework every week, there should be a pool of students with prepared written content for bottom-up interaction. Since the English language is a challenge for most Japanese college students, it is suggested that the length of most written assignments be limited to one or two pages. Advise the students to type their homework and print it out. They will then be able to have the confidence to read their homework when called upon in the classroom. At the end of the class collect the homework and grade it and try to give it back to the students by the next class. This continuous verbal and written interaction with the students in the classroom should make learning for them more motivational and effective. Finally, in addition to final exams (or final papers), recommend also having midterm exams (or midterm papers). Midterms will serve as a half-way point check for students to become more actively engaged if they are not yet at that point. Tuition in Japan is typically paid in advance of each semester and is a flat rate independent of the number of classes. Therefore, there is no financial penalty for "freeloading" any classes each semester by staying registered without attending.

Following the traditional Japanese lecture approach (high coupling/high hierarchy) in English will result in too much supply of knowledge to the Japanese students with too little understanding. A professor's teaching will quickly reach saturation without any feedback from the students through active participation. Hence, the interactive classroom is vital not only for effective learning by the students, but also for effective pacing by the professor.

U.S. engineering education can also benefit from Japanese educational norms. The top-down social influence and tight coupling in Japanese universities with respect to graduation rates, internships, and job placement provide the students structural guidance for graduating within four years with good jobs. In summary, cross-cultural opportunities for professors as well as students can benefit not only developing countries, but also developed countries so that institutions do not become too path-dependent. For future research, we plan to empirically extend this study to include teaching and learning in Korea and compare and contrast this to Japan and the U.S.

References

1. Nations, U., *Word Investment Report.* United Nations Conference on Trade and Development, 2015: p. 258.
2. Brynjolfsson, E.a.A.S., *Wired for Innovation: How Information Technology Is Reshaping the Economy.* 2010, Cambridge, Mass.: MIT Press.
3. Yin, R.K., *Case Study Research: Design and Methods.* 2013: Sage Publications.
4. Nonaka, I., *A Dynamic Theory of Organizational Knowledge Creation.* Organization Science, 1994. **5**(1): p. 14-37.
5. Brown, J.S. and P. Duguid, *Knowledge and Organization: A Social-practice Perspective.* Organization Science, 2001. **12**(2): p. 198-213.
6. Alavi, M. and D.E. Leidner, *Knowledge Management and Knowledge Management Systems: Conceptual Foundations and Research Issues.* MIS Quarterly, 2001. **25**(1): p. 107-136.
7. Hansen, M.T. and M.R. Haas, *Competing for Attention in Knowledge Markets: Electronic Document Dissemination in a Management Consulting Company.* Administrative Science Quarterly, 2001: p. 1-28.
8. Moran, P. and S. Ghoshal, *Value Creation by Firms*, in *Academy of Management Best Paper Proceedings*, Dosier, Editor. 1996, In J, B, Keys& L. N.: Georgia Southern University.
9. Nahapiet, J. and S. Ghoshal. *Social Capital, Intellectual Capital and the Creation of Value in Firms.* 1997. Academy of Management.
10. Yoo, Y., K. Lyytinen, and D. Heo, *Closing the gap: towards a process model of post merger knowledge sharing.* Information Systems Journal, 2007. **17**(4): p. 321-347.
11. Garud, R. and A. Kumaraswamy, *Vicious and Virtuous Circles in the Management of Knowledge: The Case of Infosys Technologies.* MIS Quarterly, 2005. **29**(1): p. 9-33.
12. Levinthal, D.A. and J.G. March, *The Myopia of Learning.* Strategic Management Journal, 1993. **14**(S2): p. 95-112.

13. Jones, Q., G. Ravid, and S. Rafaeli, *Information overload and the message dynamics of online interaction spaces: A theoretical model and empirical exploration.* Information systems research, 2004. **15**(2): p. 194-210.
14. Rogers, E.M., *Diffusion of Innovations*. 1983, New York: Free Press.
15. Karahanna, E., D.W. Straub, and N.L. Chervany, *Information technology adoption across time: a cross-sectional comparison of pre-adoption and post-adoption beliefs.* MIS Quarterly, 1999: p. 183-213.
16. Bandura, A., *Social Learning Theory*. 1977, New York: General Learning Press.
17. Kanawattanachai, P. and Y. Yoo, *The Impact of Knowledge Coordination on Virtual Team Performance Over Time.* MIS Quarterly, 2007. **31**(4): p. 783-808.
18. Bento, J., Jose Duarte, Manuel V. Heitor, William Mitchell, *Collaborative Design and Learning: Competence Building for Innovation (International Series on Technology Policy and Innovation)*. 2004, Westport, Connecticut: Praeger. 409.
19. Bento, R., et al., *A taxonomy of participation in online courses.* Journal of College Teaching & Learning (TLC), 2011. **2**(12).
20. Deutsch, M. and H.B. Gerard, *A study of normative and informational social influences upon individual judgment.* The journal of abnormal and social psychology, 1955. **51**(3): p. 629.
21. Brass, D.J., et al., *Taking Stock of Networks and Organizations: A Multilevel Perspective.* Academy of Management Journal, 2004. **47**(6): p. 795-817.
22. Lou, H., W. Luo, and D. Strong, *Perceived critical mass effect on groupware acceptance.* European Journal of Information Systems, 2000. **9**(2): p. 91-103.
23. Kraut, R.E., et al., *Varieties of social influence: The role of utility and norms in the success of a new communication medium.* Organization Science, 1998: p. 437-453.
24. Katz, M.L. and C. Shapiro, *Network externalities, competition, and compatibility.* The American economic review, 1985. **75**(3): p. 424-440.
25. Agichtein, E., et al. *Finding high-quality content in social media.* in *WSDM '08 Proceedings of the international conference on Web search and web data mining*. 2008. Palo Alto, CA: ACM.
26. Liang, H., et al., *Assimilation of enterprise systems: The effect of institutional pressures and the mediating role of top management.* MIS Quarterly, 2007. **31**(1): p. 59-87.
27. Liang, H. and Y. Xue, *AVOIDANCE OF INFORMATION TECHNOLOGY THREATS: A THEORETICAL PERSPECTIVE.* MIS Quarterly, 2009. **33**(1): p. 71-90.
28. **Hofstede G,** G.H., **M** Minkov, ***Cultures** and **Organizations**: Software of the Mind.* 1991, New York: McGraw Hill.
29. Weick, K.E. and K.H. Roberts, *Collective Mind in Organizations: Heedful Interrelating on Flight Decks.* Administrative Science Quarterly, 1993. **38**(3): p. 357-381.
30. Burns, T. and G.M. Stalker, *The Management of Innovation.* University of Illinois at Urbana-Champaign's Academy for Entrepreneurial Leadership Historical Research Reference in Entrepreneurship, 1961.
31. Fiedler, F.E., *A Contingency Model of Leadership Effectiveness.* Advances in Experimental Social Psychology, 1964. **1**(1): p. 149-190.
32. Bandura, A., *Social Foundations of Thought and Action: A Social Cognitive Theory*. 1986, Saddle River, New Jersey: Prentice-Hall, Inc. 617.
33. Blau, P.M., *Presidential Address: Parameters of Social Structure.* American Sociological Review, 1974: p. 615-635.
34. Wagner, J.A., *Organizations*. Encyclopedia of psychology, ed. A.E. Kazdin. Vol. 6. 2000, New York: Oxford University Press, American Psychological Association. 508.
35. Johnson, M.D., et al., *Cutthroat Cooperation: Asymmetrical Adaptation to Changes in Team Reward Structures.* Academy of Management Journal, 2006. **49**(1): p. 103-119.
36. Wagner, C., *Wiki: A Technology for Conversational Knowledge Management and Group Collaboration.* Communications of the Association for Information Systems, 2004. **13**(13): p. 265-289.
37. Johnson, E.C., B.A. Robbins, and M.C. Loui, *What Do Students Experience as Peer Leaders of Learning Teams?* Advances in Engineering Education, 2015. **4**(4).
38. Hannafin, M.J. and H.K. M., *Cognition and Student-Centered, Web-Based Learning: Issues and Implications for Research and Theory*, in *Learning and Instruction in the Digital Age*, J.M. Spector, Editor. 2009, Springer: New York. p. pp 11-23.
39. Pennings, J.M., *Structural Contingency Theory: A Reappraisal.* Research in Organizational Behavior, 1992. **14**: p. 267-267.
40. Orton, J.D. and K.E. Weick, *Loosely Coupled Systems: A Reconceptualization.* Academy of Management Review, 1990: p. 203-223.

41. Giessner, S.R. and T.W. Schubert, *High in the Hierarchy: How Vertical Location and Judgments of Leaders' Power are Interrelated.* Organizational Behavior and Human Decision Processes, 2007. **104**(1): p. 30-44.
42. Berente, N. and Y. Yoo, *Institutional Contradictions and Loose Coupling: Postimplementation of NASA's Enterprise Information System.* Information Systems Research, 2012. **23**(2): p. 376-396.
43. Weber, M., *Economy and Society: An Outline of Interpretive Sociology*, ed. G.R.a.C. Wittich. 1922/1978, New York: Bedminster Press.
44. Hollenbeck, J.R., et al., *Asymmetry in Structural Adaptation: The Differential Impact of Centralizing versus Decentralizing Team Decision-making Structures.* Organizational Behavior and Human Decision Processes, 2011. **114**(1): p. 64-74.

The role of the socio-psychological disciplines in the training of engineers (KNRTU experience)

Mrs. Nailya Sh. Valeyeva, Kazan National Research Technological University

I am a Professor, Dean of the Faculty of Socio-Technical Systems; Head of Department of Social Work, Psychology and Pedagogics; Deputy Director of Institute of Innovation Management in Kazan National Research Technological University; and Honored Scientist of the Republic of Tatarstan. I received my specialty in Physics in 1975 at Kazan State University and PhD in Pedagogics in 1990. Since 1998 I am a Doctor of Pedagogical Sciences. My professional career covers: teaching at undergraduate, graduate and post-graduate level; planning, developing and managing project in the areas of Educational Systems. My research interest include the foreign educational programs, dual degree, project training, foreign industrial practice. I am a member of Russian Academy of Natural Science and Academy of Social Education.

Mr. Roman V. Kupriyanov, Kazan National Research Technological University

I am an Assistant Professor of Kazan National Research Technological University. I received my specialty of chemical technology of high-molecular compounds in 1997. At the same time, I studied at the Faculty of additional education of KNRTU on specialty Psychology. I was a post- graduate student at the specialty 19.00.05 Social psychology in 1997 - 2000. I defended the thesis on Socio-Psychological Mechanisms of Conflict in the Dyad Teacher-Student of a University in 2000. My research interest include the process of implementation of the Bologna Agreement in the Russian Higher Education System.

Mrs. Elvira R. Valeyeva, Kazan National Research Technological University

I am an Assistant Professor of Kazan National Research Technological University. I received my specialty in Social Work in 2005 and PhD in Pedagogics in 2008. My professional career covers: teaching at undergraduate and graduate level (College, Bachelor, Master); planning, developing and managing project in the areas of Educational Systems. My research interest include the process of implementation of the Bologna Agreement in the Russian Higher Education System.

The role of the socio-psychological disciplines in the training of engineers (KNRTU experience)

Introduction. The challenges facing today's engineers are so complex that they require the involvement of teams of experts specialized in different fields. The volume of information and knowledge needed to understand and solve problems is increasing so fast that one person cannot keep up with it all. Intense specialization increases the value of the ability to function on a multidisciplinary team. As a result, the need to add humanistic subjects to the traditional professional education of engineers and other technical staff is becoming evident, which generally manifests in the form of continued "humanitarization" of Higher Engineering Education. The subject of humanitarization of Higher Engineering Education is discussed extensively in the works of Y.Vetrov, A.Ivashkin, T.V.Smirnova, V.N.Stegniy, G.V.Shevtsova, and others[7, 8, 9, 14]. There are sufficient reasons to assume that high-quality professional engineering education must comprise humanitarian component.

This opinion is shared by a number of international researchers[3, 15, 16], they agree that teamwork, team building, leadership, and conflict resolution skills are essential for a successful engineering career. This interdisciplinary approach to the training of undergraduate students helps to develop a more comprehensive vision of an engineer's professional activity.

The importance of incorporating of socio-psychological component into the content of undergraduate engineering curriculum is emphasized in the State Higher Professional Educational Standard (hereinafter – Educational Standard). For example, the Educational Standard for "Chemical Technology" program (bachelor's degree) requires that the graduates who have successfully completed the program, in addition to core professional competences, could demonstrate the following skills: the ability and willingness to develop productive working relationships with colleagues; teamwork; high motivation to perform professional activities; the ability to plan, organize, and supervise the work of subordinates; make managerial decisions.

Despite the existing difficulties in the process of adopting a competence-based approach in the Russian Education System[12], changes in the traditional educational paradigm and a revision of the content of education continue. One of the challenges in the process is to design the fundamentally new activity-focused techniques aimed at developing these competences.

As an example, we present the experience of Kazan National Research Technological University (KNRTU), where the "Workforce Psychology" course has been a part of the undergraduate engineering curriculum for several years now. Over 2,500 KNRTU students complete this course every year.

"Workforce Psychology" course is an integrative course that combines the elements of traditional academic disciplines (psychology, engineering psychology, social psychology, psycho-diagnostics). This course teaches students to recognize and appreciate the unique personality of each team member; it enables students to develop self-awareness and self-management skills that will help them reach their full potential, maintain productive relationships with other team members, and achieve effective communication. It is found that the stage of professional education includes the development of activities from educational-cognitive to training-professional and from it to a real professional[13]. In this connection, the main objectives of the course, in addition to the development and enhancement of students' theoretical and practical knowledge in team working, are to provide an understanding of the psychological processes that influence professional interactions and prepare students to handle various challenges and break down barriers to inter-team communication. To achieve these objectives, the course includes the following sections: Business Communication, The Role of Personality in the Workplace, Professional Conflicts, Socio-Psychological Climate in the Workplace, etc. The learning process is designed in a way that teaches students how to integrate academic knowledge with practical application in their future careers.

Methods of the Study. During research the following methods were used: analysis of normative documents, generalization of teaching practices, content analysis, systematization and generalization of facts and concepts, action research method, observation, questionnaire survey, pedagogical experiment.

The Experimental Base of the Study. Experimental work was carried out on the basis of Kazan National Research Technological University. 178 students participated in the experiment.

Stages of the Study. The study was conducted in three stages:

- The first stage is devoted to the analysis of current state of the studied problem in the pedagogical theory and practice.
- The second stage covers the content development of the "Workforce Psychology" course; identifies stages and defines the principles of formation of social and psychological competencies, including objectives, methods, pedagogical conditions, criteria.
- The third stage involves the systematization, interpretation and synthesis of the research results; refines theoretical conclusions; processes and represents the obtained results.

Teaching methods. Modern practice-oriented forms, methods and educational technologies replace the classical (lecture-seminar-practicum) mode. The modern education is based on workshops, business and role-playing games, problemoriented methods, context study, case study, etc. The main advantage of the new EdTechs is interactive nature, that allows stimulating better students' engagement and their active participation in the training process. Thus practical skills are acquired faster. The rate of learning is especially important nowadays because the period of studies has been reduced[5]. In our view the student's knowledge and skills in Workforce Psychology would benefit significantly if during the training the following methods are practiced: (i) socio-psychological trainings; (ii) simulation games; (iii) interactive group activities; (iv) case study; (v) brainstorming; (vi) discussion.

Karimov A. and Kazakova V. suggest that «although intellectual virtues cannot be taught directly, they can be fostered in the appropriate environment»[2]. The appropriate environment can be created through active learning methods.

In order to enhance teamwork skills, the ability and willingness to cooperate with colleagues, we developed a system of training exercises which include diagnostic and formative components. Exercises in this block are aimed at team building, helping each participant of an interaction to reach their full potential, developing the skills of conflict-free interaction. These training exercises complement the content of lectures on "The Role of Personality in the Workplace", "Socio-Psychological Climate in the Workplace" and demonstrate how to maintain an effective cooperation. Active interaction during exercises allows you to assess and create a positive socio-psychological climate in the group (and, if necessary, adjust it), which in turn affects the level of student motivation. Training exercises in this case are used as the method of achieving the didactic goals.

Simulation game is an effective method of learning and developing the supervisory, management, and decision-making skills. It helps "re-create the objective and the social content of a professional activity, simulate the system of interactions specific to a certain type of workplace"[6]. In addition, the simulation game eliminates the gap between "the abstract nature of the subject and the real nature of the professional activity" (ibid.) The experience shows that games enables the development of all mental processes of a person more effectively than other types of activities, facilitating the transition to a new, higher stage of self-development[11]. This method helps identify the main barriers to cooperation and mutual understanding, identify stressors and conflict-generating factors of teamwork, and identify ways to prevent or overcome these factors. Through participation in simulation games students have the opportunity to identify the leadership qualities, acquire group management, decision-making, and stress-management skills. Simulation games method was implemented as an addition to lectures on "Business Communication", "Professional Conflict Management".

The focus of training exercises and simulation games was on the management of interpersonal interaction.

In order to gather feedback from students on their participation in exercises/games and assignments, and learn how these activities impact the development of teamwork and communication skills, the authors developed reflective questionnaires for each block of assignments (training exercises, business games), asking students to answer questions that reveal the level of cohesion of the group, the presence and influence of the positive socio-psychological climate on the performance and the willingness to work in this group, the presence of barriers in interactions. In addition, each student had the opportunity to express their opinion about the work accomplished individually or in collaboration with others, describe their classroom experience and emotions during interaction within the team, etc.

It should be noted that the reflection was carried out after the completion of each exercise and at the end of a class. A content analysis of an exercise or class in general, creates the conditions for self-assessment, self-learning, self-development of students, giving a new meaning to the experience, information, summarizing the emotional outcomes, gathering feedback.

In addition to personal reflection, which was carried out in the form of a written essay, we conducted a group reflection, summarized the results, which helped participants to develop the ability to assess, understand, and perceive themselves and other members of the group. Students had an opportunity to find out how adequate their self-image is and how they are perceived by other members of the group.

Results. The study identified that all respondents, 178 people, acknowledge the positive impact of training exercises, business games and assignments for the development of socio-psychological and communication skills. The analysis of free-form written student essays on their classroom experience, impressions and comments showed the following.

All students are satisfied with the result of the group as a whole. The involvement of every member of a group in productive teamwork has been identified as an essential attribute of team cohesion (79% of respondents). In addition, all participants reported the significant positive affect of participation in practical sessions on team effectiveness. After students completed assignments/games, 65% of respondents said that the exercises helped them enhance attention, observation, and the ability to feel the mood and psychological state of partners.

Among the factors that influence the successful implementation of group assignments, students identified the following factors: personality and management style of a leader (89%), individual psychological characteristics of group members (67%), basic conflict resolution and prevention skills (69%), ability to develop a strategy and an action plan (84%), and so on. However, many respondents noted that they deliberately chose a position of subordination, since the leadership role has already been taken by other member of the group during the implementation of particular assignments (31%).

After assessing the communication barriers, students noted that people's behavior is influenced by past experience, personal relationships, current needs, and other factors. Exercises aimed at developing the communication competences create conditions that help to identify the motives, focus, and needs of a partner.

As identified by students, the challenges faced by the members of a group can be objective and subjective. The objective challenges include the following: lack of care and attention from classmates to the opinion of some members of the group (46%); lack of assignment implementation supervision (67%); lack of understanding (41%); lack of clear understanding of the future profession (48%), which is understandable, since all participants are first-year students.

The subjective challenges include: difficulty of self-assessment (31%); difficulty of overcoming communication barriers, uneasiness (35%); limited time and lack of the skills necessary for assignment implementation (41%), etc. Despite the difficulties in cooperation within the group, participation in interactive exercises helps students gain experience and tools for using self-reflection as a learning technique in the future.

After the successful completion of the "Workforce Psychology" course, students were asked to evaluate the importance of effective communication skills acquired during the course for their future careers. It should be noted that all of the respondents agreed on the importance of acquiring these skills. More specifically, some respondents stated that they: "realized the importance of recognizing the individual psychological characteristics of each member of a group", "acquired effective goal- setting skills necessary to organize the work of subordinates," "classes contributed to positive group dynamic," etc.

Conclusion. Today's labor market demands engineers who are multidisciplinary experts, able to learn fast and make decisions in rapidly changing environment[4]. Modern engineers are required to have special technical training, as well as the social and psychological skills to ensure the successful interaction with other team members. Integration of socio-psychological disciplines into the content of engineering curriculum is fundamental to training of engineers that can meet the requirements of today's labor market.

The experience of teaching the social-psychological disciplines and feedback received from students who completed the "Workforce Psychology» course show that the acquired skills are highly valued by engineering students and can benefit them during their training and in their future career.

The use of interactive and active learning methods of teaching encourages increased interest and involvement of students in solving practice-oriented professional problems modeled in quasi-professional environment, that despite the conventionality of the game situation, allow participants to experience the real, true emotions, resolve conflicts, both interpersonal and intrapersonal challenges , develop the necessary knowledge, skills through group and individual reflection[1, 10]. Specially organized didactic games encourage critical thinking and creativity in students; help them prepare for academic success and future careers.

References

[1] Hutorskoy, A.V. (2003) *Didakticheskaya evristika. Teoriya i tehnologiya kreativnogo obucheniya* [Didactic heuristics. Theory and Creative Learning Technology] Moscow: Publishing House in Moscow State University, 416 p.

[2] Karimov, A., & Kazakova, V. (2015) *Intellectual Virtues and Education Practice.* The Social Sciences, 10 (6): 1317-1323.

[3] Khalid, A., Chin, C.A., Atiqullah, M.M., Sweigart, J.F., Stutzmann, B., Zhou, W. (2013) *Building a Better Engineer: The Importance of Humanities in Engineering Curriculum* Paper presented at ASEE Annual Conference, Atlanta, Georgia. https://peer.asee.org/19270

[4] Kupriyanov, R.V. (2011) *Funkcional'ny' podhod k podgotovke specialistov mezhdisciplinarnyh professi' (na primere nanotehnologov i social'nyh rabotnikov)* [Functional approach to training multidisciplinary professionals (the case of nanotechnology engineers and community development workers)] Vestnik Kazanskogo tehnologicheskogo universiteta, N15: 315-318.

[5] Kupriyanov, R.V., & Gorodetskaya, I.M. (2015) *Global trends in higher education and thier impact on engineering training in Russia.* International Conference on Interactive Collaborative Learning (ICL), pp. 244-250.

[6] *Psychology. Dictionary.* (1990) Edited by Petrovsky A.V. & Yaroshevsky M.G., 2nd ed., rev. and additional. Moscow: Politizdat, 494 p.

[7] Shevtsova, G.V. (2009) *Paradigmal'nyi podhod k izucheniyu processa gumanizacii inzhenernogo obrazovaniya v Rossii* [Paradigm approach to study of the process of humanization of engineering education in Russia] Vysshee obrazovanie segodnya, N11: 50-53.

[8] Smirnova, T.V. (2009) *Vozmozhnosti vizual'nogo myshleniya v gumanitarizacii inzhenernogo obrazovaniya* [Visual thinking potential in humanization of engineering education] Vysshee obrazovanie v Rossii. N4: 93-98.

[9] Stegniy, V.N. (2013) *Gumanitarizaciya inzhenernogo obrazovaniya* [Humanization of engineering education] Vysshee obrazovanie v Rossii. N11: 55-63.

[10] Sternberg, R.J. (2012) *The assessment of creativity: an investment-based approach.* Creativity Research Journal, 24(1): 3-12.

[11] Valeyeva, N.Sh., & Mingaleyeva, A.V. (2010) *Formirovanie gotovnosti k professional'no-lichnostnomu samorazvitiyu posredstvom avtorskogo speckursa* [Facilitating personal and professional development through the author's course] Vestnik Kazanskogo tehnologicheskogo universiteta, N3: 202-206.

[12] Valeyeva, N.Sh., Kupriyanov, R.V., Valeyeva, E.R. (2015) *Results and challenges of Russia's integration into Bologna Process.* International Conference on Interactive Collaborative Learning (ICL), pp. 404-406.

[13] Valeyeva, N.Sh., Kupriyanov, R.V., Valeyeva, E.R., Nadeyeva, M.I., Guryanova, T.N., Yurtayeva, N.I., Kaysarova, Zh.E., Safina, A.A. (2016) *The Managerial Mechanism of Social Sphere Future Specialists' Professional World View Formation.* International Review of Management and Marketing, 6(S2): 135-141.

[14] Vetrov, Y., & Ivashkin, A. (2006) *Gumanizaciya i gumanitarizaciya inzhenernogo obrazovaniya* [Humanization and humanitarization of higher engineering education] Vysshee obrazovanie v Rossii, N1: 45-50.

[15] Welch, R.W. (2011) *Humanities and Social Sciences within Civil Engineering Curriculum.* Paper presented at Annual Conference & Exposition, Vancouver, BC. https://peer.asee.org/18070

[16] White, C.K., Breslow, L., Hastings, D. E. (2015) *Understanding Curricular Approaches to Communication as a Global Competency: An Interdisciplinary Study of the Teaching and Learning of*

Communication. Paper presented at ASEE Annual Conference and Exposition, Seattle, Washington. 10.18260: p.24957.

Establishing Partnerships for Global Service Learning in Engineering

Dr. Jeffrey R Seay, University of Kentucky

Dr. Jeffrey R. Seay is the PJC College of Engineering Professor and Associate Professor of Chemical and Materials Engineering at the University of Kentucky at Paducah Extended Campus. Dr. Seay joined the University of Kentucky in 2008 following a twelve year career in the chemical industry. Dr. Seay completed his B.S. and Ph.D. from Auburn University and his M.S. from the University of South Alabama, all in Chemical Engineering. His primary research area is process systems engineering focused on sustainability, green chemistry, biofuels and appropriate technology for underdeveloped regions.

Mr. John Christopher Higgins, University of Kentucky

John Higgins is an undergraduate Mechanical Engineering student at the University of Kentucky College of Engineering, Paducah Extended Campus. John works for the University of Kentucky Appropriate Technology and Sustainability (UKATS) research team as an undergraduate researcher.

Ms. Chandni A Joshi, University of Kentucky

Chandni Joshi is an undergraduate Chemical Engineering student at the University of Kentucky College of Engineering Paducah Extended Campus. Ms. Joshi joined the University of Kentucky Appropriate Technology and Sustainability (UKATS) research team in 2013 and has traveled with the group to Cameroon, Gabon, India and Uganda to implement their research.

Sarah Frances Willett, University of Kentucky

Establishing Successful Partnerships for Global Service Learning in Engineering

Abstract

Service learning experiences can be life changing events for both engineering students and faculty. These courses can provide educational context for the impacts – both positive and negative – that engineers have had and continue to have on the developing world. Non-traditional education abroad experiences in developing countries can be a challenge to organize, but can benefit greatly from strong partnerships with organizations in the host country. The University of Kentucky Appropriate Technology and Sustainability (UKATS) research group is an undergraduate, service learning and sustainability focused research group at the University of Kentucky (UK) College of Engineering Paducah Extended Campus Program. This group has established successful partnerships with Non-Governmental Organizations (NGO) in Cameroon and India as part of a non-traditional study abroad course in global service learning in engineering. The service learning projects undertaken by the UKATS research group as part of the EGR 390: Global Service Learning in Engineering course at UK are focused on the development and dissemination of technology based on the principles of appropriate technology and sustainability – particularly the societal impacts of those projects on the local communities. This contribution gives an overview on how these partnerships were developed and describes the importance of strong in-country partnerships for successful service learning experiences. In particular this contribution will illustrate how the need for community interaction is critical to the success of any project in a rural or developing community. The experiences of the UKATS group with the African Centre for Renewable Energy and Sustainable Technology (ACREST) in Cameroon and the Organization of Development Action and Maintenance (ODAM) in India will be described. In addition, lessons learned regarding properly preparing students for service learning based education abroad experiences will be described.

Keywords: Africa, Cameroon, India, education abroad

Introduction

Engineers play a critical role in society. Most people are well aware of the more visible contributions of the profession – from roads and bridges to clean water to manufactured products, pharmaceuticals, chemicals and computers. The list goes on and on. Despite these impacts however, most engineering students fail to realize the importance of their place in the global society. Global service learning courses can help educate engineering students as to the challenges in the developing world and how the engineering profession can address these challenges in an appropriate societal context. That said, in-country experiences in the developing world can be difficult for students due to both logistical and cultural barriers. A group of engineers, no matter how well intentioned, cannot simply show up in a rural area of an underdeveloped region and expect to be successful. Establishing an engaged in-country partner is critical for the chances of successfully meeting the objectives for an international service learning project

Challenges of Working in Developing Regions

International service learning projects differ significantly from traditional study abroad experiences. A traditional study abroad experience features a class or group of classes offered for a foreign university for college credit. Because of the university setting of a traditional study abroad program, room and board and classroom facilities are readily available. Although global service learning projects can be terrific learning experiences for students, the level of planning is much greater than for a traditional study abroad program. In addition, numerous other challenges are encountered when working in developing regions, as will be described.

Cultural challenges
Obviously the potential for a language barrier is perhaps the biggest challenge, but other cultural norms such as style of dress, local cuisine, religious preferences, alcohol use and gestures vary greatly from place to place. It is important not to inadvertently upset or offend residents of the host country. Often social norms, particularly with regard to the treatment of women and their traditional place in society, are very different from the experiences of the typical college student. A good in-country partner will be aware of the differences between western and local culture so that both the visiting students and the local community will be aware of differences, preparing both sides so that any misunderstandings are avoided.

Logistical challenges
Transportation in a rural region in an underdeveloped country is quite different from what students are most likely used to. Issues with clearing customs at the arrival airport are typical and having local support at the airport is invaluable. Outside the airport, travel takes longer and can be unpredictable. The means of transportation can also be quite different, ranging from autorickshaws, to motorcycle taxis to walking. Additionally, there can be safety issues for foreigners traveling in an unfamiliar region.

Lodging and General Safety
In a rural or underdeveloped region, western style lodging can be difficult to find. Indeed, part of the service learning experience is adapting to local norms with regard to lodging and general comfort. Limited electricity, limited communication options, lack of indoor plumbing and climate control are certainly common. That said, ensuring the safety of the travelers is paramount. Since it can be difficult to get information before traveling, complying with university requirement for vetting student accommodation prior to travel is a real challenge.

Community acceptance
Just because a project seems like it will be beneficial doesn't mean that it will be accepted by the local community. The project may address a problem that is not seen as important to the local community or may violate their established cultural norms. Simply having an idea does not automatically lead to a successful project. Although the objective of a global service learning in engineering course is the education of the participating students, the project is a waste of time if it does not benefit the local community. Furthermore, ensuring that the project is accepted by the community and sustainable using local resources after the students leave is also a key challenge.

Healthcare and hygiene
For global service learning programs held outside of a university setting, healthcare may be difficult to find in an emergency situation. Additionally, foodborne and insect borne illnesses are also common in many regions meaning that special precautions are needed to keep students healthy for the duration of their project. It should also be noted that students traveling to these regions should consult with a physician prior to departure regarding medications and vaccines recommended for the particular region. The U.S. Centers for Disease Control website is a good resource for students and faculty. The in-country partner should be able to ensure that the food prepared for students is effectively cleaned and cooked to proper temperatures to avoid illness. This should be a point of discussion between the faculty advisor and the in-country partner before departure.

What Makes a Good International Partner?
In addition to simply having a presence in the target country, an effective partner organization must have a number of key characteristics. Among these are:

Well established in the target country
Unlike in the United States, in many countries, foreign visitors are expected to register with local governmental authorities. This is necessary for local security as well as for the safety and well-being of the

travelers. If the in-country is well established and has a good relationship with local authorities, many problems can be avoided.

Experienced with international visitors
Strong community outreach does not always translate into the ability to host international visitors. Look for organizations who have experience hosting western visitors. This is particularly important with regard to health and hygiene. Keeping a group of visiting students from getting sick requires a good understanding of safe food handling and hygiene. And organization with hosting experience can also best prepare guests with regard to local customs and culture.

Strong local community ties
It is important for students to be briefed on local culture and customs. However, the local partner can also educate the local community about what to expect from their foreign visitors. A two-way understanding of cultural norms and expectations can help to avoid any inadvertent misunderstandings. A good partner organization should know and understand the community needs and be able to create and direct a project that is wanted, needed and accepted by the community.

Ability to handle transportation and logistical issues
Travel is often difficult is underdeveloped countries. Taxi service around large international airports has a reputation for being unsafe for foreigners. Having the in-country partner organize transportation is often the best option. In many cases prices are not fixed and haggling over cost is normal practice. The assistance of the local partner can ensure that reasonable prices are charged for transportation. In addition, the in-country partner is available in the event of unexpected travel disruptions.

Able to provide room and board for the group
Although the in-country partner may not maintain room and board facilities themselves, the partner will be the most important resource in securing housing that is safe and provides an appropriate level of hygiene. Of course it is important to realize that the accommodations in a rural or underdeveloped region will be very different from home. Ensuring the health and safety of the travelers are top priorities. A good in-country partner must provide help with selecting an appropriate location, if they do not provide this themselves.

Access to basic healthcare
Despite the best preparations, some students will get sick when visiting a foreign country. Although most illnesses will only require rest and fluids, the in-country partner must at a minimum have a plan to get to sick individuals to qualified medical professionals if the need arises. The visiting students and faculty should plan for minor medical needs such as bringing a first aid kit and medications such as a pain reliever, antibiotic ointments and medicine for stomach discomfort. It is of course recommended to consult the Centers for Disease Control prior to departure to check for recommended vaccines and a list of common illnesses, but the in-country partner should be able to inform the group of the types of previous incidences that have occurred.

Project sustainability
Once the student group returns home, it is the in-country partner who will bear the responsibility of continuing the project and addressing any problems that may arise. This underscores that the partner must be more than simply a provider of logistics services – they must be a fully engaged partner in the technology or technical solution as well. A successful partnership will only last as long as it is actively maintained. The university group and the in-country partner must establish a plan for maintaining the project after the students leave. Establishing channels for regular communication requires planning and commitment by both parties.

Organizational Partnerships

The UKATS research team has partnered with two NGOs to conduct global service learning experiences for undergraduate students. The first is the African Center for Renewable Energy and Sustainable Technology (ACREST), www.acrest.org, located in Bangang, Cameroon. The second NGO is the Organization of Development, Action and Maintenance (ODAM), www.odamindia.org, located in Tiruchuli, Tamil Nadu, India.

Recent Global Service Learning Projects

Two recent global service learning projects conducted by the UKATS research team highlight the results of two successful international partnerships. The first project was conducted in collaboration with ACREST in Cameroon and the second was conducted in collaboration with ODAM in India. Both projects were conceived based upon the principles of green chemistry, sustainability and appropriate technology (Schumacher, 1973).

Project 1: Developing a Novel Low Cost Biodiesel and Biochar Processor for Rural Cameroon Using Locally Sourced Materials and Feedstocks

The principle objective of this project was to develop a low-cost, environmentally benign technology for producing biodiesel and the associated raw materials, i.e. methanol and base catalyst, from locally available resources available in rural Cameroon in sub-Saharan Africa (Seay, *et al.*, 2014, Seay and Lumkes, 2014). To achieve this objective, three tasks were required:

- Design a low cost process for generating biochar and methanol by destructive wood distillation,
- Design a low cost process for generating biodiesel from locally sourced feedstocks,
- Complete the mechanical design and fabrication plans of a processor for producing biodiesel, biochar and methanol using locally sourced materials.

This project was initially conceived through the partnership with the ACREST. The processor that resulted from this project was designed to be simple to build and operate and to cost less than 100 USD. Locally sourced scrap materials such as discarded steel drums and junkyard automobile parts were the primary construction materials.

Although the production of biodiesel from vegetable oil and methanol from biochar production is certainly well understood, there are numerous technical challenges involved with adapting this technology for the developing world. Principle among these challenges are: ensuring that the design is robust enough to handle wide variations in feedstock quality; developing quality control safeguards for production in rural areas; and ensuring that the chemistry is well understood based on locally available resources.

In rural parts of Africa, electricity is often expensive and unreliable. Therefore the biodiesel processor needed to be designed to require no moving parts, operate without needing electricity and utilize a simple and innovative temperature control strategy. Although a loss in efficiency was expected due to the techniques employed, the student design was rugged and extremely simple to build and operate, making it ideally suited for small scale producers in the developing world.

The partnership with ACREST was critical to the success of this project. The ACREST facility is located approximately 275 km from the international airport in Douala, Cameroon. This meant that transportation would be a critical issue. In addition to organizing all in-country travel, room and board was provided at the ACREST guesthouse in Bangang, Cameroon. ACREST maintains a guest house in addition to the workshop and technology center. The guest house can accommodate up to 18 students and faculty. The ACREST staff prepares all meals and is trained to ensure proper hygiene for the visiting students. Because of the rural location the facility, it is necessary for ACREST to provide all services for visiting groups.

Although electricity and phone service in the village is limited, ACREST provides comfortable accommodations for all their guests which enhanced the student experience.

This project was funded in part by a US EPA People, Prosperity and the Planet grant award (Award number SU835303).

Figure 1. ACREST Workshop in Bangang, Cameroon

Project 2: A Green Chemistry Approach for Producing Non-Synthetic Pesticide in Under-Developed Regions

The main objective of this UKATS research team project was to develop a sustainable, low-cost method for producing a non-synthetic pesticide, wood vinegar, from already existing biochar production techniques present in rural regions of Cameroon, and Tamil Nadu, India. Collecting the wood vinegar reduces the release of volatile chemicals and tars into the atmosphere, which is the case when making biochar using current practices. As a result, this research focuses on minimizing atmospheric pollution by collecting and condensing the released chemicals into a useful and inexpensive, non-synthetic pesticide, herbicide, and fungicide known as wood vinegar. To achieve this goal, the tasks below were required:

- Design and optimize a production method for generating wood vinegar by pyrolysis of locally available biomass.
- Determine the composition of the wood vinegar and its variance as different biomasses are pyrolyzed.
- Conduct field studies in Cameroon and India for the use of wood vinegar as a pesticide, herbicide, and fungicide to enhance its effectiveness.
- Complete the mechanical design and fabrication plans for making wood vinegar using locally sourced construction materials to implement the prototype at ACREST and ODAM.

This project was conceived and conducted through the partnership with ACREST and ODAM. The description of the student team project will be to design and optimize the construction of the wood vinegar processor, and analyze its product for green chemistry use as a non-synthetic pesticide. The challenge to the student team was to fabricate this processor using only locally sourced materials present in rural regions while keeping cost below 100 USD.

To achieve this goal, the UKATS P3 research team developed a sustainable method for collecting, condensing, and transforming the waste gas produced from the production of charcoal into a non-synthetic pesticide, known as wood vinegar. The use of renewable feedstock leads to generation of safer, more sustainable pesticides, which minimize adverse effects on human health and environment. The results of this collaborative design project will combining engineering technology in a way that makes the collection and use of wood vinegar based pesticide easy and accessible to a wide array of potential users in the developing world.

As with the previous project, the partnership with ODAM was critical to the success of this project. ODAM maintains a technical facility in Tiruchuli, with kitchen and dining facilities, but does not have room to accommodate a large student group. Unfortunately there were no suitable lodging facilities in Tiruchuli. Because of this, ODAM arranged lodging at a facility in the nearby city of Aruppukottai, about 20 km away, and arranged daily transportation from the lodging location to the ODAM workshop. Due to the duration of the trip, ODAM also arranged for several excursions to nearby areas of cultural importance.

This project was funded in part by a US EPA People, Prosperity and the Planet grant award (Award number SU836115).

Figure 2. ODAM facility in Tiruchuli, Tamil Nadu, India

Conclusions

In conclusion, the successful execution of a global service learning project is dependent on many factors, but perhaps the most important is having a strong in-country partner. Finding a partner is a critical first step for faculty interested in engaging in a service learning project in a developing nation. Overcoming cultural as well as logistical hurdles is certainly most effectively done with a committed partner organization. As a result this contribution has highlighted some of the key characteristics of an in-country partner and provided examples of how having such a partnership has led to successful project implementation in India and Cameroon.

Acknowledgements

The authors wish to acknowledge the support and assistance of Dr. Vincent Kitio and Dr. Isaac Zama with the African Center for Renewable Energy and Sustainable Technology in Bangang, Cameroon and Mr. Elango Jeyaraj with the Organization of Development Action and Maintenance in Tiruchuli, Tamil Nadu, India.

References

Schumacher, E. F. (1973): *Small Is Beautiful: Economics as if People Mattered*, Harper & Row, New York, New York.

Seay, J. and J. Lumkes (2014): "Multi-University Partnership for Global Service Learning in Sub-Saharan Africa", *International Journal of Service Learning in Engineering*, Special Issue, pp 367-380, Fall 2014.

Seay, J., I. Zama and B. Butler (2012): "International Partnership Helping to Bring Appropriate Biofuel Technology to Rural Cameroon", *International Journal of Service Learning in Engineering*, Vol. 7, No. 2.

ASEE 2016
INTERNATIONAL FORUM
NEW ORLEANS, LOUISIANA / JUNE 25, 2016

Paper ID #17510

A Comparison and Evaluation of Aeronautical Engineering Learning Outcomes using an Airborne Flight Laboratory and a Flight Simulator Laboratory.

Mr. Raymond Colin Lewis, University of New South Wales at the Australian Defence Force Academy

Commenced flying career in the Australian Defence Force; became an Airline Pilot; returned to university to inform work as a Human Factors facilitator for the airline. Worked in Middle East as pilot before commencing as a Senior Lecturer for the University of New South Wales (Canberra) in 2001. As well as teaching, operated University aircraft for Aeronautical Engineering laboratories.

Dr. Matthew Garratt, University of New South Wales

A/Prof Matt Garratt is the research admissions coordinator for the school of Engineering and IT and is based in the Canberra campus of UNSW Australia. His main research areas focus on sensing, guidance and control for autonomous systems. Some of his research successes include demonstration of terrain following using vision for an unmanned helicopter, landing an unmanned helicopter onto a moving deck simulator and control of helicopters using neural networks. Some of his current research projects include achieving autonomous flight in cluttered environments using monocular cameras and range sensors, landing UAVs on moving platforms, formation satellite control and building and manipulating 3D models collected by robots for visualisation purposes. He is senior member of the American Institute of Aeronautics and Astronautics (AIAA), member of the American Helicopter Society (AHS), the Association for Unmanned Vehicles Systems International (AUVSI) and the Australian Association for Unmanned Systems (AAUS).

Comparison and Evaluation of Aeronautical Engineering Learning Outcomes using an Airborne Flight Laboratory and a Flight Simulator Laboratory.

Introduction

Traditional teaching approaches in courses covering aircraft dynamics and control typically utilize theoretical development, comprised of the derivation of governing equations followed by hand calculations of problems and/or simple simulation examples. One of the major challenges experienced by students in appreciating such teaching is the inability to visualize complicated, multi-modal aircraft motions thus leading to a decrease of students' motivation and understanding of fundamental concepts [12].

Like many other engineering schools, the School of Engineering and Information Technology (SEIT) at the University of NSW, Canberra recognises the importance of student laboratories to complement classroom theory. As engineering educators have espoused, this is because laboratory work enables students to observe the relationship between theory and practice. Importantly, students begin to gain confidence in the application of theory by observing its practical limitations [4].

For this reason, in 1998 SEIT decided to develop an airborne laboratory facility. An aeroplane was acquired and it was equipped with a suite of sensors and instruments that allowed many aeroplane flight parameters to be measured and recorded. Aeronautical Engineering students and candidate pilots carried out a flight which allowed them to investigate aspects of aircraft performance, handling qualities and stability (static and dynamic) in a 1.2 hour flight. These experiments maximized the students' experience and exposure to flight test.

After an evaluation of the effort and time that academic staff required to operate the flight laboratory, in 2010 the airborne flight laboratory was discontinued. In its place there has been developed an Aviation Studio, equipped with a fixed-base flight simulator. Similar to the work carried out at another Australian University, the engineering flight simulator has been specifically designed as a versatile and practical hands-on aid to the teaching of flight mechanics and dynamics and aircraft design [3]. Using a flight console, screens and X-Plane software, students can manipulate many aircraft characteristics.

The use of technology to simulate physical phenomena most likely originated in the "Blue Box" developed by Edwin Link in 1928 [5]. Link trainer flight simulation was used for pilot training extensively during World War II "saving millions of dollars and more than a few lives" [1]. However, flight simulation for engineering training in an academic environment is a relatively recent development [6]. It is maintained that "there is little detailed information on how flight simulators have been implemented in coursework, how effective they are in learning improvement, or how this has been assessed" [6] (p.429). Accordingly, the aim of this paper is to describe and compare the learning outcomes of two teaching paradigms of an aeronautical engineering course on aircraft dynamics and control - the airborne flight laboratory and a more 'contained' learning environment – the Aviation Studio.

The Flight Laboratories

The airborne flight laboratory was conducted in a specially instrumented Cessna 182RG light aircraft (Figure 1). In addition to the standard aircraft instrumentation, this aircraft was fitted with a variety of special instruments and sensors which included; an air data boom (Figure 2) providing airspeed, altitude, angle of attack and sideslip; an inclinometer to measure the inclination of the longitudinal axis of the aircraft; elevator, aileron and rudder control surface angular deflection sensors and pitch and roll rate gyros. Additionally, a computer-based data acquisition and control system, allowing up to 16 channels of data to be recorded at 100 Hz was installed. The fitting of these additional aircraft instruments met the requirements of Civil Aviation Regulation 35 [10 11].

Figure 1: Cessna 182RG VH-CKA

Figure 2: The air-data boom mounted on the starboard wing.

A typical flight laboratory session was conducted with the academic staff/pilot performing manoeuvres and the student recording the parameters of the aircraft as they were displayed on the fitted laptop computer (Figure 3). As the flight progressed the pilot briefed that, apart from an initial control force input, the aircraft was to be flown 'hands off' so that the flight characteristics of the aircraft could be demonstrated. For instance, after a control input, the aircraft was allowed to take up a phugoid motion (Figure 4).

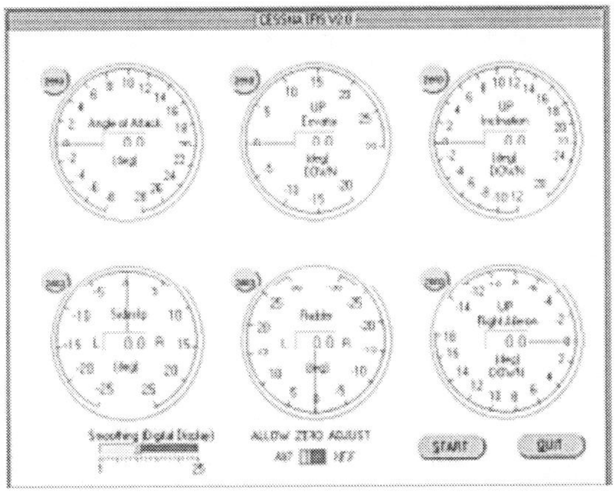

Figure 3: A typical data display using data acquisition computer.

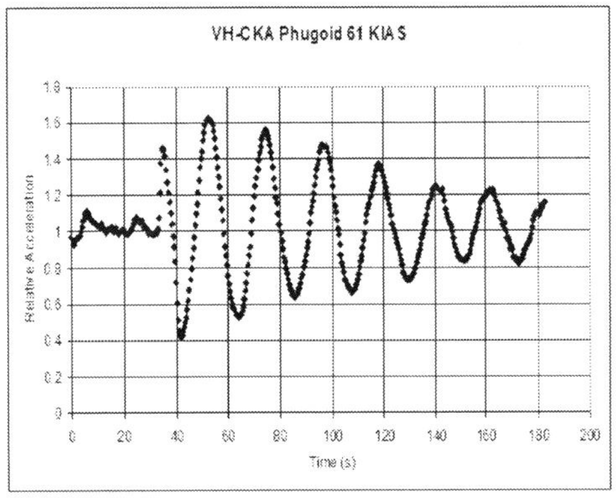

Figure 4: Long. dynamic stability: the Phugoid

During the flight laboratory experiments, two students were taken up at a time. They worked as a team to observe and record data during the flight in a flight-laboratory logbook. (Figure 5). After completing the flight, they analysed their data and submitted a report in which they were required to demonstrate an understanding of the behaviour of the aircraft during each of the flight manoeuvres.

Figure 5: Student recording data during airborne flight laboratory

An important feature of the airborne laboratory was that every effort was made by the pilot to perform low 'g' manoeuvres. This was to avoid discomfort and motion sickness and not compromise the students' ability to observe and record information. For this reason turn performance manoeuvres were discontinued as part of the laboratory as this testing often led to motion sickness problems as students tried to observe and record turn rates, bank angles and 'g' loadings during steady turns.

The Aviation Studio utilises a Precision Flight Controls console, screens and the X-Plane flight simulator software package (Figure 6). Students are able to 'fly' the simulated aircraft and manipulate many aircraft characteristics.

As in the case of the airborne flight laboratory, students record their data and submit a report demonstrating their knowledge and understanding of certain aircraft dynamics and control.

Figure 6: Aviation Studio Flight Simulator

The Aviation Studio experiments were designed to achieve the learning outcomes of the airborne laboratory: straight and level drag polar; lateral and directional static stability; longitudinal handling qualities and demonstration of longitudinal and lateral/directional dynamic modes – phugoid, Dutch roll and spiral modes.

It can be argued that both forms of the aircraft dynamics and control laboratory – airborne and studio-based – are experiential in their implied learning processes. Researchers of experiential learning paradigms maintain that experiential "learning is the process whereby knowledge is created through the transformation of experience" [8 9]. Depending on the nature of the task," experiential learning offers enormous potential for confronting students with highly complex and dynamic situations" [2] (p350). Students must analyse what is going on in the game or exercise, synthesize solutions to address the situation and evaluate their relative merits.

What is germane to the present study is the question of whether the flight experience was too experiential. As previously stated every effort was made to ensure a smooth flight – for instance, when strong westerly winds were forecast and turbulence surrounding the designated laboratory airspace could be expected, flight laboratory sessions were rescheduled. However, it can be reported that there were many occasions when students suffered and complained of nausea which was attributed to the manoeuvres performed in the aircraft.

Evaluation of Learning Outcomes

In both the airborne laboratory and the flight simulator laboratory students submitted a laboratory report some two weeks after the laboratory session. The marked and assessed laboratory reports are a component of the total assessment for a Fundamentals of Flight course. A comparison of the marks and quality of laboratory reports produced in the two teaching and learning environments provided a basis of comparison for the learning outcomes of the two teaching and learning laboratories.

It is realised that comparisons of examination results and test scores are not a valid metric when evaluating the learning outcomes of different cohorts of students. However, the present study made a qualitative assessment of the submitted laboratory reports. Students who were the authors of the laboratory reports – either airborne flight laboratory or Aviation Studio flight laboratory - were matched on the basis of academic prowess. In this way some of the methodological issues when comparing the learning outcomes of two student cohorts may be overcome. The weighted average mean result (WAM) achieved by each student at the end of the previous academic year was ascertained. Students with similar WAMs and who had carried out either the airborne lab or the studio laboratory were matched and the resultant qualitative difference in their laboratory report assessed and recorded.

The laboratory reports were graded on a scale of 1 – poor attainment of learning outcomes to 10 – complete attainment of learning outcomes. The mean of the results of 10 laboratory reports resulting from the airborne flight laboratory and the mean of the results of 10 laboratory reports resulting from the Aviation Studio flight laboratory were calculated and are presented in Table 1.

Table 1: The mean of 10 laboratory reports for the airborne and the studio flight laboratory.

	Airborne Flight Lab.	Studio Flight lab
Grading	7.25	7.75

It is apparent that the Aviation Studio flight laboratory produced a marginally better result in terms of measured learning outcomes of the respective flight laboratories. However, in this study, the relatively small sample size and the methodological issues concerning the comparison of two separate cohorts of participants means that a significant result cannot be claimed. Notwithstanding the cost savings and elimination of risk, perhaps it can be claimed that the students' appreciation of aeroplane flight dynamics is not poorer by the learning experience obtained in an Aviation Studio environment.

Discussion

It must be conceded that flying in an aeroplane was an unsettling (sometimes almost traumatic) experience for some students. Rarely were they allowed to manipulate the controls. The instructor/ pilot would set the experimental conditions and then sit hands off – allowing the aircraft to enter the phugoid or spiral dive. Confinement in a small aircraft cabin and the possible undesirable effects of motion served to distract many students from the task in hand.

The flight simulator studio has none of these distractions. Students are allowed to work at their own pace and manipulate the 'aircraft'. Flight dynamic variables such as changes in aircraft centre of gravity may be changed at will and the resultant changes to aircraft stability observed in a calm learning environment.

The outcome of the comparison and evaluation of aeronautical engineering learning outcomes using an airborne flight laboratory and a flight simulator laboratory is a positive for the flight simulator laboratory given the constraints of comparing and evaluating two cohorts of learners. This finding will serve to inform the development of other learning paradigms and research activities in the flight simulator laboratory.

Recommendations/Implications/Conclusions

To an ever greater degree, teaching institutions are moving towards a virtual world where hands-on experience of equipment and artefacts are being replaced by e-learning paradigms; YouTube video and screen presentations of engineering and associated concepts. The benefits of making a comparison and evaluation of aeronautical engineering learning outcomes using an airborne flight laboratory and a flight simulator laboratory is that it may inform further development of teaching practices in a laboratory environment to better reflect a real world experience and improve learning outcomes.

For instance, the human-machine-interfaces (HMI) of many modern aircraft and unmanned aircraft systems (UAS) are rapidly evolving. Touch screens and other input devices are being integrated into the flight decks of new generation fighter aircraft; commercial airliners and the ground control stations of UAS. The ability to easily and quickly input data may significantly improve situational awareness of flight crew and ground station controllers. The new HMIs may be evaluated and experience gained in the new interfaces during Aviation Studio flight laboratory sessions.

It is expected that the Aviation Studio learning environment will produce better learning outcomes not only for existing aircraft technology but also for future aircraft. Because of the flexibility and adaptability of apparatus and software programs, novel HMIs will be able to be assessed and explored to determine their efficacy and ease of use both from an operator and aeronautical engineering perspective.

The use of simulations in education and training are considered to be beneficial for several reasons; "economy of time and cost savings; the benefit of neutralizing risks; exposure to different experiences that can accelerate learning and the appeal of the simulated, often highly interactive, experience" [13] (p. 492). Thus simulations find a place when it becomes too expensive or too risky to allow students to learn in the real world. Students are allowed to explore, make mistakes and learn valuable lessons in virtual environments. As evidenced in this comparison and evaluation of aeronautical engineering learning outcomes using an airborne flight laboratory and a flight simulator laboratory, the use of a flight simulator laboratory may lead to enhanced learning outcomes.

References

[1] American Society of Mechanical Engineers, (2000). *The Link flight trainer*, ASME, Binghamton NY.

[2] Cannon, H. M and Feinstein, A. H. (2005). Bloom beyond bloom: using the revised taxonomy to develop experiential learning strategies. *Developments in Business Simulations and Experiential learning, Vol32,* 348-356.

[3] Done, G. and Neal, C. (2012). Use of an engineering flight simulator in the teaching of flight mechanics and dynamics and aircraft design. *Proceedings of the 50th AIAA Aerospace Sciences Meeting,* Nashville.

[4] Eley, M. G. (1995). A laboratory approach to teaching in engineering, *Research and Development in Higher Education, Vol 18*, 284-289.

[5] Feisal, L. D. and Rosa, A. J. (2005) The role of the laboratory in undergraduate engineering education. *Journal of Engineering Education* 94(1), p121-130.

[6] Gibben, P., Medagoda, E and Dumble, S. (2009) Enhancement of learning in aircraft handling qualities through variable stability flight simulation. *Proceedings of the 2009 AaeE Conference, Adelaide.*

[7] Harrap, M. J. (2007). An Airborne Laboratory for Undergraduate and Postgraduate Education, *Proceedings of the 2007 AaeE Conference*, Melbourne.

[8] Kolb, D. A., Boyatzis, R. E., and Mainemelis, C. (2000). Experiential learning theory: previous research and new directions in R. J. Sternberg and L. F. Zhang (Eds.), *Perspectives on cognition, learnig and thinking styles.* NJ. Lawrence Erlbaum.

[9] Kolb, D. A. (1984). *Experiential learning: experience as the source of learning and development.* Englewood Cliffs, NJ. Prentice Hall.

[10] Lewis, R., and Harrap, M. J. (2010). A Flight Test Laboratory for Aeronautical Engineering Education. *Journal on Education, Informatics and Cybernetics. Vol 2 (2)* 36-40, International Institute of Informatics and Cybernetics (IIIC)

[11] Lewis, R., and Harrap, M. J. (2009). A Flight Test Laboratory for Aviation Education *Proceedings of the Inaugural Conference of the Aviation Education and Research Conference.* Blenheim, New Zealand.

[12] Shankar, P., Chung, W.-T., Husman, J., and Wells, V. (2013). A novel software framework for teaching aircraft dynamics and control. *Comput. Appl. Eng. Educ., 23*: 63–71.

[13] Wood, R. E., Beckman J. F., and Birney D. P. (2009). Simulations, learning and real world capabilities, *Education and Training, Vol 51 iss5/6* 491-510.

Learning Off the Grid: Implementing Engineering Service Projects in Developing Countries to Achieve Student, Faculty, and Community Outcomes

Dr. Paul John Ackerman Jr P.E., Virginia Military Institute

Paul J. Ackerman, Jr., PhD, PE

Paul Ackerman is an Assistant Professor of Civil Engineering at Virginia Military Institute (VMI). Paul is also the Assistant Program Director and Faculty Advisor for Keydets Without Borders, VMI's international service program.

Paul has over 20 years of project management and construction engineering experience on a variety of local, state, and federal projects. In addition Paul has assisted with engineering service projects in Uganda and Bolivia.

Paul received a BSCE from Virginia Military Institute in 1993, an MSCE from West Virginia University in 1995, and a PhD in Civil Engineering from Virginia Tech in 2014.

Dr. Tim Moore P.E., Virginia Military Institute

A 1997 graduate of VMI, I earned my PhD in Environmental Engineering from Virginia Tech in 2009. I believe that research and new technologies greatly enhance the overall learning experience, and that environmental engineers have a social responsibility. Therefore, I have established new research and initiatives to advance instruction at VMI through the development of VMI Keydets Without Borders (VMI-KWB), a research and study abroad program that provides design and construction of sustainable water and sanitation facilities in developing communities. This program has been instrumental in teaching the concepts of austere, rudimentary development and operation of water and wastewater treatment systems coupled with sustainable energy concepts (solar, biomass). To date, our group has designed solar and gravity powered pumping facilities for sustainable "portable farm" aquaculture projects in Uganda and Bolivia, solar thermal energy showers using recycled construction materials in Bolivia, anaerobic waste treatment and energy generation facilities and solar devices for the capture and conveyance of potable water sources within resource poor communities of Haiti, Bolivia and Uganda. This program has also focused on the pyrolitic development of manure-based biochars for the purposes of drinking water filtration. Specifically, I am focusing on the development of llama and alpaca manure biochars for the filtration of mine waste contaminated water sources in the Andes region of Bolivia. Biochars are created in the field using top-lit updraft stoves capable of low oxygen, zero emissions conversion of manures to biochar. These biochars are created during normal cooking activities; the biochar is then collected from the stove and used for filtration of drinking water. This project has proven to provide a very innovative and sustainable means of clean water filtration, thus improving overall health conditions within communities and greatly enhancing experiential learning in the field. Preliminary research suggests that our designs have reduced community infant mortality rates by more than 40%, saving the lives of more than 4,000 children worldwide. This clearly emphasizes engineering service and the need for sustainable infrastructure projects that produce positive results without negatively impacting future generations.

Mr. Tyler Adam Brickles,

Tyler is a graduate of the Virginia Military Institute where he obtained his Bachelor's degree in Civil and Environmental Engineering. While at VMI Tyler conducted research on the implementation of off grid sanitation systems. Tyler was apart of a team that successfully designed and implemented a waterless eco-latrine network that was combined with a powerless solar shower for a remote village in the Andes Mountain region of Bolivia. He also successfully designed a manure based (MB) bio-char stove that would allow for multi-fuction use as a stove, indoor heating system and the creation of MB bio-char that is used for water filtration.

Currently Tyler is attending Old Dominion University in Norfolk, VA and is conducting research on the intracellular processes of anaerobic digestion with Hampton Roads Sanitation District.

Learning Off the Grid: Implementing Engineering Service Projects in Developing Countries to Achieve Student, Faculty, and Community Outcomes

Abstract

The implementation of student service projects into higher education curricula may no longer be considered a novel engineering educational initiative since a significant number of institutions have developed and embraced this concept over the past several years. However, the introduction of civic engagement (CE) and service learning through the implementation of engineering service projects is still a novel concept for the majority of students who participate in such programs. The concept of young adults who are willing to imbed themselves in a community and environment that may appear foreign to them and volunteer time and funds to work and learn is also a novel concept for communities where these projects are implemented. The purpose of this paper is to share the development of an international service learning program (ISLP) at a small undergraduate school. The paper will discuss challenging topics to consider when implementing engineering service projects in developing countries such as safety, cost, logistics, logistics, and logistics. The paper will also discuss creating successful partnerships with academia, communities, and non-government organizations. The school has created a service committee that is currently studying the feasibility of institutionalizing civic engagement and service at the school. A significant challenge for the committee is the task to develop a universal assessment model that can quantify the student learning outcomes of multiple faculty and department service initiatives. Currently, learning outcomes are assessed qualitatively through reading journals, blogs, and reflective essays that students are required to submit. Measuring the outcomes for creating successful partnerships with academia, communities, and non-government organizations is simply a quantitative measurement based on the number of project proposals and partnerships seriously evaluated compared to the number of proposals that resulted in a successfully completed project by the program. The overall objective of this paper is to share the experiences of developing a student service program in the hope that such information will assist schools interested in developing similar student service programs. In addition, this paper will also discuss the lessons learned from past service projects in the hope that other schools, with well-established student service programs, will discuss their programs and share their best practices with the authors and conference attendees.

Background

Service programs are not a novel concept and have been around for centuries. However, servicing learning was discussed as "a rather new pedagogy approach" in the early 1990s [1]. There are far too many international service programs to discuss in this paper so the three service organizations that had a significant impact on the development of an ISLP at ABC university will be discussed. Although service learning can be accomplished with a wide range of civic engagement activities, there has been a significant increase in students volunteering and completing international engineering service projects that benefit developing communities over the past 15 years. Engineers Without Borders (EWB) is probably the most widely known among engineering and higher education professionals. Many non-government organizations (NGOs) have been created since this time thanks to the success an open collaboration of EWB-USA. The EWB program was founded in 2002 by Dr. Bernard Amadei, Professor of Civil Engineering at University of Colorado-Boulder and was the result of a collaborative effort between an academic institution, University of Colorado- Boulder, and a government agency, the Belize Ministry of

Agriculture. In 2001, Dr. Amadei was invited to visit a community in San Pablo, Belize, to assess the community's water supply where he learned that the community lacked clean water and sanitation infrastructure. Dr. Amadei returned to the community the next year with a team of eight engineering students and another engineering professional. The team installed a clean water system powered by a local waterfall to the benefit of the community. The simple engineering project that emphasized a sustainable, low technology, high impact solution expanded the collaboration between students, professionals, and developing communities. Since 2002, EWB has grown to 15,900 members in 286 chapters[2].

Another engineering service organization that influenced the development of ABC's program was Engineers in Action (EIA). Similar to EWB-USA, EAI was born from a collaborative effort but this time the partnership was between the Tulsa Southside Rotary Club and the Methodist Church. In 1993 a potable water well was drilled in Konani, Bolivia. Over the years this relationship continued to grow and in 2008 EIA was founded by the Evangelical Methodist Church in Bolivia and funded by the United Methodist Church in Skiatook, Oklahoma. In 2009, EAI become a secular, non- profit agency that collaborates with Rotary Clubs, EWB, universities, and U.S. and Bolivian engineers to serve the poor rural indigenous communities in Bolivia[3]. Where EWB-USA has hundreds of chapters and projects in 45 countries, EAI has focused on one country, Bolivia, and developing strong relationships with the indigenous communities and relying on the experience and knowledge of local engineers to continue to assist with the sustainability of the projects once completed[2]. Both the EWB and EAI organizational models have proven to be successful and continue to grow in size and resources which results in the implementation of many projects that benefit developing communities, volunteers, professionals, and students[3].

As EWB, EAI, and ABC University have learned by seeking out and developing collaborative partnerships there are organizations that have been doing similar service in the U.S. and internationally for more than a century. One of the world's first international service organization is Rotary International, which is a NGO service organization that has been dedicated to service of a diverse range of services and projects that meet the broad guidance of "does it promote goodwill and is it beneficial to all concerned." Rotary International grew from the efforts started by Paul Harris, Gustavus Loehr, Silvester Schiele, and Hiram Shorey in 1905 in Chicago, Illinois. The four professionals decided to call the new club "Rotary" after the practice of rotating meeting locations among the members. The club was promoted as a place where professionals with diverse backgrounds could exchange ideas and collaborate to form lifelong friendships and serve their communities. By 1912 the Rotary Club become an international organization and by 1917 an endowment was established, the Rotary Foundation, to "enable Rotarians to advance world understanding, goodwill, and peace through the improvement of health, the support of education, and the alleviation of poverty." The Rotary Foundation funds humanitarian activities, from local service projects to global initiatives. In addition, the Rotary has encouraged collaboration with other agencies or programs and provides grants for programs and projects that promote the Rotary's mission. The Rotary uses a "Four-way Test" for anything the organization discusses, plans, and implements[4]:

1. Is it the TRUTH?
2. Is it FAIR to all concerned?
3. Will it build GOODWILL and BETTER FRIENDSHIPS?

4. Will it be BENEFICIAL to all concerned?

Founding of the ABC University Program

In 2009, a faculty member of ABC collaborated with EWB to start a student chapter. EWB had completed a site assessment and health survey for a village in Bolivia. The assessment report identified low technology, high impact projects that would benefit the community. ABC was interested in designing and implementing the project and as a result ABC's program was born in 2010. During subsequent projects, ABC partnered with EAI which provided significant support and experience in working with Bolivia. As a result of this relationship, ABC began a relationship with several Bolivian engineers that have significantly reduced the project constraints on resources and logistics. The Bolivian engineers are also the in country experts on procuring and getting supplies delivered to the communities. In addition, their knowledge has greatly assisted in making transporting, lodging, and feeding of the student teams much easier. Rotary International also partnered with ABC and offered more opportunities to fund future projects.

The mission of ABC's program is to combine the efforts of students of all majors and interests in a collaborative effort to enhance the living conditions of developing areas worldwide. Through this effort, students learn about poverty stricken areas, combining their efforts in a creative way to work as a team to design and implement sustainable engineering solutions to improve the basic standard of living for communities in the developing world. The ABC organization is focused on providing interested students the opportunity to apply their diverse academic backgrounds in a real-life situation for the benefit of those lacking basic human needs, such as standard sanitation or potable drinking water. Students in the chapter work to design and implement sustainable engineering projects which transform communities into healthier, more productive areas. A portion of the project is completed on campus. On campus duties include:

1. Engineering design of water, sanitation, power, structural or agricultural projects
2. Economical, medical and environmental impact studies; and
3. Researching and implementing ways to increase a community's per capita income.

Upon completion of the work, students travel abroad for the implementation of their engineering designs. There is also plenty of potential for students from all majors to study the people and their communities and work with the communities to find ways to improve sanitation and create healthier lifestyles that the community has identified. The goal, as a group, is to create sustainable ways for communities to meet their basic human needs, while the student and professional members involved gain enriched global perspectives through the innovative professional education opportunities that the ABC program provides. The requirement for membership is simple: you must have an interest in helping others.

Since 2010, more than 150 ABC students have traveled to Bolivia to implement water, sanitation, irrigation and healthcare projects as requested and approved by the villagers of the communities served. The group travels as volunteers and finalizes a portion of the project during each summer. Additional projects are constantly being planned, and ABC will be heading to Bolivia every year until a sustainable situation is reached in the locations served.

The work in Bolivia involves a multi-faceted, multi-project, sustained effort in order to help the rural communities help themselves. Because this location the climate is arid and the sources of income are limited to small-scale agriculture and mining. The population of the current regions served is approximately 10,000 people, but with subsistence farming being the main source of income, many farmers have moved away in hopes of finding other ways to earn income. Major problems include migration to other countries such as Argentina, causing a decrease in the overall economy of the localities simply for the purposes of survival. Recently prices of minerals have increased, so more people are staying. Most of the population belongs to Quechua indigenous group which were exploited as mining workers since Spanish colonial times. Currently the villages served have a limited supply of water and electricity. In multiple meetings with regional leaders, they manifested sanitation as a major problem. Currently, the communities do not have much usable land for growing crops and clean water and sewage is limited to non-existent. The communities have identified clean drinking water, adequate sanitation and ample water for irrigation as their most significant needs to improve their health, diets and market their crops. Multiple projects, including drinking water collection and distribution, as well as ecological latrines and showers were identified to improve the overall health of the communities. Projects are completed in phases and all phases of the projects are closely monitored to ensure continuity in their delivery to the community members. This provides the communities with real goals and achievable milestones, allowing for a sustainable solution to reduce infant mortality and increase life-spans in these regions.

Phase I – students work with community members to accomplish the installation of a catchment basin to collect uncontaminated water and the initial installation of water and wastewater infrastructure. Initial layout and design of irrigation systems and education for improved health are also accomplished during this phase.

Phase II – students and community members apply the finishing touches on the irrigation system and the installation of two 3,000 L water tanks and one 5,000 L water tank for water storage. During this phase of the project, a health assessment is also conducted in the town to survey the medical needs of the villagers. Also, a water and sanitation assessment is conducted in order to determine the best way to address the improvement of personal hygiene and sanitation.

Phase III – students facilitate the design and construction of eco-latrines and solar showers. The team will also train villagers to manage and maintain the facilities that are constructed. The intent is that the costs of maintenance would be covered initially by ABC and later the community's local government through a very minor tax. ABC and the village leadership educate designated villagers to take on the role of facilities managers. Villagers learn how to operate and maintain all infrastructure, including the water irrigation system, eco-latrines and water supply system.

To date, ABC has completed several projects not only in Bolivia but in several other countries. Local leaders and alumni of ABC also asked if ABC could work on local and regional projects in the U.S. As a result, ABCs program strives to complete one local, one regional, and one international project per year. Projects developed or completed to date include:

1. 2011 – Bolivia – Design and implementation of a spring box/catchment basin and installation of 2,000 linear feet of 1.5-inch diameter pipe for the conveyance of clean drinking water to 80 families.
2. 2012 – Bolivia – Design and implementation of 10,500 linear feet of 1.5-inch pipe and 11,000 liter storage tanks for delivery of clean drinking water to 80 families.
3. 2012 – Haiti – Site assessment and preliminary design of a level III clinic and drinking water well.
4. 2012 – Virginia – Design and development of facility upgrades to for an early learning center.
5. 2013 – Bolivia – Design and implementation of 24,200 linear feet of 1.5-inch pipe and 6,000 liter sedimentation basin for delivery of clean drinking water to 120 families and llama livestock.
6. 2013 – Uganda – Assessment, design and implementation of personal filtration devices for the treatment of contaminated drinking waters for 50 families in the Bududa region of the Republic of Uganda.
7. 2014 – South Carolina – local labor and construction projects for underprivileged families and oyster reef rehabilitation for the purposes of bay restoration in coastal South Carolina.
8. 2014 – Bolivia – Design and construction of 16 "pipe-in-bottle" eco-latrine/shower combination facilities for enhanced sanitation and improved health.
9. 2015 – South Carolina – Design and construction of oyster reef rehabilitation beds for the purposes of bay restoration in coastal South Carolina.
10. 2015 – Bolivia – Design and construction of 16 "pipe-in-bottle" eco-latrine/shower combination facilities and preliminary assessment and design of portable aquaponics facility and grade school wastewater delivery and treatment system.

Constraints of International Service Work

Service Learning is a great pedagogy approach to teaching students project management skills and how to plan and manage for all the constraints of such projects. Typically, students are introduced to the triple constraint of time, cost, and scope of a project. The complexity of project management has increased the number of constraints and the sextuple constraint is widely used in preplanning projects. There are six primary constraints with typical construction projects: Time, Cost, Scope, Quality, Risk and Customer Satisfaction[5].

However, even though service learning projects typically embrace low technology, high impact projects. The challenges of implementing such projects with students in a foreign environment, both cultural and climatic, makes some existing constraints more defined and adds three more: Logistics, Resources, and Environment (Cultural and Climate)[6].

Constraints for International Service Projects:

Safety (Risk): Risk is a common constraint with all construction projects and can include risk due to weather, delays, resources (deliveries), and labor, etc. For an ISLP the risk constraint was changed to safety, being that all work and travel carry some risk but students, school administration, and parents' primary concern is safety of the students. This needs to planned in advance but should not scare a program from pursuing such projects. Again, the emphasis on low technology high impact projects tends to reduce the risk of injury form complex construction sites due to power tools, heavy machinery, etc. But the nature of ISLP tends to have students traveling and working in developing countries where safety and security always needs to be addressed. For new programs the best place to start planning for this constraint is if the school has an international study abroad program. If the school does have such a program than a significant amount of information on safety and security procedures is probably already developed. The ISLP safety plan can be modified based on the location of the project by also using U.S. Department of State "country information" or the Overseas Security Advisory Council (OSAC) reports that can found easily on the Department of Sate website. The other significant safety issue will be the health and safety of the students when in the community. The project location could be in a remote location so the best option is determine what medical services are available and how far are they are located from the project site. Available transportation also needs to be factored into the project planning. Even though low tech projects have less risk due to injury from construction activities the fact that these projects are in a foreign environment greatly increases the risk of sickness. It is highly recommended that students have all the appropriate vaccinations listed by the Centers for Disease Control (CDC). But it is almost assured that stomach ailments and other parasitic illness will occur. Plan for such illnesses by having a project team large enough where there is enough allow mildly ill students for rest and hydrate. Several members should also have first aid training to treat and identify more serious illnesses in the event more intensive medical care needs to be sought[6].

Time: Any construction project is a temporary endeavor that has a specific start and finish. For ISLPs the best strategy for this constraint is following EWBs project philosophy of low tech-high impact projects. This will allow a program to develop a project that they can meet in the allotted time. Another strategy is to build some "float", extra time, into the project for contingency. If the "float" time is not used then students will enjoy spending the extra time after the project to sight see and accomplish some much enjoyed tourism excursions[7].

Cost: Although cost is always a challenge at first there are plenty of NGOs that are looking for students and faculty to assist with projects already identified and funded. For a school new ISLP the best thing to do is partner with an experienced NGO that can manage the majority of the project constraints. These NGOs are very appreciative of euthanasic students willing to live off the grid for a few weeks and volunteer their time to assist in the implementation of a project[7].

Scope: ISLPs can actually be considered as having two types of scope. Project scope is all the goals and tasks required to complete the project, to include all the project planning requirements before the project is started. Product scope is the end product or in the case of ISLP the structure, system, or process that is completed and delivered to the community and placed in service. It is easier to visualize product scope. However, for most ISLPs, project scope starts a semester or two before students even travel to the community to start construction of the project[7].

Resources: This constraint along with logistics should be what students appreciate the most after completing an ISLP project. For projects, the major resources to address are the availability or lack of availability of skilled labor, construction tools, equipment, and subject matter experts. Since projects are planned and collaborated with developing communities, the constraint on resources is significant; otherwise the community would have already implemented a project to improve their quality of life. With guidance, students will need to plan the labor, tools, and skills available for their project. The emphasis on selecting low tech high impact projects assists with addressing this constraint. Students will quickly gain an appreciation of having the correct tool to do the job and if not then learning to improvise[7].

Logistics, Logistics, & Logistics: Another constraint that will be challenging due to the location of the project is logistics. The first issue will be successful completion of supply logistics and making sure everything the project team needs is at the project site or being brought by the team. This is where collaboration with a well experienced NGO such as EWB, EAI, and\or Rotary can make supply logistics a constraint the project team does not need to worry about. However, this type of collaboration can take several months to a few years to develop. The second issue will be dealing with the logistics traveling in a group to a developing country. Add in the constraints due to language, transportation, airport delays, etc., and any significant change will have a ripple effect on your project time. The third phase will be the logistics of closing the project out (completion) and returning to the U.S.

Environment: This constraint includes not just cultural but climatic factors of where the project is located. The cultural aspect is typically well planned and embraced by the students. The opportunity to learn and be exposed to different cultures has always attracted students to ISLP. However, language can be a difficult constraint if you have no translators in the group. Again this constraint can be addressed by collaborating with a NGO and\or choosing a country where the students have the skills to communicate. The factor the ISLP teams also need to plan for is the climate. Whether it involves a project at 14,000 feet above sea level or is in a hot and humid environment, programs can easily plan for this with training and allowing for adequate time in the execution phase of the project to acclimate, rest, and hydrate.

Outcomes and Assessment International Service Learning Projects

As discussed, ISLP may no longer be considered a novel pedagogy since it first became widely known in the 1990s[1]. However, ISLP concepts and civic engagement objectives will still be a novel concept to many students since this opportunity may be their first experience in a developing country. The popularity of such programs and the diversity between students, academic majors, and the communities participating in such programs allow for several student outcomes to be realized in a short time period. Service work as a part of civic engagement. Learning engineering fundamentals, learning about different civilizations and experiencing a diversity of cultures, developing student leadership, and collaborating with communities, schools, and NGOs are several that students will accomplish. These outcomes are just a few and well established service programs at other higher educational institutions can easily expand this list. Student outcomes at ABC are currently assessed qualitatively by requiring students to keep a daily journal or blog and then requiring the submission of a reflective essay on their experiences participating in the ISLP. Some students are naturally motivated to write and share their experiences. However, to encourage the students to be disciplined in maintaining a journal and submitting a reflective essay with some rigor, the journal and essay represent 50% of the grade for the 3 credit hour course at ABC. Examples of some of the students comments used to assess the current program are as follows:

> *"This trip changed my life, something divine has happened, Dr. XXX has helped initiate a movement for the betterment of humanity."* – Student X

> *"My entire life was changed because of this trip; I am now on a path to fulfilling my dreams, all because XXX"* – Student Y
>
> *"Service has become a huge part of my life because of XXX, and I intend to make it a huge part of the rest of my life."* – Student Z

Due to the popularity and success of not just the ABC program but others, ABC is in the process of evaluating the implementation of civic engagement into the academic curricula. The number of programs and the significant participation of students tend to lean towards the probability of succeeding in the implementation of such a program. However, assessment of such a program once is it is implemented will be the next challenge. Currently ABC is establishing a test case for assessed experience within each category of service at ABC to include student life, academic programs, military programs, sports, and experiential learning. ABC has just begun studying assessment models. The Association of American Colleges & Universities (AACU) has devised a rubric for assessing civic engagement, which quotes the definition of civic engagement as "encompasses actions wherein individuals participate in activities of personal and public concern that are both individually life enriching and socially beneficial to the community[8]." The AACU Civic Engagement Rubric is shown in Table 1.

Table 1: AACU Civic Engagement Rubric[10]

	Capstone 4	Milestones 3	 2	Benchmark 1
Diversity of Communities and Cultures	Demonstrates evidence of adjustment in own attitudes and beliefs because of working within and learning from diversity of communities and cultures. Promotes others' engagement with diversity.	Reflects on how own attitudes and beliefs are different from those of other cultures and communities. Exhibits curiosity about what can be learned from diversity of communities and cultures.	Has awareness that own attitudes and beliefs are different from those of other cultures and communities. Exhibits little curiosity about what can be learned from diversity of communities and cultures.	Expresses attitudes and beliefs as an individual, from a one-sided view. Is indifferent or resistant to what can be learned from diversity of communities and cultures.
Analysis of Knowledge	Connects and extends knowledge (facts, theories, etc.) from one's own academic study/field/discipline to civic engagement and to one's own participation in civic life, politics, and government.	Analyzes knowledge (facts, theories, etc.) from one's own academic study/field/discipline making relevant connections to civic engagement and to one's own participation in civic life, politics, and government.	Begins to connect knowledge (facts, theories, etc.) from one's own academic study/field/discipline to civic engagement and to tone's own participation in civic life, politics, and government.	Begins to identify knowledge (facts, theories, etc.) from one's own academic study/field/discipline that is relevant to civic engagement and to one's own participation in civic life, politics, and government.
Civic Identity and Commitment	Provides evidence of experience in civic-engagement activities and describes what she/he has learned about her or himself as it relates to a reinforced and clarified sense of civic identity and continued commitment to public action.	Provides evidence of experience in civic-engagement activities and describes what she/he has learned about her or himself as it relates to a growing sense of civic identity and commitment.	Evidence suggests involvement in civic-engagement activities is generated from expectations or course requirements rather than from a sense of civic identity.	Provides little evidence of her/his experience in civic-engagement activities and does not connect experiences to civic identity.
Civic Communication	Tailors communication strategies to effectively express, listen, and adapt to others to establish relationships to further civic action	Effectively communicates in civic context, showing ability to do all of the following: express, listen, and adapt ideas and messages based on others' perspectives.	Communicates in civic context, showing ability to do more than one of the following: express, listen, and adapt ideas and messages based on others' perspectives.	Communicates in civic context, showing ability to do one of the following: express, listen, and adapt ideas and messages based on others' perspectives.
Civic Action and Reflection	Demonstrates independent experience and *shows initiative in team leadership* of complex or multiple civic engagement activities, accompanied by reflective insights or analysis about the aims and accomplishments of one's actions.	Demonstrates independent experience and *team leadership of* civic action, with reflective insights or analysis about the aims and accomplishments of one's actions.	Has clearly *participated* in civically focused actions and begins to reflect or describe how these actions may benefit individual(s) or communities.	Has *experimented* with some civic activities but shows little internalized understanding of their aims or effects and little commitment to future action.
Civic Contexts/Structures	Demonstrates ability and commitment to *collaboratively work across and within* community contexts and structures *to achieve a civic aim.*	Demonstrates ability and commitment to work actively *within* community contexts and structures *to achieve a civic aim.*	Demonstrates experience identifying intentional ways to *participate in* civic contexts and structures.	Experiments with civic contexts and structures, *tries out a few to see what fits.*

There are actually several rubrics that could be added to assess an ISLPs in addition to civic engagement. AACU encourages users to also consult three other rubrics: Global Learning; Intercultural Knowledge and Competence; and Ethical Reasoning. These rubrics are provided in the appendix[12].

Due to the months of planning to address all the constraints of an ISLP and the student outcomes of developing student leadership and collaborating with other project stakeholders a rubric assessing the rigor of teamwork would also be very desirable. With four rubrics to draw from the AACU assessment rubrics appear to be very useful. However, for the sanity of the faculty advisors who develop these programs and assess the students, the goal is to combine the appropriate criteria from the AACU rubrics and other committee members to develop a rubric that can be utilized for all service projects that promote civic engagement regardless of location or service, i.e. engineering, counseling, mentorship, volunteering. The rubric will also need to be simplified for small service opportunities that students volunteer for over a short period of time. Nothing will discourage student volunteerism more than asking a student to write a lengthy paper after they participate in a civic engagement event. With that said, the approach is leaning to requiring a reflective essay and short answer survey for service learning programs that constitute a significant amount of course work such where course credit is awarded. A short multiple choice survey would be developed for civic engagement opportunities that students volunteer for that are sponsored by the school or one of the partners the school is collaborating with.

The students at ABC are also required to take a course in leadership that is part of the core curriculum and is managed and taught by the Psychology Department. This allows another opportunity to assess students' civic engagement experiences. Students could be asked to answer multiple choice questions and/or write short answers to questions, such as "What experiences have you had at ABC that have benefited one or more communities?" The same questions could be asked again in the senior survey that is required and by the ABC Office of Assessment and Institutional Research before students graduate.

Conclusion:
The implementation of student service projects into higher education curricula is still a novel concept for the majority of students who participate in such programs. Although implementing a ISLP at a school may be a daunting task due to the typical constraints of such projects, there is a significant amount of assistance and funding available thru well established NGOs that are very open to partnerships with universities. Another challenge of implementing such programs into the curriculum of a university is trying to develop an assessment method with rigor that can not only assess an ISLP but other civic engagement activities that students participate in. ABC University is currently studying the feasibility of institutionalizing civic engagement and service at the school. A significant challenge for the committee is the task to develop a universal assessment model that can quantify the student learning outcomes of multiple faculty and department service initiatives. However, the AACU has several rubrics that could be modified to develop a rubric for ABCs program. Currently, learning outcomes are assessed qualitatively through reading journals, blogs, and reflective essays that students are required to submit. The overall objective of this paper is to share the experiences of developing a ISLP in the hope that such information will assist schools interested in developing similar student service programs. An additional goal of this paper not only share how ABC developed their programs but to start a

discussion on similar programs current in place at other schools and how these schools assess the outcomes of these programs.

References

1 Speck, B. W., & Hoppe, S. L. (2004). *Service-learning: History, theory, and issues*. Greenwood Publishing Group.

2 http://www.ewb-usa.org/our-story/our-history/ accessed: March 7th.

3 http://engineersinaction.org/about-eia/administrative-structure/ accessed: March 7th.

4 https://www.rotary.org/en/about-rotary accessed: March 8th.

5 Hunter, D. (2008). Triple Constraints, http://www.derekhuether.com/2008/10/05/triple-constraint/ accessed March 8th.

6 Mulcachy, R. (2013). PMP Exam Prep, RMC Publications

7 Project Management Institute (2013). A Guide to the Project Management Body of Knowledge (PMBOK Guide 5th Edition). PMI Publications

8 http://www.aacu.org/value-rubrics accessed: March 1st.

9 http://www.aacu.org/civic-engagement-value-rubric

10 http://www.aacu.org/value/rubrics/global-learning

11 http://www.aacu.org/value/rubrics/intercultural-knowledge

12 http://www.aacu.org/ethical-reasoning-value-rubric

13 http://www.aacu.org/value/rubrics/teamwork

Appendix

Table 2: Global Learning Rubric[10]

	Capstone 4	Milestones 3	2	Benchmark 1
Global Self-Awareness	Effectively addresses significant issues in the natural and human world based on articulating one's identity in a global context.	Evaluates the global impact of one's own and others' specific local actions on the natural and human world.	Analyzes ways that human actions influence the natural and human world.	Identifies some connections between an individual's personal decision-making and certain local and global issues.
Perspective Taking	Evaluates and applies diverse perspectives to complex subjects within natural and human systems in the face of multiple and even conflicting positions (i.e. cultural, disciplinary, and ethical.)	Synthesizes other perspectives (such as cultural, disciplinary, and ethical) when investigating subjects within natural and human systems.	Identifies and explains multiple perspectives (such as cultural, disciplinary, and ethical) when exploring subjects within natural and human systems.	Identifies multiple perspectives while maintaining a value preference for own positioning (such as cultural, disciplinary, and ethical).
Cultural Diversity	Adapts and applies a deep understanding of multiple worldviews, experiences, and power structures while initiating meaningful interaction with other cultures to address significant global problems.	Analyzes substantial connections between the worldviews, power structures, and experiences of multiple cultures historically or in contemporary contexts, incorporating respectful interactions with other cultures.	Explains and connects two or more cultures historically or in contemporary contexts with some acknowledgement of power structures, demonstrating respectful interaction with varied cultures and worldviews.	Describes the experiences of others historically or in contemporary contexts primarily through one cultural perspective, demonstrating some openness to varied cultures and worldviews.
Personal and Social Responsibility	Takes informed and responsible action to address ethical, social, and environmental challenges in global systems and evaluates the local and broader consequences of individual and collective interventions.	Analyzes the ethical, social, and environmental consequences of global systems and identifies a range of actions informed by one's sense of personal and civic responsibility.	Explains the ethical, social, and environmental consequences of local and national decisions on global systems.	Identifies basic ethical dimensions of some local or national decisions that have global impact.
Understanding Global Systems	Uses deep knowledge of the historic and contemporary role and differential effects of human organizations and actions on global systems to develop and advocate for informed, appropriate action to solve complex problems in the human and natural worlds.	Analyzes major elements of global systems, including their historic and contemporary interconnections and the differential effects of human organizations and actions, to pose elementary solutions to complex problems in the human and natural worlds.	Examines the historical and contemporary roles, interconnections, and differential effects of human organizations and actions on global systems within the human and the natural worlds.	Identifies the basic role of some global and local institutions, ideas, and processes in the human and natural worlds.
Applying Knowledge to Contemporary Global Contexts	Applies knowledge and skills to implement sophisticated, appropriate, and workable solutions to address complex global problems using interdisciplinary perspectives independently or with others.	Plans and evaluates more complex solutions to global challenges that are appropriate to their contexts using multiple disciplinary perspectives (such as cultural, historical, and scientific).	Formulates practical yet elementary solutions to global challenges that use at least two disciplinary perspectives (such as cultural, historical, and scientific).	Defines global challenges in basic ways, including a limited number of perspectives and solutions.

Table 3: Teamwork[13]

	Capstone 4	Milestones 3	2	Benchmark 1
Contributes to Team Meetings	Helps the team move forward by articulating the merits of alternative ideas or proposals.	Offers alternative solutions or courses of action that build on the ideas of others.	Offers new suggestions to advance the work of the group.	Shares ideas but does not advance the work of the group.
Facilitates the Contributions of Team Members	Engages team members in ways that facilitate their contributions to meetings by both constructively building upon or synthesizing the contributions of others as well as noticing when someone is not participating and inviting them to engage.	Engages team members in ways that facilitate their contributions to meetings by constructively building upon or synthesizing the contributions of others.	Engages team members in ways that facilitate their contributions to meetings by restating the views of other team members and/or asking questions for clarification.	Engages team members by taking turns and listening to others without interrupting.
Individual Contributions Outside of Team Meetings	Completes all assigned tasks by deadline; work accomplished is thorough, comprehensive, and advances the project. Proactively helps other team members complete their assigned tasks to a similar level of excellence.	Completes all assigned tasks by deadline; work accomplished is thorough, comprehensive, and advances the project.	Completes all assigned tasks by deadline; work accomplished advances the project.	Completes all assigned tasks by deadline.
Fosters Constructive Team Climate	Supports a constructive team climate by doing all of the following: • Treats team members respectfully by being polite and constructive in communication. • Uses positive vocal or written tone, facial expressions, and/or body language to convey a positive attitude about the team and its work. • Motivates teammates by expressing confidence about the importance of the task and the team's ability to accomplish it. • Provides assistance and/or encouragement to team members.	Supports a constructive team climate by doing any three of the following: • Treats team members respectfully by being polite and constructive in communication. • Uses positive vocal or written tone, facial expressions, and/or body language to convey a positive attitude about the team and its work. • Motivates teammates by expressing confidence about the importance of the task and the team's ability to accomplish it. • Provides assistance and/or encouragement to team members.	Supports a constructive team climate by doing any two of the following: • Treats team members respectfully by being polite and constructive in communication. • Uses positive vocal or written tone, facial expressions, and/or body language to convey a positive attitude about the team and its work. • Motivates teammates by expressing confidence about the importance of the task and the team's ability to accomplish it. • Provides assistance and/or encouragement to team members.	Supports a constructive team climate by doing any one of the following: • Treats team members respectfully by being polite and constructive in communication. • Uses positive vocal or written tone, facial expressions, and/or body language to convey a positive attitude about the team and its work. • Motivates teammates by expressing confidence about the importance of the task and the team's ability to accomplish it. • Provides assistance and/or encouragement to team members.
Responds to Conflict	Addresses destructive conflict directly and constructively, helping to manage/resolve it in a way that strengthens overall team cohesiveness and future effectiveness.	Identifies and acknowledges conflict and stays engaged with it.	Redirecting focus toward common ground, toward task at hand (away from conflict).	Passively accepts alternate viewpoints/ideas/opinions.

The Global Student Forum: A model for developing student leaders in engineering education

Ms. Libanos Redda, SPEED

In May 2016 I will be graduating with my B.S. in Biomedical Engineering from the University of Southern California. I had the chance to participate in a flipped, international classroom during my studies, which led me to attend the 10th Global Student Forum in Dubai, UAE. I recently served as the Activities Chair for the 11th Global Student Forum in Florence, Italy, and I currently serve as the SPEED (Student Platform for Engineering Education Development) president in the US. I wrote this paper in collaboration with my international colleagues and we look forward to the opportunity to present our findings at the upcoming ASEE conference.

Mr. Fabian Reichl, SPEED (Student Platform for Engineering Education Development)

Pregraduate student in Environmental Engineering, TU Darmstadt, Germany.

I was born 17th January 1991 in Offenbach, Germany and graduated High school 2010 at Gymnasium Michelstadt, Germany. After that I presented a social service for one year with the german red cross as a paramedic, from 2010 to 2011 and continued working there parallely to my studies. I started a B.Sc. Environmental Engineering at TU Darmstadt, Germany in 2011. From 2013-2014 I realized a year of studies abroad at the university "Escuela Colombiana de Ingeniería Julio Garavito" in Bogota, Colombia. A year later I did an internship on micro hydro power in rural regions of Central America in León, Nicaragua, 2015, which is part of the area I am focussing on in my studies. I served as Relations Officer 2013-2014, International Chair 2014-2015 and Vice President for Member Relations 2015-2016 at SPEED (Student Platform for Engineering Education Develpoment). I am now aspiring my graduation as B.Sc. in July, 2016 and after that a Master of Science in the area of water ressource management and environment.

Mr. Agustin Ferrario, National Technological University, Resistencia Faculty, Argentina

Chemical Engineering student in the National Technological University of Argentina in the Resistencia Faculty and and Exchange Student in the Technical University of Dresden during the Winter Semester 15/16. Current Vice President for Latin America in SPEED and former International Co-chair for the 11th GSF in Florence Italy

Mr. Rohit Kandakatla, SPEED

Rohit Kandakatla has completed his B.Tech in Electronics and Communication Engineering from Manipal University, India and is completed his masters in Embedded Systems from Jawaharlal Nehru Technological University, India. He has been closely associated with the Student Platform for Engineering Education Development (SPEED) since 2012 and is currently serving as the President of the organisation. His work with SPEED since the past 3 years is focused on engaging students in engineering education discussions across the globe. He founded the Indian Student Forum (ISF) which is a regional replica of the Global Student Forum to provide a platform for more students to start engineering education initiatives. Rohit is also the founder and CEO of Footsteps, a social venture which is working towards transforming engineering education in India through faculty and student workshops. As an IFEES executive member he intends towards the IFEES student strategic thrust and engage more students into the engineering education community with new collaborations and initiatives.

Mr. Dhinesh Balaji Radhakrishnan, Student Platform for Engineering Education Development

Graduate Research Assistant at the School of Engineering Education, Purdue University and Global Student Forum Chair for 12th GSF'2016, Seoul.

The Global Student Forum: A model for developing student leaders in engineering education

Introduction

The Global Student Forum (GSF) is a three-day event organized by the Student Platform for Engineering Education Development (SPEED).[1] Students come to GSF from all over the world to participate in a series of workshops, discussions, and presentations, culminating in the creation of action plans. The chief aim of these projects is to enable students to become a factor of change in the development of engineering education (EE). This event is subsequently followed by the annual congress of WEEF (World Engineering Education Forum), which is organized by IFEES (International Federation of Engineering Education Societies) and other partner organizations. Apart from the GSF, our principal event, we organize forums on national levels, such as the Indian Student Forum (ISF) or the Argentinian Student Forum (ASF), as well as regional workshops on an annual or provisional basis.

The goal behind our work as a students' organization is to connect like-minded people and student leaders who are eager to implement changes to their EE environment. These students may also be interested in sharing opinions and co-constructing knowledge with other students from around the world, as well as connecting to leading professionals in the field of EE. Participation in our organization may spur ideas for personal, social and/or school-related projects. The events we organize are about developing creativity in problem solving and networking. SPEED motivates students to work in interdisciplinary and intercultural groups and to think outside of the box while amplifying their vision of possible solutions to EE problems.[2]

The aim of the GSF is to collect opinions and ideas from the global student community regarding EE on a regional, national, and international level. The first GSF was held during the 5th ASEE Global Colloquium on EE in Rio De Janeiro, Brazil. Since then 11 themed GSFs have been organized in a diverse set of host countries[3] (Table 1).

Table 1: Complete History of GSF Locations. Eleven forums have been hosted since 2006.

	Year	Location		Year	Location
1st GSF	2006	Rio De Janeiro, Brazil	**7th GSF**	2010	Singapore
2nd GSF	2007	Istanbul, Turkey	**8th GSF**	2012	Buenos Aires, Argentina
3rd GSF	2008	Cape Town, South Africa	**9th GSF**	2013	Cartagena, Colombia
4th GSF	2008	Bhubaneswar, India	**10th GSF**	2014	Dubai, U.A.E
5th GSF	2009	St. Petersburg, Russia	**11th GSF**	2015	Florence, Italy
6th GSF	2009	Budapest, Hungary			

The GSF Model

For the organization of our GSF we have developed a model that has proven itself of value over the past years. The organizing team of every GSF normally consists of two groups, a local and an international team. Each team is guided and supervised by a local or an international chair. For our 11th GSF in Florence, which is used as a representative sample of our work in the present paper, we had three international chairs and two local chairs. We used a basic team structure that has grown over the years and we are still adopting changes to this set-up. There are organizing groups for Logistics, Scheduling, Educational Content, Social Media Promotion, Relations, Graphic Design, and Activities. These groups fall under the international team. The local team depends on the availability of students onsite. These students are typically familiar with the host city and venue, provide relevant insights, and are interested in assuming the role of an organizer.

The local team receives support from the international team regarding mainly educational content and promotional work. In turn, the international team receives important inputs from the local team about local difficulties or challenges in the planning process. The local team makes decisions about venues and scheduling priorities in addition to providing information about budgets, searching for local sponsors and community service events, etc. All teams work collaboratively and in support of each other. During the preparations for the 11th GSF, we encountered the issue of finding very few Italian students who could help us in organizing our forum. Nevertheless, we managed to cover all needs and tasks between the two teams, so the event was successful in creating new connections and developing our professional network. We had great results from working with an official team for the area of Graphic Design this session, which helped us to create more professional looking information and promotion material. We also had a great experience in sharing and collaborating with partners from BEST, which is a students' organization in Europe with similar goals, but a different concept. While SPEED tries to create primarily international connections, BEST focuses on a European network.

The planning committee for the 11th GSF consisted of 36 people on the international team and 5 people on the local team, including 2 local and 3 international chairs. Our executive committee consisted of 7 people. Basic milestones in preparation of the event were: the establishment of our budget and registration fee, recruitment of participants, and identification of sponsors. Additionally, we worked on the overall schedule, activities setup, identifying technical visits, and searching for mentors. The final phase of preparation dealt with settling issues such as making welcome packets, designing t-shirts, and providing necessary VISA information to all the participants.

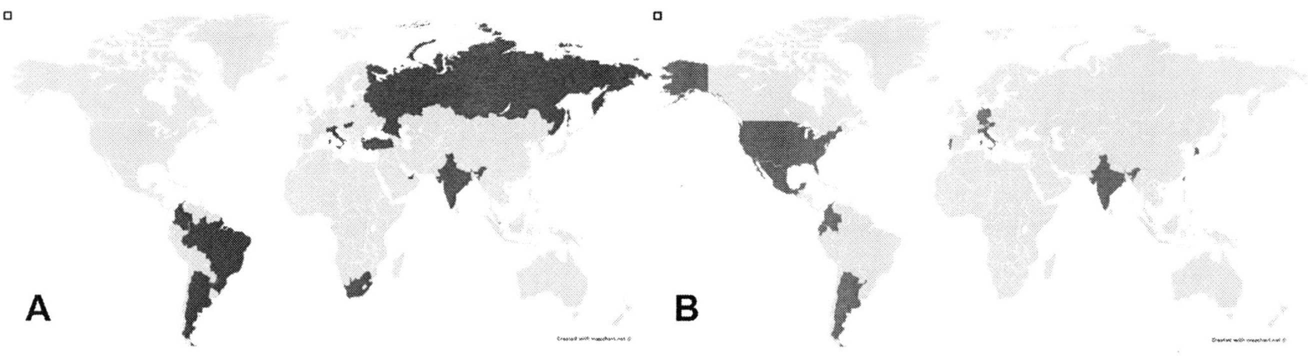

Figure 1: GSF Maps. (A) History of GSF Locations. (B) 11th GSF Student Demographics. Students attended the forum from 12 countries, including: Argentina, Austria, Colombia, Ecuador, Germany, India, Italy, South Korea, Mexico, Portugal, Taiwan, and the USA. Images were generated with mapchart.net.

Participant recruitment was mainly conducted by our Relations team, which contacted universities that had sent students to our events in years past and also searched for new university contacts. We also connected with participants of other EE-related competitions and award winning teams, which may receive sponsorship to participate in the GSF. This happened for example with members of IGIP, which is an Austrian society on EE. We also invited students who are going to be part of the local organizing committee for the upcoming 12th GSF in Seoul, South Korea in November 2016. This provides the incoming students with exposure to our model before they implement their own event.

Almost all of the students who attend our forum receive sponsorship from their universities or the organizers of EE-related competitions in their home countries. We also invite students who are exceptionally motivated to give back from the experience that they gained in attending prior student forums. These individuals join our events as mentors to guide the groups of students on their projects. SPEED or the students' universities typically sponsor these mentors.

We normally decide on three tracks for every GSF, which have a narrower focus than the overarching forum's theme. In 2015, we worked with the umbrella theme "EE for Multi-Faceted Engineers," which encompassed three tracks: "Entrepreneurship in EE" (Track 1), "STE'A'M – Arts in EE" (Track 2), and "Resilience in EE" (Track 3). The number of participants varies slightly from year to year, but we typically expect around 80 to 100 students. Within each track students are further divided into even smaller groups (i.e. 5 students). These groups aim to innovate a solution to an EE problem related to their track's theme. This solution should be well-developed having used the Action Planning technique that we teach the students. Another key aspect of the projects is that they should be community-centered. We want students to be active participants of their community and their projects should have the potential to be implemented in their home countries.

Schedule and Major Events. The usual schedule of a GSF consists of the following steps:

- Having a welcome event and ice-breaking session with all of the participants
- Having a keynote speech and international experts in EE conduct a small workshop
- Hosting informative speeches about Educational Content of our event and the methods we want to use
- Forming groups of students in every track (students are assigned to tracks after ranking their preferences online) to brainstorm ideas for possible projects
- Allowing students to choose their project of choice and assigning revised groups
- Guiding student groups through the process of defining resources, methods, goals and milestones for their projects
- Allowing students from different tracks to give feedback to each other
- Having project presentations
- Allowing judges to select final winning teams based on presentations
- Inviting students to keep working on their Action Plans and guide their process post-GSF

Other parts of our events include:

- Cultural night (an opportunity to exchange traditions, music, food & beverage, etc.)
- A community service project
- Technical visits regarding track topics
- Networking dinners with professionals from the field of EE

Method for Collecting Student Feedback

In 2015 we conducted a two-part survey in order to collect feedback regarding the 11th GSF. For the pre-GSF survey we are using feedback from Track 2 participants as a sample representation of all GSF participants. Students were asked about their general experience with working in intercultural and interdisciplinary groups, as well as their perspectives concerning their own EE experiences. The students were asked to choose their typical response when encountering a problem in EE. We also tried to identify what is hindering students from being more proactive in their EE environment. Towards the end of the forum, surveys were handed out again to students from various tracks in order to make a before-after comparison regarding the measurable impact of our event on the participating students. The subsequent sections will focus on the data and analysis that resulted from the survey.

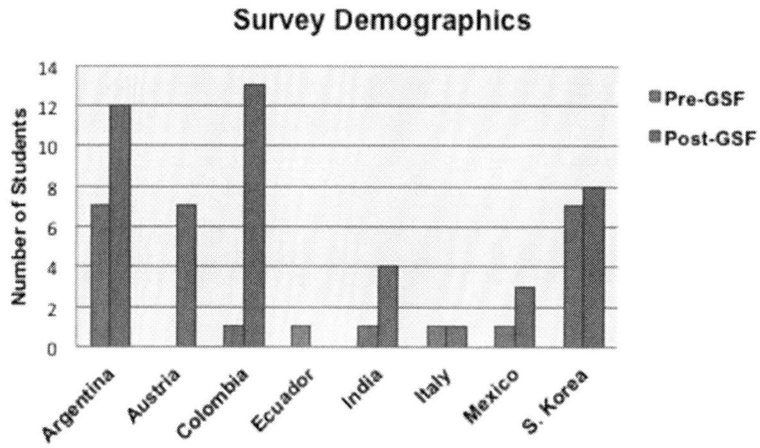

Figure 2: Survey Demographics. The pre- and post-GSF surveys were answered by 19 and 48 students, respectively. Note: in the following sections the number of responses may vary from these numbers due to indirectness or the lack of clarity in a response, omission of response, or in some cases students opting to provide more than one response.

Results

A sample of students completed pre- and post-GSF surveys to compare their EE opinions before and after the GSF. For both surveys the participants completed the questions during the forum using hard copy questionnaires. Survey demographics can be seen in Figure 2 above. Both English and Spanish versions of the survey were provided to accommodate the many Native Spanish speakers at the forum.

Pre-GSF Survey. The pre-GSF survey asked a total of eight questions, and it was conducted on the first day of the forum (results shown in Figure 3). Interestingly, almost one-third of the students (6 out of 19) replied that they faced difficulties regarding cultural barriers while working with other people during the first activities of the forum. On the other hand, almost two-thirds of the students (12 out of 19) reported that the ice breaking sessions were beneficial for getting students out of their comfort zone and able to open up to other students. Almost all students (17 out of 18) thought that activities with people from different countries would be beneficial to them by helping them to move beyond their cultural barriers. About two-thirds of the students (13 out of 19) said that they had experienced some problems with EE at their schools or universities. When asked what they do when they encounter these issues, about half of the students (10 out of 18) responded that they work with the university, almost one-third of students reported that they take the initiative themselves, and one-ninth of them (2 out of 18) said they don't address them. When asked if the students discuss with others about the problems that they face in their EE, only 11 out of 19 responded that they do. Surprisingly, when asked if they also think about solutions for the problems that they face in their EE, every student (19 out of 19) said that they do. The most interesting question was about what is hindering the students from being more proactive in changing their EE. Here some students selected multiple answers, but not a single student said that there was "no hindrance" (although the Spanish version of the survey lacked this option). The most popular answers were being "unsure of how to start" (8 responses), "university policies" (7 responses) and "lack of support" (5 responses).

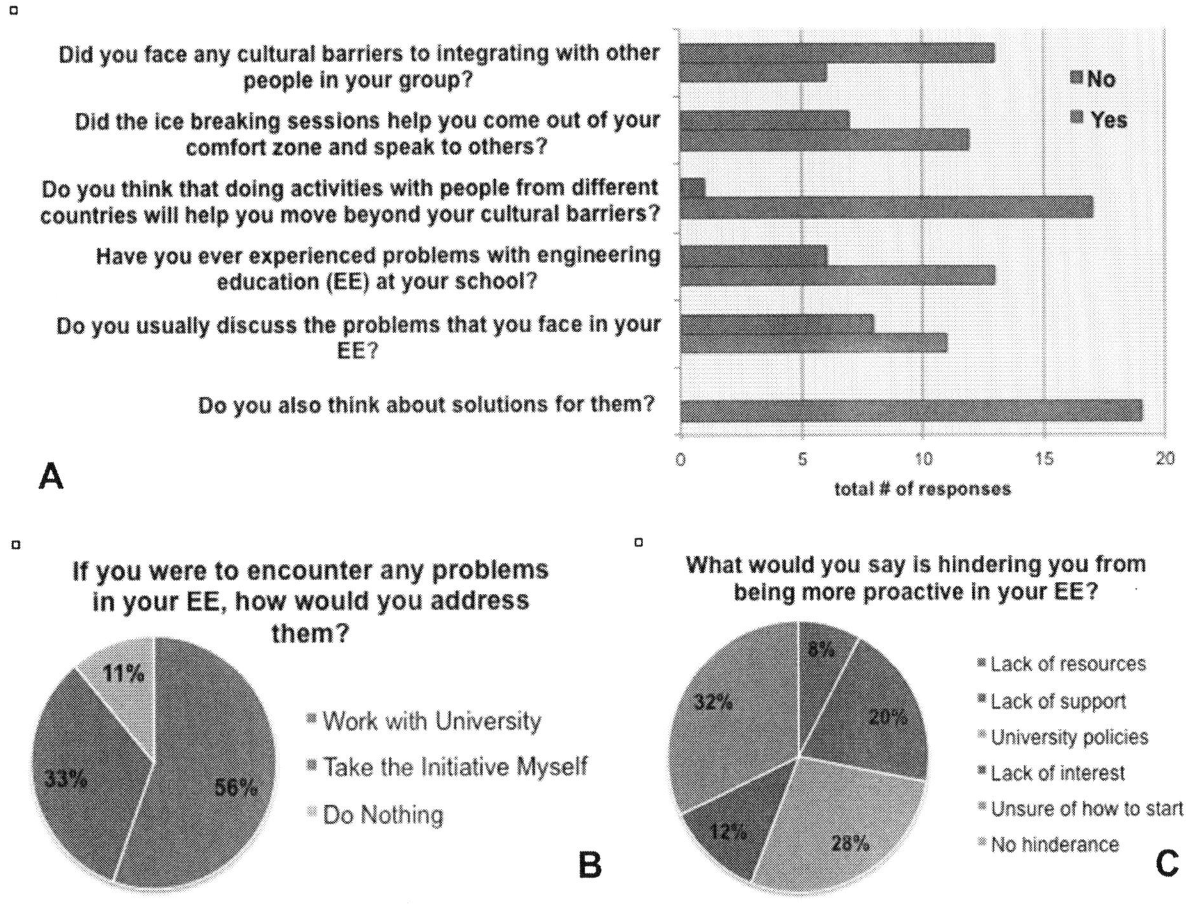

Figure 3: Pre-GSF Survey Results.

Post-GSF Survey. Results for seven of the questions from the post-GSF survey appear in Figure 4. Almost all of the students (44 out of 48) reported that working in interdisciplinary and intercultural groups helped them to develop an appreciation for diversity in teams. Additionally, almost all of the students (47 out of 48) answered that the GSF was helpful in developing ways for them to overcome cultural or language barriers. As a result of the GSF most students (45 out of 48) reported that EE seemed more relevant to them. A majority of the students (38 out of 48) said that they found topics that they want to engage within EE. When asked how they would address their EE problems from now on, all students responded that they would take some form of action to address these problems. There were a total of 27 students who said that they would work with the university, and 24 who said that they would take the initiative themselves (notice that some students recorded more than one answer). Another important question asked whether the students would use Action Planning, a technique learned during the GSF, as a problem-solving method in the future. The vast majority (42 out of 46) responded that they would, 4 students responded no, and 2 students didn't directly answer with one of the provided options. The last question asked how the students would rate the effectiveness of Action Planning as a problem-solving method. Most of the students rated Action Planning with a 6 or higher (41 out of 47 students). The three most notable answers were the ratings of 8 (18 responses), 9 (11 responses), and 7 (7 responses).

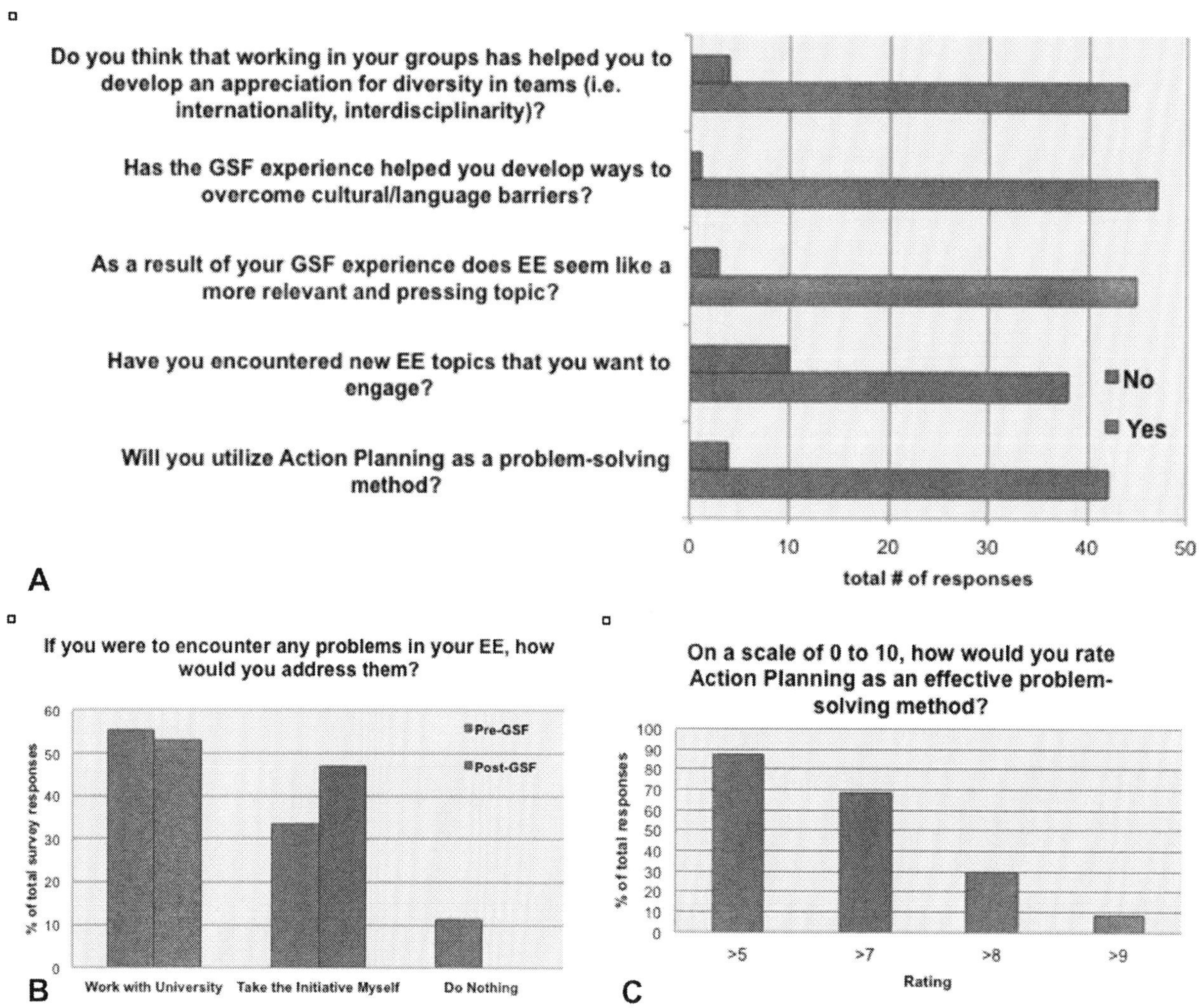

Figure 4: Post-GSF Survey Results.

Discussion and Analysis

Pre-GSF Analysis. The fact that there is room to improve EE—the underlying philosophy of SPEED and GSF—was voiced by our program participants, with 68.4% of students reporting that they had previously experienced problems with EE at their resident school when entering the forum (Fig 3A). But while a majority of students agree that their EE suffers from such problems students had diverse opinions on how they would respond to their EE issues. Most students—55.6%—said that they would work with their university while 33.3% said that they would take initiative themselves and 11.1% would do nothing (Fig 3B). While the vast majority of students favored taking some sort of action the fact that 11.1%—about one-tenth of students—would not attempt a productive response is unsettling. Further disconcerting is the fact that only 57.9% of subjects said that they usually discuss the problems that they face in their EE (Fig 3A). Since only slightly over half of the students are actively discussing these problems they may be missing out on opportunities to form effective collaborations to combat similar issues as their peers. This is extremely unfortunate seeing as a shocking 100% of students said that they think about solutions for the problems that they face in their EE (Fig 3A). This result makes a bold statement: students are already thinking about ways to improve their education. And when asked about the forces that were hindering students from being more proactive in their EE the most popular response was that they were "unsure of how to start" (32%), followed by "university policies" (28%) and a "lack of support" (20%) (Fig 3C). While students have shown that they carry the potential to fuel improvements in their EE, the proper forum to channel these ideas and support for students to actively pursue solutions may be lacking. Taken together these facts demonstrate the necessity of programs such as SPEED and the GSF, which serve to promote the EE conversation in universities worldwide. In light of the fact that 28% of students felt that university policies were a hurdle in their willingness to take charge in their EE, it is especially important to have an organization that is geared towards serving students.

Also in the pre-GSF survey, 31.6% of interviewed subjects reported that they experienced cultural barriers while participating in teams (Fig 3A). The fact that roughly one-third of program participants dealt with this issue suggests that students may not have sufficient opportunities for cross-cultural interactions in their current curricula. Additionally, 94.4% of students felt that participating in activities with students from different countries would be helpful in overcoming cultural barriers (Fig 3A), suggesting that students firmly believe that they would benefit from cross-cultural activities. One model of successful cross-cultural activities would be the ice-breaking sessions included in the forum, with 63.2% of surveyed students reporting that the ice-breaking sessions were helpful in bringing them out of their comfort zone and facilitating their engagement with other students (Fig 3A).

Post-GSF Analysis. Overall, GSF appears to have had a positive impact on its participants. At the conclusion of the forum 91.7% of interviewed subjects agreed that working in groups during the forum helped them to develop an appreciation for diversity in teams (Fig 4A) and 97.9% said that GSF was helpful to them in developing methods to overcome cultural/language barriers (Fig 4A). The combination of these results show that GSF had a strong influence on its participants in terms of learning how to gain an appreciation for multicultural teams and overcoming the challenges that may be encountered when working in such a diverse environment.

GSF also helped to put EE on the radar for its participants. At the end of the forum 93.8% of students reported that EE seemed like a more relevant and pressing topic directly as a result of their participation in the forum (Fig 4B). Furthermore, 79.2% of participants stated that they encountered new topics in EE that they would like to engage (Fig 4A). This data set indicates that students seemed to gain an interest in EE and an enhanced understanding of its relevance. Action Planning, one of the problem-solving methods that students used in the forum, was also reviewed positively by many students. At the end of the forum 91.3% of students said that they would utilize action planning as a problem-solving method in the future (Fig 4A) with 68.1% of students rating the effectiveness of action planning as an 8 or higher on a scale of 0 to 10 (Fig 4C). These results suggest that students may be open to learning or adopting new problem-solving methods and a program such as GSF may be the appropriate forum to introduce such tools to students. Lastly, in the post-GSF survey students were once again asked how they would respond to a problem that they could potentially face in their EE. At the end of the forum 52.9% of students said that they would work with their university while 47.1% said that they would take the initiative themselves (Fig 4B). It is also interesting to note that the 11.1% of students that said that they would "do nothing" in the pre-GSF survey dropped to 0% in the post-GSF results. There was also a 13.7% increase in students who would take the initiative themselves between the pre- and post-GSF results (Fig 3B & 4B). In short, GSF seems to have boosted the willingness of students to take action in their EE.

Conclusion

The GSF has developed into a well-established model for uniting a diverse group of students for the common cause of improving EE. With approximately one decade of GSF experience behind us it is a critical point in time for us to reflect on ways to further bolster our principal event. In the future we would like to develop more creative ways to preserve the strong sense of motivation that students develop for EE over the course of the forum, even after the conference is over. Perhaps this can be achieved by keeping students in touch with their GSF project teams, finding a sponsor to help fund and kickoff their projects, and assigning teams to SPEED mentors for ongoing collaboration and project development after the GSF. Our survey also provides future directions for our organization. The pre- and post-GSF survey results helped to characterize present gaps in EE—on a global scale—as well as evaluate the impact of GSF as a whole. We learned that students are already thinking about solutions to EE challenges, but not knowing where to start, university policies, and a lack of support hinder them. We should enhance our university relations to act on these findings. Furthermore, we should continue to survey students with a greater sample size in future forums. Once we collect more data it may even be possible to identify regional differences in student responses. In closing, as argued by Rugarcia et al., the 21st century engineer must be equipped to handle a wide realm of challenges that are not strictly in the form of technical engineering topics, such as globalized markets and social responsibility.[4] Fortunately, the GSF model revolves around inclusion of these principles by motivating a diverse group of students to collaborate and enact positive change in their communities. The GSF provides a framework for international student collaboration, which can be applied to EE and beyond.

References

[1] D. Delaine *et al.* (2009) "The Student Platform for EE Development (SPEED) Empowering the Global Engineer" SEFI Annual Conference, Rotterdam, Netherlands, August 2009.Boyer, E. L. (1996).

[2] R. Kandakatla, R. Dhinesh Balaji, "Impact of Action Plan workshop in enhancing Indian EE", Twelfth LACCEI Latin American and Caribbean Conference for Engineering and Technology (LACCEI' 2014) Guayaquil, Ecuador, July 2014.

[3] R. Kandakatla, A.K.R. Venkathapathy, A. Seeman, J. Cano, J. Packhem, "Reimagine Student Perceptions of Barriers in EE through the 10th Global Student Forum", 43rd SEFI Annual Conference, Orléans, France, June 2015.

[4] A. Rugarcia *et al.*, "The Future of Engineering Education" *Chemical Eng. Educ.,* vol. 34, no. 1, pp. 16-25, 2000.

ASEE 2016
INTERNATIONAL FORUM
NEW ORLEANS, LOUISIANA / JUNE 25, 2016

Paper ID #17540

Preparing globally competent and competitive STEM workforce of the 21st century in the Global STEM Classroom®

Dr. Vitaliy Popov, 1. University of San Diego. 2. The Global STEM Education Center

Vitaliy Popov is a Post-Doctoral Research Scholar at the Mobile Technology Learning Center, University of San Diego, USA. In addition, Dr. Popov is a consultant at the Global STEM Education Center. For his PhD dissertation, he designed and implemented studies in which over 500 students from more than 55 countries worked together with the help of online collaborative technologies. As a postdoc he is currently conducting a cross-sector research study involving the cross-pollination of ideas between K-20 education and private enterprise in STEM-related disciplines as they implement mobile learning in teaching and learning. Vitaliy has published articles in international peer-reviewed scientific journals such as, Computers in Human Behavior; The Internet and Higher Education; Technology, Pedagogy and Education; International Journal of Intercultural Relations.

Mrs. Larisa K. Schelkin, Global STEM Education Center, Inc.

Larisa K. Schelkin CEO, President and Founder, Global STEM (Science, Technology, Engineering and Math) Education Center, Inc. 501(c) (3) nonprofit corporation; She is also a Founder of the Global Technology and Engineering Consortium (GTEC) and a co-founder of the DOME Foundation. Larisa Schelkin is the author of the Global STEM Classroom® teaching and learning model . She held executive positions in academia and STEM global corporations (WPI, WIT, Tufts University; TYCO Electronics Global Corporation). Larisa is a Fellow for Education Policy, Rennie Center for Educational Research and Policy & Institute for Education Leadership (IEL), Washington, DC (Class 2015); Larisa studied Global Education at Harvard University Graduate School of Education "Think Tank" (Class 2016); she is a Fellow and Board member of Massachusetts Academy of Science. Larisa serves on The STEM-Space Project at SETI Advisory Board; on IUCEE Advisory Board and on Massachusetts Metro North Regional Employment Board; She was recently appointed a Co-Chair for the Global STEM Education Sub-Committee and also serves on Diversity in STEM Education Sub-Committee, MA Governor's STEM Education Advisory Council. Larisa received a Catalyst Award from the Science Club for Girls and Distinguish Service Award New England Regional ASEE. She holds BS and MS in Petroleum Engineering & MS in Computer Science. Prior to moving to Massachusetts, USA, where she and her family residing now, they have lived in Russia, India, and Africa.

Dr. Russell Faux, DSRA

Russell Faux, Ed.D. studied philosophy at the Universities of Chicago and Wisconsin before turning to education and completing his doctorate at Boston University. He taught language and literature for six years at the University of Padua in Italy. After a year as a Spencer Scholar at Stanford University, Dr. Faux returned to Italy as a research consultant for Bolt, Beranek, and Newman. Dr. Faux held senior research positions at TERC and Riverdeep Interactive (perhaps best known for the Carmen Sandiego software) before starting Davis Square Research Associates in 2001. His work encompasses qualitative and quantitative methods, including grounded theory, instrument development and validation, quantitative modeling, and social network analysis. DSRA clients have included the National Science Foundation, the National Institutes of Health, Purdue University, the University of Maine, Tufts University, the Boston Museum of Science, Boston University, the University of Massachusetts, the U.S. Department of Education, Houghton Mifflin, Verizon, and the Corporation for Public Broadcasting.

CONFERENCE PAPER FOR ASEE INTERNATIONAL FORUM 2016

"Preparing globally competent and competitive STEM workforce of the 21st century in the Global STEM Classroom®" -

Abstract

Technological advancements over the last decade have dramatically changed the reality of the STEM professional world, the way STEM professionals work, and the skills that are now required. One of the major changes is that STEM professionals are working in teams, generally multicultural teams in the United States or global teams collaborating through technology. Today's students—tomorrow's workers— without ability to learn and work with people from diverse cultural backgrounds, and skills to function productively in an interdependent world community will not be able to improve the United States' economic competitiveness, and provide leadership in innovation and creativity. The Global-STEM Classroom® reflects these rapid changes, addresses the new skill requirements, and provides innovative technology experiences for students and teachers by modeling authentic, multi-cultural, multi-disciplinary collaborative, STEM professional teamwork processes.

The Global STEM Education Center, Inc 501 (c) (3) nonprofit founded in 2008 with a mission to "increase participation in STEM disciplines by all students and, especially girls and students from underrepresented groups, to meet the needs of an innovation-driven globally competitive workforce". It involves a collaborative learning environment specific to Science, Technology, Engineering, and Math (STEM) projects. The Global STEM Classroom® implies creating virtual environment of one joint classroom where students from two or more schools in different countries receive instructions from one teacher as if they are in the same classroom (using both synchronous and asynchronous teaching and learning approaches). The participating schools are located in the United States, United Kingdom, France, Russia, Ukraine, Mexico, Norway and the Netherlands (Australia, Ireland, India will be joining next academic year). The Global STEM Classroom® program specifically addresses secondary education with students ages 12-17 years old (more than 50% girls and students from underrepresented groups) giving students a learning advantage in preparation for higher education, raising an awareness of and "sparks" the interest in engineering professional and studying STEM disciplines and expands cultural and technical competency for the global workforce. The Global STEM Classroom® participants typically complete real-life STEM (both research and hands-on) projects with an online virtual global student team. The present study was undertaken to answer the following research question: *What are the attitudinal effects of participation the Global STEM Classroom® in terms of students' motivation for participation, levels of engagement, and interest in continuing to participate in similar projects?* The research model was a within-group, change-over-time model and data collection relied largely upon pre-post survey findings to determine whether the Global STEM Classroom® was effective at meeting its declared objectives. The methods included an extensive online survey comprising the constructs embedded in the research question. Findings from the survey showed that:

- Students chose to participate because they were motivated by the content, the possibility of collaborating internationally, or because they thought it would be good for their academic and future careers.
- Students most enjoyed the interactive aspects of the project.
- Students tended to use relatively simple and ubiquitous communication tools (e.g., Skype).
- Teachers were motivated by the professional development opportunities and by the idea that the project would be good for their students.
- Teachers also enjoyed the interactive aspects of participation, including the guest speakers (professional engineers and other STEM professionals)

The above findings, however preliminary, show that the Global STEM Classroom® holds a great deal of promise for a larger-scale rollout. The Global STEM Classroom® is able to tap into students motivations and provide them with meaningful and innovative educational experiences that appear to have exercised significant effects on learning, attitudes, and behavior. There are important aspects of future research that will need to be addressed, however, including: a more complete sample with a full description of the participants, an assessment tool for measuring both fidelity of implementation and extent of exposure, and more work on assessments of learning that are tied to the actual content of the STEM projects.

Introduction

Technological advancement over the last decade has dramatically changed the reality of the professional world of engineers, the way they work and the skills they now require. One of the major changes is that many engineers now work in teams, generally multicultural, either in the United States or internationally, cooperating through technology. Based on the data on international collaboration that were collected in 2006 by SETAT, one in six scientists and engineers in the United States reported working with individuals located in other countries. These scientists and engineers used telephone, e-mail, and Web-based or virtual communication as main means of communicating during their international collaboration. Thus, it is becoming increasingly imperative for today's students— tomorrow's workers— not only to have the opportunity and skills to use various means of communication but to be able to work with people from diverse linguistic and cultural backgrounds.

This growing demand for engineers with global experience requires that K-12 schools expand the number of graduates competent in Science, Technology, Engineering and Math (STEM) and who are interested in pursuing engineering as a career path. However, there is a large disparity between the number of K-12 students who enroll in school each year (the National Center for Education Statistics reported approximately 50.1 million K-12 students enrolled in 2015) and the amount number of students who pursue engineering courses (the National Academy of Engineering reported that from 1990-2009 only about six million K-12 students received any formal engineering education). For the few students who had engineering instruction, the curriculum did not always portray the diversity and excitement of modern technological and collaboratively global nature of 21st Century engineers. This may lower the interest and engagement for students and leave them less prepared if they do pursue an engineering career path later on (Fralick et al., 2009; Knight & Cunningham, 2004).

There were a few studies, mostly at the level of higher education, which investigated the benefits of study abroad collaborative engineering programs (Streiner et al., 2015). The students who participated in these studies acquired a greater understanding of the current issues that engineers face in a global and social context (Jesiek et al., 2014). However, it may be unrealistic and financially difficult for K-12 students to study abroad and experience the benefits of international learning environments. The new online collaborative technologies that we advance can provide more accessible opportunities for students to gain international experiences by simulating globally dispersed engineering teams, which could potentially lead to increased interest in engineering.

A well-documented body of research now exists on the effectiveness of a form of collaborative learning that uses a network of computers, namely computer-supported collaborative learning (CSCL) (Stahl, Koschmann, & Suthers, 2006). CSCL is a coordinated, synchronous and/or asynchronous activity between two or more learners who collaborate via the computer to solve problems or co-construct knowledge. Learning occurs by means of the articulation and sharing of ideas to be built upon by the collaborating partners (Stahl, Koschmann, & Suthers, 2006). Collaborative learning is an effective instructional approach that can better prepare students as prospective engineers because it fosters working in teams. The skill that is highly valued in the engineering profession. In this study, CSCL with culturally diverse middle/high school students is referred to as Global STEM Classroom®. The Global STEM Classroom® participants complete real-life STEM/engineering (both research and hands-on) projects with an online virtual global student team. In addition, the Global STEM Classroom® involves creating virtual environment of one joint classroom where students from two or more schools in different countries receive instructions from one teacher as if they are in the same classroom (using both synchronous and asynchronous teaching and learning approaches).

Therefore, we believe that the inherent practices of 21st century engineers should inform the way students work and learn about engineering in school. The introduction of this knowledge into, specifically intercultural, collaborative learning environments will allow us to mirror the contemporary internet-based and intercultural workplace of professional engineers in the field. Middle and high school students will not only acquire STEM skills, but also will potentially be inspired to pursue a career in engineering. There is university level research regarding the benefits of international collaborations (in the form of study abroad programs) on the development of beliefs and attitudes towards becoming engineers and scientists (Besterfield-Sacre et al., 2015). However, little is known about how the Global STEM Classroom® (computer supported intercultural collaborative learning) may affect middle/high school students who work on authentic engineering projects and whether this approach can influence students' attitudes and knowledge about engineering and their self-efficacy around novel engineering collaborations. The present study was undertaken to answer the following research question: *What are the attitudinal effects that can be reasonably attributed to participation in the Global STEM Classroom®?* We particularly investigated students' motivation for participation, levels of engagement, the use of various technological tools, problems encountered, attitudinal gains, and interest in continuing to participate in similar projects.

Theoretical framework

Attitude can be defined as a predisposition towards a stimulus acquired through a person's past experiences (Ajzen, 2001). These past experiences whether they are positive or negative, can determine how an individual will react towards future situations, concepts, people or objects. According to Myers (1993) a combination of emotional, cognitive and behavioral attributes shape the outcome of attitude. According to Social Cognitive Career Theory, attitudes along with values and abilities are three important aspects that determine how someone's career path might develop (Lent & Brown, 2002). There is a positive correlation between a student's familiarity with STEM careers and his/her interest in pursuing STEM professions (Clewell & Campbell, 2002; Robinson & Kenny, 2003). The shortage of students enrolling in engineering courses could be linked to their insufficient knowledge of the actual tasks engineers do at their jobs on a daily basis.

Previous research suggests that interest and positive attitudes towards engineering need to be developed as early as middle school, so that students will desire to enroll in high school courses centered around science and math (Akos et. al., 2007). Without initiating this essential academic path, students are less likely to be academically competent and may lose interest in pursuing college programs and future careers as engineers. Akos and his colleagues (2007) reported that an overwhelming majority of middle/high school students make decisions about their coursework based on preparation for a future college degree or career (Akos et al., 2007). There are more negative attitudes towards science in older students compared to younger ones (Neathery, 1997). Therefore, it is imperative to positively engage students as early as middle school in STEM subjects or they are less likely to enroll in the required high school courses that would be necessary in pursuing future careers in engineering.

How can we facilitate the process of formation of positive attitude towards engineering among middle and high school students? According to Epistemic Frame Theory, developed by Shaffer (2006), the goal of a professional is to create an epistemic frame (values, knowledge, epistemology, identities) through their professional experiences. We believe, of the limited amount of middle/high/high school students who receive engineering instruction, they rarely have access to practical hands-on professional experiences that would form an epistemic frame of engineering. This may explain why some lose interest and quit engineering programs. One approach to this problem is implementation of online collaborative learning environment, so called the Global STEM Classroom® here, that will enable us to mirror the contemporary internet-based and intercultural workplace of professional engineers in the field. By doing so, middle and high school students will work on authentic engineering projects and gain more understanding of the engineering profession; therefore, they might be more likely to form a positive attitude toward engineering.

From a theoretical perspective, socio- constructivist learning theories (Vygotsky, 1978) assert that knowledge emerges from the collaborative problem-solving process in learning communities with information distributed among its members. Collaborative activities such as asking questions and giving explanations are thought to change participants' cognitive structures (King, 1997). The personal and cultural backgrounds of learners are viewed as particularly important because they influence the manner in which the learner attains knowledge (Wertsch, 1998). Heterogeneous groups and culturally heterogeneous ones, in particular, can benefit from the sharing of different perspectives and different background knowledge (Popov et al., 2014; Popov et al., 2013; Weinberger et al., 2007). From an educational policy perspective, moreover, many schools and universities are using collaborative learning technologies to better prepare students for the world of work. Balakrishnan (2015) has found that the use of CSCL had positive impact on engineering students' learning in terms of their learning achievement and positive attitude towards online collaborations. These findings seem to be consistent with other research on the use of CSCL in engineering education (see for details Rubio et al., 2010; Fuller & Moreno, 2004).

Methods

The study setting

The students from 8th-9th grade (N=112 in total) from US (n=56), Russia (n=28), France (n=14) and UK (n=14) have participated in a Global STEM Classroom® collaborative project on NASA GRACE and climate change in Spring Semester 2013 (January to June). NASA GRACE is a joint international mission and stands for "Gravity Recovery and Climate Experiment".

From one class in one public school in the US (Dennis Yarmouth HS in Massachusetts) and one class from a public school in UK (Finham Park in Coventry, Mid-Land). We formed 7 culturally mixed student groups (in each team two students from US and two student from UK). The collaborative partners met/worked in their groups synchronously two times per week during regular class hours in US -and after school hours in UK. Each team needed to chosen a research topic under the "umbrella" of a major theme "NASA GRACE and Space Exploration". Each team have conducted a collaborative research, developed a Google Drive PPT presentation and collaboratively reported (synchronously via Google Hangout or Skype), the results to the joint UK-US teaching team, other science teachers from both US and UK, other students and parents during the final presentation day. Similar setup was done with two more classes from Moscow, Russia (Lyceum "Vtoraya shkola" and Gymnasium #53) and one more class from France (Ballan-Mire) with the US students as their collaborative partners.

During the Global STEM Classroom®, the students engaged in hands-on collaborative activities devoted to NASA GRACE data collection and data visualization, data processing, plotting and interpretation; ice sheet/water content changes for specific regions: Antarctica, Greenland, CA valley, 16 basins of Greenland (each team have chosen a different global region). Guidance for teachers was provided by one of the NASA GRACE mission specialists. The students collectively studied the materials provided by NASA GRACE mission http://www.nasa.gov/mission_pages/Grace/#.VtB7wOaauNo and had a DVD disk provided by the American Museum of Natural History http://www.amnh.org/explore/curriculum-collections/grace. There were four staff members from each school (US, UK, France, Russia/Russia) 20 staff member (one lead-science teacher, one assistant science teacher, IT Director and one IT technician or, one Foreign Language and one IT technician.

Sample, data collection and analysis

The sampling frame for the 2013 Global STEM Classroom® student survey was determined in consultation with the GTEC leadership. Participating students were provided the link to the SurveyMonkey site and completed the survey online. The final respondent sample for the survey included 30 students (28 from the US and 2 from Russia) from the classrooms of three teachers). Students ranged in ages from 13 to 18, with the average age of about 15, and about 2/3 of the respondents being female. The respondent composition by class and gender is presented in the following table:

Teacher	Boy	Girl	Total
2	1	5	6
3	7	11	18
4	2	4	6
Total	10	20	30

The survey instrument itself was created through the collaboration of the project with the external evaluator (DSRA) and explored the participant perceptions of program components and the resultant attitudinal effects of participation (the survey can be found here https://www.surveymonkey.com/r/BV8332G). The Cronbach Alpha reliability value for the survey was calculated at a very high 0.930, providing some indication that the survey was consistent in its measurements. No construct validity analysis was undertaken as the sample size was too small.

The rather long survey included 120 items, including 10 open responses. The six central domains of the survey included motivation for participation (e.g., 'I am interested in collaboration with students from other countries'), levels of engagement ('I am interested in science and engineering components of the program'), the use of various technological tools, problems encountered, attitudinal gains, and interest in continuing to participate in similar projects (e.g., 'I believe that this program is highly beneficial for me in general'). The research model was a within-group, change-over-time model and data collection relied largely upon pre-post survey findings to determine whether the Global STEM Classroom® was effective at meeting its declared objectives. Subsequent analyses and visualizations were done either in SPSS or R, with the bulk of the statistics being non-parametric. The overarching research question for the student survey included:

What are the attitudinal effects of participation the Global STEM Classroom® in terms of students' motivation for participation, levels of engagement, and interest in continuing to participate in similar projects?

Results

Findings from the survey covered a wide range of points, and these will be summarized in the following bullet points:

- Students chose to participate because they were motivated by the content, the possibility of collaborating internationally, or because they thought it would be good for their academic careers.
- Students most enjoyed the interactive aspects of the project.
- Students tended to use relatively simple and ubiquitous communication tools (e.g., Skype).
- Teachers were motivated by the professional development opportunities and by the idea that the project would be good for their students.
- Teachers also enjoyed the interactive aspects of participation, including the guest speakers

The following series of density plots show examples of the distributions some of the reported pre-post gains in the target constructs. The constraints of space limit what can be presented, though the reader will no doubt glean some insights into the relative effectiveness of the Global STEM Classroom®.

The first set of density plots show the statistically significant ($p<.05$, Wilcoxon) shift in the students' attitudes toward the study of engineering in this context.

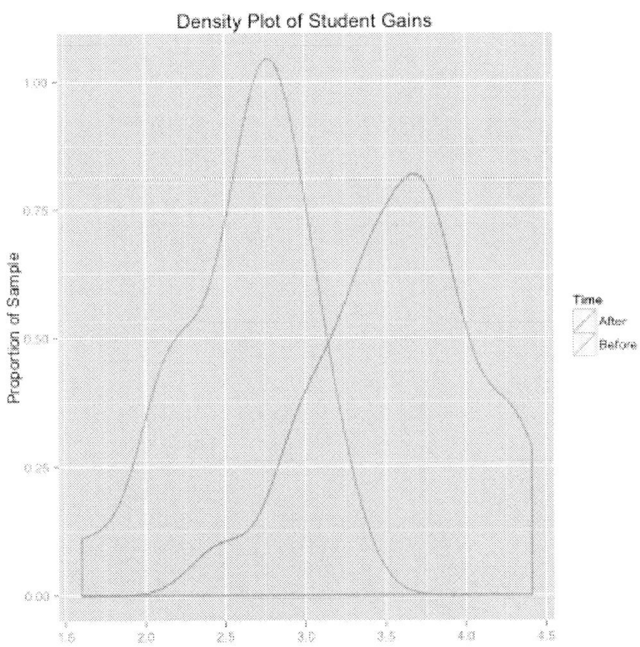

The following density plots show the statistically significant ($p<.05$, Wilcoxon) shift in teachers' engineering talk.

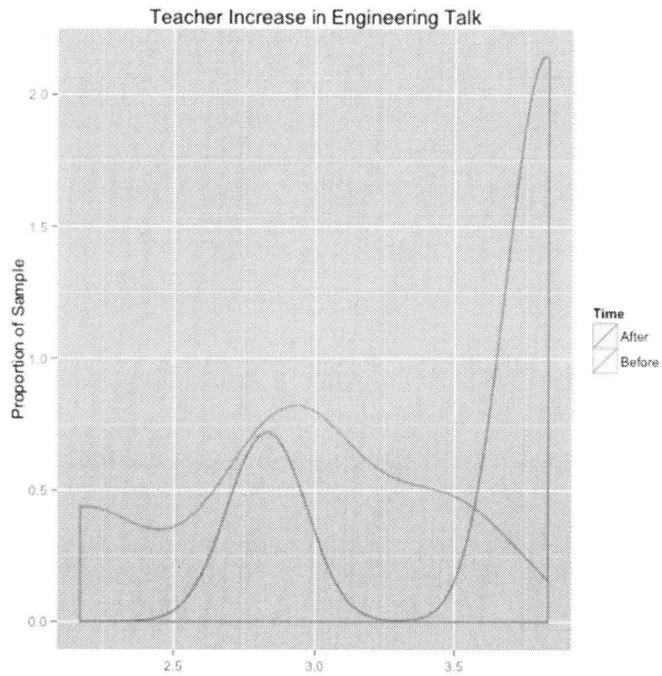

A final pair of density plots shows the statistically significant pre-post increase ($p<.05$, Wilcoxon) in the teachers' reported levels of self-efficacy in teaching engineering.

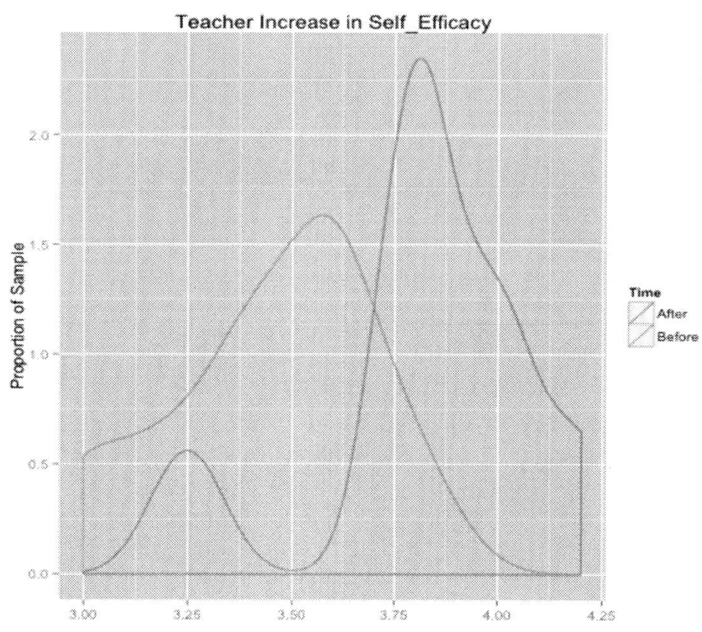

Conclusion

The above findings, however preliminary, show that the Global STEM Classroom® holds a great deal of promise for a larger-scale rollout. Thus study was able to tap into participant motivations and provide participants with meaningful and innovative educational experiences that appear to have exercised significant effects on learning, attitudes, and behavior. There are important aspects of future research that will need to be addressed, however, including:

- A more complete sample with a full description of the participants
- An assessment tool for measuring both fidelity of implementation and extent of exposure
- More work on assessments of learning that are tied to the actual content of the project
- Additional work on the validation of the project research instruments, perhaps expanding these with a mode developed track of qualitative inquiry.

Previous research findings indicate that if engineering students are properly prepared for working in an international environment prior to the experience, which includes being clear about their own expectations and how they are connected to the experience, they will be able to optimize their engineering learning goals and become more globally adept (Ragusa et al., 2014). Simply placing students from different cultures together in a group does not appear to lead to successful learning experiences (Lin & Kinzer, 2003). It is necessary, rather, to emphasize the importance of preparation for the intercultural group projects and reflection activities after their international collaborations, in order to recognize and acknowledge the impact of these experiences on their cognitive, attitudinal, social and behavioral development. Future research studies may investigate a number of interventions that could help students reap the benefits such learning experiences. These could include: reflection activities through discussions for increased meaning, and guided facilitation of collaborative activities (e.g., through collaboration scripts (Popov et al. 2013) and teachers' active leadership roles in cultivating students' attitudes, knowledge, and skills.

Overall, based on these preliminary findings Global STEM Classroom® hold promise for the STEM education. The Global STEM Classroom® model, considered as a kind of pedagogical "container space," could well support any number of other content areas, or even trans-disciplinary inquiry. Building on the participants' interest in international contacts and technology, the Global STEM Classroom® is in an excellent position to help students extend their social and cognitive networks, help them to experience the building of social capital, and assist them in the derivation of considerable learning and attitudinal benefits in the process.

Reference:

Ajzen, I. (2001). Nature and Operation of Attitudes. *Annual Review of Psychology 52,* 27–58.

Akos, P., Lambie, G. W., Milsom, A., & Gilbert, K. (2007). Early adolescents'' aspirations and academic

Balakrishnan, B. (2015), Online computer supported collaborative learning (CSCL) for engineering students: A case study in Malaysia. Computer Applications in Engineering Education, 23, 352–362.

Besterfield-Sacre, M., Shuman, L., Ragusa, G., Matherly, C., & Benson, L. (2015). "Assessing the Spectrum of International Undergraduate Engineering Educational Experiences", 122nd ASEE Annual Conference & Exposition, June 14-17, 2015, Seattle, WA

Fralick, B., Kearn, J., Thompson, S. & Lyons, J. (2009). How middle schoolers draw engineers and scientists," *Journal of Science Education and Technology, 18*(1), 60-73.

Jesiek, B. K., Haller, Y., & Thompson, J. (2014). Developing Globally Competent Engineering Researchers : Outcomes-Based Instructional and Assessment Strategies from the IREE 2010 China Research Abroad Program. *Advances in Engineering Education, 4*(1), 1–31.

King, A. (1997). ASK to THINK-TELL WHY: A model of transactive peer tutoring for scaffolding.

Knight, M., & Cunningham, C. M. (2004). Draw an Engineer Test (DAET): Development of a tool to investigate students' ideas about engineers and engineering. *Proceedings of the 2004 American Society for Engineering Education Conference & Exposition, Salt Lake City, UT.*

Myers, J.E. (1993) Personal Empowerement. Ageing International, XX, 3-8.

Neathery, M. F. (1997) Elementary and Secondary Students' Perceptions toward Science and the Correlation with Gender, Ethnicity, Ability, Grade, and Science Achievement. *Electronic Journal of Science Education, 2*(1).

Popov, V., Biemans, H. J. A., Kuznetsov, A., & Mulder, M. (2014). Use of an Interculturally Enriched Collaboration Script in Computer-Supported Collaborative Learning in Higher Education. *Technology, Pedagogy and Education*, 23(3), 349- 374.

Popov, V., Biemans, H. J. A., Brinkman, D., Kuznetsov, A., & Mulder, M. (2013). Facilitation of Computer-Supported Collaborative Learning in Mixed- versus Same-Culture Dyads: Does a Collaboration Script Help? *The Internet and Higher Education*, 19, 36-48.

Shaffer, D. W. (2006) Epistemic frames for epistemic games. *Computers & Education, 46*(3), 223-234.

Stahl, G., Koschmann, T., & Suthers, D. (2006). Computer-supported Collaborative Learning: An Historical Perspective. Cambridge University Press, Cambridge, UK.

Streiner, S., E. McCave, M Besterfield-Sacre, G. Ragusa, L. Shuman, C. Matherly, L. Benson, S. Levonisova, R. Savage. (2015). "An Inductive Qualitative Analysis of Student Interviews on Engineering Global Preparedness", 2015 ASEE Annual Conference and Exposition, Seattle, WA, June 2015

Vygotsky, L. S. (1978). *Mind and society: The development of higher mental processes*. Cambridge, MA: Harvard University Press.

Weinberger, A., Clark, D. B., Hakkinen, P., Tamura, Y., & Fischer, F. (2007). Argumentative Knowledge Construction in Online Learning Environments in and across Different Cultures: a collaboration script perspective. *Research in Comparative and International Education, 2*(1), 68-79.

Wertsch, J. (1998). Mind as action. New York: Oxford University Press.

Multination Research Programmes: The UNESCO UNITWIN in Humanitarian Engineering Outreach Case Study

Dr. Jane Goodyer P.E., Massey University

Jane Goodyer is an Associate Professor in the School of Engineering and Advanced Technology at Massey University. Regarded as an expert in manufacturing systems improvement she had led numerous research projects working with organisations, including Aston Martin, Jaguar and Caterpillar. Jane is also the New Zealand Coordinator for the UNESCO (United Nations Educational, Scientific and Cultural Organisation) University Twinning network in Humanitarian Engineering working with other Universities from Tanzania, Malta and the UK.

Until recently Jane was Director for Undergraduate Teaching and Learning where she led an initiative to design a new engineering curriculum. Her work has been acknowledged internationally with Fellowships from the UK's Institute of Engineering and Technology and the Institution of Professional Engineers New Zealand. Jane continues to be recognised for her work in engineering education and is currently engaged with advising NZ's Tertiary Education Commission on introducing degree apprenticeships.

Dr. Lizzie Miles, Coventry University UK
Dr. Anh Lan Ho Tran, Coventry University

Dr Anh Tran is a Senior Lecturer of Humanitarian Engineering at Coventry University. Her research interests are in the area of Appropriate and Humanitarian Technology, particularly in water, sanitation and hygiene (WASH), sustainable waste management and renewable energy, humanitarian applications of 3D printing and unmanned aerial vehicles (UAVs), social entrepreneurship and engineering education. Her current research projects are in 1) the use of fluorescence spectroscopy to assess water quality in areas of poor sanitation and disaster relief 2) floating biodigesters – a renewable energy, waste and sanitation solution for communities living on the water and the social enterprise models to bring this technology to market, 3) humanitarian UAVs for disaster relief - thermal imaging for search and rescue, 4) global competencies and intercultural awareness of engineers and 5) raising aspiration and widening participation in science, technology, engineering and mathematics (STEM) through Humanitarian Engineering outreach programs.

Multination Research Programmes: The UNESCO UNITWIN in Humanitarian Engineering Outreach Case Study

Abstract

This paper describes the building of partnerships involving multiple nations through the United Nations Education, Scientific and Cultural Organisation's university twinning and networking scheme (UNESCO UNITWIN) in Humanitarian Engineering. The originators of the UNESCO UNITWIN span the globe: the network's coordinator is Coventy University, based in the UK, and is supported by the Institute of Accountancy Arusha, Tanzania; the University of Malta and Massey University in New Zealand.

The main objectives of the UNESCO UNITWIN is to promote an integrated system of research, training and information on humanitarian engineering and to enhance cross-cultural learning and sharing. We believe that humanitarian engineering is about having a forward thinking view: investing in what adds value to society, looking at how to use our engineering capability to apply technology appropriately into communites' self-identified needs.

This paper reviews how the UNESCO UNITWIN was formed and describes a case study of an outreach programme; addressing one of the network's goals to raise public awareness about how engineering can help society. Using these two 'lenses' we highlight the challenges of partnering with multiple nations using documented theories of transdisciplinary working. Challenges such as inexperience and communicating over large time-zone differences replicates what other transdisciplinary programmes experience. What we have learnt is that knowledge about multination collaboration is tacit. The challenge for the engineering education community is to make this knowledge explict so that we can equip ourselves and the next generation of engineers to effectively practice across disciplines and cultures.

Introduction and Rationale

Currently, engineering education globally is going through a tough time. Firstly there is a shortage of young people wishing to embark on an engineering career. The Royal Academy of Engineering[1] reports that the UK will require "1.28m new science, engineering and technology professionals and technicians by 2020." In Australia, a Senate inquiry[2] was set up to address the shortage of engineering and related employment skills. The American Immigration Council[3] are calling for new policies to improve Science, Technology, Engineering and Mathematics (STEM) training throughout the U.S. educational system in the interest of their economy's recovery. To compound this problem current engineering education programmes may be regarded as too abstract, out of touch, hard work and boring that focus on theoretical foundation rather than student-centred practice[4]. In addition there is an under representation of females practicing engineering where (in countries such as the UK and Australia) women bearly reach 10% of the profession's cohort[5,6].

There are many initatives government, educational and industry sectors are implementing to address these issues, e.g. developing curricula that is more aligned for the challenges students will face in their working career (such as the Conceive-Design-Implement-Operate movement[7]) and the resurgence of apprenticeships[8]. A suggestion, in UNESCO's commissioned landmark report[4] on transforming engineering education purports that the:

"promotion of public and policy understanding and interest in engineering, will happen with the better appreciation of the vital contribution of engineering to development, sustainability and poverty reduction.......activities such as "Engineers Without Borders" .....are attractive to students.... Such initiatives help enrolment, public awareness and policy implementation of the importance of engineering in social, economic, international and humanitarian development."

We believe that by being very explicit about connecting engineering with its benefits to society it will help address both the gender disparity and increase general interest (irrespective of gender) in engineering. This is one of the reasons why a key objective of our UNESCO UNITWIN programme is developing effective outreach programmes that relate to humanitarian engineering.

If we analyse the definition of the humanitarian engineer[9] "the artful drawing on science to direct the resources of nature with active compassion to meet the basic needs of all – especially the economically poor, or otherwise marginalized" it forces engineers to embrace their social, cultural and ethical skills. Skills that may be underdeveloped, in many practicing engineers, due to their education and professional experience. This leads to a call to converge concepts, disciplines, and epistemologies to solve complex problems that are bound by larger social systems and its interdependencies, i.e. leading engineers to work with various disciplines from agriculture, medicine, earth sciences, social sciences and education[9]. This transdisciplinary sytems approach also supports the thrust of the UNESCO UNITWIN programme to support projects that are interdisciplinary and intersectoral. This scheme, established as a result of a UNESCO resolution adopted in 1991, serves as a prime means of building the capacities of higher education and research institutions through the exchange of knowledge and sharing, in a spirit of international solidarity.

It is appreciated that UNESCO sanctioned the formation of the UNITWIN network in Humanitarian Engineering. The UNESCO UNITWIN has purposely sought to provide a transdisciplinary network through its founders. We have a convergence of humanities and international relations (through the University of Malta), business and economics (through the Institute of Accountancy Arusha, Tanzania), engineering education (through Massey University, New Zealand) and humanitarian engineering (through Coventry University, UK). Through its international platform, the purpose is to examine key areas where engineering can be harnessed to improve humanitarian and development outcomes. These areas are to: develop resource materials that are appropriate, culturally-relevant, that promote effective practice; deliver an international postgraduate course specifically in the area of humanitarian engineering; develop outreach programmes at all levels that encompasses pre-school through to industrial professional development.

This paper describes the initial process of the UNESCO UNITWIN's formation followed by a description of how two of the collaborators have operated during the development of one of its objectives i.e. the development of outreach programmes. We shall use Stokols et al. working model[13] of transdisciplinary scientific collaboration to benchmark our collaborative effort. This model is favoured as it has been extensively cited[14]. We will also reflect on our ability to collaborate between a disparate body of researchers as we wish to add to the debate on international collaborations; particularly as transdisciplinary networks, that span major geographic and cultural boundaries, are extremely complex[10]. As general literature points to a number of factors that inhibit the development of successful collaborations such as expertise, language, cultural values, belief and norms, management, time and resources, age/seniority, negotiation style, trust, personal connection[10,11,12].

UNESCO UNITWIN Formation

The UNESCO UNITWIN in Humanitarian Engineering was signed in May 2013. Two years of the initiative have transpired and the first report has not yet been submitted to UNESCO, therefore the data summarised in this paper should be treated very much as preliminary data and interpreted as such.

When initially setting up the UNESCO UNITWIN strong parallels could be drawn between its operation and the Tuckman Model for 'Developmental Sequence in Small Groups'[15]. With the actual development and writing of the proposal simulating the Forming Stage cultivating a positive and polite atmosphere between partners. This has rapidly been followed by the Storming Stage of Tuckman's Model where members start to push at boundaries and look for their positions and roles to be clarified. These challenges are magnified when looking to work in a multination transdisciplinary project such as the UNESCO UNITWIN.

In their paper designed to evaluate transdisciplinary science, Stokols et al. presented a conceptual framework[13] that took three 'antecedent conditions' (i.e. personal, physical and bureaucratic and structural issues), reviewed the 'processes' that these antecedent conditions produced and the resultant 'outcomes' that influence the prospects for a successful transdisciplinary project. This was then strengthened by the work of Morgan et al. who, after reviewing data provided by several transdisciplinary centres, identified several overarching themes that presented challenges in this type of global cross cutting work[16]. When examined in detail it can be said that these overarching themes present themselves as 'processes' in Stokol's conceptual model allowing for the development of an 'outcome' to improve the operation of the transdisciplinary project. This is summarised in Table 1.

Table 1. Adaptation of Stokols et al. Model of Transdisciplinary Scientific Collaboration[13]

Antecedent	Overarching Theme (Process) – *Brief explanation*	Example 'Outcomes'
Bureaucratic & Structural	1. Universities inexperience with transdisciplinary work – *Universities are rarely prepared to make transdisciplinary collaborations run smoothly*	Group Coaching, pre project team training, clarification of project mission
Personal, Physical	2. Opportunities for face-to-face interaction – *spatially distant locations limit face to face interaction*	Structured communication strategy
Bureaucratic & Structural	3. Tenure and merit review procedures – *few universities offer incentives to encourage transdisciplinary collaborations*	Incentives to engage with the process
Personal	4. Departmental chauvinism and disciplinary disrespect – *debates and tensions surrounding social science versus technical science*	Structured training programmes
Personal, Physical	5. Collaborative successes and progress towards intellectual integration – *can progress towards innovative work bridging multiple fields be demonstrated*	Publications
Personal, Physical	6. Critical experiences and 'milestone' events in the development of effective collaboration – *certain meetings have been identified as playing a pivotal role in simulating transdisciplinary collaboration*	Structured communication strategy
Bureaucratic & Structural, Personal	7. Readiness to collaborate – *an initial lack of readiness of some partners to collaborate*	Group Coaching, pre project team training
Bureaucratic & Structural	8. Support for working models and conceptual themes – *joint efforts to link concepts and methods drawn from two or more fields*	Structured training programmes

Over the preliminary 18 months of operation of the UNESCO UNITWIN several generic challenges have been identified which can be mapped against the Table 1 and seen to be a 'process' resulting from an antecedent condition. By using both Stokols et al's and Morgan et al's models[13,16], constructive 'Outcomes' have been developed to support the progress of the UNITWIN. Tables 2, 3 and 4 summarise the challenges resulting from or a lack of an antecedent condition.

Table 2. Antecedent Condition – Personal

Highlighted Challenge	Themes
Differing understandings of the term 'coordination' some cultures interpreted this as 'organisation' and 'instruction of activities' – compounded by lack of direction from UNESCO	1, 2, 7
Lack of knowledge of each other's strengths and areas of expertise, what each partner brought to the table and what each partner wanted from the Network	1, 2 4, 7
Some partners already had longstanding working relationships or deep understanding of each other's cultural differences that resulted in them moving their Networks forward faster than the newer members	7, 8
Cultural issues – 4 very different universities – which was at the start perceived as a strength – but in operation can also act as a weakness as each has very different humanitarian problems that they wish to address as highlighted by the Outreach Case Study	1, 4, 5, 7

Table 3. Antecedent Condition – Physical Environment

Highlighted Challenge	Themes
It was always known that the time difference between countries would be a challenge but this has proved more difficult than initially anticipated	1,2,5,6,7,8
Incompatible online video conferencing technologies made it difficult to conduct virtual face-to-face meetings and webinars critical for collaborative working practices over geographically separate locations	1,2,7

Table 4. Antecedent Condition - Bureaucratic and Structural Issues

Highlighted Challenge	Themes
The initial paperwork required by UNESCO required each Head of University (Vice Chancellor / Rector) to sign in person and due to a lack of logistical reliability across countries on one occasion the paper work went missing.	1, 7
Changes in internal structuring with in some of the partner organisations. One organisation of the UNESCO UNITWIN has resided in 3 different departments in its 2 years of operation, none of which have the same operational procedures.	1
Changes in key personnel or personnel's working hours resulting in no clear key point of contact as many of the initial negotiations took place at a very senior level and these are not the members of staff delivering the outputs for the partnership.	1
Each partner has different modes of operation for term times, teaching loads and other duties of staff external to the UNITWIN. This means it is difficult to get a time that all partners are able to dedicate time to the project.	1,5
Some partners have no financial targets in their Institutional Research Targets and have active strands that specify deliverables in Multi-Disciplinary Research where as other partner academics were under pressure to use the Network to generate income and had targets in the areas of Internationalisation, Student Satisfaction and Curriculum Development.	1, 3, 5, 7, 8

As a result of analysing these challenges the following Outcomes have been proposed and are in the process of being implemented to enhance the operation of the UNITWIN:

- Development of a 'New Partners agreement' for new institutions joining the UNITWIN. This will incorporate a method of recording clear expectations on both sides, reporting methods and timings.
- Better planning of subject specific meetings with exact regularity under negotiation.

- Development of potential retreats and workshops (some on line some linked to conferences or academic exchange programmes)
- Expansion of the use and population of the website.
- Requirement of every partner to have a minimum of a single dedicated link member of staff with encouragement of this to be an early career academic to challenge Stokol's observations that junior members of staff are reluctant to engage with transdisciplinary initiatives.
- Calendar and map to be added to website to record term times, key meetings, geographical locations and time differences of partners.
- Technology policy developed to give to new partners of what technology is ideal and how to engage if that technology is not available.

Case Study: Outreach Programme in Humanitarian Engineering

We strongly support the suggestion made by Beanland & Hadgraft that by demonstrating the social benefits of engineering to young people (girls and boys) it will raise their aspirations to pursue a career in engineering[4]. This linkage of engineering to social benefit is also highly pertinent to girls and women. Research[17] shows women are more inclined to study subjects they find 'socially relevant', and more likely to be motivated by the idea (to study engineering) that they would 'make a difference'[18]. Due to the potential that humanitarian engineering can raise the aspirations of young people to pursue a career in engineering, one of the key objectives of the UNESCO UNITWIN is to develop outreach programmes that can be disseminated internationally.
During 2014 the UNESCO UNITWIN's Coordinator, Liz Miles at Coventry University, specifically asked each of the parties to determine a work plan of cooperation during the four years of the programme. During this phase there were face-to-face meetings to discuss each parties' priorities with regards to contributing to the objectives of the UNESCO UNITWIN. The result of this phase is that Coventry University (UK) and Massey University (NZ) decided to intially focus on the outreach objective. As the Institute of Accounting in Arusha has no engineering provision it was decided that they would focus initially on engagement with the Academic and Student Exchange strand of the programme and as Malta University is the only University on the Island they chose to observe this initial phase of dissemination and are in discussions as to the use of the outreach work specifically to address gender balance in their engineering programmes. Using Stokols et al. model[13] the following discusses the collaboration of these two partners, and describes the antecedents that were present and questions if these built strong or weak processes that are conducive for positive outcomes.

Antecedent Conditions – Personal Environment

Stokols et al. identify that personal factors such as values, expectations, goals and experience have an affect on transdiciplinary collaboration[13]. The key people leading the outreach programmes are Liz Miles at Coventry University and Jane Goodyer at Massey University. An important factor is that Liz and Jane have a long-term professional and personal relationship. Jane worked closely with Liz at Coventry University for 6 years, prior to her relocating to NZ. They have a positive working relationship that is also quite informal. They thoroughly understand each other's nuances at a cultural and personal level. They also have quite similar work experiences and are at similar stages in their academic careers. They are also of similar age and both women. Interestingly gender has been studied in terms of its influence on collaboration with findings suggesting that females collaborate more with females than male researchers do[19].

In terms of values, goals and expectations they have similar views about humanitarian engineering and how it can influence young people in a positive way. They both agree that it can influence those who are typically alienated from learning about STEM. The only difference in their goals relates to who they target the outreach programme to. Coventry University is striving to disseminate humanitarian engineering to 14-21 year olds, irrespective of gender, and is including young people who are either excluded or at-risk of school exclusion, newly arrived to the UK refugees or asylum seekers. Massey University is targeting young girls aged between 10-13, who will come from lower decile schools (i.e. have a higher proportion of students from low socio-economic communities) and will include girls that come from regions in NZ that have a higher proportion of Maori and Pasifika population.

It is quite clear that the personal antecedents of the two partners mirrored each other, making collaboration much easier. Also, it is important that there is flexibility in how partners pursue the goals to aid collaborative effort.

Antecedent Condition – Physical Environment

The UK and NZ are at opposing sides of the planet. The extreme lack of spacial proximity has created many challenges. Many of the challenges encountered by the UK and NZ are not unique to this section of this article and have been discussed in relation to Stokols' model earlier in the paper. However we have tried to overcome these by the extensive use of Skype and email communication methods. The time difference (typically 11-13 hours) does mean that both partners have to be willing to engage in Skype meetings outside of normal work hours, sometimes late into the evening or early in the morning.

To cope with such physical displacement we suggest that collaborators must be willing to use technology to communicate and are able to work flexible hours.

Antecedent Condition – Bureaucratic and Structural Issues

Massey University acknowledges the importance of developing innovative, collaborative, multi-institutional partnerships to sustain world-class research and scholarship in its Research Strategy. However, Massey University does not have systems nor policies in place that focus on establishing and nurturing collaborations, particularly international. This is fundamentally different to Coventry University that has recently introduced 'Internationalisation' into its merit review system as a compulsory requirement for all academic staff. There is a broad spectrum of activities that can be categorised as 'Internationalisation', which can range from internationalisation of curriculum through case studies through to taking students overseas to study. The university has provided large scale investment to encourage academics to engage in this activity and it will be interesting to review the impact of this change in the merit system on engagement with the UNTWIN by Coventry academic staff.

Coventry University has also received support from its bureaucratic structure (i.e. the Faculty of Engineering and Computing) to release resource to develop the humanitarian outreach programme. They have developed ten outreach workshops with approximately £18,000 funding from the Royal Academy of Engineering, with assistance from industry (Siemens), academia (University of Bristol), private organisations (ReciproBoo) and non-govermental organisations (Serve On, Practical Action). Each workshop contains facilitator support materials, links to the UK's national curriculum, kitsets, student support materials and 'train-the-trainer' support.

At Massey University there are institutional drivers that create support for this work. For example, in NZ there is a national performance-based research fund (PBRF), which funds universities. This is measured by publications, research funding, peer esteem, etc. Currently the NZ government, through its Ministry of Business, Innovation & Employment[20] have launched a $2 million fund to excite and engage harder-to-reach young people (e.g. indigenous Maori or people from low socio-economic communities) to take part in STEM. Jane Goodyer is bidding for $130k to adapt the UK programme into a national free afterschool club, called the Hello Café. If the bid is successful the research outputs, i.e. related to womens' engagement with STEM, will contribute to Jane Goodyer's PBRF rating and the university's goal to increase the research funding from PBRF.

This collaborative effort could be described as an organic and voluntary collaboration rather than one which has been developed by an administrative, bureaucratic system, which supports the argument that collaboration works best as a voluntary process[21]. However, the strategic goals and performance measures of universities does play an important role in allowing time and resource to be released to undertake collaborative work.

Conclusions

The future challenges and complexity the practice of engineering will encounter, such as population growth and a global economy, increases the importance of the engineering education community to develop a skill base that can effectively work with others within the engineering community, other disciplines and other cultures. However collaboration across differing nations and disciplines is difficult. The UNESCO UNITWIN is an excellent programme that brings together intersectoral and interdisciplinary partners, however there is an assumption that these partners have the skills and systems to effectively collaborate. Our experience has shown that the four founding partners of the UNESCO UNITWIN in Humanitarian Engineering had differing capability and experience to effectively collaborate from day-one of the network's formation. It has taken some time for the partners to come to grips with working together. Writing this paper has provided us an opportunity for reflection about how we are operating the UNESCO UNITWIN and by engaging with literature about the 'science of team science' has unveiled a myriad of ideas and theories that we should have taken account of at the network's inception. Our experience mirrors what other transdisciplinary programmes experience, what isn't apparent is the mutual learning that can occur by systematically sharing our experiences more widely. As an engineering education community we need to ask ourselves how we're equiping engineers of the future to effectively collaborate at a global level and also what professional development is required to help ourselves advance in the 'science of team science'.

References

[1] RAEng. (2015). Engineering for Growth. Retrieved 10 March, 2016, from http://engineeringforgrowth.org.uk.

[2] Senate. (2012). *The shortage of engineering and related employment skills*. Canberra, ACT, Australia: Senate Printing Unit.

[3] AIC. (2013). *Always in Demand: The Economic Contributions of Immigrant Scientists and Engineers*. Washington, DC.

[4] Beanland, D. & Hadgraft, R. (2013). UNESCO Report *Engineering Education: Transformation and Innovation*. RMIT University Press, Melbourne, Australia.

[5] Women's Engineering Society. (2014). *Useful Statistics*. Retrieved 11 February, 2016, from http://www.wes.org.uk/statistics

[6] Engineers Australia. (2013). *The Engineering Profession: A Statistical Overview*. Tenth Edition. September 2013, Institution of Engineers Australia, Barton ACT.

[7] Bankel, J., Berggren, K.-F., Blom, K., Crawley, E.F., Wiklund, I., & Ostland, S. (2013). The CDIO syllabus: a comparative study of expected student proficiency. *European Journal of Engineering Education*, 28 (3), 16.

[8] Goodyer, J. & Frater, G. (2015). *Stepping into One Another's World: Apprenticeships – Transforming Engineering Technologist Education in New Zealand*. Palmerston North, New Zealand: Massey University Printery.

[9] IJSLE. (2012). *Convergence: Philosophies and Pedagogies for Developing the Next Generation of Humanitarian Engineers and Social Entrepreneurs*. International Journal for Service Learning in Engineering: Humanitarian Engineering and Social Entrepreneurship. USA.

[10] Wang, L-C. & Beasley, W.A. (2014). International collaboration in higher education between the United States and China: Differences in cultural perspectives [online]. *International Journal of Continuing Education and Lifelong Learning*, Vol. 7, No. 1, Dec 2014: 47-57. Availability:<http://search.informit.com.au/documentSummary;dn=805270935005833;res=IELHSS> ISSN: 1997-7034. [cited 15 Feb 16].

[11] Johnson, I. M. (2009). International collaboration between schools of librarianship and information studies: Current issues. In *Asia-Pacific Conference on Library & Information Education & Practice*.

[12] Bozeman, B., Fay, D., & Slade, C. P. (2013). Research collaboration in universities and academic entrepreneurship: the-state-of-the-art. *The Journal of Technology Transfer*, *38*(1), 1-67.

[13] Stokols, D., Fuqua, J., Gress, J., Harvey, R., Phillips, K., Baezconde-Garbanati, L., ... & Morgan, G. (2003). Evaluating transdisciplinary science. *Nicotine & Tobacco Research*, *5*(Suppl 1), S21-S39.

[14] Wagner, C. S., Roessner, J. D., Bobb, K., Klein, J. T., Boyack, K. W., Keyton, J., ... & Börner, K. (2011). Approaches to understanding and measuring interdisciplinary scientific research (IDR): A review of the literature. *Journal of Informetrics*, *5*(1), 14-26.

[15] Tuckman, B.W. (1965). Developmental sequence in small groups. *Psychological Bulletin*, 63, 384-399.

[16] Morgan, G.D. Kobus, K., Gerlack, K.K., Neighbors, C., Lerman, C., Abrams, D.B., & Rimer, B.K. (2003). Facilitating transdisciplinary research: The experience of the transdisciplinary tobacco use research centres. *Nicotine & Tobacco Research*, 5 (1), 11-19.

[17] National Research Council. (2006). *To Recruit and Advance: Women Students and Faculty in Science and Engineering*. The National Academies Press. Washington, D.C.

[18] Ministry of Women's Affairs. (2012). *Does Gender Matter?* Ministry of Women's Affairs, Wellington, New Zealand.

[19] Bozeman, B., & Corley, E. (2004). Scientists' collaboration strategies: Implications for scientific and technical human capital. *Research Policy*, 33(4), 599–616.

[20] MBIE. 2016. *Unlocking Curious Minds*. Retrieved 15 February, 2016 from http://www.curiousminds.nz/discover/article/4/30/unlocking-curious-minds

[21] Chompalov, I., Genuth, J., & Shrum, W. (2002). The organization of scientific collaborations. *Research Policy*, 31(5), 749–767.

ASEE 2016
INTERNATIONAL FORUM
NEW ORLEANS, LOUISIANA / JUNE 25, 2016

Paper ID #17480

The STEM Loop: Undergraduate Engineering Students Create a STEM Children's Book

Dr. Leslie Seawright, Texas A&M University at Qatar

Leslie Seawright is an Assistant Professor of English at Texas A&M University at Qatar. She has published several journal articles and book chapters related to pedagogy, Engineering education, intercultural communication, and notions of identity through literacy. Her research interests include technical writing, discourse analysis, community literacy practices, and transnational education.

Prof. Ibrahim Hassan P.E., University of Texas, Austin

The STEM Loop:
Undergraduate Engineering Students Create a STEM Children's Book

Abstract

This paper documents an innovative project in which undergraduate mechanical engineering students created and produced a children's book about combustion engines. Funded through a grant provided by Texas A&M University at Qatar, students researched, designed, and wrote a children's book intended to promote interest in STEM fields. The book, written in both English and Arabic, will be used in Qatari public schools and in the Texas A&M University at Qatar's STEM Outreach Program. The interdisciplinary project was co-led by a mechanical engineering professor and an english professor. Students gained valuable transferable skills while working on this project including creative thinking, audience awareness, teamwork, technical writing, visual design, and communication. The authors of this paper, who include the students that worked on the project, aim to promote and encourage the idea of undergraduate students actively engaging and creating STEM programs and initiatives for K-12.

Introduction

The idea of STEM education has been around since the early 1900's. Originally called, SMET (science, mathematics, engineering, and technology), educators focused on instructing students in skills that would benefit their future societies[1]. The idea of enriching these fields by creating students well versed in them quickly gained popularity. Currently, STEM education is recognized in many countries as a way to raise educational standards and produce a generation of ingenious individuals. In the Middle East, where this study was conducted, STEM education and engineering programs specifically, are currently the fastest growing majors. Professionals in engineering and science are in high demand in the region as countries like UAE, Qatar, Kuwait, and Saudi Arabia use their abundance of petroleum revenues on State development and new technologies.[2]

Texas A&M University at Qatar offers engineering programs for Qatari and international students. The university advances STEM pedagogy and incorporates transformative educational experiences into its curriculum. These effort correspond to the current belief that engineering education is outdated and does not prepare students for the challenges they will face in industry. David Goldberg in his book, *A Whole New Engineer*, suggests that current engineering education is misaligned with 21st century needs. He argues that change in engineering education "can't be accomplished in the old paradigm under the old assumptions about how education change happens, but in the right atmosphere, the change flows organically from the students themselves" [3] . It was this kind of thinking about a new way of educating future engineers that spurred the idea for this project.

The idea to create a children's book came from a discussion that took place in the spring of 2015 in a Business and Technical Writing course at Texas A&M University at Qatar. The discussion centered on a student's prospective audience for his upcoming descriptive paper. The student jokingly suggested creating a pop-up book on combustion engines for an audience of children. This remark resulted in a discussion about the possibility of doing a children's book series focused on engineering concepts that would inspire children to enter the engineering field.

English professor, Dr. Leslie Seawright, partnered with a mechanical engineering professor, Dr. Ibrahim Hassan, to pitch the idea to the administration of the university. They were awarded a Transformational Education Experience grant of $12,500 in order to produce a children's book on combustion engines.

The University and Students

Texas A&M University at Qatar (TAMUQ), is an international branch campus more than 8,000 miles away from the main campus of Texas A&M University in College Station, Texas. Despite its location in the Middle East, students at TAMUQ are required to complete identical coursework as their peers on the main campus, including courses on American politics and American history. The student body is comprised of students from around the world, but primarily students come to TAMUQ from the Arab Gulf region and Asia. TAMUQ offers Bachelor of Science degrees in chemical, electrical, mechanical and petroleum engineering. The university maintains an enrollment of around 550 students, of which 58% are male and 41% are female[4] . Nearly all of the students at TAMUQ speak English as their second or third language. The faculty of the university is as diverse as the students, with faculty from the United States of America, Jordan, Lebanon, Egypt, Greece, France, and elsewhere.

TAMUQ is located in Education City and funded by the Qatar Foundation, a non-profit organization founded in 1995 by the then ruling Emir of Qatar, Sheikh Hamad bin Khalifa Al Thani. Education City hosts the branch campuses of five other American universities, including Georgetown School of Foreign Service, Carnegie Mellon University, Northwestern University, Virginia Commonwealth, and Weill Cornell Medical College. Education City is a bustling community of students and faculty in all disciplines[5].

The Project

Once Drs. Hassan and Seawright received the grant for the children's book, they went to work finding students for the project. Since the assignment was not part of a class, the students voluntarily committed to the project and spent time outside of class to work on the book. Many of the students came from the mechanical engineering program, though not all. Of the ten students involved in the project, six were in mechanical engineering, two in electrical engineering, and one each in petroleum engineering and chemical engineering. The project not only had program diversity but gender diversity as well, as four of the students in the project were female along with one of the project's advisors, Dr. Leslie Seawright.

The students and advisors meet weekly starting in the Spring semester of 2016 to discuss ideas for the children's book. Many important challenges and opportunities were discussed and debated in these project meetings. The first decision to be made was the target audience and the main character of the book. Other decisions such as narrative style, illustrations, number of pages, type of book paper, and other fundamental aspects of the project were raised. Mohammad AlGammal, a technical laboratory coordinator, agreed to tackle the precise mechanical drawings necessary for the book. Two students volunteered to create and draw the main character and background. In addition to the students and advisors, two students were hired as part-time workers for the book project. These students communicated and planned the student meetings, worked with publishers to obtain quotes for various types of children's books, and collected data on the project.

The division of labor and scope of the project was decided by the students themselves. They divided into two groups: an illustration group and a dialogue/text group. Despite these divisions, the students actually met as a whole on most occasions and discussed the book project as a team. As the students reached the mid-point in the semester, it became clear that the book would not be accomplished using only student art. An illustrator was hired in order to take the students' concept drawings and turn them into the actual book pages.

Findings

This project was funded through a Transformative Educational Experiences Grant awarded by Texas A&M University at Qatar. The idea of transformative educational experiences (TEE) has been adopted by many colleges and universities in order to encourage undergraduate research and meaningful student-centered work. These types of experiences are especially meaningful for engineering students, as their future careers will require them to work in teams and think critically about project decisions. Equally as important are the children that these engineering students hope to reach. As Carlson and Sullivan note, "Preparing children with the skills necessary to flourish in an increasingly technological world becomes more challenging every day…engineering is about building things to help people and society."[6] This project allowed engineering students to reach out to young primary school children and encourage them to learn more about the field of engineering.

The first decision to make was the type of children's book and the target age group. The students and advisors brought in books and discussed the options. It was decided that the group would like to make a book with flaps on each page in order to make the book more interactive for children. The students were adamant about wanting to connect with the children reading the book, and they also discussed the possibility of adding QR codes, videos, and an app connected to the book that could provide additional information. Ultimately, the students had to discard these ideas as the short time span of 5 months simply wasn't enough to incorporate all of their ideas. Even the flap idea had to be abandoned after the students discovered that the flaps nearly doubled the cost of the book and created complications related to illustration and design. The students struggled with the decision to drop the page flaps for several months. Their decision to move on with a more traditional-style children's book was driven by time and money concerns, both issues they will face again and again as engineers. The real-world scenario of book production forced them to make hard decisions that affected their overall design and aspirations for the book since these goals did not correspond to the amount of time and money available.

The age range for the children's book was set at 6 to 10 years old. This decision was made by studying the book industry and the common age ranges for children's books. While it appeared a minor decision to make, the students soon realized that this factor would ultimately affect the vocabulary, style, illustrations, and organization of the entire work. As the dialogue group began writing the story, they often thought about the words and examples children 6 to 10 would know and understand. Also, this age group created a dilemma in thinking about the level of writing that should be used. Should the book be able to read by 6 year olds, making it very simple and perhaps too childish for the 10 year olds, or should it be written at the top end of the age group challenging the 10 year olds and requiring an adult or sibling to read it to the 6 year olds? The students decided to aim for the upper end of the age range in regards to reading level. This decision would allow the book to "grow" with the 6 year olds and be informative and challenging for the 10 year olds.

The creation of this children's book also required a great deal of care in the selection of the main character, plotline, and gender/culture/language considerations. The students agreed in the first meeting that the main character of the book should be female. When the advisors asked them why they decided to make the main character female, several students explained that boys would be immediately drawn to the idea of a book on combustion engines while a girl may not. They wanted to encourage female readers to learn about combustion engines by having a female lead character. This impressed the advisors, as the majority of students in the group are male but the students all quickly decided that females should be encouraged to enter STEM fields and the creation of a female lead character was one way to do this. What should she look like? How old should she be? What should she wear? The students spent several minutes discussing the main characteristics of the female character. One student suggested that the girl be a bit older than the 6 to 10 age range because children look up to older children. The students decided that she would be 12 or 13 years old. They chose her name, Sara, because this is a name found in both English and Arabic-speaking regions of the world.

The advisor, Dr. Seawright, reminded the students that it was important how they chose to dress and characterize the girl. The students quickly agreed that she should have black or dark brown hair and wide anime-style eyes. The eye selection was driven by the student illustrator who preferred this style and by the fact that anime is an Asian art style. The students felt that this style would speak to the children in the Qatar region, as anime is popular in the region. The character's clothes were also discussed in several meetings. Students wanted the girl to look comfortable in a story about combustion engines. The initial sketch was of the girl in a pair of mechanic's overalls and a baseball hat. In the Middle East, Muslim women commonly cover their hair in order to respect their cultural and religious beliefs. The girl was given a hat in a nod to this tradition. Later, however, the sketch was changed and the girl had her shoulder length hair down. This was done in order to make it more believable that she had just removed her go-cart racing helmet (a plot detail). The student illustrator (a female) explained that the girl in the story didn't need to cover her hair because she was 1) young enough to not have to cover and 2) had short hair so it was not necessary to cover it in the student's opinion of her own culture and religion. The students and advisors agreed with this new choice.

The plot remained the most important part to work out in terms of getting started with the book. By this point, the students had decided on the age range, the number of pages, and the main character's details. One student suggested that the girl ask her father while riding in a car how the car's engine worked. The advisor pushed back asking the students to help the girl feel empowered in a journey to discover more about combustion engines. Soon the students were excited by the idea of the girl as a go-cart racer whose engine has broken down. The story of how combustion engines work evolves as she tries to find out what happened to her engine. It was important for the students and advisors that they give the girl agency in researching combustion engines.

Several weeks into the project, it became clear that student-only illustration was not going to be possible. None of the advisors or students had experience illustrating and producing a children's book, so an artist was hired to illustrate the book based on the conceptual drawings for each page and the character sketches of the main figure. In addition to the illustrator artist, a publishing expert was used to put together the final pages of the book combining the art, mechanical drawings, and text into one cohesive document. While students had initially wanted to produce two completely separate books, one in Arabic and one in English, they ultimately had to settle for a book in English in the first half and Arabic in the second half. This was done in order to cut costs and production time. The benefit of having the two languages in one book will also help young children learning to read in both languages, a common situation in Middle Eastern countries like Qatar.

The idea for this children's book came from a student and became a student-led project. An ethic of selflessness was prevalent throughout the project. Brainstorming, writing, and designing were all done in a team environment that sharpened students' skills in critical thinking, writing, designing, and working as a team. Because students were in charge from the beginning, they felt an immediate sense of responsibility and commitment. Many of them now claim that because they were allowed primacy in decision making, they felt more dedicated to this project than their required coursework. The students saw the children's book as something they had ownership in, something they had created rather than an assignment they simply completed.

Discussion

ABET requires engineering students to be competent in writing, communication, presentation, and cultural awareness in order to be truly competitive and ready to perform as engineering professionals[7]. The students involved in creating this children's book were actively acquiring these critical skills while building cultural and gender awareness. At each phase of the process, students were grappling with creating a complete picture of the audience in order to select appropriate vocabulary and analogies. Developing audience awareness is one of the central aspects of all good technical writing pedagogy. This is a difficult skill to teach, however, because in nearly all cases students are merely writing for their teacher and trying to figure out what one person requires of their writing. In this project, students had to write for a real audience of potentially thousands of children with varying English abilities, nationalities, and scientific knowledge. In addition, the engineering students had to select a plot, illustrations, and scenarios to best fit the needs of their readers. The development of presentation skills resulted from the decisions the students had to make in regards to illustration style, color saturation, text fonts, and other choices.

Perhaps the most interesting ABET requirement achieved in this project was cultural awareness. Students developed a deeper understanding of their own (mainly Arab) culture while reaching across boundaries to better understand other cultures. The students discussed customs, styles of dress, architecture, cultural norms (especially those related to gender) of their own and other cultures in order to create a book that could be easily accepted by many cultures and peoples. This is a skill that is difficult to teach and difficult for students to acquire in more traditional engineering courses.

It is the hope of these students and faculty advisors that this project will turn into a book series. Each year a new group of students could explore a new topic and create a children's book that fits within the book series. These books could then be distributed to local Arabic and English schools, as the book is printed in both Arabic and English. Engineering students at TAMUQ could also invite local primary students to campus in order to engage them in the stories and lead them in hands-on engineering activities.

Allowing students to lead this project was challenging and rewarding in many ways. If faculty can find ways to let students come up with projects and support their implementation, they can begin to change engineering education one project at a time.

References

[1]"STEM EDUCATION: Preparing Students for a Growing Field", *connectionslearning*, 2016. [Online]. Available: http://www.connectionslearning.com/Libraries/PDFs/STEM_Primer_PreparingStudents.pdf. [Accessed: 16-Apr-2016].

[2]A. Durrani, "Arab Region Engineering Degrees Address Industry Demands", *US News & World Report*, 2015. [Online]. Available: http://www.usnews.com/education/best-arab-region-universities/articles/2015/07/07/arab-region-engineering-degrees-address-industry-demands. [Accessed: 19- Apr- 2016].

[3]D. Goldberg, M. Somerville and C. Whitney, *A whole new engineer.*

[4] "Enrollment Profile Fall 2015," Texas A&M University, Nov. 2015.

[5]R. Hamilton, "After 10 Years in Qatar, A&M is Ready for 10 More, by Reeve Hamilton", *The Texas Tribune*, 2016. [Online]. Available: https://www.texastribune.org/2012/05/25/after-10-years-qatar-m-looking-forward-10-more/. [Accessed: 19- Apr- 2016].

[6]L. Carlson and J. Sullivan, "Int. J. Engng Ed", *Hands-on Engineering: Learning by Doing in the Integrated Teaching and Learning Program*, vol. 15, no. 1, pp. 20-31, 1999.

[7] "Criteria for Accrediting Engineering Programs, 2016 – 2017," *ABET Criteria for Accrediting Engineering Programs 2016 2017 Comments*, 16-Oct-2015. [Online]. Available at: http://www.abet.org/accreditation/accreditation-criteria/criteria-for-accrediting-engineering-programs-2016-2017/. [Accessed: 17-Apr-2016].

Acknowledgements

A special thank you to the students that participated in this project:
Ali Alashar, Lolwa Al-Muhannadi, Shaikha Al-Qahtani, Omar Al Sharif, Doaa Awad, Muhammad Faizeen, Saad, Moazam, Mohamed Safan, Shaza Shehab, and Yamen Zaza

Students that participated in the writing of this article include:
Ali Alashar and Yamen Zaza

Increasing Diversity through Internationalization of Engineering Education

Prof. Rose-Margaret Itua, Ohlone College

Rose-Margaret Ekeng-Itua is an innovative Engineering Professor based in the Silicon Valley Area in California, USA with over 10 years of experience in Higher Education globally. Before moving to the USA she was the Chair of Engineering and Senior Lecturer at the University of West London, London, UK .She graduated from the Federal University of Technology Owerri, Nigeria with a B.Eng. (Hons) in Electrical/Electronic Engineering majoring in Telecommunications. She gained an MSc. in Mobile and Satellite Communication from the University of Surrey, Guildford, UK and currently her Doctorate in Cybernetics at the University of Reading, UK . She has been internationally recognized by UNESCO, BBC, the IEEE for her contributions towards youth empowerment and campaigning for equity for African Americans and Females in STEM education and Career. She has research interests in the Internet of Things, Wireless Sensor Networks, Sustainable Energy and Appropriate Technologies for Emerging Economies and bridging gender and race divides in STEM education and careers. Rose-Margaret has implemented various initiatives to encourage social entrepreneurship amongst African Youths and to encourage African girls to pursue STEM education/careers.

Increasing Diversity through Internationalization of Engineering Education based on the Pedagogy of Humanitarian Engineering

Rose-Margaret Ekeng-Itua[1, 3], Sharnnia Artis [2,3]

[1]Engineering Department, Ohlone College, Fremont California,

[2] The Henry Samueli School of Engineering, University of California, Irvine, California

[3] Center for Energy Efficient Electronics Science, University of California, Berkeley, California

Abstract

The lack of diversity in engineering classrooms is a cause of concern for both academia and industry, as industry looks to academia to provide the much needed diverse workforce in engineering. Though there exist various suggestions on how to increase diversity in engineering classrooms, not many exist that creates intrinsic motivation for students. This paper discusses how internationalization of the engineering education through the pedagogy of humanitarian engineering could increase diversity in engineering classrooms through creating intrinsic motivations in students to remain in engineering education. This paper is based on a five semester long study of how internationalization embedded in the Introduction to Engineering Curriculum increased retention and success of underrepresented students in engineering. Teaching pedagogies applied and results are also discussed in this paper. We also discuss our NSF-UC Berkeley funded collaboration on Context-Based Learning and the IEEE-CPMT/SCV Section's support for Humanitarian Engineering Students Projects at Ohlone College.

Female and Underrepresented Ethnic Group Students in Engineering

To continue advancement in energy science and research and to thrive in a global economy, the U.S. will have to rely on scientists and engineers to develop innovative and high-value-added products and services, as well as improve productivity through the use of technology-based tools.[1] This pipeline of scientists and engineers, with its under-representation of women and underrepresented minorities (African Americans, American Indians or Alaskan Natives, and Hispanic Americans), is a critical concern for the U.S.[2] In 2010, the National Academies of Science reported that underrepresented minorities "embody a vastly underused resource and a lost opportunity for meeting our nation's technology needs".[3] With today's society facing global challenges in energy that are essential to sustaining our current way of life, it is even more critical for 4-year institutions to reach out to pools of students traditionally underrepresented in science and engineering programs.

One pool of such students is community college students pursuing math, chemistry, and physics courses that are transferrable to baccalaureate programs. In a Strategy for American Innovation, the Obama administration stated that "President Obama is taking continuous steps to improve our educational system ... and to promote student achievement and careers in STEM fields", and "the Administration is committed to restoring America's global leadership in college graduation rates ...., making investments in community colleges ...."[4,5] With more resources being allocated to community colleges to stimulate student achievement in STEM fields, and community colleges serving many ethnic and racial minorities, community college students are uniquely positioned to fill the pipeline of STEM professionals.

With enrollment in the nation's community colleges hitting an all-time high, students from these institutions are a rich source of the nation's recipients of undergraduate and graduate degrees in STEM fields. The community college transfer pathway is particularly important for African American, Hispanic and Native American STEM degree recipients, as well as low-income students due to its low cost ($36/unit).[6] In this time of high unemployment and economic crisis, as in earlier recessions, community college enrollment has surged. During the 2009-2010 academic years, California alone enrolled 2.7 million students. The California Community College System (CCCS) is the largest community college system in the U.S., serving 25% of the nation's community college students.[7] Of the students enrolled for 2009-2010, 40% were from NSF-categorized underrepresented minority backgrounds. In fall 2009, nearly 50,000 CCCS students transferred to University of California (UC) or California State University (CSU) campuses[8]. One year earlier, nearly 20% (3,344) of all UC B.S. degrees in STEM fields were earned by community college transfer students, but only 11% (356) of these transfer graduates were from underrepresented minority backgrounds; 40% were women.[6] In 2015 fifty-five percent of community college students are people of diverse ethnic backgrounds and roughly 53 percent are female.[6]

Despite the relatively significant number of female and ethnic minority students in the CCCS we see a disproportionately low number in the STEM fields. Therefore, there is the need to adapt our classroom pedagogies to engage these demographic groups of students.

Context Based Learning Pedagogy

The Obama Administration has forecasted that over the next decade, the U.S. economy needs approximately 1 million more STEM professionals than the U.S. will produce at current rates.[9] In February 2012, the President's Council of Advisors on Science and Technology (PCAST) issued an undergraduate STEM education report indicating fewer than 40% of students who enter college intending to major in a STEM field complete a STEM degree. Increasing the retention of STEM majors from 40% to 50% would alone generate three-quarters of the targeted 1 million additional STEM degrees over the next decade. With the first two years of college being the most critical years for the recruitment and retention of STEM majors, it is imperative to focus actions on methods that influence the quality of STEM faculty and adopt teaching methods supported by evidence derived from experimental learning research.

Research has found that high-performing students frequently cite uninspiring introductory courses as a factor in their choice to switch majors.[10] Empirical evidence about how people learn and assessment of outcomes in STEM classrooms all point to a need to improve teaching methods to enhance learning and student persistence.[10] However, a significant barrier to broad implementation of evidence-based teaching approaches is that most faculty lack experience using these methods and are unfamiliar with the vast body of research indicating their impact on learning. This gap in experience has resulted in STEM education lacking relevance to a student's life, where STEM classes traditionally teach sets of scientific and engineering principles and concepts with little course time devoted to exploring the application of principles and concepts in real-world context or technology implications.[11] Additionally, scientific principles are typically taught in scientific discipline silos, without real-world application that can span more than one scientific discipline.

Responding to the challenges identified in the PCAST report, UC Berkeley created a research experience program for community college faculty, called **RET in Engineering and Computer Science Site: UC Berkeley's Context-Based Research Experience for Community College Faculty** *(also referenced as the UCB Context-Based RET Site in this article)* that offered community college faculty a nine-week experience that integrated individual hands-on research with team-based curriculum development to enable new research concepts to be introduced in community college classrooms in the *context-based approach.* Closely aligned with project-based learning and inquiry-based science education, the context-based approach was selected to provide the participants a pedagogical method that brings their research experience alive in the community college classroom by tying the teaching with applications that students can relate to in their lives.[12] The *context-based approach* has been shown through assessments to enhance the students' interest in STEM and has been applied successfully in college teaching, particularly to the teaching of chemistry[10,12]. Empirical evidence has also shown that context-based education helps students see and appreciate more clearly the links between science and everyday lives.[13,14,15] The context selected for this RET program was technology applications that offer societal benefits and have employment opportunity potential. The premise is that using such context will increase community college student engagement in STEM studies, thereby contributing to student persistence and eventually, improved retention rates of students in STEM.

Pedagogy of Humanitarian Engineering in the Introduction to Engineering Class

Engineering education could be said to be at a crossroad, with educators and researchers calling for new ways to understand engineering's social role. Two approaches to engineering education that are relevant to the social context of engineering have emerged. First, there is a growing trend for engineering education to address issues of human development. Vesilind suggests that, historically, engineers have been employed as "hired guns, doing the bidding of both political rulers and wealthy corporations[21]. However, he says, there is a new kind of engineering emerging, one "rooted in the greater ideas and aspirations of engineering as a service to all humanity."[21] Robbins also predicts that we are at the beginning of an emerging field in engineering.[22] Second, there is a need to embed globalization within the engineering curriculum to provide students with the knowledge and skills to respond to globalization issues and to work in a global context. The importance of globalization of the engineering curriculum has been highlighted by various researchers. Some of these researchers suggest that showcasing engineering within a global context is necessary for global competitiveness, cultural inclusivity, and sustainable design.[23,24,25]

Vandersten closely links presenting engineering in a global context to the humanitarian engineering pedagogy.[26]

Humanitarian Engineering (HE)

Humanitarian Engineering (HE) as a discipline was founded in 2003, when the William and Flora Hewlett Foundation funded the creation of a minor program at Colorado School of Mines (CSM).[27] Muñoz describes the new discipline as "a wave that's passing through the world among young people that are bent on trying to improve the lives of humans on the planet in a sustainable way."[28]

Though Munoz describes humanitarian engineering as a discipline and established it as a full program at the School of Mines, this paper showcases the integration of humanitarian engineering in an existing engineering course/curriculum.[27] The argument is made that where having humanitarian engineering as a program may be challenging due to time and resource constraints, Colleges can integrate humanitarian engineering in already existing courses. This integration provides students the opportunities to be globally competitive, but more so to better appreciate cultural diversity which has a boomerang effect on creating an inclusive learning environment for female and underrepresented students in Engineering.

Burnham highlights that compared to other professions, Engineers seems to have immense power and responsibility and should therefore be afforded the opportunities to channel this desire for positive change around the world.[29] Engineering Education should therefore prepare students to make lasting positive impact in the lives of people globally.

Munoz also defines Humanitarian Engineers as Engineers who "try to balance technical excellence, economic feasibility, ethical maturity, and cultural sensitivity" through a set of specially designed technical, humanities, and social science classes, as well as a Design Experience.[30] This could arguably be said to really be the definition of an engineer, not just a humanitarian engineer.

Munoz also went further and defined the Humanitarian Engineering Program by breaking down the definition as:

"Humanitarian: an artifact, process, system, or practice promoting present and future wellbeing for the direct benefit of underserved populations."[31]

"Engineering: designing and creating a component, subsystem, or system under physical, political, cultural, ethical, legal, environmental, and economic constraints."[31]

"Humanitarian engineering: design under constraints to directly improve the wellbeing of underserved populations."[31]

HE seeks to augment technical engineering expertise with understanding of the social, political, and economic realities that define the systems in which engineers live, work, and for which they design solutions.[29] HE looks to address, in a comprehensive way, the needs of students, faculty, industry, the global community, and governmental as well as non- governmental organizations, to work toward solutions to the basic problems plaguing the human experience on earth.[32]

Humanitarian Engineering is quite popular with many engineering students, as there is a need for the students to see that their future work or career would contribute to making a positive impact in the world. In many US universities, there is an increase in student interest in service learning, community engagement, and global issues, with undergraduate and graduate programs developing to support this interest.[33] Claes Helgesson notes that, "many engineering students and professional engineers are frustrated at being tied up with solving problems connected to people in the wealthy part of the world."[34] This new movement towards active social responsibility in engineering has a desire to ensure that products, systems and processes designed by engineers are not just meeting functional standards but are also meeting societal standards.

HE similar to context based learning pedagogy helps the student contextualize their work, see the direct value and applications, create accurate mental models and as such become intrinsically motivated to pursue their engineering education. This popularity of HE is even more evident amongst female and underrepresented students.

Various researches suggest that one of the most important plus of Humanitarian Engineering is that it can attract and engage a more diverse pool of individuals.[35] There is a growing understanding that, for women and for students from some underrepresented groups, a major factor in career choice has to do with making a difference.[35] "This has not been a widespread image—engineers as members of one of the 'caring professions.' But it's an image that we can build, both for our students and for the community as a whole."[36]

A number of Colleges with HE programs are beginning to see greater diversity in their engineering cohort of students. Purdue and Colorado School of Mines have reported that over twice the percentage of women are interested in HE than in the average engineering department.[37] This is encouraging news given the fact that "the U.S., as a nation, has made no progress in diversifying engineering in some key dimensions."[38]

Through showcasing the ideas of HE, engineering schools can attract more minorities, such as women and non-technically oriented students, pushing diversity in engineering to new levels, and widening the range of perspectives needed to solve problems. The concepts at the core of the Humanitarian Engineering (HE) movement are not utopian fantasies, but rather a realistic analysis of where engineering has been, where it is now, where it is heading, and what engineering students need to learn today in order to be prepared for the future.[38]

Ohlone College Engineering for Female and Underrepresented Students (EFUS) Project through the Introduction to Engineering Course

Research has shown that the need to nurture in female students is one of the reasons why they pursue careers in healthcare and the arts. Hence the EFUS project aims to provide the opportunity for female students to 'nurture' through engineering and as such engage them in seeing that engineering provides a lot of opportunities to create solutions that 'nurture'. Engineering for Humanitarian needs will likewise create the feeling of being valuable and relevant in solving problems around the world, as gain research has shown that being valuable is important for underrepresented students to stay engaged in a course/career.

The EFUS project was first implemented in the Spring Semester of 2014. The IEEE-CPMT/SCV funded the project with $2500, donated to the Ohlone College Foundation. The project was embedded into the Introduction to Engineering Course, which has the flexibility and scope to support the EFUS Project. Students were very excited and motivated just to hear that their course/project is being supported by the IEEE-CPMT/SCV, it gave them a sense of belonging to a larger community of Engineers and also made them feel valued. This in no small way was instrumental to the intrinsic motivation evidenced by all the students on the course while facing design and team work challenges during their projects. The IEEE-SIGHT group emphasizes the importance of global humanitarian engineering, and provided students with initial project ideas.

The project gave the students a taste real life Engineering as they went through the whole Engineering Design cycle and Problem Solving Sequence on their respective projects. The female and underrepresented students in the class where motivated to take on leadership roles in their teams and contributed to identifying humanitarian needs in the world and formulating probable solutions. Furthermore the female and underrepresented students felt valued and had a sense of purpose as they saw how they could change the world positively and engage in a good cause through engineering (this was particularly true for the female students).

Many of the students from underrepresented ethnic groups felt valued as they identified and discussed humanitarian needs that they or a family member or friend was experiencing/had experienced. They also highlighted to their team mates the appropriate technologies that could be used to solve those humanitarian problems. Feedback from the students concerning the EFUS Project was very positive. Here are some quick statistics from the students' feedback.

- 100% of the students felt that the EFUS project had motivated them to stay on course with their Engineering Education and Careers.
- 100% of the Female students mentioned that through the Project they saw how they could change the world positively through Engineering.
- 100% of students who were undecided about Engineering as a career (at the beginning of the semester), decided to commit themselves to an Engineering Education (80% of the female students decided to take summer math classes to speed up their Engineering Education).

The students also presented their Projects at the Ohlone STEM day on May 9, 2014 to 60 Middle School Students. This had a domino effect as many of the middle school students were very interested in the projects and wanted to design and build the sort of prototypes they saw. The EFUS Project students felt a sense a fulfillment as they were also engaged in the process of encouraging the next generation of Engineers, through Service Learning. The EFUS project also featured the June edition of the Tri-City Newspaper (http://www.tricityvoice.com/displayPages.php?issue=2014-06-24&page=5)

Implementing the RET UC Berkeley Experience in the Introduction Engineering Course at Ohlone College

The RET experience was an invaluable experience that supported the pedagogical methodologies of humanitarian Engineering and Context Based Learning. Community College Faculty engaged in research on Sustainable Manufacturing which was easily embedded in the Introduction to Engineering Course as a topic under Engineering Ethics and Engineering Design. Students had to include different aspects of sustainability in their Capstone Projects which was underlined by humanitarian engineering theme. The research experience at UC Berkeley provided Faculty with an in depth understanding of global sustainability issues which faculty discussed in the class room and provided students with a better understanding of these issues.

Again the retention of female and underrepresented students is significant as many have made a commitment to engineering as a career. Feedback from some students were;

"The Humanitarian Engineering Project really made me see the difference I can make as an Engineer;"

"I now see that Engineering is also about caring for the world- which I really want to be part of."

"My opinions as a female student were valued as we had to really understand the social injustice that people face- somehow my team of male students valued my contribution and I took leadership role".

In conclusion this paper highlights the role that humanitarian engineering and context based learning has in retaining/increasing the number of female and underrepresented students and shows how HE can be effectively incorporated in the Engineering Curriculum via the Introduction to Engineering Course. This paper also highlights the importance of Programs like the UC Berkeley RET program for Community College Faculty in propagating HE and CBL.

Acknowledgements

This work was partially supported by the RET in Engineering and Computer Science Site: UC Berkeley's Context-Based Research Experience for Community College Faculty, a project funded by NSF Award EEC-1405547. Additionally, the authors would like to extend a special thank-you to all of the UC Berkeley Transfer-to-Excellence Research Experiences for Undergraduates students for their hard work, their mentors for their time and patience, and the program staff for their organizational efforts and support.

References

1. Babco, M., Chubin, D., & May, G. (2005). Diversifying the engineering workforce. *Journal of Engineering Education,* 94 (1): 73-86.
2. National Action Council for Minorities in Engineering (NACME). (2008). *Confronting the "New" American Dilemma. (2008).* Retrieved from July 2, 2011, from http://www.cpst.org/NACME_Rep.pdf
3. Committee on Underrepresented Groups and the Expansion of the Science and Engineering Workforce Pipeline; Committee on Science, E., and Public Policy; Policy and Global Affairs; National Academy of Sciences, National Academy of Engineering, and Institute of

Medicine (2010). Expanding Underrepresented Minority Participation: America's Science and Technology Talent at the Crossroads. Washington, D.C., The National Academies Press. Retrieved, November 15, 2012 from http://www.cossa.org/diversity/reports/Expanding_Underrepresented_Minority_Participation.pdf

4. Obama, B. (September 2009). A Strategy for American Innovation: Driving Towards Sustainable Growth and Quality Jobs. Retrieved November 4, 2012, from http://www.whitehouse.gov/administration/eop/nec/StrategyforAmericanInnovation/
5. Obama, B. (November 4, 2012). A Strategy for American Innovation: Securing Our Economic Growth and Prosperity. Retrieved from November 4, 2012, from http://www.whitehouse.gov/innovation/strategy.
6. California Community College Chancellor's Office (2011). *Key Facts.* Retrieved, August 11, 2011, from http://californiacommunitycolleges.cccco.edu/PolicyInAction/KeyFacts.aspx
7. California Post Secondary Education Commission, Enrollment (2010). *Fall Transfers to Public Institutions by Discipline, for Fall 2009.* Retrieved, August 10, 2011, from http://www.cpec.ca.gov/OnLineData/GenerateReport.ASP, accessed 8/10/11.*
8. Bransford, J. (2000). *How people learn: Brain, mind, experience, and school.* Washington, DC: National Academy Press. Available online at http://www.nap.edu/catalog.php?record_id=9853.
9. Executive Office of the President President's Council of Advisors on Science and Technology (2012). Report to the President – Engage to Excel: Producing One Million Additional College Graduates with Degrees in Science, Technology, Engineering, and Mathematics. Retrieved October 1, 2013, from http://www.whitehouse.gov/sites/default/files/microsites/ostp/pcast-engage-to-excel-final_2-25-12.pdf.
10. Bennett, J. (2005). The effects of context-based and Science-Technology-Society (STS) approaches in the teaching of secondary science on boys and girls, and on lower-ability pupils. Retrieved August 23, 2013, from http://eppi.ioe.ac.uk/cms/LinkClick.aspx?fileticket=SqD-s-48RCY%3d&tabid=329&mid=1242.
11. Hulleman, C. S., and Harackiewicz, J. M. (2009). Making Education Relevant: Increasing Interest and Performance in High School Science Classes. *Science, 326,* 1410-1412.
12. Bennett, J., and Holman, J. (2002). Context-Based Approaches to the Teaching of Chemistry: What are They and What Are Their Effects?. In Gilbert, J. (Ed.), *Chemical Education,* pp. 165-184).
13. King, D., and Ritchie, S. M. (2012). Learning science through real-world contexts. In Fraser, B. Tobin, K. G., and McRobbie, C. J. (Eds.) *Second International Handbook of Science Education,* pp. 69-72

14. Hofstein, A., Kesner, M., and Ben-Zvi (2000). Student perceptions of industrial chemistry classroom learning environments. *Learning Environments Research, 2,* 291-306.

15. Ramsden, J. M. (1997). How does a context-based approach influence understanding of key chemical ideas at 16+? *International Journal of Science Education, 19,* 697-710.

16. Remarks by the President in State of the Union Address (2013). Retrieved September 25, 2013, from http://www.whitehouse.gov/the-press-office/2012/01/24/remarks-president-state-union-address

17. Executive Office of the President President's Council of Advisors on Science and Technology (2011). Report to the President on Ensuring American Leadership in Advanced Manufacturing, p.9. Retrieved September 25, 2013, from http://www.whitehouse.gov/sites/default/files/microsites/ostp/pcast-advanced-manufacturing-june2011.pdf

18. The Boston Consulting Group (2013). Majority of Large Manufacturers Are Now Planning or Considering 'Reshoring' from China to the U.S. Retrieved September 25, 2013, from http://www.bcg.com/media/PressReleaseDetails.aspx?id=tcm:12-144944

19. Sirkin, H. L., Zinser, M., and Rose, J. (2013). The U.S. as One of the Developed World's Lowest-Cost Manufacturers. The Boston Consulting Group. Retrieved September 25, 2013, from https://www.bcgperspectives.com/content/articles/lean_manufacturing_sourcing_procurement_behind_american_export_surge/

20. Foroohar, R., and Saporito, B. (2013). Made in the U.S.A., *Time Magazine.* Retrieved September 25, 2013, from http://business.time.com/made-in-the-u-s-a/

21. P.A. Vesilind, "Peace engineering," J. Prof. Issues in Engineering Education and Practice, pp. 283–287, 2006.

22. P.T. Robbins and B.B. Crow, "Engineering and development: Interrogating concepts and practices," J. Int. Development, vol. 19, pp. 75–82, 2007.

23. P. Vohra, R. Kasuba, and D. Vohra, "Preparing engineers for a global workforce through curricular reform," Global J. Engineering Education, vol. 10, no. 2, pp. 141–148, 2006.

24. S. Johnson, "Towards culturally inclusive global engineering," Euro. J. Engineering Education, vol. 26, no. 1, pp. 77–89, 2001.

25. C. Johnston, D. Caswell, and G. Armitage, "Developing environmental awareness in engineers through Engineers Without Boarders and sustainable design projects," Int. J. Environmental Studies, vol. 64, no. 4, pp. 501–506, 2007.

26. J.D.J. VanderSteen, "Humanitarian engineering in the engineering curriculum," Ph.D. thesis, Queen's University. Kingston, Ont., Canada, 2008.

27. Muñoz, D.R., & Colorado School of Mines (2008, March 12) Humanitarian Engineering Program, promotional video. Retrieved April 13, 2008, from http://www.youtube.com/watch?v=m4hgdwhj_Zc

28. Burnham, M.G. The 'Systems Approach' to Human Problems: How Humanitarian Engineering Can Help
29. Muñoz, 2006 & 2008.
30. Skokan, C., Simoes, M., & Crocker, J. (July, 2006). Designing Humanitarian Engineering Classes. Paper T3-H presented at the 9th International Conference on Engineering Education, San Juan, PR.
31. Helgesson, C.I. (2006). Engineers Without Borders and Their Role in Humanitarian Relief. IEEE Engineering in Medicine and Biology Magazine, May/June 2006, pp. 32-35.
32. Tinto, V. (1993). Leaving College: Rethinking the Causes and Cures of Student Attrition. University of Chicago Press, Chicago, Ill.
33. Seymour, E. & Hewitt, N. (1997). Talking About Leaving: Why Undergraduates Leave the Sciences. Westview Press, Boulder, CO.
34. Noddings, N. (1992) Gender and Curriculum, in Handbook of Research on Curriculum, P. W. Jackson (ed.), Macmillan, New York.
35. Rosser, S.V. (1990). Female-Friendly Science. Pergamon Press, Elmsford, N.Y.
36. Rosser, S.V. (1995). Teaching the Majority: Breaking the Gender Barrier in Science, Mathematics, and Engineering. Teachers College Press, New York.
37. Matyas, M.L., & Malcolm, S. (1991). Investing in Human Potential: Science and Engineering at the Crossroads. American Association for the Advancement of Science, Washington, D.C.
38. Oakes, J., Gamoran, A., & Page, R.N. (1992) Curriculum Differentiation: Opportunities, Outcomes, and Meanings, in Handbook of Research on Curriculum, P.W. Jackson (ed.), Macmillan, New York.

ASEE 2016
INTERNATIONAL FORUM
NEW ORLEANS, LOUISIANA / JUNE 25, 2016

Paper ID #17553

What Underrepresented Minority Engineering Majors Learn from Co-Ops & Internships

Dr. Terrell Lamont Strayhorn, The Ohio State University

Dr. Terrell Strayhorn is a professor of higher education and director of the Center for Higher Education Enterprise (CHEE) at The Ohio State University. Author of 10 books, more than 50 book chapters, and over 100 journal articles and scientific abstracts, Strayhorn is a former NSF CAREER grant recipient, reviewer for the Journal of Engineering Education, and one of the nation's leading diversity scholars.

Dr. Royel Montel Johnson, Center for Higher Education Enterprise

Dr. Royel M. Johnson is Policy Analyst for the Center for Higher Education Enterprise (CHEE) at The Ohio State University (OSU), where he is also an affiliate in the Criminal Research Justice Center. Recently, he earned a doctorate in the Higher Education and Student Affairs (HESA) program at OSU with a cognate in Race and Social Policy; he also holds a B.A. in Political Science and Ed.M. in Educational Policy Studies from the University of Illinois at Urbana-Champaign (UIUC).

What Underrepresented Minority Engineering Majors Learn from Co-Ops & Internships

INTRODUCTION

Higher education in the United States (U.S.) is an expansive enterprise, comprised of more than 4,300 colleges and universities and approximately 21 million college students enrolled part-time or full-time for study. Countries vary in the size and scope of postsecondary education systems (Altbach, Berdahl, & Gumport, 1999), making the U.S. system the largest in the world. Women represent the numerical majority on most U.S. college campuses and nearly 40% of U.S. college students identify as underrepresented ethnic minorities (URMs) including Blacks/African Americans, Hispanics, Asian Pacific Islanders, Native Americans, and bi/multiracial (non-White) students. Despite their growing presence in higher education, URMs still represent a small proportion of students majoring in high-demand fields such as engineering. The National Science Foundation (NSF) reports that 39% of URMs "plan to major" in science and engineering fields as freshman, but less than half of these earn an undergraduate engineering degree; rates are startlingly lower for Blacks and Hispanics.

One way to increase retention and graduation rates in engineering for all students is to enhance their overall skills and readiness for engineering-related work by way of cooperative education (i.e., co-op) and internships[1] that offer students "real-life, hands on" experience in their major subject. The Center for Postsecondary Research at Indiana University identifies such experiences as a "high-impact practice" that likely produces significant learning gains for students, although research is sparse on learning that accrues from internships and co-ops in engineering (Linn, Howard, & Miller, 2004). Still, it is estimated that thousands of students participate in either co-op or internships annually, 67% of recent college graduates completed at least one internship while in school, and 56% of employers expect to hire more interns in the future. Despite their widespread use and popularity, we know relatively little about the influence of co-ops and internships on URM students' learning in engineering. This is the gap addressed by this study.

---

[1] There is consistent evidence that *all countries* are concerned about work preparation and provide "hands on experience" during training through internships, practicum, or apprenticeships.

Using a blend of survey and interview data, we will investigate the role that co-ops and internships play on enhancing URM students' learning in engineering in two primary areas (i.e., professional competencies, problem solving) as well as their professional identity. Survey data were collected using an online, web-based survey administered to URM engineering majors at predominantly White institutions in the southeast and Midwest regions of the country. Interview data were collected using a semi-structured protocol that asked questions about the nature of the co-op or internship, elicited stories about students' experiences, and what the student believed they gained from the opportunity. Implications for policy, practice, and teaching will be highlighted for the US and similarly-situated countries or related fields/disciplines.

REVIEW OF LITERATURE

To conduct a study on learning that accrues from co-ops and internships to URM students majoring in engineering, it was necessary to review the existing literature in a few areas. First, a long line of inquiry supports the role that experiential learning plays in the cultivation of talent. Experiential learning holds that experience (i.e., practice) is the source of learning and development (Kolb, 2014). Of course, experiential learning represents a number of opportunities ranging from service-learning to co-ops and internships (Stanton, 1992).

Prior research on service-learning internships provides ample support for the power of co-ops and internships to produce learning in professional fields such as engineering. For instance, Rehling (2000) focused on students in business and communications and found that their skills and understanding were enhanced through such internships, as well as their sensitivity to the needs of others. Industry-sponsored internships were deemed valuable as part of career preparation.

Many industries or fields use hands-on experiences such as co-ops and internships to prepare trainees for high-quality performance in a particular field or job. For instance, internships are highly effective in principal preparation programs. Thus, educators strive to expose principals-in-training to competent models—practicing administrators—from whom they can learn (Cottrill, 1994). And internships are a core component of ABET-approved engineering programs (ABET., 2010).

Just as engineering education programs are major- or discipline-specific (e.g., environmental, computer), so too are co-ops and internships in such fields. Although only 10% of construction education programs include field internships nationally, there is a long-standing belief that knowledge, skills, and abilities are enhanced through authentic involvement in co-ops and internships in construction engineering and internships produce learning that cannot be attained any other way. Tener (2004) studied the Purdue Construction Engineering Management Internship program and found that students learn time management strategies from weekly site meetings, effective communication skills working on a team, and the challenges of leadership as they observe interactions between the superintendent and subcontractors, for example.

There is other evidence supporting the notion that industry-based internships constitute significant learning experiences for students in engineering or technical fields. For instance, Biasca and Hill (2011) evaluated the Paper Science and Engineering program (PSEN) at the University of Wisconsin-Stevens Point that typically offers sophomore students employment at a paper mill for eight months. They found that students gained a sense of accomplishment from helping the mill meet its "bottom line" goals, time management skills as they worked to prioritize daily tasks, and communication strategies from reflective journaling and speaking "on the job." There was less evidence that students learned much about the organization itself, industry issues, corporate values, or aspects of their own professional identity. Not only have researchers shown that internships produce learning but Fogg and Putnam (2004) provided data indicating that internships often lead to higher earnings once in a professional job.

Despite this evidence base, there are important limits to what we know about internships in engineering and the role they play in student success. Scholars who have focused on internships and co-ops tend to discuss them generally using learning theory as a frame (e.g., Linn et al., 2004; Stanton, 1992), while those who focus on industry-based internships in engineering education programs rarely focus on URMs or the role that race/ethnicity might play in creating differential learning opportunities (e.g., \Tener, 2004 #6662}. This study was designed to address this gap in knowledge specifically and we strongly believe that the analysis and conclusions in this study may well be applicable to experiential learning in other disciplines, fields, for different student populations and countries beyond the United States that are concerned about workforce preparation.

METHODS

This study is part of a larger federally-funded research study designed by the lead author (Strayhorn) as part of a five-year CAREER grant from the U.S. National Science Foundation (NSF). The objectives of the larger study were to understand students' pathways to and through science, technology, engineering and math (STEM) fields with a principal focus on ethnic minority males. Although the larger study is comprised of national data, surveys, interviews, and observations, this work-in-progress paper is based on survey and interview data only.

Sample

The survey sample was comprised of 1,150 students majoring in engineering or a related subfield at a NSSE-participating institution. Sixty-four (64%) percent of the sample were women and over 57% were Black/African-American or Latino. Consistent with the nature of engineering and ABET-approved programs, 29% of respondents had "done" or completed a co-op or internship, 51% "plan[ned] to do" one, and interestingly approximately 20% "did not plan" to do so or hadn't decided at the time of survey. A fuller description of the analytic sample will be included in the final paper.

Interview participants (n = 10) were mostly minority males, given the lead authors' interest in this subpopulation, and over 50% were sophomores or juniors majoring in engineering or a subfield. All interview participants were bachelor degree-seeking and had the ability to participate actively (and without assistance) in a one-on-one interview with a member of our team. First-year, first-semester students and those living with severe disabilities were excluded from the study as the former are not likely to have internship experiences yet (or to have declared a major necessarily) and the latter may not be able to work in an internship without audio, visual, or other assistance. Table 1 presents a summary of the interview participants.

[Table 1 about here]

Data Collection

Part One: Survey. Survey data were collected in a national administration of the National Survey of Student Engagement (NSSE) sponsored by the Center for Postsecondary Research at Indiana University. The NSSE is a survey instrument designed to measure the quality and quantity of students' engagement in educationally purposeful college activities. Items relate to participation in various co-curricular (e.g., clubs and organizations) and curricular (e.g., internships) programs and activities. Additionally, the NSSE questionnaire includes a set of items that elicit information about students' perceived learning gains, making it suitable for the present study. For instance, one item asked: "To what extent has your experience at this institution contributed to your knowledge, skills, and personal development in analyzing quantitative problems?" Response options ranged from 1 (*very little*) to 4 (*very much*); other items asked about communication skills, working on a team, and ethics, to name a few. To date, more than 600 colleges and universities have participated in the national survey.

A web-based approach was employed since response rates are frequently low to mailed, hardcopy surveys (Crawford, Couper, & Lamia, 2001). Specifically, students attending NSSE-participating institutions were invited in the Spring semester to participate in the study via email. Electronic invitations included a hyperlink to the URL of the website on which the NSSE questionnaire was located. University registrars forwarded the invitation to potential participants using a randomly selected list of eligible students. No strategies were used across-the-board to encourage student participation, although evidence, to date, suggests that NSSE response rates range from 20 to 45%, which is considered good for web-based surveys (Crawford et al., 2001).

Part Two: Interviews. Interview data were collected from Fall 2008 through Spring 2015 by the authors or members of a formal research team led by the principal investigator. Qualitative data were collected via in-depth, one-on-one interviews, which is a widely-accepted technique for eliciting such information (Kvale, 1996). The purpose of interviewing is to "find out what is in and on someone else's mind" (Patton, 1990, p. 278). Interviews were conducted in such a way to elicit stories from each participant about his academic experiences generally but also what he derived from co-ops and internships specifically. Interviews averaged approximately 45 minutes, ranging from 23 to 85 minutes.

In short, a semi-structured interview protocol was developed, tested, and used in this study that included questions about each participant's demographic background, prior schooling experiences, aspirations for college, as well as the quality and nature of his engineering field experiences. Questions were designed to elicit recollections of challenges, people, and circumstances related to what was learned from co-ops and internships. Where necessary, probes were used to prompt remembrance of pivotal moments or to clarify meaning (e.g., "Can you tell me about a time when..." or "Can you say that differently..."). Use of probes and follow-up prompts are highly recommended in qualitative studies (Patton, 2002). All interviews were digitally recorded and professional transcribed verbatim for later analysis. Transcripts were stored electronically in a data information system.

Data Analysis

Given the study's objectives, data analysis consisted of three major tasks. First, survey data analysis proceeded in three stages. Descriptive statistics were calculated to observe any patterns among data points. Then, exploratory correlation analyses were conducted to test the magnitude and direction of relationships among variables; these are not included in the paper due to page limits but will be presented in the conference session. Third, hierarchical linear regression techniques with a nested design were used to estimate the influence of internship/co-op experiences and background variables on URM engineering majors' perceived learning gains, controlling for a battery of potentially confounding factors. Hierarchical regression analysis is "a method of regression analysis in which independent variables are entered into the regression equation in a sequence specified by the researcher in advance" (Vogt, 1999, p. 129). This approach yields more conservative estimates of statistical relationships, thereby reducing the chances of making Type 1 errors—that is, in this case, claiming a relationship between internships/co-ops and learning when there is none.

Interview data were analyzed using a three-stage constant comparison method described by Strauss and colleagues (Strauss, 1995; Strauss & Corbin, 1998). First, transcripts were read and re-read to generate initial categories of information or codes that represented "an initial plot of the terrain" (Miles & Huberman, 1994, p. 69); this is known as open coding (Strauss & Corbin). Next, codes were collapsed by grouping categories that seemed to be related to each other while leaving intact those that stood separate from all others. This smaller list of categories was used to generate "supercodes" or categories (Strayhorn, 2011). Discrete codes were then sorted and organized into a smaller list of 10 categories. Lastly, categories were compared and contrasted to understand the degree to which they were similar; closely related categories were combined into themes and renamed, where needed, so that the whole named reflected the sum of its parts.

FINDINGS

Hierarchical linear regression (HLR) tests were used to evaluate the relationship between URM engineering majors' participation in internships/co-ops and perceived learning gains as measured by the NSSE. Results were statistically significant ($F[4,923] = 7.46$, $p < 0.01$, $R = 0.18$, $R^2 = 0.03$, Adj. $R^2 = 0.027$), indicating that URM engineering majors who participate in internships/co-ops report greater learning gains than their same-race peers who do not work in internships/co-ops. Other significant predictors of URM engineering majors' perceived learning generally and problem-solving learning specifically were age (b = - 0.002) and transfer status (b = -0.139). Collinearity statistics suggest that multicollinearity was not a problem for this investigation and both part- and partial correlations reinforce interpretations from the HLR analysis.

Consistent with the NSSE analysis, interview participants spoke at length about the ways in which their internship or co-op experiences constituted significant learning as they were authentically involved in the "actual work of the field," as one participant put it. Generally, URM engineering majors were satisfied with their internship/co-ops experiences and found them rewarding. Consider the following:

> From co-op to co-op, ever since my first experience it was great. It was like my first real job—it was with Honda. It was kind of everything I imagined it would be. I was in the test department and I was testing brand new cars and the people I worked with were great, my mentor was great, my boss was great so I had a great experience for six months. (Derrick)

> Actually my best learning experience wasn't in the classroom, it was from my internship. I finally got a chance to learn and apply stuff from class to real-life problems. Once I had a good experience, I wanted to stay. (Homer)

Not only did participants speak, at length, about their satisfaction with internships and co-ops but they provided insights about what they derived from their experiential field experience in terms of learning, professional development, and understanding of their professional identity. Consider the following:

> You learn a lot—not just real-life engineering issues, like what you'll do as an engineer but other stuff like working with people. At [my internship] I worked with pretty young like 30's people and then people on the top like my boss. I had to work with American managers and Japanese managers and clients all over [the world]. (Miguel)

> On my resume I have knowledge of AutoCAD and, yes, I learned that from my engineering degree [program] but I *really* (emphasis added) learned it from my summer internship when I worked with my mentor doing some CAD projects with AutoCAD mechanical. When you're doing an AutoCAD design a lot of shapes are already built into the program so you're creating the design for the electrical person to follow. It's cool because I learned how to build the basic components myself. (Alexis)

While participants shared countless examples of what they learned from internships and co-ops in terms of communication, interpersonal, and technical skills such as CAD use, relatively few offered clear examples of what they learned about the profession, the organization where they interned, engineering issues, ethics, or their own professional identity. Some students shared that they felt like mentors and co-op supervisors gave them "stuff to do" but didn't always take the time to explain why they're doing it, how it fits within the larger scope of work, or prepares them for the workplace.

Interestingly, the vast majority of participants—both men and women—spoke candidly about the ways in which race/ethnicity seemed to shape their internship experience. For example, one participant explained feeling looked over, ignored, or invisible at his internship site as a Black male. Some women shared how mentors seemed to expect less of them (and, consequently, assign them menial tasks) based on the mere fact that they were undergraduate women. Findings suggest critically important implications for research, teaching, and practice.

CONCLUSION

Recall that the purpose of this study was to investigate the role that co-ops and internships play on enhancing URM engineering majors' learning engineering in two primary areas (i.e., professional competencies, problem solving) as well as their professional identity. Result from analysis of NSSE data provide persuasive evidence supporting the conclusion that engineering majors engagement in internships and co-ops produce significant learning gains in terms of problem-solving, communication, and learning more about work.

Insights from one-on-one interviews support these conclusions using stories shared by students themselves. Clearly, engaging URM engineering majors in internships and co-ops is one way to facilitate learning in core areas (e.g., problem-solving), soft skills (e.g., working on a team), and career learning. From internships students learn how to communicate professionally, skills for managing time and prioritizing tasks, and technical skills such as CAD design, computer programming, and management/leadership.

Much more needs to be done to educate students about engineering as a profession, the infrastructure and goals of engineering industries and organizations, as well as students' professional ethics and values. For instance, engineering faculty might incorporate these topics specifically in undergraduate engineering courses or professional seminars. Internship coordinators might follow the example of Biasca and Hill (2011) to develop a rubric that requires students to complete reflective essays and web-based surveys after each "work day" at an internship site that asks about ethical decisionmaking, professional values, and what they've learned about the organization.

REFERENCES

ABET. (2010). Criteria for accrediting engineering programs. Baltimore, MD: Accreditation Board for Engineering and Technology.

Altbach, P. G., Berdahl, R. O., & Gumport, P. J. (1999). *American higher education in the twenty-first century: Social, political, and economic challenges*. Baltimore, MD: Johns Hopkins University Press.

Biasca, K. L., & Hill, S. (2011). *Assessment of ABET student outcomes during industrial internships*. Paper presented at the American Society of Engineering Education (ASEE) Forum, Washington, DC. http://www.engr.uky.edu/~aseeched/papers/2011/1678.pdf

Cottrill, D. (1994). Internships: Learning the most from the best in the least amount of time. *NASSP Bulletin, 78*(559), 72-74.

Crawford, S. D., Couper, M. P., & Lamia, M. J. (2001). Web surveys: Perceptions of burden. *Social Science Computer Review, 19*, 146-162.

Fogg, N., & Putnam, M. (2004). Considering the needs of different stakeholders: The impact of co-op job quality on post-graduation earnings. In P. L. Linn, A. Howard & E. Miller (Eds.), *Handbook for research in cooperative education and internships* (pp. 229-250). Mahwah, NJ: Lawrence Erlbaum Associates.

Kolb, D. A. (2014). *Experiential learning: Experience as a source of learning and development*. New York: Prentice Hall.

Kvale, S. (1996). *InterViews: An introduction to qualitative research interviewing*. Thousand Oaks: Sage.

Linn, P. L., Howard, A., & Miller, E. (Eds.). (2004). *Handbook for research in cooperative education and internships*. Mahwah, NJ: Lawrence Erlbaum Associates.

Miles, M., & Huberman, A. (1994). *An expanded sourcebook: Qualitative data analysis* (2nd ed.). Thousand Oaks, CA: Sage Publications.

Patton, M. Q. (1990). *Qualitative evaluation and research methods* (2nd ed.). Newbury Park, CA: Sage.

Patton, M. Q. (2002). *Qualitative research & evaluation methods* (3rd ed.). Thousand Oaks, CA: Sage.

Rehling, L. (2000). Doing good while doing well: Service learning internships. *Business Communication Quarterly, 63*(77), 77-89.

Stanton, M. (1992). Internships: Learning by doing. *Occupational Outlook Quarterly, 36*(2), 30-33.

Strauss, A. (1995). *Qualitative analysis for social scientists*. Cambridge, UK: Cambridge University Press.

Strauss, A., & Corbin, J. (1998). *Basics of qualitative research: Techniques and procedures for developing grounded theory* (2nd ed.). Thousand Oaks, CA: Sage.

Strayhorn, T. L. (2011). Singing in a foreign land: An exploratory study of gospel choir participation among African American undergraduates at a predominantly White institution. *Journal of College Student Development, 52*(5), 137-153.

Tener, R. K. (2004). Using theory in Research I: Understanding the learning experienced in structured internships in construction engineering. In P. L. Linn, A. Howard & E. Miller (Eds.), *Handbook for research in cooperative education and internships* (pp. 315-336). Mahwah, NJ: Lawrence Erlbaum Associates.

Vogt, W. P. (1999). *Dictionary of statistics and methodology: A non-technical guide for the social sciences*. Thousand Oaks: Sage.

Paper ID #17591

Cross Cultural Diversity in Engineering Professionals—Russia, India, America

Julia Ziyatdinova,

Julia Ziyatdinova graduated from Kazan State Pedagogical University in 1999. Her major areas of study were foreign languages and she finished her University course with honors and qualification of teacher of English and Turkish. She continued her training and obtained PhD in Education degree in 2002. The topic of her PhD study was titled "System of Character Education in the US Schools: Current State and Trends for the Development". She also received additional minor degrees in Management (1998) and Psychology (1999) in Kazan State Technological University.

Julia joined the team of Kazan State Technological University as an instructor at the Department of Foreign Languages and the School of Foreign Languages "Lingua" in 1999 and was rapidly promoted to the position of Associate Professor at the Department of Foreign Languages in 2003. Her teaching career was perfectly balanced by the experience of a translator and an interpreter. She is a well-known person at Kazan international conferences and other events for her high quality consecutive and simultaneous interpreting, such as interpreting for the Academy of Sciences of the Republic of Tatarstan.

The new milestone in Julia's career was the position of the Chair of Department of Foreign Languages for Professional Communication in 2007, when she took over all the responsibilities related to foreign language training at Kazan State Technological University. The teaching and research priorities of her department were then focused on professional and intercultural communication for students in a technical university, professional translation and creation of foreign language environment at a university.

Because of her talents and activities, Julia became one of key figures in university international life. When Kazan State Technological University obtained the new status of a National Research University and joined the list of Top 30 Russian universities, Julia was offered a position of a Head of University International Office. She took over this position in April 2011 and rapidly gathered a strong team of professionals to face the challenges of the new university status and transformed International Office into University International Affairs with two offices covering all the aspects of internationalization.

In addition to her intensive career, Julia is also the Director of Center for Intercultural Communication – a company within the University structure offering excellent language training services for students and adult employees.

Julia is the author and co-author of over 85 publications including monographs, journal articles and study guides.

Dr. Artem Bezrukov

Artem Bezrukov graduated from Kazan National Research Technological University (KNRTU) in 2007. His major program was Chemical Engineering. He has also a minor degree in Translation for Professional Communication. Artem Berukov received his PhD in chemistry at the same university in 2010. His is the head of Protocol Office at International Affairs and an associate professor at the Department of Physical Chemistry at KNRTU. His activity areas include internationalization of higher education, collaboration with universities and research organizations in the USA, Europe, and Asia. Artem Bezrukov is the author and co-author of over 50 publications including papers in peer-reviewed journals, proceeding of international conferences, workbooks and monographs.

Dr. Phillip Albert Sanger, Purdue University, West Lafayette

Dr. Sanger is a professor in the School of Engineering Technology in the College of Technology of Purdue University. His focus and passion is real world, industry based, senior capstone experiences both domestically and internationally. He has successfully developed this area at Purdue and at Western Carolina University. Prior to his career in academia, Dr. Sanger had a successful 30 year career working in and with industry managing and participating in broad range technology development and commercialization.

Dr. Petr Osipov

Cross Cultural Diversity in Engineering Professionals—Russia, India, America

Abstract

In this global world, today's engineer is likely to have to work in global international teams with colleagues from other nationalities. The challenge for many engineering curricula is how to include, in a realistic way, this global dimension and increase the student's awareness of the issues that are encountered. However as curricula begins to be developed, it would be beneficial to study what the differences might be between cultures.

To expose the issues that may be encountered for future multidisciplinary teams made up of students from USA, Russia and India, the Miville Guzman Universal Diversity Scale (MGUDS-S) survey and form which assesses cross cultural diversity was used to measure differences between American and Indian engineers. Data using the MGUDS-S survey was collected from a population of Indian professional business people and entrepreneurs attending a personal development program and compared their responses to data collected for a similar survey with predominantly Russian business people and entrepreneurs.

This latter comparison group was made up of experienced professional business people and entrepreneurs from around the world including a large population of academics from Russia, plus entrepreneurs from Mexico, Luxembourg, Australia, South Africa, Uzbekistan, Brazil, and Finland. This international group has quite similar demographics to the Indian group and is being used as a reasonable comparison for project-minded professionals. Both groups should have very similar attitudes as expected. However significant and interesting differences have been found and are discussed in the paper.

A strikingly higher emotional discomfort level was found in Indian male participants for cross cultural diversity. In addition this study studied the differences between male and female responses across cultures. Finally the results of these two studies were compared to the results of surveys with engineering students from U.S.A., Germany, Poland and Russia from previous cross cultural diversity studies conducted by the authors.

Background

In the age of global grand challenges, engineers are likely to work in international teams so as to find good opportunities to be realized together. The challenges have been defined by different organizations, including UNO[11], Oxford Martin School[9], The Millenium Project[10], and etc. A special attention is given to the grand challenges for engineering listed by different organizations, among them the US National Academy of Engineering[5]. This paper does not make an attempt to review these challenges, however, they form a sufficient background for defining the attributes and competencies of a global engineer whose cross-cultural skills are of crucial importance for his professional activities.

A global engineer's competencies have been investigated in different surveys and papers by a number of international authors, and included various sets of attributes, thus Parkinson (2009) emphasized an appreciation of other cultures, an ability to communicate across cultures and in other languages, an ability to work in cross cultural teams and deal with ethnic and cultural diversity and having an international work/educational experience[7], Osipov (2013) gave priority to self-directed development capabilities[6], others distinguished critical thinking[14], social competence[8], intercultural competence[12], foreign language proficiency[1].

We summarize these attributes in three general categories: a global mindset, continuous self-directed development skills, and readiness for innovations. At the root of these attributes is the heightened awareness of various global cultures and a sensitivity to the cross cultural nature of an engineer's profession. The challenge for many engineering education programs is how to include this global dimension in the curriculum. The most common and traditional approach is study abroad programs, however, a limited number of students can afford them due to financial and other reasons. Thus, universities try to find other opportunities for the so-called 'internationalization at home'[3] e.g. cross-cultural communication trainings[2]. A comparative analysis of the engineering curricula at the US and Russian leading research universities shows that only few of them succeed in developing a global mindset of their graduates. This process requires a new generation of faculty who are globally engaged themselves[13].

An approach described in our previous paper[16] focuses of international capstone projects which give good results. For those, however, who were not lucky to go through international projects at their home university, other more internationalized institutions offer special short time professional development programs. This paper will focus of one of the most successful among them, Global Village for Future Leaders of Business and Industry at Lehigh University Iacocca Institute[4]. This is a five week intensive program which gathers around 100 diverse participants from different countries every summer to teach them entrepreneurship and leadership skills and to create a global network of future leaders. The program does not finish after the five week course, and its graduates together with their business partners also participate in a short 10 day Global Village of the Move (GVOTM) program, hosted in different countries every year. In the last 15 years the program was held in Peru (2000, 2011), Spain (2003), Western Australia (2005), Middle East (2006), Malaysia (2010), Italy (2012), China (2013)

In 2014, GVOTM was hosted by Kazan National Research Technological University (Russia), where participants from Russia prevailed, and in 2015 moved to Viva College (India) with predominantly Indian people. The rule for participation in GVOTM is to gather people with real world international experiences who have already worked or are working for international companies, and who are ready to share and develop their global professional networks. GVOTM often results in joint international businesses carried out by its participants.

This study surveys and compares the cross-cultural attitudes of GVOTM participants of 2014 and 2015, whose attitudes are expected to be similar. The data from these groups are considered as the professional baseline to be compared to for regular engineering students from the US, Germany and Poland, and those who participated in international capstone projects.

Population

Data were collected from a population of Indian professional business people who participated in GVOTM 2015. This group included professionals of Indian business and international participants from the USA, Great Britain, Spain, Australia, Israel, Colombia, Bulgaria, Mexico, Puerto Rico etc., in other words, it represented a strong multicultural environment, it was used as the reference group.

A similar demographics was demonstrated by GVOTM 2014: Russian entrepreneurs plus their colleagues from America, Europe and Australia. Comparison of these groups is of special interest as both of them represent multinational and multicultural communities of business professionals. The results of these studies were compared to the results of surveys with engineering students from the U.S., Germany and Poland and students who participated in international capstone project.

Table 1 summarizes the survey demographics. The total number of the survey participants is 315 with non-uniform distribution among different groups. The academic level of the participants is mostly undergraduate, followed by Masters and PhDs.

The survey also provided us with the percentage of international students in the groups and the gender statistics, allowing for a more detailed data processing. Although the majority of the survey participants were male, the percentage of female participants was sufficient for making important conclusions.

Finally, the survey data provide an overview of statistics about foreign languages the students can speak (additional to English) and their study abroad experience. 85% indicated that they have at least certain skills in another language or use it regularly.

Source		Academic		International		Gender		Language		Study Abroad	
All	315	All	315	All	315	All	315	All	315	All	315
American*	30%	Undergraduate	82%	Yes	13,1%	Male	70%	Yes	85%	Yes	33%
GVOTM India	11%	PhD	3%	No	87,3%	Female	28%	No	15%	No	67%
GVOTM Russia	7%	Master	14%								
Int Capstone*	2%	Juris Doctor	0%								
German*	13%	Other	0%								
Polish*	37%	Specialist	1%								
		CPA	0%								

Table 1. Demographics of the Survey.

*Added for comparison from previous cultural diversity studies conducted by authors

Results of the Survey

The survey data were compared using three subscales of the Miville Guzman Universality-Diversity Scale (M-GUDS)-S. Each subscale summarized statistics of 5 questions in the survey. The numerical values for the emotional comfort subscale were stated in reverse response value (higher values mean less emotional comfort).

Table 2 provides comparison of three subscales for two Global Villages on the Move with similar demographics. Each value in the sum of points from 5 questions (each 6 points maximum). The highest value of a subscale is 30 points indicating the highest diversity of contact, relative appreciation and emotional **dis**comfort (in the latter case).

As it was stated above, these two groups are quite similar in demographics and represent similar multinational and multicultural groups. It is not a surprise, that the values are quite similar and these groups demonstrate broad interest in participating in diverse social and cultural activities (high diversity of contact and recognize that diversity could have positive impact on their self-understanding and personal growth (relative appreciation). Their degree of emotional comfort with individuals of a different culture is also high.

	GVOTM India	GVOTM Russia
Diversity of Contact	26,12	26,27
Relative Appreciation	25,53	24,73
Emotional Comfort	10,38	10,09

Table 2. Comparison of survey results of two GMOTM professional development programs.

A more detailed analysis, however, reveals interesting differences even in these similar groups. Table 3 summarizes subscale points for these groups without international participants (only Russians and Indians, but still young entrepreneurs with strong international experience and multicultural awareness). Once again, groups in total demonstrate insignificant differences. However, a strikingly higher emotional discomfort was found in Indian male participants for cross-cultural diversity (almost 30% higher than in female participants). For Russian participants the results are excitingly opposite: now female participants reveal higher emotional discomfort, although the different is not so evident.

	Russian	Indian
	All	All
Diversity of Contact	26,80	25,35
Relative Appreciation	24,93	24,94
Emotional Comfort	10,60	10,76
	Russian	Indian
	Male	Male
Diversity of Contact	25,17	25,00
Relative Appreciation	25,50	23,89

Emotional Comfort	8,50	12,22
	Russian	Indian
	Female	Female
Diversity of Contact	27,40	25,75
Relative Appreciation	24,70	26,13
Emotional Comfort	11,50	9,13

Table 3. (M-GUDS)-S subscales by gender for Russia and India.

The results of GVOTM studies were compared to the results of the same surveys with engineering students from USA, Poland and Germany. The results are shown in Table 4.

Global villagers reveal the highest commitment to diverse social and cultural activities. High scores are also demonstrated by International Capstone Project participants. American, German and Polish students are very similar in their attitude to diversity of contact, while the differences between them and GVOTM groups are significant in all the subscales.

	GVOTM India	GVOTM Russia	American*	German*	Polish*	Int Capstone*
Diversity of Contact	26,12	26,27	19,81	19,12	19,99	20,86
Relative Appreciation	25,53	24,73	22,16	19,90	21,65	22,43
Emotional Comfort	10,38	10,09	9,54	16,05	12,51	9,71

Table 4. A comparison of attitudes to cultural diversity from different national groups.
*Added for comparison from previous cultural diversity studies conducted by authors

Table 5 summarizes the answers to multiple questions of the (M-GUDS)-S as a detailed comparison of these different groups. The GVOTM India group with many international entrepreneurs was selected as the reference one and mean values of scores were put in its column. The score differences were taken for other groups for comparison with the reference group and between them. A positive difference means that the mean scores are lower that for the reference group. A negative difference indicates higher scores.

	GVOTM India	GVOTM Russia	American	German	Polish	International Capstone
Count	34	22	92	42	117	7
Item	Mean	Difference	Difference	Difference	Difference	Difference
1	5,82	0,10	1,35	1,70	1,74	1,11
2	5,12	-0,16	0,47	1,05	0,65	-0,03
3	1,91	-0,13	0,18	-3,06	-0,06	0,05
4	5,00	0,00	1,50	1,79	1,29	1,57
5	5,03	0,21	0,42	0,89	0,70	0,89
6	2,24	0,78	0,62	-0,41	-0,11	0,66
7	4,47	-0,44	1,02	0,80	0,36	0,76

8	4,88	0,66	0,83	1,26	0,81	0,88
9	1,88	0,38	0,27	-0,43	-0,35	-0,12
10	5,65	0,19	1,15	1,34	1,28	1,08
11	5,26	0,26	0,69	1,29	0,90	0,84
12	2,62	-0,93	-0,10	-0,93	-1,00	0,05
13	5,18	-0,01	1,29	1,37	1,46	0,75
14	5,24	-0,17	0,96	1,14	0,82	0,52
15	1,74	0,19	-0,12	-0,84	-0,62	0,02

Table 5. The answers to multiple questions of the (M-GUDS)-S by different groups.

The results of the GVOTM Russia International Capstone Group have no significant differences from the reference group. The most significant differences for American, German and Polish groups can be found in the answers to questions 1,4,10, and 13: they are less likely to join an organization which emphasizes getting to know people from different countries and get intercultural dancing and music experience. They demonstrate less interest in learning about cultures that have existed in the world and getting to know people from different racial backgrounds.

It is quite interesting that all these questions belong to the first group "Diversity of Contact", so these groups demonstrate narrower interest in participating in diverse social and cultural activities such as music, dance, celebrations and organization which focus on behavior. At the same time, individual answers to the questions from the relative appreciation and emotional comfort subscales are less different from the reference group and only their sums provide visible deviations (Table 4).

Conclusions and Suggested Approaches

This study discusses a specific part of a large research project in cross-cultural diversity problems in different groups and countries. The authors have already published some results of surveys using the Miville Guzman Universality-Diversity Scale (M-GUDS)-S method [14]. More countries and cultures are expected to be put in the list, while the results in this paper will contribute to developing a methodology roadmap for further research in this project.

1. Although modern students demonstrate self-understanding of diversity importance and reduced emotional barriers in front of different cultures, engineering universities still face such internationalization challenges[16] as insufficient commitment of students to diverse cultural activities. Increased awareness in cross-cultural diversity can be achieved by early analysis of student groups and integrating respective intercultural aspects to their curriculum.

2. A good option for senior students to become less strained in intercultural relationships is participation in international capstone projects offering immersion to intercultural

environment accompanied by intensive study activities to gain useful skills for further career in international companies.

3. Professional development programs can be offered to engineers and businessmen, but a truly multinational and multicultural environment should be an indispensable component of such programs (immersion to other country in a group of professionals with no predominant nationality or culture). A successful example of such a program is Global Village on the Move by Lehigh University and its global partners.

Acknowledgement

The research was funded by Russian Foundation for Humanities grant, project # 15-16-16003.

Bibliography

[1] Bezrukov, A., Ziyatdinova, J. (2014) Internationalizing engineering education: A language learning approach. Proceedings of 2014 International Conference on Interactive Collaborative Learning, ICL 2014, P.229-302.

[2] Gorodetskaya, I.M.,Shageeva, F.T.,Valeeva, E. (2015) Cross-cultural communication training for future engineers - A model developed at the Kazan National Research Technological University to prepare students for mobility programs and the global market place. ASEE Annual Conference and Exposition, Conference Proceedings Paper ID 11381.

[3] Knight, J. (2013) The changing landscape of higher education internationalisation – for better or worse? Perspectives: Policy and Practice in Higher Education. Vol. 17:3. P. 84-90.

[4] Lehigh University Iacocca Institute Global Village http://www.iacocca-lehigh.org/Iacocca/globalvillage/index.shtml

[5] National Academy for Engineering Grand Challenges for Engineering in the 21st Century URL: http://www.engineeringchallenges.org/cms/challenges.aspx

[6] Osipov, P.N. (2013) Training competitive specialists as the priority of modern education International Conference on Interactive Collaborative Learning, ICL 2013, INSPEC Accession Number: 13873590

[7] Parkinson, A., (2009), The Rationale for Developing Global Competence, Online Journal of Global Engineering Education, 4(2)

[8] Salyakhova, G.I., Valeeva, R.A. (2015) Pedagogical stimulation of university students' social competence development by means of interdisciplinary integration. Review of European Studies, Vol.7, No.5, P.186-192.

[9] The 17 Great Challenges of the Twenty-First Century URL: http://www.elon.edu/docs/e-web/predictions/17_Great_Challenges.pdf

[10] The Millenium Project. URL: http://www.millennium-project.org

[11] United Nations. Global Issues. URL: www.un.org/en/globalissues/

[12] Valeev, A.A., Valeeva, L.A.,Sirazeeva, A.F. (2015) University students' intercultural competence development in foreign language teaching by means of axiological approach. Review of European Studies, , Vol.7, No.5, P.178-185.

[13] Valeeva, E.E.,Kraysman, N.V. (2014) The impact of globalization on changing roles of university professors. Proceedings of 2014 International Conference on Interactive Collaborative Learning, ICL 2014, P.934-935.

[14] Valeeva, L.A., Valeeva, R.A. (2013) Development of future engineers' critical thinking in foreign language teaching. International Conference on Interactive Collaborative Learning, ICL 2013 6644619, pp. 438-439

[15] Ziyatdinova, J.N, Ivanov, V.G., Bezrukov, A.N., Osipov, P.N., Sanger. P.A. (2015) Going Globally as a Russian Engineering University. 2015 ASEE Annual Conference and Exposition, Conference Proceedings. Paper ID 12035.

[16] Ziyatdinova, J., Sanger, P.A., Kropiwnicki, J., Nguyen, P.V. (2015), Changing Attitudes in Cross Cultural Diversity Through International Senior Capstone Projects. Proceedings of the 2015 Annual Conference and Exposition 122nd ASEE. Seattle, WA, Paper. P.12045

Building Global Infrastructure for Diversity and Inclusion in Engineering Education

Dr. Autumn Marie Reed, University of Maryland, Baltimore County

Dr. Reed coordinates campus-wide initiatives designed to enhance and support faculty diversity at UMBC. Working collaboratively with the Leadership Team in the Office of the Provost, the Deans, and Faculty Leaders from the Executive Committee on the Recruitment, Retention, and Advancement of URM Faculty, the ADVANCE Executive Committee, the STRIDE Committee, and the faculty leaders of UMBC's Community-Based Faculty Groups, Dr. Reed develops, implements, and evaluates programmatic interventions designed to recruit, retain and advance diverse faculty at UMBC. Dr. Reed also routinely disseminates best practices learned from UMBC's diversity initiatives at national and international venues. Dr. Reed is on the advisory board for the Mid-Atlantic Higher Education Recruitment Consortium.

Dr. Renetta G. Tull, University of Maryland, Baltimore County

Renetta Garrison Tull is Associate Vice Provost for Graduate Student Professional Development & Postdoctoral Affairs at the University of Maryland, Baltimore County (UMBC: An Honors University in Maryland), where she is the Co-PI and Founding Director for the National Science Foundation's PROMISE: Maryland's Alliance for Graduate Education and the Professoriate (AGEP) for the 12 institutions in the University System of Maryland, and Co-PI Louis Stokes Alliance for Minority Participation (LSAMP) Bridge to the Doctorate at UMBC. Dr. Tull has worked with thousands of students from Alaska to Puerto Rico, and in Latin America through graduate school preparation workshops that have been sponsored by AGEP, The National GEM Consortium, National Society of Black Engineers, Society for Hispanic Professional Engineers, Society for the Advancement of Chicano, and Native American Scientists, American Indian Science and Engineering Society, and the Alliance/Merck Ciencia Hispanic Scholars Program. She has presented workshops on graduate school admissions, "The Success Equation," STEM initiatives, and PhD Completion in Panama, Mexico, Ecuador, Colombia, Puerto Rico, and schools around the United States. Tull is on the board of advisors for the PNW-COSMOS Alliance to increase the number of American Indian/Alaska Native (AI/AN) students who complete STEM graduate programs, and is a speaker on "GRADLab" tour with the National GEM Consortium, giving talks across the US on Saturday mornings during the Fall. Tull researched speech technology as former member of the faculty at the University of Wisconsin-Madison. She has co-authored several publications on achievement in STEM fields, and is a mentoring consultant for Purdue, Carnegie Mellon, Cornell, MIT, and other schools. She co-leads the "ADVANCE Hispanic Women in STEM" project in Puerto Rico, and the Latin and Caribbean Consortium of Engineering Institutions' (LACCEI) "Women in STEM" forum. Tull was a finalist for the 2015 Global Engineering Deans Council/Airbus Diversity Award, and has presented on diversity in the US, Latin America, Europe, Australia, India. She is a Tau Beta Pi "Eminent Engineer."

Dr. David A. Delaine, Universidade de São Paulo

David A. Delaine is a progressive engineer who has strong interests in the intersections of engineering, education, and society. He has obtained a Ph.D. in electrical engineering from Drexel University, in Philadelphia, USA. He is currently serving his second term as an executive member of the International Federation of Engineering Education Societies (IFEES). In this role he serves as a Vice President, representing Diversity and Inclusion. He is currently performing research as a Fulbright Scholar postdoctoral researcher at the Escola Politécnica da Universidade de São Paulo for his project titled "An Action Research of Boundary Spanning Intervention on University-wide STEM Educational Engagement" where he will attempt to optimize community/university relations for broadening participation in the STEM fields. David is a co-founder and past president of the Student Platform for Engineering Education Development (SPEED). He has ambitions to significantly broaden the global pipeline of STEM talent and to unify the needs of the engineering education stakeholders in order for engineering education to more accurately reflect societal needs. Diversity and inclusion, university/community engagement, educational research methods, action research, and student led initiatives fall within the scope of his growing expertise.

Dr. Darryl N Williams, Tufts University
Ms. Rovani Sigamoney, UNESCO

Rovani Sigamoney is a chemical/environmental engineer from South Africa who started in the platinum refinery/mining sector and then moved on to researching bioenergy systems and biofuels for Africa.

She joined the United Nations Educational, Scientific and Cultural Organisation (UNESCO) HQ in Paris, France in 2007 and later ran the Chemistry programme and International Year of Chemistry 2011 and thereafter the Engineering programme. The Engineering Programme is working with countries, international partners and program experts to strengthen engineering education through curricula development, hands-on training and capacity building. In line with UNESCO's global priorities on Africa and Gender Equality, it focuses on women and Africa. Rovani is passionate about women in engineering and encouraging more youth to pursue careers in engineering.

She previously worked at the United Nations Environment Programme (UNEP), Paris on a biofuels strategy and also at the Wuppertal Institute of Climate Change in Germany on a policy document for the European Parliament on the security of energy supply.

Building Global Infrastructure for Diversity and Inclusion in Engineering Education

Introduction

In the 21st century, the global knowledge economy faces numerous challenges. In order to surmount these challenges, we must encourage the full participation and equal regard of all diverse talents. In engineering education this broadening of participation is of paramount importance, as this field grapples first-hand with how to effectively address many global challenges. Unfortunately, STEM fields, engineering, in particular suffer from a lack of diversity due to gendered, ethno-racial, and cultural biases, often implicit, that circumvent the full inclusion of members from underrepresented groups [1]. This lack of diversity threatens to impede engineering's ability to tackle these pressing societal challenges. As such, the field of engineering is at a critical crossroads, at which it is imperative that thought-leaders from multiple nations across academia, industry, and the government, come together to build a global infrastructure that increases diversity and inclusion in engineering education.

This paper describes one such diversity effort that took place at the 2015 World Engineering Education Forum (WEEF) in Florence, Italy, that resulted from prior meetings and conversations at the 2014 WEEF meeting in Dubai, UAE, the 2015 ASEE meeting in Seattle, WA, USA, and at UNESCO, which in concert expand on the International Federation of Engineering Education Societies' (IFEES) increased emphasis on inclusion. This new effort, the September 2015 WEEF special session: "Diversity & Inclusion in Global Engineering Education- Initializing Global Scale Collaboration," was designed to educate and spur a diverse international audience of engineering students, professors, deans, and corporate representatives from Nigeria, Turkey, South Africa, Switzerland, the United Kingdom, the United States, and other countries, to collective action toward fostering practices for the diversification of engineering. In two interrelated sections, this paper 1) reviews the educational component and 2) shares the resulting outcomes and recommendations of this session. The first section summarizes the framing of the issue through literature and data on diversity and inclusion, followed by a description of an interactive training on explicit and implicit biases at the session. Section two is action-oriented and builds upon the educational awareness and training as outlined in section one. This second section describes our novel use of Structure-Behavior-Function (SBF), an approach from systems engineering, as an empowering tool that leaders can implement to broach the subject of diversity, and foster actions that lead to respect, appreciation, equity, and inclusion of individuals from underrepresented backgrounds, especially ethnic and racial minorities, within their respective countries. As a continuation of conversations from the special session at WEEF in Italy, this paper offers recommendations that can assist global stakeholders in countries outside of the U.S. to develop locally supportive climates for people from all backgrounds (students, faculty, and staff) who seek and share engineering education. Further, faculty within the U.S. can use these recommendations as they train their students to be culturally competent, and globally relevant engineers.

Framing the issue

As we examine primarily issues that affect people from various racial and ethnic backgrounds, it is important to define how diversity is being used in the context of this paper. We are using the designation "underrepresented" to describe racial groups that the National Science Foundation from the United States has defined as racial and ethnic groups historically underrepresented in STEM disciplines: African Americans, Hispanic Americans, American Indians, Alaska Natives, Native Hawaiians, and Native Pacific Islanders. Underrepresentation in engineering in the United States can be observed at several points along the engineering education pipeline, e.g., undergraduate college and university enrollments and graduation rates, postgraduate enrollments in master's and doctoral programs. An example of the disparity can be seen in the numbers of undergraduate students who are enrolled in engineering programs in the U.S. In 2011, there were 471,920 undergraduate students enrolled, and of those, 16.2% were from racial groups that are underrepresented in STEM, based on NSF's definition [2]. These types of disparities have been highlighted at international engineering education conferences to demonstrate underrepresentation in the U.S. context. Table 1 was presented in a plenary session at the 2016 International Conference on Transformations in Engineering Education in India by one of the authors of this paper to show that the number of doctorates conferred in engineering over the years has not reached parity. Comparing the number of people with doctorates from The Survey of Doctorate Recipients' longitudinal biennial survey, to the number of people in the population, Table 1 shows that collectively, based on 2013 data, 46,535 more engineers with doctorates from Hispanic, Black, and American Indian groups would be needed to reach parity with their respective population demographic [3]; [4]; [5].

Table 1: The total number of U.S. citizens with doctorates in engineering in the U.S. from a longitudinal study as of 2013, compared to the number of citizens in the U.S. population

	Engineering Doctorates N=150,600	United States Population N=316.5 million
Hispanic (Latino)	2.9% n=4,400	17% Need 25,602
Black (African-American)	1.92% n=2,900	13.2% Need 19,879
American Indian (Native American)	0.13% n=200	0.7% Need 1,054

Societal Issues and Bias

It is unfortunate that some people from underrepresented groups have experienced discrimination and have been targeted because of their racial and ethnic backgrounds. The U.S. has a longstanding history of institutionalized discrimination, including the enslavement of African Americans, the removal of Native Americas from their ancestral lands, the removal of Hispanic/Latinos from their properties in the Southern U.S. states, and immigration quotas directed at non-white immigrants. Indeed, it was not until the second-half of the 20th century, with the passage of the Civil Rights Act, and the relaxation of immigration restrictions, that these overt discriminatory practices toward ethno-racial minorities became illegal. Notwithstanding this progress, issues of discrimination have (re)surfaced, for example, police brutality directed toward African Americans [6], the increased deportation of Hispanic/Latinos, and heightened surveillance and profiling of individuals from Middle Eastern or South Asian countries. Ethnic and racial discrimination is not limited to the United States. Reports from Australia revealed that race debates escalated following discriminatory insults directed at Adam Goodes, former star player of the Sydney Swans football team. Debates about the treatment of Goodes, an indigenous Australian, raised the issue of treatment of indigenous, aboriginal groups [7]. The presence of discrimination has not been limited to police brutality in the streets, nor insulting comments on the sports field. Racial, ethnic, and gender biases that undermine the talent and performance of people from underrepresented groups have also been experienced in academic spaces and in the workplace [8]; [9]. However, Twitter campaigns such as #ILookLikeAnEngineer, and research articles such as Carlone & Johnson's (2007) work on STEM identity have sought to address negative perceptions about the abilities of people from underrepresented groups by addressing stereotypes [10]; [11].

Explicit and implicit bias: Definitions and applications

After defining our use of underrepresented minorities and sharing examples of racial, ethnic, and gendered bias, within the U.S. and Australian contexts, we facilitated interactive bias awareness training. We included training within our workshop to equip the audience with a shared language to name, recognize, discuss, and challenge everyday instances of unintended bias pervading engineering academic spaces. As our overall goal for the session was to initialize a global collaboration to advance diversity and inclusion in engineering education, it was crucial that our audience be empowered to continue these conversations following the session. This training consisted of two parts: 1) a definition of biases, explicit and implicit, including examples of consequences of implicit biases within a variety of contexts and 2) an interactive activity through an anonymous PollEveryWhere.com survey in which audience members shared their experiences with implicit bias within engineering spaces.

In part one of the training, we defined bias as habits of mind--preferences, inclinations, or patterns of thought. Biases play a pivotal role in shaping how we navigate judgments in our everyday lives. Some biases are *explicit,* in that we are fully aware (conscious/intended) of the systematic patterns/preferences behind our decisions. In hiring an engineering faculty member, for example, search committee members have an explicit bias, a systematic preference, that eligible candidates must have a PhD in engineering. In their review of resumes, search committee members consciously exclude candidates without a PhD from consideration. Other biases are *implicit,* in that we are unaware (unconscious/unintended) of the habits of mind we employ in our judgments. We described implicit biases as habits of mind that are 1) ordinary, 2) learned from culture, 3) pervasive, and 4) often conflicting with consciously endorsed beliefs [12]. For the purpose of our workshop, we were primarily concerned with implicit biases, in particular implicit bias related to gendered, ethnic, racial, national, cultural, sexual orientation, socioeconomic, religious, and ability identities. We aimed to model for our audience how these implicit biases contribute to the underrepresentation of certain groups in engineering. We highlighted the importance of cultural contexts in the formation of biases. We argued for example that underrepresentation in a Chinese engineering academic space, and the implicit biases involved, may differ radically from South African or U.S. engineering academic space. We stressed that as increasingly globalized field, it is critical that we, as engineers, are attuned to the locally relevant contexts of underrepresentation in our spaces, yet become globally competent in drawing connections and parallels between our local and global contexts of engineering academic spaces. .

Discussions around bias and prejudice can be contentious and individuals often struggle with 1) acknowledging their own implicit biases, and 2) how to have conversations with colleagues and students about this topic. To overcome this obstacle, we reiterated to the audience that having biases do not make us bad people, because as humans we are imperfect beings, whose judgments are not always correct. Additionally, we asserted that although understanding the unconscious processes behind our unintended biases mitigates some of the personal blame we may feel about having such biases, we nevertheless have a shared responsibility to reduce its effects [13]. To further reduce possible discomfort audience members may experience as we progressed in our discussion of this topic, we provided examples of concrete consequences of implicit biases reproducing inequality in various institutions. Using the United States as an exemplar, we cited literature documenting how implicit ethnic, and racial biases toward African Americans and Hispanic/Latinos, the two largest ethno-racial minority populations in the United States, led to disparities in their medical treatment [14], longer and disproportionate sentencing in the criminal justice system [15], their association with crime due to media portrayals of African Americans and Hispanics as criminals [16], underperformance in higher education [17], and their underrepresentation in the labor market [18], most notably in STEM academic spaces [19]; [20].

After providing examples of the material consequences of implicit biases, we returned to and adapted our original hiring scenario to illustrate the consequences of implicit bias in engineering academic spaces. We proposed that within a U.S. faculty search, implicit bias occur if search committee members unintentionally excluded candidates who received engineering PhDs from historically black colleges or universities (HBCU). Here the implicit bias is the search committee members' unconscious stereotype of HBCUs as less prestigious institutions as opposed to an Ivy League or Research Intensive University. Or perhaps it was an South African faculty search, implicit bias may occur in a search committee member preferred the resumes of male engineers or female engineers based on gender cues in the resume and cover letter. Here the implicit bias is the assumption that men are more qualified engineers than women. Again, in the U.S. and South African examples, the search committee member may explicitly share their commitment to diversity and inclusion in the hiring process, but their actions do not align with their intentions due to implicit bias.

Implicit bias in engineering brainstorm

Having laid a foundation, we shifted to the interactive Implicit Bias Brainstorm using Polleverywhere.com to invite audience members to share their experience with implicit bias. We asked the following question: "What type(s) of implicit bias have you experienced and/or witness in engineering. Table 2 provides a summary of the examples of bias anonymously shared by the audience.

Table 2: Implicit Biases Experienced or Witnessed by the Audience

Identity	Implicit Bias Example
Gender	Use of the pronoun "he" to generalize engineering profession; crediting a woman's idea to a man; boys receiving additional support in class, complimenting a woman on her looks and not the content of her presentation; mistaking a man for the boss when the boss was really the woman; suggesting that an all-female team needed adult (male) supervision

Race	Denial of job opportunities; racist jokes in the office; being told grant award was not earned, but due to equal opportunity laws
Gender/Race	Believing that browned-skinned women are "exotic" rather than smart
Ability	Assuming a student with a physical disability also has a mental disability
Geo-Spatial/Cultural Location	Assuming the women cannot work in the Middle East; failure to recognize that cultural norms vary around the world and using politically incorrect comments that do not make sense or offend in local contexts
Language	Assuming someone with an accent or who speaks slowly is not smart and/or unable to complete an assignment

As a large group we discussed our experiences and learned that implicit biases related to language, gender, race, ability, and geo-spatial location are a shared theme in engineering education, but the manifestations of these biases differ depending on the local context in which they occur. Overall, this activity helped us develop a common language for discussing this issue, connecting with this issue on global and local levels, and becoming more aware of the detrimental effects of implicit biases in engineering education.

Structure-Behavior-Function (SBF): A tool to reduce implicit bias in engineering education

After the audience identified examples of implicit bias in engineering education, it was crucial to share tangible actions and steps that audience members can take to reduce the effects of implicit bias in engineering education. Given the diversity of the audience, however, it was first necessary to frame the discussion from a practical, uniform, and familiar perspective. Again, for our work to be successful, it was critical that we develop a shared language to communicate about this topic. Consequently, we adapted the structural, function, and behavioral (SBF) model [21] an approach from systems engineering, commonly used to consider future design considerations for complex systems. In our translational adaptation of the framework, engineering education figures as the complex system that requires a (re)design. Specifically, SBF enables us to systematically pinpoint those emergent patterns that promote or hinder broader participation of diverse groups in engineering.

We initiated this segment of the conversation by categorizing some of the structural, behavioral, and functional components within engineering education. For example, there do exist structural components (e.g. departments, social networks, formal learning environments) that may encourage or discourage diversity and inclusion. The function (or purpose) of the engineering education system and its respective components that reflect its function may be identified as the types of resources, tools, supports, and people (e.g. professors, students, staff) that allow the system to operate and which may facilitate or obstruct diversity and inclusion. With this said, these structural and functional components can shift or be manipulated to change the system, but may or may not produce the desired effect of promoting and sustaining diversity and inclusion. It is not until the entire system identifies the behavioral components (e.g. beliefs, dispositions, preferences, or understanding among constituent groups) and acknowledges the influence these behavioral components play in the full execution of change that the system can reliably undergo a redesign. For engineering education to make significant changes to influence the composition and broaden the participation of diverse groups in the field of engineering, it will require a hard look at the intersection between the structural, functional, and behavioral factors that perpetuate the challenges of implementing effective diversity and inclusion strategies. Once appropriately characterized, it is feasible that various strategies could be implemented using evidence-based best practices.

Applying SBF to act and support others in overcoming implicit biases

After providing an overview of SBF, and modeling its application to diversity and inclusion efforts within the system of engineering education, the collaborators facilitated a second interactive activity. Unlike the previous activity, in which the audience identified implicit bias, the purpose of this activity was to move the conversation from thoughtful reflection to more action-oriented steps that can positively impact diversity and inclusion in engineering education. Due to time constraints, this activity was narrowly focused on identifying some of the behavioral components that perpetuate biases in the participants' respective contexts. We then asked the audience to describe the ways in which they could take action and be supportive in addressing the identified bias challenges that currently exist in the structural and functional makeup of the system of engineering education. Here, we reminded the audience that while implicit bias does call attention to the fact that everyone has biases, that does not give allowance or level the playing field for marginalized groups within the system of engineering education. A higher form of vigilance and advocacy is necessary to significantly reduce the negative impact of biases. Having provided this reminder, we asked the audience to respond to two specific questions:

1) In your daily function within engineering education, what ways can you **ACT** to overcome biases affecting your environment?
2) In your daily function within engineering education, what ways can you **SUPPORT** others in overcoming the biases affecting them?

We then organized the audience into groups of 3-4 individuals for the activity. Each group discussed the first question for 10 minutes, followed by a 10 minute large-group debrief, followed by a 10 minute discussion of the second question, and a final 10 minute large-group debrief. In the follow-up discussion a common theme around the need for additional training for leaders and management on how to tactfully and constructively broach and act on the subject of implicit biases. A second theme emerged around the need to be more inclusive in the formation of research collaborations and mentoring relationships, and to develop intercultural competence to work across differences in these globalized professional relationships.

This activity proved powerful in offering an international community of engineering education stakeholders a lens through which to view and begin changing behavioral issues related to bias in the context of their particular professional roles. The biggest takeaway from this activity, however, is the urgent need and desire for more hands-on training on how to confront implicit bias and create an inclusive climate in engineering education. Although an initial step, we believe this activity offers an example platform by which individuals from across international boundaries and professional functions can have thoughtful and productive discussions about diversity and inclusion in engineering education.

Recommendations

This work provided a platform to open the conversation about biases, experienced and observed. This paper serves as a reference to the workshop, but more importantly, it is designed to serve as a resource to address and explore issues of racial bias, without the overhead of discomfort. We provide the following recommendations for conversations about racial bias in engineering:

1. ***Have an open discussion about diversity challenges in order to increase awareness.*** This paper, or other resources can be used as a catalyst for the conversation.
2. ***Allow open and honest dialog, where all responses are valued.*** Plan to let the data speak for itself and inspire action to achieve equity.
3. ***Identify systemic constructs that prevent or obscure inclusion.*** Multi-modal approaches, e.g., oral conversation, anonymous electronic polls, anonymous comments on uniform notecards, give all participants an opportunity to contribute.
4. ***Carefully reflect and choose colleagues who will adopt "champion" roles to support inclusion.*** Be sure that the audience includes people in positions of power in order to effect change in policies as appropriate.
5. ***Encourage the champions to be comfortable with their role.*** Champions can encourage of collegial behavior modification among those who are consciously or unconsciously stalling inclusive excellence in engineering at all levels (e.g., undergraduates, graduate students, postdoctoral trainees, faculty hiring, research collaborations)
6. ***Request a level of commitment from participants.*** Participants can commit to being advocates, catalysts and change agents who take actions against bias and for inclusion within the department, at the university, on regional/national/global levels. Remind colleagues of written statements and commitments to develop a more inclusive engineering profession from leading international engineering organizations such as ASEE's letter of commitment from engineering deans, the Global Engineering Deans Council/Airbus diversity initiative, and SEFI's (European Society for Engineering Education) conference theme of "Diversity in Engineering Education."

Acknowledgements

The authors would like to thank the following organizations and institutions for the support of this on-going work: The American Society for Engineering Education, the International Federation of Engineering Education Societies, the World Engineering Education Forum, UNESCO, the National Science Foundation (Collaborative Research: AGEP - T: PROMISE AGEP Maryland Transformation # 1309290), Airbus/Global Engineering Deans Council (GEDC) Diversity Award, the Escola Politécnica da Universidad de São Paulo, the Tufts University Center for STEM Diversity, and the Office of the Provost ADVANCE Program and Faculty Diversity Initiatives at the University of Maryland, Baltimore County (UMBC).

VII. References

[1] National Academy of Sciences (US), National Academy of Engineering (US), and Institute of Medicine (US) Committee on Maximizing the Potential of Women in Academic Science and Engineering. (2007). Beyond bias and barriers: Fulfilling the potential of women in academic science and engineering. Washington (DC): National Academies Press (US); 4, Success and Its Evaluation in Science and Engineering. Available from: http://www.ncbi.nlm.nih.gov/books/NBK9811/

[2] National Science Foundation, National Center for Science and Engineering Statistics. (2013). Survey of doctorate recipients. 2013.http://ncsesdata.nsf.gov/doctoratework/2013/. Retrieved January 6, 2016.

[3] National Science Foundation, National Center for Science and Engineering Statistics. (2013). Survey of doctorate recipients. 2013.http://ncsesdata.nsf.gov/doctoratework/2013/. Retrieved January 6, 2016.

[4] National Science Foundation, National Center for Science and Engineering Statistics. (2013). Survey of doctorate recipients. 2013.http://ncsesdata.nsf.gov/doctoratework/2013/. Retrieved January 6, 2016.

[5] U.S. Census Bureau. 2014. State and county quickfacts, http://quickfacts.census.gov/qfd/states/00000.html. Retrieved September 17, 2014.

[6] Wells, C. (August 15, 2015). Protesters march against police brutality on anniversary of Ferguson unrest, The Baltimore Sun.

[7] BBC. (July 31, 2015). Adam Goodes AFL race case sparks discrimination debate. BBC. http://www.bbc.com/news/world-australia-33729797.

[8] National Academy of Sciences (US), National Academy of Engineering (US), and Institute of Medicine (US) Committee on Maximizing the Potential of Women in Academic Science and Engineering. (2007). Beyond bias and barriers: Fulfilling the potential of women in academic science and engineering. Washington (DC): National Academies Press (US); 4, Success and Its Evaluation in Science and Engineering. Available from: http://www.ncbi.nlm.nih.gov/books/NBK9811.

[9] Gupta, S. (January 26, 2015). Study: 100% of women of color in STEM experience bias, Fortune. http://fortune.com/2015/01/26/study-100-of-women-of-color-in-stem-experience-bias/

[10] Carlone, H.B. & Johnson, A. (2007). Understanding the science experiences of women of color: Science identity as an analytic lens. Journal of Research in Science Teaching, 44 (8), 1187-1218.

[11] Dickey. M. (August 3, 2015). #ILookLikeAnEngineer Aims To Spread Awareness About Diversity In Tech, TechCrunch.com. http://techcrunch.com/2015/08/03/ilooklikeanengineer-aims-to-spread-awareness-about-gender-diversity-in-tech/

[12] Carnes, M., & Fine, E. (February 2016). Breaking the bias habit: a workshop to promote racial equity in hiring and clinical practice. Workshop presented at the 8th Understanding Interventions Conference, Philadelphia, PA.

[13] Carnes, M., & Fine, E. (February 2016). Breaking the bias habit: a workshop to promote racial equity in hiring and clinical practice. Workshop presented at the 8th Understanding Interventions Conference, Philadelphia, PA.

[14] Blair, I. V., Havranek, E. P., Price, D. W., Hanratty, R., Fairclough, D. L., Farley, T., et al. (2013). Assessment of biases against Latinos and African Americans among primary care providers and community members. American Journal of Public Health, 103(1), 92-98.

[15] Papillon, K. (2013). The court's brain: Neuroscience and judicial decision making in criminal sentencing. Court Review, 49, 48-62.

[16] Dixon, T. L. (2008). Crime news and racialized beliefs: Understanding the relationship between local news viewing and perceptions of African Americans and Crime. Journal of Communication, 58, 106–125.

[17] Steele, C. (2010). *Whistling vivaldi: How stereotypes affect us and what we can do.* New York, New York: W.W. Norton & Company, Inc.

[18] Bertrand, M., & Mullainathan, S. (2004). Are Emily and Greg more employable than Lakisha and Jamal? A field experiment on labor market discrimination. The American Economic Review, 94(4), 991-1013.

[19] Guy, S. (2012). Making diversity a core value in engineering education.SWE Magazine, 58(4), 34.

[20] Moss-Racusin, C. A., Dovidio, J. F., Brescoll, V. L., Graham, M. J., & Handelsman, J.. (2012). Science faculty's subtle gender biases favor male students. Proceedings of the National Academy of Sciences of the United States of America, 109(41), 16474–16479.

[21] Goel, A. K. (1996). Adaptive modeling. In *Qualitative Reasoning: Papers from the 10th Annual Workshop, Technical Report WS-96-01* (pp. 67-73).

Understanding International Perspectives in Science and Engineering Ethics

Dr. Thomas M. Powers, University of Delaware

Thomas M. Powers is the founding director of the Center for Science, Ethics, and Public Policy (CSEPP) at the University of Delaware. He holds appointments as Associate Professor in the Department of Philosophy and in the School of Public Policy and Administration, and resident faculty at the Delaware Biotechnology Institute. His research concerns ethics in science and engineering, the philosophy of technology, and environmental ethics, and his publications range from topics in artificial intelligence and robotic ethics to the ethical aspects of design. Powers received aB.A. in philosophy (College of William and Mary) and a Ph.D. in philosophy (University of Texas at Austin) for a dissertation Immanuel Kant. He has been a DAAD-Fulbright dissertation-year fellow at the Ludwig-Maximilians-Universität, Munich, a National Science Foundation postdoctoral fellow in the School of Engineering andApplied Science at the University of Virginia, and a visiting researcher at the Laboratoire d'Informatique (LIP6) at the Université Pierre et Marie Curie in Paris, France.

Understanding International Perspectives in Science and Engineering Ethics

Abstract

In 2014 an interdisciplinary team of researchers from the University of Delaware began a 5-year project with the Online Ethics Center (OEC) to gather international perspectives on science and engineering ethics, broadly construed.[1] We seek instructional and scholarly materials from international institutions and individual foreign scholars and teachers. This paper describes the rationale and outlines the theoretical foundation of this project. The central issue to be addressed here is: if there is indeed a benefit to the "internationalization" of science and engineering ethics, how can that benefit be understood from the more general standpoint of the purposes of science and engineering ethics *per se*?

It is now widely understood that contemporary science and engineering are increasingly international and collaborative. These complex practices are subject to difficulties introduced by differences in languages, cultures, social norms, education, religion, political systems, laws, resource and infrastructure availability, and other factors. Despite these variations, there is an important common factor in the production of scientific and engineering knowledge: it is not just descriptive, explanatory, and predictive, but also normative. This means that there are explicit and implicit rules and guidelines, regulations, and norms of practice and sanction regarding how one *ought to* behave as a scientist or engineer. This normative part concerns how to do research, but also how to orient one's research to a world that expects to benefit from it.

Given this character of science and engineering as in part normative, a better appreciation of these norms worldwide will be necessary to understand actual practices, as well as to motivate critical appraisals of practice. I explain the rationale for collecting those views, and making them widely available for further study.

In the final part of the paper I discuss a "contractarian" theory of the good of publicizing differences in norms. This approach, which is inspired by the political philosophy of John Rawls, can be seen as an alternative to a "conventionalist" understanding of science and engineering ethics.

Introduction

Ethical issues associated with the integrity of science and engineering include the falsification of experimental results, fabrication of data, conflicts of interest in industry-funded research, plagiarism, and improper attribution of authorship for collaborators, to name just a few. All of these factors threaten the secure accumulation of knowledge. Additionally, revelations of misconduct can lead to the perception of a lack of professional integrity and a corresponding erosion of public support and trust in science and engineering. These threats could be mitigated by recognition and enforcement of ethical norms and expectations for research conduct. However it is difficult to develop useful ethical guidelines that are applicable across different academic and industry areas, as well as in different cultural contexts. For instance, conventions differ on the criteria for authorship and fair use of the published work of others. One ethicist has even gone so far as to claim that the field of research ethics is itself incoherent.[2] Compounding these challenges is the fact that science and engineering research often involves competition, so individuals and groups have an incentive to seek advantage by violating ethical norms of research and thus to free ride on the general compliance with these norms by competitors. As the economist Kenneth Arrow notes, *"an ethical code is useful only if it is widely accepted. Its implications for specific behavior must be moderately clear, and above all it must be clearly perceived that the acceptance of these ethical obligations by everybody does involve mutual gain."* [3] While the benefits to science and society of better programs of ethics instruction and inculcation seem obvious, it is far from obvious how this ought to be achieved.

While popular media outlets sometimes report research ethics scandals, critical ethical inquiry into science and engineering is usually limited to university study in the various areas of applied ethics. Topics such as pollution and the use of non-renewable resources in science and engineering, STEM in the defense industry, the proper use of taxpayer dollars to fund controversial research, and the role of information technology in threats to individual privacy are included in curricula in environmental ethics, computer ethics, and bioethics. But these issues are not merely academic. Scientists and engineers must eventually face them when asked to justify the use of taxpayer monies to fund their research, to secure future investment funding, and in some instances to respond to complaints from interest groups or politicians. In recent years, funded research *and* ethics instruction have faced scrutiny from members of the U.S. Congress.

The work described here can be considered as an extension of efforts begun in the U.S. to improve online access to ethics resources in science and engineering. An important milestone in those efforts was the 2007 America COMPETES Act (ACA). In implementing the ACA, the National Science Foundation (NSF) acknowledged a need for the involvement of educators in science, technology, engineering, and mathematics (STEM) fields, as well as the involvement of ethicists, other humanists, universities and research organizations and the funding agencies themselves. In the wake of the ACA, the NSF funded three online ethics sites to assist individuals and institutions in meeting the mandate of the ACA: the Ethics in Science and Engineering National Clearinghouse at UMass Amherst, the Online Ethics Center at the National Academy of Engineering, and the National Professional and Research Ethics Portal, which evolved into Ethics CORE (Collaborative Online Resource Environment) and incorporates materials from the UMass project. Together, these online sites have contributed greatly to the resources available in research ethics and the responsible conduct of research (RCR) in science and engineering. The offerings include case studies, syllabi and other curricular aids, ethical and professional codes, original research on and news about research ethics, exercises, videos, and best practices. Past and ongoing efforts have not, however, focused on resources from outside of the U.S. This is the focus of the team of philosophers, social and natural scientists, and engineers that I lead at the University of Delaware.

We began our efforts by first recognizing the difficulties inherent in the subject matter, irrespective of global context. There are two basic challenges for understanding and improving the ethics of science and engineering. The first is the wide range of practices over which ethics is to be considered. Science and engineering include inquiries into the living and the dead, at many different scales, by academics, public sector employees, and industry researchers alike. How they *ought* to behave when they are practicing their respective crafts is often thought to be relative to each craft. The second challenge concerns variation *within* each practice, due to differences in mores that are manifest in laboratories, universities, journals, professional meetings, organizations, and academy-industry collaborations. Both challenges are made more difficult by what we might call the "anthropological" realities of *global* research, namely, the cultural, national, and linguistic diversity of researchers that complicates the matter of their collective ethics in many cases of collaboration. There is ample evidence that different normative traditions of research and practice have now begun to collide.[4] [5] [6] [7] Yet the benefits of doing international and collaborative science and engineering are such that no reasonable participant in these practices would want to retreat to an era in which collaboration was limited to colleagues in close proximity, and of the same nationality.

An analogy with the advantages of the Internet might be helpful here. There was a time when information exchange was slow, cumbersome and costly, and now it is fast, easy, and cheap. It is true enough that a networked world has also brought us spam, computer viruses, and the Distributed Denial-of-Service attack. But there is no going back—or at least there is no way back that does not involve tremendous loss. Similarly, national, cultural and linguistic diversity might complicate international science and engineering, but it is also part of its strength. So our main question can be restated: how can we better use the international character of science and engineering to understand and improve its ethics?

Overcoming the two challenges mentioned above, in light of the "anthropological" difficulty brought about by the scope of these practices, presents opportunities to develop a new, critical enterprise in an area of applied ethics. Instead of bemoaning the normative complexity introduced by global science and engineering, ethicists can use the diversity of norms to lead richer discussions of the ethics of science and engineering on the whole. This "windfall" of difference and contrast is something that is cherished in the ethics classroom. The best discussions of ethical issues almost always occur when students (respectfully) disagree. The ethics of science and engineering can also benefit from some disagreement—under the assumption that this is part of a process that is headed towards the convergence and acceptance of norms of research and practice, and not the opposite, towards divergence and chaos.

An anthropology of science and engineering ethics

As indicated above, contemporary science and engineering as a whole might be given an anthropological treatment, so that it could be studied as a collaborative set of complex behaviors, undertaken by individuals and institutions that vary by language, culture, social norms, education, religion, political systems, laws, media institutions, resource sufficiency, infrastructure, available technology, and national goals or agendas of the research enterprise. Cutting across all of these variations is a dominant but often overlooked factor in the production of scientific and engineering knowledge: it is thoroughly normative, with explicit and implicit rules and guidelines, regulations, and "folkways" of practice and punishment regarding what it is to do science and engineering. Our focus as part of the OEC project is on these normative factors, as they are manifest outside of the U.S., according to the contributions of foreign practitioners of science and engineering, including those who do science and engineering research, but also those who study, teach, and critique the practices. The prior projects funded by the NSF have focused on domestic resources in the same topical areas: responsible conduct of research, research ethics, the ethics of individual sciences and engineering fields, as well as the social responsibility of science and engineering. The scope of the expanded OEC, and hence of our international supplement to it, now covers areas such as sustainability, privacy and security, law and public policy, risk, and other normatively influenced topics. Clearly, this is a very large endeavor.

A distinction may help to organize what is being sought in such a collection of resources. Normative factors in science and engineering can be understood in narrow or broad scope. Narrowly, "ethical and responsible conduct of research" might focus just on the *processes* of research and practice themselves—e.g., how data ought to be gathered, represented in publication, or "managed" during and after publication; how inclusion and order of authorship is determined; what constitutes proper attribution (or failure thereof) concerning ideas and other intellectual contributions; and how scientific misconduct ought to be investigated.

Institutions and practices to regulate these narrow process issues have been or are being created in most developed nations with high levels of research,[8] in part in response to the perception that scientific fraud is a growing and international phenomenon that must be addressed. (The question whether scientific fraud is actually growing is quickly beset by a puzzle: since when?) There have been considerable international activities to address the narrow process issues of science and engineering; four meetings of the World Conference on Research Integrity have taken place since 2007 (Lisbon, Singapore, Montreal, and Rio de Janeiro). In 2011 the InterAcademy Council, through the InterAcademy Partnership (IAP) and its Committee of Research Integrity, began an international stakeholder project to forge "an international consensus on responsible conduct in the global research enterprise" and one year later issued their first report.[9] The IAP has now published a guidebook for responsible conduct of research, with a focus on global perspectives. [10] The narrow process factors of science and engineering ethics are increasingly under scrutiny.

Normative factors can also be understood more broadly to include a critical evaluation of the *products* of science and engineering research such as new knowledge, new technologies, or even negative outcomes such as inequities, inefficiencies, or social injustices that result from (or at least are not addressed by) such research.[11] The products of research may offer new opportunities to respond to global challenges such as climate change or world hunger, but they may also present new challenges of their own, such as bioterrorism, political instability, or the erosion of individual privacy. To date, these broad "product" issues have been less widely studied from an international and comparative perspective.

In considering both the (narrow) processes and (broad) products of science and engineering, the study of their normative factors can be undertaken as a descriptive inquiry in which norms, guidelines, and conventions are set out and clarified. As a description, the contributions of our international supplement to the OEC are designed to help populate the anthropology of science and engineering mentioned earlier.

But our goals go beyond description. Our larger aspiration is that the collection and publication of such materials can aid in a critical enterprise—one through which conventions and norms are deliberately and continually subject to challenge and refinement. In both cases—as description and as critical enterprise—the international supplement is intended to support pedagogic goals of breadth, balance, and the consideration of foreign perspectives in developing courses, syllabi, and instructional materials.

In thinking about ethics, many non-specialists will select from a variety of collections (physical and online libraries, available textbooks and other printed media), or may take the less-formal but still time-honored approach: ask a mentor or colleague. An online or physical collection—or any personal conversation, for that matter—runs the risk of inadvertent bias in favor of the ethical views of a particular community (linguistic or geographical), discipline, or nation. There is also the possibility of bias in developed world vs. developing world conceptions of science and engineering. After all, the most highly-cited research, as well as most of the textbooks, come from nations with highly-developed research infrastructures. The U.S. is the prime example. There is also the fact of cheap and fast access to those views that have been made available online. Contrast, here, the dearth of online resources from developing countries where Internet access and infrastructure is limited.

In terms of university pedagogy, it would be nice to think that considerations of difference and contrast in ethics will be useful. Moreover, it seems desirable to impart not just knowledge about science and engineering ethics, but also a kind of wisdom that that will serve students in the international, cross-disciplinary, and cross-cultural contexts through which their careers may take them. Since conventions differ markedly, it seems important to teach students how to reason critically through quandaries and disputes that are both value based and constitutive of behavioral norms. While the benefits to society of better programs of ethics instruction and inculcation seem obvious, it is far from obvious whether formal instruction can do all the work. Having good, critical online resources available for anyone to use should support the cause of formal ethics education.

Though it might be tempting to think that we can establish ethical standards governing research merely by adopting and enforcing conventional standards of conduct, these standards cannot be justified without examining the proper aims of research. Consider, for instance, that if one aim of research is to increase the number of publications, instead of the quality and depth of scientific knowledge, some level of plagiarism might well be justified. So the convention against plagiarism does not stand alone; it requires us to ask what we want out of science. Though conventions have their place in fostering collaboration, they require examination and testing. Students, faculty, and practitioners of science and engineering make up a diverse audience, and they will increasingly need to grapple with hypothetical cases, abstract principles, and engage in ethical reasoning. Reporting conventions, alone, cannot do it all.

The best print resources to help this audience will become rapidly outdated as we are in a quickly changing and global world of research. Print media are also costly to translate and to distribute widely. Finally, they include too few—and primarily Anglophone—viewpoints. Hence print resources in science and engineering ethics are likely to be of limited use and to run the risk of inadvertent bias towards the views of those who have easiest access to them.

To summarize, our basic challenges for understanding and improving the ethics of science and engineering are also our best assets. Variations in norms over cultural, national, and linguistic communities can the foundation of a new, critical ethics enterprise. This diversity provides opportunities for more fruitful discussion of the ethics of science and engineering on the whole.

A contractarian theory of science and engineering ethics

There is a deeper motivation behind the collection of international resources in science and engineering ethics that I would like now to sketch. To the extent that collecting resources in applied ethics requires a theory, I want to argue for one here that rejects a descriptivist or conventionalist understanding of science and engineering ethics. On conventionalist views, when a practitioner of science or engineering conforms to an ethical practice (such as refraining from plagiarizing), he or she does so because that is just "the way people behave" in the relevant context. If the convention were otherwise, the practitioner would just as easily conform to the contrary convention.

On the theory offered here, comparing and contrasting of international norms of science and engineering has a contractarian basis—one that follows, by analogy, the conception of the hypothetical social contract of "justice as fairness" made famous in the 20th century by the philosopher John Rawls. [12] On the contractarian view, there is a strong "publicity" requirement for airing of differences of opinion about norms of conduct within any practice that hopes to become more just. The same can be said of science and engineering: the opportunity to negotiate better norms of practice first requires a publicizing of differences. After, by means of a process that Rawls called "reflective equilibrium," practitioners are able to converge on an improved understanding of a norm. And, importantly, they will have done so through an exercise of "public reason" and therefore should be more inclined to endorse the outcome as one they would have adopted under ideal conditions. The goal, then, of internationalizing science and engineering ethics is both an eventual convergence of norms, and also a deeper internalization of them through reasoned consensus.

The development of an ethics by way of "reflective equilibrium" is intended to emphasize the public nature of agreements over principles of shared governance, the rational ability to find compromise between abstract principles (codes of ethics) and considered judgments about established practices, and ultimately the fairness of the results that we jointly consider to be reflective equilibria. For Rawls, the procedure was intended ultimately to justify our distribution of rights, liberties, opportunities, and other primary social goods. For scientists and engineers, reflective equilibrium can serve as a procedure to adjust their considered judgments about the rules and principles of research and practice to other, somewhat different views. As the bioethicist David Resnik has noted, the procedure of reflective equilibrium can serve to justify ethical rules and principles, allow for the resolution of conflicts between them, and avoid the kind of moral relativism that is often associated with outcomes validated merely by group deliberations.[13]

The idea of generating helpful contrasts with international ethics resources follows the pedagogy of problem-based learning, studies of which have suggested that it serves as an effective way to achieve cognitive as well as motivational effects.[14] In addition, the process of coming to consensus through discussion and debate should reinforce the fairness of the resulting norms of science and engineering. It is our hope that the collection of international science and engineering ethics perspectives will be a down payment on those goals.

References

---

[1] The project is a supplement to an NSF-funded expansion of the Online Ethics Center (OEC), with the National Academy of Engineering: NSF SES (1355547) "Becoming the Online Resource Center for Ethics Education in Engineering and Science" 2014-2019. The Principal Investigator is Rachelle Hollander of the OEC.

[2] Pimple, K.D. Six domains of research ethics: A heuristic framework for the responsible conduct of research. *Science and Engineering Ethics* 8 (2): 191-205, 2002

[3] Arrow, K. Social responsibility and economic efficiency, Public Policy 21: 303-317, 1973

[4] Altman, L.K., and Broad, W.J. Global trend: more science, more fraud. *New York Times*, December 20, 2005

[5] Kochan, C.A., and Budd, J.M. The persistence of fraud in the literature: the Darsee case. *Journal of the American Society for Information Science* 43 (7): 488-93, 1992

[6] Reich, E.S. *Plastic Fantastic: How the Biggest Fraud in Physics Shook the Scientific World.* Palgrave: Macmillan, 2009

[7] Tanimoto, T., et al. Research misconduct and scientific integrity: a call for a global forum, *The Lancet*, Volume 382, Issue 9896, 14–20 September 2013

[8] Scott-Lichter, and the Editorial Policy Committee, Council of Science Editors. *CSE's White Paper on Promoting Integrity in Scientific Journal Publications.* Reston, Va., Section 3.2.1: 2006

[9] IAP-IAC Committee of Research Integrity. Responsible conduct in the global research enterprise: A policy report. InterAcademy Council, IAP, 2012; http://www.interacademycouncil.net/24770/24032/28355.aspx

[10] InterAcademy Partnership, *Doing Global Science: A Guide to Responsible Conduct in the Global Research Enterprise*, Princeton University Press, 2016

[11] Schienke, et al. The role of the National Science Foundation Broader impacts criterion in enhancing research ethics pedagogy. *Social Epistemology* 23:317-336, 2009

[12] Rawls, J. A Theory of Justice. Harvard University Press, 1971

[13] Resnik, D. Philosophical foundations of scientific ethics, in M. Thomsen (ed.), *Proceedings from the workshop on ethical issues in physics* Ypsilanti, MI: Eastern Michigan University, 2004

[14] Dolmans, D., and Schmidt, H. What do we know about cognitive and motivational effects of small group tutorials in problem-based learning? *Advances in Health Sciences Education* 11: 321-36, 2006

Enhancing Multiple Thinking through the Engineering Design Process

Dr. Jaby Mohammed, Petroleum Institute

Jaby Mohammed is a faculty at The Petroleum Institute, Abu Dhabi, UAE. He received his PhD in Industrial Engineering from University of Louisville (2006), masters in Industrial Engineering from University of Louisville (2003) and also a master's in business administration from Indira Gandhi National Open University (2001). His research interests include advanced manufacturing, design methodologies, six sigma, lean manufacturing, and engineering education. He previously taught at Indiana Purdue Fort Wayne, IN and at Morehead State University, KY. He is a member of IIE, SME, ASQ, ASEE, and Informs.

Ms. Mary Ragnhild Hilja Hatakka, Petroleum Institute

Dr Mary Hatakka has over 20 years of experience teaching academic literacy skills, English for Specific Purposes to engineering students and study skills in general. She has recently completed her doctorate at the university of Exeter, UK, and is currently working on improving first year students' transition from school to university. Her current research interests are academic literacy skills,engineering habits of mind and developing workplace skills for engineering students.

Enhancing Multiple Thinking through the Engineering Design Process

Abstract

The impact of the quality of engineering students' education on the industry of a country cannot be underestimated. While there are numerous studies on the types of qualities that graduate engineers should have, the extent to which undergraduate engineering students choose to acquire these qualities or skills depends greatly on the type of epistemic beliefs that they have. The purpose of the current study was to investigate the epistemic beliefs of first year engineering students. It is the first part of a study aiming to establish both the general epistemic beliefs [1] of engineering students, as well as if and how these beliefs relate to students' Engineering Habits of Mind [2]. The primary instrument used in this first phase of the study was Schraw, Dunkle, and Bendixen's (2002) Epistemic Belief Inventory (EBI) [3]. The instrument is in the form of a questionnaire and it was administered to 39 female students in their freshman and sophomore years in an engineering college in the Arabian Gulf. The results showed that there was very little variation between the two levels of students, indicating that the issue of developing students' epistemic beliefs needs to be addressed more thoroughly. This could be done possibly in the introductory engineering courses or as curriculum infused in other Science, Technology, Engineering and Mathematics (STEM) courses.

Keywords—epitemic beliefs; inventory; engineering students; female; Arabian Gulf

Introduction

Several studies have been done on the kinds of qualifications, knowledge and skills that the 21st century engineer should have. To illustrate an example of such skills reflecting the needs of industry, Rajala [4] defines the areas of competence necessary for a global engineering professional as global competence, technical competence and professional competence. However, she points out that the attributes which are associated with each of these competences present a challenge for engineering educators regarding both the context in which they should be introduced, as well as "determining what can be accomplished within the constraints of a university education" [4]. From an educational stance, the Accreditation Board for Engineering and Technology (ABET) [5], states engineering students' minimum learning outcomes (a) through (k) as shown in Table 1:

Table 1: ABET learning outcomes

a) an ability to apply knowledge of mathematics, science and engineering
b) an ability to design and conduct experiments, as well as to analyze and interpret data
c) an ability to design a system, component, or process to meet desired needs within realistic constraints such as economic, environmental, social, political, ethical, health and safety, manufacturability, and sustainability
d) an ability to function on multidisciplinary teams
e) an ability to identify, formulate, and solve engineering problems
f) an understanding of professional and ethical responsibility
g) an ability to communicate effectively

h) the broad education necessary to understand the impact of engineering solutions in a global, economic, environmental, and societal context
i) a recognition of the need for, and an ability to engage in life-long learning
j) a knowledge of contemporary issues
k) an ability to use the techniques, skills, and modern engineering tools necessary for engineering practice.

The lists of competences, attributes and learning outcomes are based on much research and have been modified over time to ensure that engineering education develops along with industrial needs and college learning outcomes are adjusted accordingly. Feedback is also frequently sought from industries [4], which have become more and more multinational, another aspect that affects the nature of learning outcomes in engineering education.

However, the extent to which undergraduate engineering students choose to acquire the desired qualities or skills depends greatly on the types of epistemic beliefs that they have. Epistemology is the philosophic study of where one believes knowledge comes from and what it is like [6]. Epistemic beliefs can be related to research, personal knowledge or professional knowledge among other things. As far as college education is concerned, the person who is recognized as the pioneer in developing the first framework of the kinds of knowledge that undergraduate students have is William Perry [1]. According to Perry [7], undergraduate students can go through four hierarchical levels of knowledge development as follows; the first stage is dualism (knowledge is either right or wrong as determined by a figure of authority); stage two is multiplicity (knowledge is about differing opinions); stage three is called relativism (knowledge is dependent on context) and stage four, which Perry called 'commitment' whereby knowledge is the ability to make decision based on informed opinions. In studies done on undergraduate engineering students using Perry's model in the USA in the 1980s, 1990s and at the start of the 21st century, "most engineering undergraduates complete college in the lower classifications of either dualism or multiplicity" [1]. According to [8], Perry's model is the most common model used to measure an understanding of where knowledge comes from in engineering education. However, Schommer was the first to come up with a quantitative way to measure epistemological beliefs [1] and the Epistemic Belief Inventory (EBI) [3] used in the current study is based on Schommer's model.

Background

Context of the Current Study

Students in the Arabian Gulf region opt to study engineering majors for a variety of reasons, including parental pressure. A job in any field of engineering, but specifically in the field of Petroleum Engineering is seen as demanding and highly respected, guaranteeing a very acceptable status in society. This is due to the fact that oil related industries generate the most money in the region and provide the main income of most Gulf States [9], contributing to both the growth of the economy and to national security [1]. As future employees in local petroleum engineering companies, students should be fully aware of the implications of the consequences of the quality of their work [9], which is guided by their basic epistemological beliefs [1]. These in turn affect the students' work ethics and the way they view the discipline of engineering. The aim of the current study is to understand what kinds of epistemological beliefs Gulf Arab female students have and whether they develop during their undergraduate studies.

Framework within which the study is situated

The beliefs can be measured at a general level using the scales in the Epistemic Belief Inventory EBI [1, 3]. The extent to which students have adopted engineering habits of mind, as established by the Royal Academy of Engineering in the UK [2], will be related to the current study in a subsequent follow up study. Moreover, the results of the female and male students will be compared to find out if there are any significant differences between the two types of students. The results of the study could have practical implications in the engineering college where the data were collected as well as in other similar institutes of higher education.

Methodology

The data for the current study were gathered from an engineering higher education institute (HEI) in the Arabian Gulf region. The case study consisted of a total of 39 female students enrolled in the ENGR 101 freshmen year engineering course and the sophomore Basic course on engineering design called STEPS 201. Most of the students involved in the study were from the Gulf region a minority of international students. Both courses were taught by the same instructors.

Epistemological beliefs were measured with the Epistemological beliefs inventory [3]. This quantitative measuring instrument is designed so that individuals respond using a 5-point Likert-type rating scale from strongly disagree (1) to strongly agree (5) to items concerning their beliefs about education and learning. The inventory was developed to measure five underlying constructs: Certain Knowledge, Innate Ability, Quick Learning, Simple Knowledge, and Omniscient Authority. The inventory, abbreviated as EBI, consisted of 32 Likert-type questions designed to measure five subscales of different types of knowledge. The subscales for the five categories included simple knowledge, which was set to define how complex knowledge is, certain knowledge regarding how tentative knowledge is, omniscient knowledge (how knowledge is acquired through authority), quick learning (how quickly knowledge is obtained), and innate ability (one's innate ability to gain knowledge) [3]. Table 2 below shows the categories of questions.

The EBI questionnaire, which was written in English, was piloted before it was officially administered. The students who participated in the study were non-native speakers of English, so based on the pilot study, it was decided that during the administration of the official survey students would be allowed to inquire about vocabulary, if they were uncertain about its meaning. This was done in order to further validate the results of the EBI study.

TABLE 2: Categories of questions

Simple Knowledge	1. It bothers me when instructors don't tell students the answers to complicated sustainable problems. 9. If a person tries too hard to understand sustainability, they will most likely end up being confused. 10. Too many theories just complicate things. 11. The best ideas are often the simplest.

	13. Instructors should focus on facts instead of theories on sustainability. 18. Sustainability is simpler than most professors would have you believe. 22. Sustainability is easy to understand because it contains so many facts. 24. The more you know about sustainability, the more there is to know.
Certain Knowledge	2. Sustainability means different things to different people. 6. Absolute moral truth does not exist. 7. Parents should teach their children all there is to know about sustainability. 19. If two people are arguing about sustainability, at least one of them must be wrong. 25. What is true about sustainability today will be true tomorrow. 31. Sometimes there are no right answers to debated sustainable problems.
Omniscient authority	4. People should always obey guidelines for sustainability. 14. I like teachers who present sustainability ideas and let their students decide which is best. 20. Children should be allowed to question their parents' belief in sustainability 23. The moral beliefs in sustainability that I live by apply to everyone. 27. When someone in authority tells me what to do, I usually do it. 28. People who question sustainability are careless.
How quickly knowledge is obtained	3. Students who learn sustainability are the most ones who appreciates sustainability 16. If you don't learn sustainability quickly, you won't ever learn it. 21. If you haven't understood an aspect about sustainability the first time through, going back over it won't help. 29. Working on a sustainability problem with no quick solution is a waste of time. 30. You can study sustainability for years and still not really understand it.
Innate ability to gain knowledge	5. Some people will never believe in sustainability no matter how much they know 8. Really smart students don't need a lot of guideline to follow sustainability 12. People can't do too much about how sustainable they are. 15. How well you perform in society depends on how much you comply to sustainability. 17. Some people just have a knack for sustainability and others don't. 26. Sustainable believers are born that way. 32. Some people are born with care and passion to future generation

Results and Discussion

After administering the EBI questionnaire to the 39 female students, the data analysis was completed using MiniTAB. MiniTAB is a tool for statistical analysis which was developed at Pennsylvania State University in the USA by [10]. The following two tables show the results for the freshman students (Table 3) and sophomore students (Table 4).

Table 3: Survey results for ENGR 101

Subscale	Range	Mean	SD
Simple Knowledge	2.92- 3.83	3.54	.335
Certain Knowledge	2.5- 3.83	3.26	0.45
Omniscient authority	2.92-3.67	3.17	0.14
How quickly knowledge is obtained	1.83-3.08	2.60	0.5
Innate ability to gain knowledge	2.42-4.00	3.00	0.58

Table 4: Survey results for STPS 201

Subscale	Range	Mean	SD
Simple Knowledge	3.22-4.28	3.65	.36
Certain Knowledge	2.72– 4.11	3.24	0.57
Omniscient authority	2.28-3.61	3.07	0.18
How quickly knowledge is obtained	1.56-3.39	2.28	0.77
Innate ability to gain knowledge	1.50– 3.78	2.65	0.77

As can be seen in Table 3 above, the students were strong in their opinions about simple and certain knowledge, as well as to a great extent on omniscient authority. The responses regarding latter issue can possibly be explained as being related to the strong, paternally authoritarian, close-knit family structure, which has managed to retain some traits of Bedouin tribalism in it, too. Moreover, as many children choose to live with their families until they are in their twenties, the authority of the father continues for some well into the children's college years. The responses coincide with the notion of dualism, the most basic form of credible knowledge on Perry's scale [7]. They are also is in accordance with [8], whereby most engineering undergraduate students fall into this category with regards to their views on knowledge. It is not, therefore, surprising that there is a wider range in the students' responses regarding the two latter categories, with the widest range being in their views on one's innate ability to gain knowledge. It could be because of the culturally appropriate somewhat preordained manner in which some people tend to perceive their existence and, as a result, the amount of control one has over matters such as one's ability to gain knowledge.

The results for the second year students are shown in Table 4. It is interesting to see that they are very similar to the results in Table 3. This is especially true in relation to the first three categories of epistemic knowledge and again coincides with the results in studies [7] and [8]. In other words there is very little difference between the epistemic views of Freshman and sophomore students. Considering the amount the students have studied and the level of difficulty they reach in their sophomore year, it can only be deduced that students continue to rely on their instructors for the right answers. It would be interesting to find out why they do not venture into more uncertain areas or towards higher order thinking skills as defined in Bloom's revised taxonomy [11] with all the knowledge they have acquired.

However, it is interesting to see that the standard deviation is slightly higher in the last two categories, as can be seen in Table 5. TABLE 5: Compilation of the results for the EBI

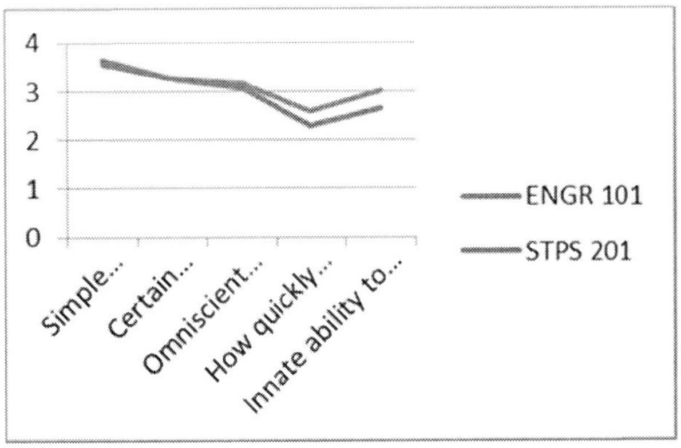

Fig (1) comparison between ENGR 101 and STPS 201

It would be interesting to interview second year students regarding this matter in future to develop a further understanding about this particular issue and to find out if there are ways in which to develop students' epistemic beliefs regarding uncertainty [7, 8] and higher order thinking [11].

Conclusion

Engineers use critical and creative thinking to solve all kinds of open-ended problems. [2] list the six main engineering habits of mind as systems thinking, problem-finding, visualizing, improving, creative problem-solving and adaptability. All these ways of thinking include an element of uncertainty in them. However, as shown in the small scale above, students are not prepared to reach into the unknown as readily as perhaps college instructors would like to think. As a result the matter becomes a curricular issue, because these kinds of thinking skills are required in industry and business to sustain economies. This is why is it vital that engineering students are able to develop these skills at college.

According to Rajala [4], engineers need to have strong technical skills, but they also need to be able to "demonstrate effective communication, creativity, entrepreneurial thinking, teamwork and understand business –all in a global context" [4]. The current study shows that students do not come to HEIs prepared with such skills, so it is the onus of engineering HEIs to instill such skills in students in order to empower them to be able to survive and flourish at work after graduating. This is why students need to be exposed to more and more real life problem-finding and -solving situations [2], as well as experiential learning where concrete results reveal the extent of their thinking [11]. Such skills can be taught using project-based learning, which also develops teamwork skills and gives rise to communicative situations in which students require higher order thinking skills. In addition students' epistemic beliefs could be expanded on using curriculum infused methods in a variety of subjects taught in engineering HEIs, as long as faculty are aware of the importance of placing the emphasis on such thinking skills. Further studies in the development of students' epistemic beliefs and their beliefs about engineering will hopefully shed more light on the situation and hopefully we will be able to provide practical recommendations within the scope of the research.

References:

[1] B. A. King, and S.Magun-Jackson, "Epistemological beliefs of engineering students," Journal of Technology Studies, pp. 56-64, 2008

[2] Royal Academy of Engineering, Thinking like an engineer. 2014.

[3] G. J. Shraw, L. D. Bendixen, M. E. Dunkle, "Development and validation of Epistemic Belief Inventory (EBI). In B. K. Hofer and P. R. Pintrich (Eds.), Personal Epistemology: the psychology of beliefs about knowledge and knowing (pp. 261-276). Mahwah, NJ: Erlbaum, 2002.

[4] S. Rajala, "Beyond 2020: Preparing engineers for the future, Preceedings of IEEE, vol. 100, pp. 1376-1383, May13th, 2012.

[5] Accreditation Board of Engineering and Technology (ABET). ABET 2014.

[6] J. Hyun Yu and J. Strobel, "Instrument development: engineering-specific epistemological, epistemic amd ontological beliefs, Proceedings of the Research in Engineering Symposium, Madrid, 2011.

[7] W. G. Perry Jr, Forms of Intellectual and Ethical Development in the College Years: A Scheme. Jossey-Bass Higher and Adult Education Series. CA: Jossey-Bass Publishers, 1999.

[8] R. Felder, and R. Brent, "Understanding student differences," Journal of Engineering Education, 94, pp. 57-72.

[9] T. Walters, A. Kadragic and L. M. Walters, "Miracle or mirage: Is development sustainable in the United Arab Emirates?" Middle East Journal of International Affairs, 10(3), pp. 77-91, 2006.

[10] T. A. Ryan, B. L. Joiner, & B. F. Ryan, Minitab™. John Wiley & Sons, Inc., 2004.

[11] L. W. Anderson, D. R. Krathwohl, P. W. Airasian, K.A. Cruikshank, R.E., Mayer, R.E., P. R. Pintrich, J. Raths, and M. C. Wittrock, A taxonomy for learning, teaching, and assessing: A revision of Bloom's taxonomy of educational objectives. New York: Longman, 2001

An empirical, comparative approach to engineering ethics (education) in international and cross-cultural contexts A study concerning Chinese engineering students' knowledge of and views concerning contents and concepts related to engineering ethics

Dr. Rockwell Franklin Clancy III, University of Michigan-Shanghai Jiao Tong Joint Institute

Rockwell F. Clancy is a lecturer in engineering ethics and philosophy at the University of Michigan-Shanghai Jiao Tong Joint Institute, Shanghai Jiao Tong University, and has acted as a long-term educational, setting up a course and writing a corresponding textbook with Heinz Luegenbiehl on global moral issues for engineers. His research and teaching interests include engineering ethics, philosophy of technology, Chinese philosophy, political philosophy, and science, technology, and society studies. Rockwell completed his PhD at Purdue University, West Lafayette, MA at the Katholieke Universiteit, Leuven, Belgium, and BA at Fordham University, New York.

Dr. Gang Zheng, University of Michigan-Shanghai Jiao Tong University Joint Institute

Dr. Gang Zheng is the Associate Dean for Undergraduate Education of the UM-SJTU Joint Institute. He is also a faculty member in Electrical and Computer Engineering. He has been leading and/or participating in curriculum development, program assessment, international joint programs, establishment of quality control and improvement mechanism, and creation of academic procedures and policies. Dr. Zheng has received the Most Popular Teacher Award and the "Candel Light" teaching award from the Shanghai Jiao Tong University.

Prof. David L.S. Hung, University of Michigan-Shanghai Jiao Tong University Joint Institute

Dr. David Hung is an associate professor of the UM-SJTU Joint Institute with a dual appointment as a professor in the Institute of Automotive Engineering, School of Mechanical Engineering at Shanghai Jiao Tong University. He received his Ph.D. in mechanical engineering from Carnegie Mellon University. His research focuses on advanced powertrain systems, engine measurements, and optical diagnostics. He has been teaching upper level and graduate subjects in automotive engineering and mechanical engineering laboratory courses, training students on engineering skills and team work through interactive learning.

An empirical, comparative approach to engineering ethics (education) in international and cross-cultural contexts

A study concerning Chinese engineering students' knowledge of and views concerning contents and concepts related to engineering ethics

An empirical, comparative approach to engineering ethics (education) in international and cross-cultural contexts: A study concerning Chinese engineering students' knowledge of and views concerning contents and concepts related to engineering ethics

Abstract: A study was conducted concerning Chinese engineering students' knowledge of and opinions about contents and concepts related to engineering ethics. The study was motivated by concerns regarding 1. the ethical standards of Chinese industries/engineers and their increasingly global presence 2. the nature of engineering ethics research/education in China and feasibility of introducing Western curricula for engineering ethics, and 3. general effectiveness of education in ethical training, based on findings in behavioral ethics and moral psychology. To address these concerns, a survey was adapted from a study previously conducted by Robert McGinn, at Stanford University, and then administered to different types of Chinese engineering students, at Shanghai Jiao Tong University, so as to compare the response data of students from these universities. As with McGinn's study, it was discovered that substantial confusion exists regarding the nature of ethical issues and conflicts. Based on findings from this data and experience educating Chinese and non-Chinese engineering students on engineering ethics in international and cross-cultural contexts, these responses are explained and tentative recommendations are made regarding how to improve such education in these contexts.

Keywords: engineering ethics; empirical approach; Chinese students; cross-cultural and international education; behavioral ethics; moral psychology

Acknowledgements

This work was made possible by an Innovative Education Grant from Shanghai Jiao Tong University in 2015-2016, a National-Level Practical Innovation Project Grant (IPP8160) in 2013-2014, an Innovative Education Grant from Purdue University in 2016, and a leave from teaching for Rockwell Clancy from the University of Michigan-Shnaghai Jiao Tong Joint Institute during the Spring 2016 semester. Additionally, the authors gratefully acknowledge the support of professors Yu Hui, Jian Yang, Liu Wenjing, and Yang Yang, who allowed them to administer the surveys for this study in their courses, and Wang Yuchen and Leng Xubo, who worked as research assistants to help translate and administer the survey, as well enter data into spreadsheets.

Introduction

This paper discusses a recent study we conducted concerning Chinese engineering students' knowledge of and opinions about contents and concepts related to engineering ethics. To do so, this paper is divided into three parts.

First, we introduce the background of and motivations for our study – an approach to engineering ethics (education) in increasingly international, cross-cultural educational and working environments. Second, we describe the nature of and rationale for our study – a comparative approach based on a study previously conducted at Stanford University. Finally, we introduce and analyze the response data to one question, on this basis making tentative recommendations for how to improve engineering ethics education both in China and abroad. As Chinese now comprise increasingly large percentages of student bodies in engineering programs and practicing engineers worldwide, and potential difficulties associated with training Chinese in engineering ethics also apply for other non-US/-Western students and practicing engineers, these recommendations would be applicable to curricula in engineering ethics in international, cross-cultural contexts in general.

I. The background of and motivations for our study

In this first section, we outline a broad research and educational agenda related to engineering ethics in global contexts. First, we discuss public safety concerns related to Chinese industries and engineers, both in China and abroad. Next, we describe the nature of engineering ethics (education) in China and why available foreign curricula would be inappropriate. Third, we discuss disciplines whose research could be brought to bear on engineering ethics (education), noting not only problems with ethics education in general and engineering ethics specifically, but also possible solutions to these problems.

A. Chinese industries and engineers in China and abroad: Public safety concerns

In the last thirty years, China has developed more and more quickly than any society in human history. This rapid development has had positive effects, raising millions from poverty, as well as negative effects related to public safety, economic corruption, and ethical integrity. Tragedies involving building, transportation, and food safety have raised concerns regarding the professional and ethical standards of Chinese industries and practitioners. The 2011 Wenzhou high-speed rail collision, for example, proved to be not only a tragedy and an embarrassment, but also a national economic setback: As a result of this incident, China was not allowed to bid on the construction of Brazil's high-speed rail network. These incidences have caused distrust in Chinese industries, leading to potentially lower economic revenues. With China's rapid development and growing global position, however, these are no longer exclusively Chinese concerns.

In addition to national tragedies and disasters within China, international cases – for example, that of tainted dog food produced in China and sold in Petco and PetSmart in the US – have also tarnished the reputations of Chinese companies and China abroad. These concerns are increased by the facts that growing numbers of Chinese students with degrees from China are choosing to complete their graduate degrees and work abroad[1], and Chinese companies are becoming more global and involved with projects outside of China. For these reasons, the ethics of Chinese engineering students and practicing engineers have economic and social consequences: If foreigners distrust Chinese engineering students, practicing engineers, and companies[2], then China risks loosing economic revenues, educational opportunities, and cultural influence. In addressing these concerns, the ethical conduct of individuals would be of importance, supported by legal and ethical norms, and ethical training and education.

Since becoming President in 2013, Xi JinPing has initiated and led a deep and widespread crackdown on corruption within Chinese industries, the government, and the Chinese Communist Party. These actions contribute to better ethical conduct by addressing large, macro-social structures, attempting to restore faith in rule of law, trust in the government, etc. To further improve ethical conduct within China, it is also necessary to address small and medium, micro-/meso-social structures, such as the behaviors of individuals, through ethical education and training.[3] Insofar as engineering is involved in all facets of modern life, ethical behavior by engineers would be a necessary condition of public safety. However, unlike countries as diverse as the US, Japan, France, and Canada, for example, courses on ethics do not figure prominently in Chinese engineering curricula.

B. Engineering ethics (education) in China, and problems with Western curricula

Students can take courses on engineering ethics, specifically, at four universities in China.[4] To date, there are scholars carrying out research in the field of engineering ethics at Tsinghua University[5], Wuhan University[6], Kumming University of Science and Technology[7], and Shanghai Jiao Tong University[4], as well as those associated with "5TU," a cooperative of five technological universities in China that address issues at the intersection of technology, society, and philosophy, including engineering ethics, which is comprised by the Dalian University of Technology, Beijing Institute of Technology, Northeastern University, Shenyang, Southeast University, Nanjing, and South China University of Technology, Guangzhou.

Although less emphasis is given to research in engineering ethics in China than internationally, recent efforts by faculty at the above universities are beginning to change this, with the publication of edited volumes[9] and organization of conferences: In 2012, the biannual conference of the Forum on Philosophy, Engineering, and Technology (fPET) was held at the Graduate University of the Chinese Academy of Sciences, Beijing, and in 2015, the biannual conference of the Society of Philosophy and Technology (SPT) was held at Northeastern University, Shenyang.

Prominent scholars include, for example, Hong Xiaonan, Professor and Director of the Programs of Philosophy of Science and Technology at the Dalian University of Technology, and Jian Wang, Professor in the Program of Science, Technology, and Society Studies at Northeastern University, Shenyang. Current efforts within the field tend to focus on the history of engineering within China[10], as well as the contributions that Chinese thought – such as Confucianism and Daoism – can make to contemporary engineering practices.[11] In Chinese, Bo Yu and Yong Fan give an overview of these research trends in engineering ethics within China.[7] Although characteristics of the contents of these efforts are unique to China, in large part, their form is the same as that of international trends within engineering ethics.

As a discipline, engineering ethics developed in the United States[12] and has been based on applying abstract, universal ethical principles – for example, those belonging to consequentialism, deontological, and virtue ethics – to concrete cases of engineering disasters and the decisions that individuals made, for example, the case of the Ford Pinto, Hyatt Regency Walkway Collapse, etc.

The field has come to include understandings of engineering ethics from the perspective of professional role responsibilities, specific duties engineers have that follow from their roles as professionals, in conjunction with an emphasis on the organizational structures of engineering work environments, for example, the ways that NASA's funding scheme, group think, etc.

contributed to the Challenger Space Shuttle disaster. There are four main texts frequently used in courses on engineering ethics.[13, 14, 15, 16]

Given its Western origins, this approach to ethics is potentially unsuitable to contemporary engineering practices, which increasingly occur in global and cross-cultural contexts.[17,18] Insofar as courses on engineering ethics should be sensitive to the professional and work environments, cultural values, and national circumstances of their students, simply using the contents of courses taught in other countries would be inappropriate.[4] These materials could be ineffective from an educational perspective. Insofar as China has a unique history and culture, it is unclear that the same norms of and education regarding professional and ethical conduct applied in Western countries could or should be applied in China.[19, 20] Unlike other engineering courses, courses in engineering ethics should, in some sense, be specific to China.

C. Disciplines with insights for engineering ethics (education): Problems and solutions

The fields of comparative applied ethics, Chinese thought, moral psychology, experimental philosophy, and behavioral ethics address problems with ethics education in general and engineering ethics specifically, and insights from these disciplines can be brought to bear on improving engineering ethics (education), not only in China but also abroad.

1. Comparative applied ethics

An increasing body of literature now exists regarding comparative approaches to applied ethics in business[21], law[22], and medicine[23] in China, although relatively little has been written on engineering ethics. Additionally, in general, little research on engineering ethics has been dealt with from an empirical perspective, although there have been notable and growing numbers of exceptions.[24, 25, 26, 27] Carrying out similar research in the field of engineering ethics is important for two main reasons.

First, doing so increases the representativeness of social scientific data in general and, therefore, the accuracy of conclusions that can be drawn from such data. Most of the data for social scientific studies are based on WEIRD (White Educated Industrialized Rich Democratic) samples. Given that these samples are potentially unrepresentative, some have raised doubts about the universality of conclusions drawn from them.[28, 29] Collecting information from Chinese engineering students and practicing engineers begins to make up for this deficiency, discovering what they know and think about engineering ethics. This would be essential to developing more effective curricula in engineering ethics.

Second, doing so allows one to develop specialized, more effective contents and materials for curricula. Some have argued that ethics cannot be taught, such that the best educators can do is develop and hone psychologically innate and previously acquired ethical commitments and processes of ethical reasoning.[26] Assuming this is the case, the first step would consist in determining the nature of these commitments and processes.

Although growing bodies of research within the interdisciplinary fields of moral psychology and behavioral ethics – further discussed below – exist, again, their findings are based primarily on the above-described WEIRD participants. Further, some of the most important commitments within engineering education – for example, commitments to life-long learning and performing in one's area of competence – are far removed from innate and previously acquired ethical commitments and processes of ethical reasoning. Engineering students have rated such commitments as having less ethical importance than, for example, telling the truth.[26] To develop more effective curricula in engineering ethics, it would be important to know more about what students know and think.

2. The merits of ethical reasoning within Chinese thought

In texts central to traditional Chinese thought, such as the *Analects*, *Dao De Jing*, and *Mencius*, questions of right and wrong are decided with regard to specific instances of persons and situations deemed to be good, making generalizations on these bases.[30, 31] Additionally, some have argued that Confucianism is a kind of role ethics.[30, 32]

These insights are important for two reasons: First, deciding questions of right and wrong with reference to abstract, universal principles applied to concrete cases – as occurs in Western approaches to engineering ethics – is largely foreign to many cultural and social traditions.[31] Second, the organization of occupations into professional groups and the characteristics that follow from this organization – the basis of another approach to engineering ethics – are, to a certain extent, unique to Western practices.[33]

Insights from traditional Chinese ethics can address these problems: First, approaches to ethics in Confucianism and Daoism are similar to bottom-up, case studies approaches to engineering ethics, which emphasize active learning and increase understanding.[34] Michael Davis has outlined such an approach, which has been employed with good results in courses composed of national and international students at Purdue University, West Lafayette, IN and Shanghai Jiao Tong University (SJTU), Shanghai, China – further discussed below. Second, despite the foreignness of a professional occupational tradition within China, orienting engineering ethics in terms of the centrality of role responsibilities with reference to Confucianism could be beneficial to the education of Chinese engineering students.

3. Moral psychology: Neither character nor autonomy

Contemporary approaches to ethics are generally more concerned with determining criteria of right and wrong than educating students to act rightly. To a large extent, this results from the influences of the liberal tradition, where the good consists in personal autonomy – persons deciding for themselves – and attempts to decide the good for others would infringe on this autonomy and, therefore, are bad.[35] However, findings in moral psychology call the conceptions of personhood on which this position is based into question.

Actions are determined more by cues within the environment than one's "character," the innate and acquired moral dispositions constituting parts of one's personality.[36] Hence, persons are not always or generally aware of why they make the decisions they do.[37] These findings call into question not only the coherence of character and personal autonomy as ontological notions but also ethical systems in which they would play central role. Given the centrality of these notions and systems to engineering ethics (education) to date, going forward, research in engineering ethics (education) should be cognizant of these findings, developing educational programs and organizing and administering engineering work environments with these in mind.

4. Behavioral ethics

As with moral psychology, the field of behavioral ethics focuses on the reasons persons act the ways they do and what can be done to insure that they act ethically.[38] Although the fields of business[39, 40], law[41], and medicine[42] have taken interest in and engaged with developments in behavioral ethics, engineering ethics has not. However, given the tremendous influence of engineers on the contemporary world – and for the sake of public safety – insights from the field of behavioral ethics can and should be brought to bear on research and education in engineering ethics.

II. The rationale for and nature of our study

In this second section, we explain the rationale for and nature of the study we conducted, orienting it in terms of the research and educational agenda described above. To do so, first, we discuss the broader rationale of our study. Second, we explain the nature of the survey itself. Third, we describe the courses in which we administered the survey and, thus, the make-up of the students surveyed, as well as how we organized and analyzed the data.

A. Rationale of the study: Anticipated outcomes

Our project consisted in creating, administering, and reviewing a survey for engineering students at SJTU, to gauge their current knowledge of and views regarding contents and concepts related to the domain of engineering ethics. This was done with the aim of better understanding Chinese students' perspectives on engineering ethics to determine the following:

1. the extent to which efforts to integrate units on the ethical and professional responsibilities of engineers into technical courses have been successful. This has been an ongoing effort on the part of civil engineering at SJTU, although it has just begun in mechanical engineering.

2. similarities and differences between their views on engineering ethics and those of their Western counterparts. We would expect that these would be different, which would be important to developing appropriate curricula, as described in section one.

3. correlations that exist between students' major fields and their cognizance of issues related to ethical and professional responsibilities. One of the hypotheses of our broader research agenda is that certain fields of engineering more easily lend themselves and give rise to reflection on and awareness regarding the ethical and professional responsibilities of engineers. In the cases of civil and mechanical engineering, for example, the negative consequences of acting unethically working on bridge and engine projects, for instance, might be more obvious than in computer engineering working on cyber security.

On this basis and in conjunction with other research, our goal is ultimately to make recommendations about how to improve engineering ethics curricula, not only in China but also contexts where Chinese and other non-US and Western students comprise large percentages of student bodies. These are increasingly becoming the norm rather than the exception.

B. The nature of the survey

So as to be able to take a comparative perspective, the survey we administered was based on one used and described by Robert McGinn in "'Mind the Gaps': An Empirical Approach to Engineering Ethics, 1997-2001."[28] In 1997, 1999, and 2001, McGinn administered a survey with questions about engineering ethics contents and concepts to 696 Stanford University engineering students.

An undergraduate engineering student working as a research assistant on our project first translated the survey into Chinese. A Chinese faculty member who speaks English at a near-native level then checked and corrected that version, and it was translated back into English to insure continuity in meaning with the original.

That version of the survey was then administered to a group of engineering majors taking "Introduction to Philosophy of Religion," a course I taught during the spring 2015 semester. The students in that course read, write, and speak English at a professional working level. The survey took, at most, twenty minutes to complete. After soliciting student feedback, slight changes were made to the Chinese translation. These were made to insure that students understood the survey questions and that the Chinese questions were soliciting the same information as those from McGinn's English original.

In addition to fixed-response questions, as with McGinn, our Chinese survey included free-response questions. These were included, first and foremost, to insure continuity between the two surveys and comparability of results. Additionally, and more interestingly, as various problems exist with administering fixed-response questions in value-related surveys in cross cultural contexts[19], the free-response questions provided Chinese students with the opportunity to express issues, values, and concerns in a freer manner than would responses to the fixed-response questions. Although the nature of these responses were, thus, qualitative in nature, so as to be able to more easily work with the data – drawing general conclusions and looking for correlations between the different variables – the answers to the free-responses questions were subsequently coded, divided into categories based on shared characteristics within the responses.

C. Courses in which the survey was administered

We selected courses in which to administer the survey based on the personal and professional relations of our research team members, educational levels of the enrolled students, and number of students enrolled in the courses.

First, as members of our team specialize in mechanical, electrical, and computer engineering, and have relations with faculty members working in these disciplines at SJTU, we began by looking for courses in these disciplines to administer our survey. Second, as the survey asks questions regarding what students have learned in and from their technical engineering courses and instructors, we next looked for upper-level courses to survey, ones in which the students surveyed would already have taken technical engineering courses. Third, as we sought to administer the survey to the largest number of students possible, we finally looked for courses with high enrollment caps.

After identifying ideal courses, we contacted the respective instructors to ask permission to administer the survey during their classes. Although we contacted many instructors, we received few replies. For that reason, we administered the survey to 172 students taking courses in civil, electrical, and computer engineering. Unfortunately, we did not administer the survey to students taking courses in mechanical engineering.

The survey was administered in four engineering courses taught at SJTU in the spring 2015 semester. These included "Basic Circuits of Communications" on April 23 taught by Yu Hui, "Variational Principles and Finite Element Methods" on May 18 taught by Jian Yang, "Basis of Design Large IC" on May 26 taught by Liu Wenjing, and "Machine Learning" on May 29 taught by Yang Yang.

For the sake of input and analysis, all of the response information was coded. This was relatively simple with regard to the fixed-response questions but more complicated with regard to the free-response questions. The free-response questions were dealt with in the way described above, where answers with similar characteristics were grouped and tallied together.

III. Introduction and analysis of one question, and preliminary recommendations

In this third and final section, we introduce and discuss the response data to one particular question, making tentative recommendations about how to improve engineering ethics education in global contexts on this basis. We begin by introducing the question we discuss and our reasons for having chosen this question here. We move on to state the response data to this question. Third, we compare this data with that from McGinn's study, attempting to explain these differences and, finally, on the basis of these explanations, we make recommendations about how to improve engineering ethics education in global contexts.

A. The question for discussion here and our reasons for choosing it

The question we discuss here is "当你面对一个问题时,会是什么让你觉得这是一个道德问题,而不是一个法 律,美学或别种问题?" Which is most literally translated into English as "When confronted by a problem, what would make you think this is a moral problem, rather than one of law, aesthetics, or another type of problem?" and corresponds to the following question from McGinn's survey: "What, as you see it, makes an issue or a conflict one that falls within the domain of ethics (as opposed to, say, one of law)?"

We chose this question for a variety of reasons. First, it is a free-response question and, thus, we anticipate that the responses to this question would be more authentically representative of the views of the Chinese engineering students we surveyed. Second, of all the questions and response data from McGinn's study, he highlights this one and presents the greatest amount of free response data, such that it can be compared with the response data from our survey. Given the disparate nature of responses he received from the survey participants, McGinn hypothesizes that this result points towards a problematic relativism with regard to ethical reasoning by engineers. Third, since this question asks about criteria used to determine ethical quandaries and, thereby, cuts to the heart of moral reasoning, focusing on this question would be the most fruitful when addressing engineering ethical concerns in international and cross-cultural contexts

B. The response data from our survey

Of the 172 students surveyed, 140 provided an answer to this question. The answers fell into roughly ten categories, the breakdown of which was as follows:
1. Non-legal 55 (39.3%)
2. Related to values, including personal, social, and professional values 37 (26.4%)
3. Concerned with safety, including potential risks to life and property 23 (16.4%)
=115 (82.1%)

4. Related to consequences in terms of risks/cost-benefits 5 (3.6%)
5. Gave specific examples 6 (4.2%)
6. Non-engineering 4 (2.9%)
7. Related to emotions 3 (2.1%)
8. Related to expertise, including professional knowledge and social experiences 3 (2.1%)
9. Concerned with illegality 2 (1.4%)
10. Do not know 2 (1.4%)

As with many of the other free-response questions, here some (46) students responded with more than one answer. The breakdown of these is as follows:
1. Concerned with safety, including potential risks to life and property 21 (45.7%)
2. Related to values, including personal, social, and professional values 18 (39.1%)
3. Related to conflicts in consequences in terms of risks/cost-benefits 2 (4.3%)
=41 (89.1%)

4. Gave specific examples 2 (4.3%)
5. Related to emotions 1 (2.2%)
6. Related to expertise, including professional knowledge and social experiences 1 (2.2%)
7. Do not know 1 (2.2%)

In response to the question of what would make an issue an ethical one, the highest percentage of students answered some variation of "a non-legal one." These responses could be explained in terms of presentation bias, conceptual confusion, cultural background, or some combination of the three, all of which we consider in relation to McGinn's findings and analyses.

This was the only free-response question the answers to which McGinn analyzed in his article. For that reason, it is also the only data from the free-response questions that can, at present, be analyzed from a comparative perspective.

According to McGinn, the 591 coherent/salient responses he received to this question fell into roughly ten categories. (McGinn reports the total number as being 592, although the number of respondents from the different categories only adds up to 591.) The breakdown of these was as follows:

1. Consequences/Harm 194 (32.8%)
2. Morality/Morals 170 (28.8%)
3. Right/Wrong 67 (11.3%)

= **431 (72.9%)**

4. Feelings 35 (5.9%)
5. Examples 30 (5.1%)
5. Values 27 (4.6%)
6. "Not Determined by Law" 26 (4.4%)
7. "Infringement of Rights" 19 (3.2%)
8. "No Clear Answer" 12 (2.0%)
9. Justice 11 (1.9%)

In his analyses, McGinn concludes from the disparate responses to this question that the engineering students he surveyed lacked a clear and unified sense of what constitutes an ethical issue or conflict, and that this would be problematic to practicing engineers. Based on the responses to our survey, we draw a similar conclusion.

First, as with responses to McGinn's survey, responses to ours were disparate in nature, comprising ten distinct categories. Second, insofar as forty-six of the 140 (32.9%) students who responded to this question did so with multiple answers, this seems to show that the students we surveyed did not have a clear sense of what constitutes an ethical issue. Third, fifty-five respondents (39.3%) said that what makes a conflict or an issue an ethical one is that it is "non-legal." This rate is considerably higher than for any of the other categories, and it is higher than the rates for any of the categories reported by McGinn. This might lead one to conclude a relative convergence regarding what constitutes an ethical issue or conflict, although we think this would be a mistake. Rather than indicating a convergence, we think this high rate actually indicates greater confusion.

In preparing this study, as mentioned above, we administered a draft of the Chinese translation of McGinn's survey to a group of undergraduate Chinese engineering students, asking them if they found any of the questions confusing. The greatest number of students mentioned this question, that they did not understand what the question was asking/how they were expected to respond. Specifically, they mentioned the "而不是一个法 律,美学或别种问题?" (="rather than one of law, aesthetics, or another type of problem?") part of the question. Students were unclear about the relationship between ethical problems, and aesthetics and law.

Additionally, students suggested that we change this question to a fixed-response format, providing options from which to choose. Although this change would have made it easier for students to select an answer, it would also have limited their potential responses, bypassing the work involved in thinking through and articulating criteria for determining ethical issues and conflicts. However, based on the responses we received, many of the students seemed to have done so anyway.

The high rate for "non-legal" issue seems to indicate students were unsure of what makes an issue or a conflict an ethical one, and thus responded based on the question itself. For these reasons, it seems as though there is as much confusion and indecision regarding what makes a conflict or an issue an ethical one among our respondents as McGinn's. In the case of our Chinese respondents, this could be explained from the perspective of moral reasoning within traditional Chinese thought, mentioned in the first part of this paper.

Again, the process of moral reasoning within traditional Chinese thought begins with examples of appropriate conduct, arriving at principles on this basis, rather than beginning with principles and applying them to particular examples. This would be a bottom-up process of ethical reasoning, where deciding issues of right and wrong with regard to abstract ethical principles is largely foreign. For this reason, one might expect that defining ethics in terms of abstract criteria – as the question prompts students to do – would be an unfamiliar and difficult task. Although we feel the high rate for "non-legal" issue is best understood as indicating confusion/indecision regarding what constitutes an ethical question, it might be explained in terms of competing processes of ethical reasoning within traditional Chinese thought.

Historically, a dichotomy has existed between the social roles of ritual within Confucianism and law within legalism. Insofar as legalists have been critical of ritual as promulgated within Confucianism – as an ineffective means of social regulation – for those who could be described as "Confucian," laws and legality could appear as a counterpoint to/the antithesis of ritual or morality.[30, 31, 43] To determine if this played a role in students' responses, in the future, it would be necessary to collect more data and conduct short interviews with students, a mix-method quantitative and qualitative study.

Comparatively, safety and the consequences of one's actions both ranked relatively high among the two groups, and the percentage of respondents who gave examples rather than listing criteria was also roughly equivalent. If we were to understand the Chinese respondents from our group as equating values with morality, then morality could would also rank high among both groups.

The percentage of our respondents who listed emotions as a criterion for determining an ethical issue was much lower than McGinn's, and they responded with right and wrong as a criterion only secondarily. Perhaps unsurprisingly, given the nature of traditional Chinese culture, none of the respondents to our survey listed either "rights" or "justice" as criteria for deciding whether problem were ethical or not.

Despite perceived differences between Westerners and Easterners with regard to social and cultural values and, therefore, ethics, the response data to this question – at least among engineering students – seems to indicate confusion regarding what constitutes the domain of ethics among both. Identifying this confusion has important implications for the development of curricula in engineering ethics, especially in cross-cultural and international environments.

C. Initial recommendations

Based on these finding, problems and the fields of inquiry discussed above, and experience educating Chinese and non-Chinese engineering students on engineering ethics in international and cross-cultural contexts, the following tentative recommendation can be made regarding how to structure curricula. (We are still in the process of identifying and analyzing correlations within the response data. Once this task has been completed, we will be in a better position to make more substantial recommendations.)

1. We recommend that curricula devote significant attention to making clear what constitutes an ethical issue within engineering. This is important for a number of reasons:

First, as the survey data shows, disagreement and confusion exists regarding what constitutes an ethical issue/conflict in general, potentially inhibiting ethical action within engineering contexts. As McGinn notes, this confusion is linked to a broader relativism regarding ethics[24], and such relativism probably has a variety of sources, for example, descriptive ethical relativism: the fact that different persons and peoples have different understandings of the nature of right and wrong, correct and incorrect actions, etc. This would be especially true in international and cross-cultural contexts, such as those where engineering and engineering education increasingly occur.

Second, as touched on above, the nature of ethical issues within the sphere of engineering can be different from and non-intuitive from perspectives of everyday or personal ethics.[26] For example, to maintain competence – and, thus, safeguard public safety – engineers should engage in life-long learning. This is a professional and ethical duty, and mentioned in various codes of engineering ethics. However, unlike honesty in stating claims, for example, a responsibility for life-long learning has no close equivalent within the spheres of everyday and personal ethics, or – at the very least – the basis, implications, and consequences of life-long learning in these two contexts are different.

Third, the specific criteria used to determine whether issues are ethical issues could vary, depending on the field of engineering.[44] What constitutes an ethical question within civil engineering, for instance, might not be the same as what constitutes an ethical problem within the domains of biomedical and computer engineering. For these reasons, it would be important to devote considerable attention to discussions and exercises regarding the nature of ethical issues, criteria for determining ethical questions.

However, given the plethora of possible views and potential disagreements, attempts to describe and defend any one particular perspective could appear as arbitrary and, therefore, remain contested or result in a backlash. Further, as mentioned above, theoretical approaches to ethics are, to some extent, unique to the "Western world." To cast a wide net, and remain realistic about criteria for ethical behavior students can be expected to understand and endorse, and with which they can easily work, we have used the following definition in teaching ethics to engineering students in international and cross-cultural contexts: "ethics is about actions that have the potential to have a serious impact on the lives of others."[45]

On the one hand, this definition is broad enough to encompass characteristics associated with and implied by ethical positions relevant to and unproblematic for engineering ethics, for example, consequentialism or the role ethics of Confucianism. A broad definition is important, to accommodate different types of engineering and cultural perspectives in engineering contexts.[46] On the other hand, this definition is narrow enough to exclude characteristics associated with and implied by other ethical positions irrelevant to and problematic for engineering ethics, for example, divine command theory or normative cultural relativism. A narrow definition is important, to limit the sphere of ethics to issues that would be of immediate and pressing concern for engineers, especially in cross-cultural and international contexts. Although it is important that engineers be aware of cultural norms, not all such cultural norms would be directly relevant to ethical issues in engineering.[46]

Again, however, given the variety of possible views and potential disagreements, this perspective could appear as arbitrary and, therefore, unconvincing. These possibilities lead to our second recommendation.

2. We recommend that curricula employ a bottom-up approach as described above, oriented around case studies. To specify and motivate the broad criterion of ethics just described, engineering ethics curricula should employ case studies that touch on those areas and topics that give rise to confusion regarding ethical issues.

Based on the previous discussion, these include situations related to: first, international and cross-cultural contexts, where different social and cultural values are present; second, general engineering contexts, where ethical issues that are of particular concern to engineers exist; third, contexts specific to different fields of engineering, where ethical issues that are of particular concern to these specific fields are present. Taking this approach allows educators to not only specify the contents of "actions that have the potential to have a serious impact on the lives of others," but also motivate the importance of ethical and reach better consensus among students regarding the nature of ethics in engineering.

Towards these ends, in courses taught at SJTU, Shanghai, China and Purdue University, West Lafayette, IN, instructors have developed and used case studies on, for instance, the Ueberlingen Mid-Air Collision in Germany and Qihoo 360's P1 Wireless Router in China. These cases are related to international/cross-cultural environments, as well as contexts of engineering in general and specific engineering disciplines.[46] Regardless of cultural, social, and educational backgrounds, when examining these cases, students can agree something went wrong and should be done to prevent such incidents from occurring.

Following a case-study procedure, raising various ethical issues, identifying relevant facts, applying broad, previously justified engineering-ethical principles,[45] etc., cultivates greater consensus among students from different backgrounds. Such an approach has been employed with good results in courses composed of national and international students at SJTU and Purdue University.

Conclusion

Contemporary engineering practice increasingly occurs in international and cross-cultural environments. Differences in cultural and social values can give rise to confusion regarding what constitutes ethical actions in engineering, putting public safety at risk. To address and mitigate difficulties associated with this changing landscape, we propose an empirical and comparative approach to engineering ethics (education). This approach pulls on findings from the fields of comparative applied ethics, Chinese thought, moral psychology, and behavioral ethics. The first step in this project is a study aimed at determining what Chinese engineering students know and think about contents and concepts related to engineering ethics, comparing this response data with that from an early study conducted at Stanford University. On this basis, one can make more effective recommendations about how to teach engineering ethics in international and cross-cultural contexts.

References

1. Heinze, J., & Zhu, J. (2012). "Current Research on Chinese Students Studying Abroad." *Research in Comparative and International Education* vol. 7 no. 1.

2. Wang, J. (2015). "Chinese the most dishonest, Japanese and British the least, study finds." *South China Morning Post*. Tuesday, November 17.

3. Bocong, L. (2012) "From a Micro-Macro Framework to a Micro-Meso-Macro Framework." In Christensen, S., Mitcham, C., Li B., & An, Y. (Eds.). *Engineering, Development, and Philosophy: American, Chinese, and European Perspectives*. Dordrecht: Springer.

4. Clancy, R. & Zhang, G. (2014). "Orienting Engineering Ethics in terms of China: Curricula Shortcomings and Case Studies on China." *American Society of Engineering Education 2014 International Forum*.

5. Su, J. and N. Cao. (2007) "A Study of Professional Code of Ethics Based on an Investigation into Bylaws of Registered Professional Engineer and Constitution of Engineering Society in China." (in Chinese) *Journal of Huazhong University of Science and Technology (Social Science Edition)* 2007 21.

6. Yi, Y. (2013) "College Students' Engineering Ethics Education Survey Report." (in Chinese) *The Guide of Science and Education* 1.

7. Bo, Y. & Fan, Y. (2014). "A Overview on Engineering Ethics Research in China." (in Chinese) *Journal of Kunming University of Science and Technology*. Vol. 14, No. 3 Jun.

8. Ming, X., Douglas, D., Gurzawska, A., & Brey, P. (2015). "Ethics Assessment in Different Countries (China)." *Ethical Assessment of Research and Innovation: A Comparative Analysis of Practices and Institutions in the EU and selected other countries. Satori.* http://satoriproject.eu/media/D1.1_Ethical-assessment-of-RI_a-comparative-analysis.pdf

9. Christensen, S., Mitcham, C., Li B., & An, Y. (Eds.). *Engineering, Development, and Philosophy: American, Chinese, and European Perspectives*. Dordrecht: Springer.

10. Hong, X. & Ma, Li. (2012). "Engineering and Development in Modern China: Challenges and Responses." In Christensen, S., Mitcham, C., Li B., & An, Y. (Eds.). *Engineering, Development, and Philosophy: American, Chinese, and European Perspectives*. Dordrecht: Springer.

11. Wang, Q. & Zhu, Q. (2012). "Traditional Chinese Thinking and Its Influences on Modern Engineering and Social Development." In Christensen, S., Mitcham, C., Li B., & An, Y. (Eds.). *Engineering, Development, and Philosophy: American, Chinese, and European Perspectives*. Dordrecht: Springer.

12. Davis, M. (1995). "An Historical Preface to Engineering Ethics." *Science and Engineering Ethics* vol. 1. 33-48.

13. Harris, C., Pritchard, M., Rabins, M., James, R., & Englehardt, E. (2013). *Engineering Ethics: Concepts and Cases 5th ed.* Boston: Wadsworth Publishing.

14. Martin, M., & Schinzinger, R. (2000). *Introduction to Engineering Ethics 2nd ed.* New York: McGraw Hill.

15. Van de Poehl, I., & Royakkers, L. (2011). *Ethics, Technology, and Engineering: An Introduction.* Malden: Wiley-Blackwell.

16. Whitbeck, C. (2011). *Ethics in Engineering Practice and Research.* New York: Cambridge University Press.

17. Luegenbiehl, H. (2004). "Ethical Autonomy and Engineering in a Cross-Cultural Context." *Techné: Research in Philosophy and Technology*, vol. 8, no. 1.

18. Luegenbiehl, H. (2002). "Issues in the Internationalization of Engineering Ethics." *First Annual Conference of the Japanese Society for Science and Technology Studies, Tokyo.*

19. Kulich, S. and Zhang (2010). "The Multiple Frames of 'Chinese' Values: From Tradition to Modernity and Beyond." In M. Bond (Ed.) *The Oxford Handbook of Chinese Psychology.* New York: Oxford University Press. 241-278.

20. Nisbett, R. (2010). *The Geography of Thought: How Asians and Westerners Think Differently and Why.* New York: Free Press.

21. Whitcomb, L, Erdener, C. & Li, C. (1998). "Business Ethical Values in China and the U.S." *Journal of Business Ethics*, vol. 17, no. 8 Jun. 839-852.

22. McMorrow, J. (2010). "Professional Responsibility in an Uncertain Profession: Legal Ethics in China." *Akron Law Review* 43: 1081-1105.

23. Fox, R. & Swazey, J. (1985). "Medical Morality Is Not Bioethics: Medical Ethics in China and the United States." *Perspectives in Biology and Medicine* 27 (4) Spring 336–360.

24. McGinn, R. (2013). "'Mind the Gaps': An Empirical Approach to Engineering Ethics, 1997-2001." *Science and Engineering Ethics* 9, 517-542

25. Kuczenski, J. (2013). "Student Ethics in Engineering: A Comparison of Ethics Survey Results from Undergraduate Engineering Students at Three Different Engineering Programs and Institutions." *American Society for Engineering Education.*

26. Stappenbelt, B. (2012). "Ethics in Engineering: Student Perceptions and their Professional Identity Development." *Journal of Technology and Science Education*, 3 (1) 86-93.

27. Kisselburgh, L., Zoltowski, C., Beever, J., Hess, J., & Iliadis, A. (2014). "Effectively engaging engineers in ethical Kisselburgh reasoning about emerging technologies: A cyber-enabled framework of saffolded, integrated, and reflexive analysis of cases." *121st ASEE Annual Conference and Exhibition.* Indianapolis.

28. Arnett, J. (2008). "The neglected 95%: Why American Psychology Needs to Become Less American." *American Psychologist.* vol 63 (7), Oct. 602-614.

29. Henrich, J., Heine, S., Norenzayan, A. (2010). "The Weirdest People in the World?" *Behavioral and Brain Sciences.* vol 33 (2-3), Apr-Jun, 61-83.

30. Hall, D., & Ames, R. (1987). *Thinking Through Confucius.* Albany: State University of New York Press.

31. Slingerland, E. (2014). *Trying Not to Try: The Art and Science of Spontaneity.* New York: Broadway Books.

32. Ames, R. (2011). *Confucian Role Ethics*. Honolulu: University of Hawaii Press.

33. Didier, C. (2010). "Professional Ethics without a Profession: A French View of Engineering Ethics." In I. Van de Poel & D. Goldberg (Eds.) *Philosophy and Engineering. An Emerging Agenda*. Dordrecht: Springer. 161-174.

34. Prince, M. (2004). "Does Active Learning Work? A Review of the Research." *Journal of Engineering Education,* 93(3), 223-231.

35. Clancy, R. (2015). *Towards a Political Anthropology in the Work of Gilles Deleuze: Psychoanalysis and Anglo-American Literature*. Leuven: Leuven University Press.

36. Doris, J. (2005). *Lack of Character: Personality and Moral Behavior.* New York: Cambridge University Press.

37. Wilson, T. (2004). *Strangers to Ourselves: Discovering the Adaptive Unconsious.* Cambridge: Harvard University Press.

38. Bazerman, M. & Tenbrunsel, A. (2013). *Blind Spots: Why We Fail to Do What's Right and What to Do about It.* Princeton: Princeton University Press.

39. Treviño, L., Weaver, G., & Reynolds, S. (2006). "Behavioral Ethics in Organizations: A Review." *Journal of Management* December. 951-990

40. Bazerman, M. & Gino, F. (2012). "Behavioral Ethics: Towards a Deeper Understanding of Moral Judgment and Dishonesty." *The Annual Review of Law and Social Sciences* 8. 85-104.

41. Robbennolt, J. (2015). "Behavioral Ethics Meets Legal Ethics." *The Annual Review of Law and Social Sciences* 11, 75-93.

42. Boscardin, C. (2015). "Reducing Implicit Bias Through Curricular Intervention." *Journal of General Internal Medicine* 30 (12). 1726-8.

43. Fung, Y.L. (1997). *A Short History of Chinese Philosophy*. New York: The Free Press.

44. Fleddermann, C. (2000). "Engineering Ethics Cases for Electrical and Computer Engineering Students." *IEEE Transactions on Education*, vol. 43 August.

45. Luegenbiehl, H. (2010). "Ethical Principles for Engineers in a Global Environment." *Philosophy and Engineering: An Emerging Agenda.* Van de Poel, I. & Goldberg (Eds.). Dordrecht: Springer.

46. These are discussed at greater length in Luegenbiehl, H. & Clancy, R. (forthcoming) *Global Engineering Ethics*. Elsevier.

The Design and Implementation of Engineering Leadership Programs: A Comparative Study

Miss Hu Yu, Shanghai Jiao Tong University

Yu Hu is a graduate student at the Graduate School of Education in Shanghai Jiao Tong University. She obtained a B.S. in biotechnology from Hebei Normal University. Her current interest focuses on engineering leadership, cognitive development of engineering graduate and undergraduate students, the assessment of teaching and learning in graduate education.

Dr. Jiabin Zhu, Shanghai Jiao Tong University

Jiabin Zhu is an Assistant Professor at the Graduate School of Education at Shanghai Jiao Tong University. Her primary research interests relate to the assessment of teaching and learning in engineering, cognitive development of graduate and undergraduate students, and global engineering. She received her Ph.D. from the School of Engineering Education, Purdue University in 2013.

The Design and Implementation of Engineering Leadership Programs: A Comparative Study

Abstract

Changing technological world requires future engineers not only show excellent technical skills, but also possess multiple skills and attributes, in particular leadership skills. Against this context, some universities and engineering colleges have developed pioneering engineering leadership programs to promote leadership among engineering students. This paper compares twenty-one engineering leadership programs around the world by analyzing the differences and similarities across these programs via text analysis. By exploring the outcomes and implementation of these engineering leadership programs, this paper found that visioning/setting goals, thinking or reasoning skills, interpersonal skills and ethics are the key attributes emphasized in these engineering programs. Lectures, field trips and practical experiences are the most common implementation platforms to develop engineering leadership. This paper aims to offer a comprehensive understanding of current practices and offer suggestions for designing similar engineering leadership programs in the future.

Introduction

Fast-paced technological transformation and innovation demands not only technical expertise from future engineers, but also multiple soft skills, in particular leadership [1]. Crumpton-Young indicated that engineering leadership is the influence and ability to lead other engineers and technical staff to design, create, innovate, implement and evaluate and assess services and products [2]. MIT-Gordon engineering leadership program describes engineering leadership as a set of capabilities and values that enable a person to accomplish a multi-disciplinary complex project by leading teams instead of functioning as an individual technical contributor. Engineering leadership is a process to encourage leaders and team members to achieve common goals that cannot be accomplished otherwise by working alone [3]. According to Nation Society of Professional Engineers, leadership consists of a number of capabilities which are crucial at a professional level to contribute to public health, safety and welfare [4]. In summary, engineering leadership is a set of integrative capabilities covering diverse aspects such as effective communication, active participation in teamwork and capabilities to influence others.

Considering the crucial role of engineering leaders in affecting change, leadership has become a new standard in cultivating future engineers. Engineers with leadership skills gain a competitive edge in labor market [5]. Engineering leadership is essential for shaping enterprise culture, optimizing the organizational structure and increasing productivity [6]. In addition, increased global cooperation and collaboration in industry calls for engineers with strong interpersonal communication skills, teamwork skills and an awareness of diverse culture [5]. Engineering leadership education has also effected on accelerating engineering students' professional development [7]. It is said that "an engineer is hired for her or his technical skills, fired for poor people skills, and promoted for leadership and management skills" [8].

With increased importance of engineering leadership, new criteria are proposed for engineers. National Academy of Engineering published *The Engineer of 2020 - Visions of Engineering in the New Century*, which indicates that engineers in 2020 need to develop analytical skills, practical ingenuity, creative capability, communication skills, concepts of business and management, leadership, ethical standards and sense of professionalism [9]. The Royal Academy of Engineering in United Kingdom published *Educating the Engineers for the 21 Century - the Industry view,* highlighting future engineering graduates need to be equipped with creativity, innovation and leadership to lead the industry to succeed [10]. A Plan for Educating and Training Outstanding Engineers published by Chinese Ministry of Education proposed a list of competencies to be possessed by engineering undergraduates upon graduation, including good professional ethics, relating mathematics, natural science and economic management knowledge, basic engineering knowledge and skills, creativity and innovation skills, communication skills, organization and management skills, global vision and intercultural cooperation ability [4]. This plan requires that future engineers should not only become technical experts in their respective fields, but also need to be taking the leading role in the changing technological world by possessing multiple skills and attributes, in particular leadership skills. Engineer who possess leadership have the ability to lead an engineering team to create, design, develop, implement and evaluate products [5]. The development of an industry needs engineering talents who possess leadership ability. Engineering leadership plays a critical role in improving industry culture, optimizing the organization structure and increasing productivity [6].

Developing leadership has become an essential part for engineering students' skill development. Engineering students who are lack of leadership ability may encounter troubles in team activities during their professional career [7]. Universities and engineering colleges have assumed great responsibility for developing engineering students' leadership ability. Engineering leadership education programs have emerged in different universities in many countries. In this study, we compare twenty-one engineering leadership education programs in different countries by analyzing outcomes and implementation process. By so doing, we identify the similarities and differences of these programs.

Literature Review

It has been said that "an engineer is hired for her or his technical skills, fired for poor people skills, and promoted for leadership and management skills" [8]. Leadership development constitutes a critical aspect for engineers in industry. A survey data showed 77 percent of engineers indicated they played some leadership roles when they worked in an engineering team, and most of the engineers admitted that they lacked leadership and/or management skills [9]. Despite some efforts in launching engineering leadership programs, however, currently most current engineering schools still focus on students' technical skills with little attention on developing students' leadership or management skills [9] [10]. It is of great significance for universities and engineering colleges to develop leadership skills among engineering students. Future effective engineers, especially engineering leaders must possess soft skills including written and oral communication, self-initiative, teamwork abilities, customer relations and decision making [5]. To achieve these skills, universities and engineering colleges, even industries should develop engineering leadership education programs to promote engineering students to develop leadership skills.

The programs of developing engineering leadership may have various forms. Some existing universities and engineering colleges which have already developed engineering leadership programs through co-operated courses, minors, integrated curriculum, scholarship programs, certificated courses and any other ways to help students acquire leadership knowledge and principle, team work skills, communication skills, creativity and innovation. MIT-Gordon engineering leadership program (ELP) is developed through cooperating with Sloan Business School, and it is an integrated curriculum to develop MIT engineering students' leadership ability [11]. The Engineering Leadership Development Minor (ELDM) of Peen State University is a minor program that engineering students complete this minor course through taking related leadership classes and obtaining the corresponding credits [12]. The engineering leadership program (ELP) in Iowa State University is structured in two phases to enable engineering student obtain eight established learning outcomes: in phase 1 (the first academic year), engineering students are devoted to community-building, and in phase 2 (rest of the academic year), engineering students are provided with an individualized program [13]. Royal Academy of Engineering in UK established Engineering Leadership Standard/Advanced Award programs in which engineering students are encouraged to participate in a series of outdoor activities for leadership training [14]. Considering the diverse formats and unique characteristics of different leadership programs, it is useful to conduct a thorough comparative study to understand the program outcomes, curriculum design and activities across the programs.

Methods

The study was conducted mainly by text analysis. Text materials related to each leadership program were collected mainly through websites of each program. Text analysis is a process which organizes and dissects qualitative data such as words, documents and forms [14]. Since the researchers themselves are research tools in qualitative research, text analysis mostly relies on the judgment and insight of researchers [15]. ATLAS.ti 7 was used as a tool to analyze text materials. In this study, we collected text materials of twenty-one engineering leadership programs from five countries, including U.S., Canada, U.K. and so on. Outcomes for different programs and their implementation processes were analyzed and compared across programs. Differences and similarities in the program outcomes and their formats of implementation were summarized based on the analyses.

Results

Program Outcome

Based on the text analysis, we found similarities and differences between engineering leadership programs. In terms of learning outcomes, most students developed professional skills, ability to adapt to the changing environment and gained international experiences either from exchange programs or internships in global enterprises. In other words, most engineering leadership programs actively incorporated leadership skills or attributes into in-class and out of class activities. We found that vision, thinking or reasoning skills, interdisciplinary knowledge, ethics were the most outstanding features of outcomes of engineering leadership programs. (Figure 1). Detailed descriptions and comparisons of outcomes across different programs are shown in Table 1.

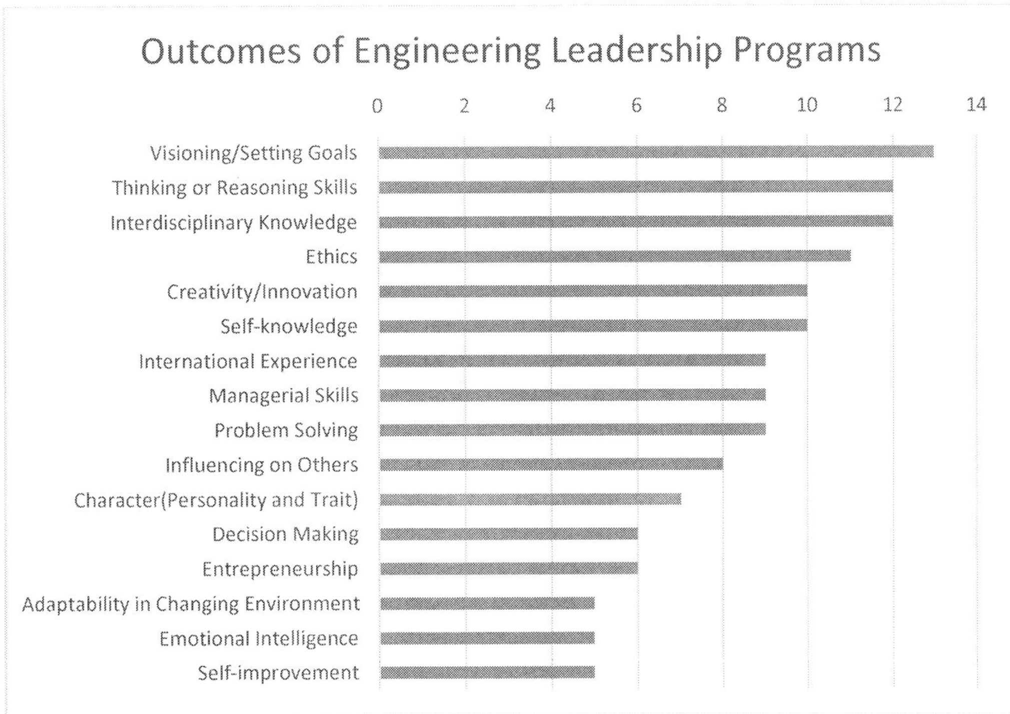

Figure 1 Outcomes defined in engineering leadership programs

Most programs paid special attention to helping engineering students develop communication/interpersonal skills and a teamwork spirit. As a result, curriculum design of engineering leadership programs aims to ensure graduates equipped with excellent communication skills to excel in their future careers. A number of programs laid emphasis on students' technical and professional excellences with regard to the learning outcomes. This is because engineering graduates who lack professional knowledge cannot become competent engineers, let alone engineering leaders. Some programs were oriented towards fostering students' vision and goal-setting skills which are vital for planning and decision-making as engineering leaders. Considering the importance of leadership foundation, many programs emphasized on students' leadership knowledge, including leadership concepts, theories and styles. Thinking or reasoning skills, such as logical reasoning, systematic thinking and analytical thinking skills, sharp observation, were highlighted by most programs. Interdisciplinary knowledge, including business, politics, law, economy and culture was also one of the key areas for these universities and engineering colleges. Additionally, we found that most programs took self-knowledge, problem solving and managerial skills into account concerning outcomes of engineering leadership programs.

Figure 1 demonstrates a list of outcomes and their frequencies across different engineering leadership programs. We also conducted a close examination of detailed outcomes across these programs (Table 1). With a combination of Figure 1 and Table 1, we found that universities, including California University of San Diego, Cornell University and Toronto University, showed at least 10 main outcomes in terms of developing engineering leadership.

Considering the differences of these programs, as mentioned before, different programs presented unique aspects. Cornell University leadership program helped students develop time management skills by introducing timeline scheme. MIT leadership program highlighted the ability of sense-making, by which engineering students should learn to analyze concrete situation when implementing project tasks and problems. Five of these programs indicated the importance of macro-ethics, which refers engineers' training should take social accountability and take sustainability into consideration. Three engineering leadership programs from Iowa State University, MIT and Rensselaer Polytechnic Institute proposed the necessity for engineering leaders to have diverse perspectives. Special outcomes may also represent the unique strengths and features of these programs. It is essential to highlight the key features and unique strength for a new leadership program whiling referring to other existent engineering leadership programs.

	1	2	3	4	5	6	7	8	9	10	11	12	13	14	15	16	17	18	19	20
California University of San Diego			√					√	√	√	√		√	√	√			√	√	√
Cornell University	√	√	√	√	√	√		√	√	√	√		√	√	√	√	√	√	√	√
Iowa State University		√	√		√			√		√	√	√	√	√				√	√	√
Kansas State University			√				√	√			√			√					√	
Kyoto University				√						√	√				√			√	√	√
Lehigh University	√		√	√	√					√					√		√	√	√	
Loughborough University			√						√	√			√	√		√		√		
MIT		√	√	√	√		√	√		√				√	√	√	√	√	√	√
Monash University	√		√	√			√	√		√		√	√				√	√		
Ohio State University		√	√			√				√			√			√		√		
Penn State University			√	√			√	√			√	√					√	√	√	√
Purdue University			√	√				√	√		√								√	
Queensland University of Technology			√					√			√	√		√					√	
Rensselaer Polytechnic Institute	√		√		√	√			√			√	√		√		√	√	√	√
Rice University	√		√				√					√	√				√	√	√	√
Royal Academy of Engineering			√	√		√			√		√		√					√		√
Southern Methodist University		√	√							√			√		√			√		√
Toronto University		√			√	√	√		√	√	√		√	√			√	√		√

	1	2	3	4	5	6	7	8	9	10	11	12	13	14	15	16	17	18	19	20
University of Central Florida			√	√					√			√	√		√		√	√	√	√
University of Colorado		√		√					√	√	√	√	√	√	√	√	√	√		
University of Michigan			√								√	√						√	√	√

Table 1 Outcomes for students as defined in the engineering leadership programs of different universities

1 Adaptability in changing environment; 2 Character (personality and trait); 3 Communication/Interpersonal skills; 4 Creativity/Innovation;5 Decision Making; 6 Emotional Intelligence; 7 Entrepreneurship; 8 Ethics; 9 Influencing on others; 10 Thinking or Reasoning skills; 11 Interdisciplinary Knowledge; 12 International Experience; 13 Leadership Knowledge; 14 Managerial Skills; 15 Problem Solving; 16 Self-improvement; 17 Self-knowledge; 18 Teamwork ; 19 Technical/Professional Excellence; 20 Visioning/Setting goals

Implementation

In order to facilitate the development of engineering students' leadership, universities and engineering colleges implemented a variety of methods. For example, capstone courses and practical projects were adopted to improve engineering students' ability of team building. Most programs conducted field trips, internships, business activities, workshops, lectures, seminars, forums and experiential programs to develop engineering leadership. A number of universities, such as California University of San Diego, Lehigh University and Monash University employed at least six different methods to conduct engineering leadership programs. Both commonalities and differences exist in the implementation of engineering leadership programs.

Many programs share some commonalities in their implementation. The first most common method to develop engineering leadership among most programs is still lecturing. Engineering students learnt leadership knowledge, interdisciplinary knowledge, engineering skills, leadership experience through lectures. Field trips (visiting industry and company, leadership journey and residential community) is the second most common method for developing engineering leadership. Through field trips, students learnt how to cooperate with team members to solve complex problems and obtained leadership and management skills. The third most common method is through practical experience, such as internships. When doing internships in enterprises, students often had opportunities to communicate with industrial specialists to gain real-world experiences. By experiential learning and practicing in an engineering environment, the goal for students is to develop problem-solving skills, communication skills, networking skills and gain knowledge in engineering and leadership.

Meanwhile, these programs also present different features in their implementation. MIT-Gordon engineering leadership program adopted a leadership model, that is, the Four-Capabilities Model, which consists of four key capabilities of future engineers, and used it as a theoretical guidance to develop students' leadership in MIT. Penn State University, University of Central Florida and Purdue University all developed engineering leadership minors for students. In addition, special sessions were arranged to develop specific skills among students. For example, to improve engineering students' management skills, Loughborough University leadership program offered a 4-day outdoor management course to cover essential fundamentals on management. Moreover, some professional activities were also organized for leadership development. For example, Rensselaer Polytechnic Institute leadership program hosted an annual one-day conference co-sponsored by industrial partners. The annual conference aimed to provide opportunities for engineering students to communicate with engineering professionals on vision, culture and value related to corporations. Cornell University leadership program had a special focus on women leaders. By offering students with courses related to gender diversity, it aimed to encourage female students to be engineering leaders.

	Business/ Entrepreneurship Program	Capstone Course	Case Study Analysis	Certification Course	Conference /Meeting/ Session	Lecture	Experiential Program	Field Activities	Group Activities	Mentoring/ Training	Practical Experience	Seminar
California University of San Diego			√			√	√			√	√	√
Cornell University				√		√	√	√	√		√	
Iowa State University		√		√		√		√			√	√
Kansas State University						√		√		√		
Kyoto University					√	√		√				√
Lehigh University			√			√	√	√		√	√	√
Loughborough University						√		√			√	
MIT						√		√	√		√	
Monash University					√	√		√	√	√	√	√
Ohio State University			√									√
Penn State University	√	√				√		√			√	√
Purdue University	√			√		√	√	√				√
Queensland University of Technology						√		√		√	√	
Rensselaer Polytechnic Institute					√	√		√		√		√
Royal Academy of Engineering						√		√	√		√	
Rice University				√		√		√			√	
Toronto University						√	√				√	√

	Business/ Entrepreneurship Program	Capstone Course	Case Study Analysis	Cer:ification Course	Conference /Meeting/ Session	Course	Experiential Program	Field Activities	Group Activities	Mentoring/ Training	Practical Experience	Seminar
Southern Methodist University							√	√		√		√
University of Central Florida				√	√	√			√		√	√
University of Colorado						√				√	√	√
University of Michigan	√	√				√		√			√	√

Table 2 Comparison of implementation measures or platforms across engineering leadership programs of different universities.

Conclusion

This paper summarized the differences and similarities of twenty-one engineering leadership programs across five countries. It presents the main characteristics and trends of engineering leadership programs by comparing their outcomes and implementation. Most programs emphasize the development of students' interpersonal skills, communication skills, visioning, influencing on others, problem-solving skills, reasoning and thinking skills, ethics and so on in their program outcomes. Meanwhile, most programs adopt varied platforms in their implementation, including field trips, entrepreneurship program. By a survey of the outcomes and implementation of these different leadership programs, this paper provides an overview of current working programs for engineering leadership development. By so doing, it offers a comprehensive list of practices for designing similar engineering leadership programs in the future.

Acknowledgement

This research was supported by Chinese Ministry of Education, Humanities Social Science Study Program (15YJC880147). Researchers would like to thank all of the interviewees who contributed to this study.

Bibliography

[1] DeLisle, P. A. (2000). Engineering leadership. *Antennas and Propagation Magazine, IEEE, 42*(1): 131-34.

[2] Crumpton-Young, L., McCauley-Bush, P., Rabelo, L., Meza, k., & Ferreras, A. (2010). Engineering leadership development programs: A look at what is needed and what is being done. *Journal of STEM Education: Innovations and Research: 11*(3/4): 10-21.

[3] *Bernard M. Gordon MIT engineering leadership program developing tomorrow's engineering leaders.* Retrieved on Apr. 1 2016 from https://gelp.mit.edu/

[4] Musselman, C. (2010). Leadership in engineering. Why is that important in engineering education? *National Society of Professional Engineers, 29.* Retrieved on Apr. 1 2016 from http://www.nspe.org/resources/blogs/pe-licensing-blog/leadership-engineering-why-important-engineering-education

[5] Lin, J., & Xu, Z. D. (2014). International engineering leadership education and its implications to China (Part 1). *Research in Higher Education of Engineering, 2*: 7-17.

[6] Morrison, R. E., & Ericsson, C. W. (2006). *Developing Effective Engineering Leadership*. London: The Institution of Engineering and Technology.

[7] McGuigan, A. P. (2010). Beyond graduation — the importance of continued leadership development.*Engineering Leadership Review, 1*(1): 29-44.

[8] Russell, J. S., & Yao, J. T. P. (1996). Education conference delivers initiatives. *Journal of Management in Engineering - ASCE, 12*(6): 17-26.

[9] Clough, G. W. (2004). *The engineer of 2020: Vision of engineering in the new century*. Washington:

National Academy of Engineering.
[10] Spinks, N., Silburn, N., & Birchall, D. (2006). *Educating the engineers for the 21st century- the industry view*. London: The Royal Academy of Engineering.
[11] Graham, R., Crawley, E., & Mendelsohn, B, R. (2009). Engineering leadership education: A snapshot review of international good practice. Cambridge, MA: Bernard M. Gotdon – MITEngineering Leadership Program.
[12] Schuhmann, R. J. (2010). Engineering leadership education the search for definition and a curricular approach. *Journal of STEM education: Innovations and research, 11*(3/4): 61-69.
[13] Athreya, K. S., & Kalkhoff, M. T. (2010). The engineering leadership program: A co-curricular learning environment by and for students. *Journal of STEM Education: Innovations and Research, 11*(3/4): 70.
[14] Popping, R. (2013). Analyzing open-ended questions by means of text analysis procedures. *Bulletin of Sociological Methodology / Bulletin de Methodologie Sociologique, 128*(1): 23-39.
[15] Fink, A. S. (2000). The role of the researcher in the qualitative research process. A potential barrier to archiving qualitative data. *Forum Qualitative Sozialforschung / Forum: Qualitative Social Research, 1*(3). Retrieved on Apr. 1 2016 from http://www.qualitative-research.net/index.php/fqs/article/view/1021

The Attributes of Future 2030 Engineers in Qatar for Innovation and Knowledge Based Economy

Dr. Mahmoud Abdulwahed, Qatar University

Dr Mahmoud holds BSc, MSc, and PhD in electrical, control and systems engineering; he completed his postgraduate studies in Germany, Sweden, and the UK. His main expertise is in Innovation, Transfer, and Education where he is ranked 14th Globally in the field of Engineering Education in last three years in terms of indexed publications in Scopus (2013-2016 period). Mahmoud worked in the UK before joining the College of Engineering at Qatar University in Fall 2011 as a faculty member, he was seconded from the College to the Office of Vice President and Chief Academic Officer in Spring 2013 where he worked as Adviser and then Manager of Strategic Initiatives & Innovations; his role focused on inception, coordination, and/or management of several strategic initiatives and innovation based practices across campus and in selected colleges. These included "Foundation of Excellence Initiative", "First Year Experience Implement Initiative", "Students At-Risk Initiative", "Technology Enhanced Learning Initiative", etc.; he also drove inception/approvals of several entities including the "Office of Strategic Initiatives & Innovations" with the VP & CAO office, the "Instructional Technology Support Unit" with OFID, and worked with the College of Engineering on inception/approval of the "Technology Innovation & Engineering Education (TIEE) Department ". Mahmoud received several fellowships, from University of Technology Sydney, University Science Malaysia, and from USA Department of State (DoS) Professional Fellows program. He published 60+ peer-reviewed conference and journal articles, and attained a number of industry research funds, academic recognitions, awards, and best papers distinctions. He published on aspects related to Internet of Things (IoT), digitally enabled learning, innovation, entrepreneurship, leadership, design, ethics, constructivism, competencies, Knowledge Based Economy/Society (KBEs), etc.

The Attributes of Future 2030 Engineers in Qatar for Innovation and Knowledge Based Economy

Abstract:

Recent studies emphasize the needs of a wider set of skills engineers require than ever thought. These studies continue to emphasize that technical content knowledge and competencies are essential for any engineer; however, in addition to being well-grounded in mathematics and science, 21st century engineers should be well-shaped in broader knowledge-base and diverse personal/ interpersonal key-skills.

Qatar is increasingly investing in knowledge development within its boundary, either through home grown manpower or through collaborative efforts from research centres from around the world.

Interviews with students, faculty, and representatives from the major industrial sectors in Qatar have been conducted. Quantitative analysis relied on comprehensive survey with multiple stakeholders from industry and academia, while qualitative analysis is constructed based on content analysis and grounded theory approaches.

In this paper, authors report on the empirical study investigating the needs of skills in engineering in Qatar in the future for a knowledge based economy. Gaps have been identified, conceptualization is developed, and recommendations have been drawn upon.

Introduction

In facing today's globalized environment, the focus on maintaining or transforming into a knowledge-based economy has been significant in many countries. Innovation driven economy is considered as the highest level of economic development, where countries focus on the development and use of cutting edge technology, acquiring the best talent, grooming them and retaining them to excel in value addition through innovation within and outside the nation. The use of cutting edge technology and expertise refers to what we call the emphasis on knowledge-based economy (KBE).

The cornerstone role of technology and engineering in driving KBEs, the industry needs, and the emergence of engineering and technology profession has led to significant number of studies, efforts, and investigations for identifying the needed engineering talent in the 21st century (NAE, 2004 & 2005; Jamieson *et al.*, 2009; Mena *et al.*, 2012; Hundley *et al.,* 2012; Knight, 2012; Spinks *et al.*, 2006; Bourn and Neal, 2008). In the Middle East, the GCC, and Qatar an investigation similarly has not been conducted, hence this study sheds an important perspectives bridging an empirical gap in the literature and contributing to better design of future engineering education in Qatar.

There are four main dimensions that are driving Qatar's economy into a KBE: 1- Qatar National Vision 2030, 2- World Cup 2022 and the associated mega projects, 3- Instability of Oil and Gas prices, and 4- Regional growth and competition for skills, products, and investments in the GCC region mainly led by UAE and Saudi Arabia.

Qatar is increasingly investing in knowledge development within its boundary, either through home grown manpower or through collaborative efforts from research centres from around the world. The country's investment in research and development, education and information and communication technology is expected to create the environment for knowledge based economy formation. The promotion of new economy is also supported through the establishment of international universities, science and technology parks, tax-exempt free zones, and relaxed employment recruitments in the tax free zones.

Since engineering and technology are the very heart of innovation and transformation in a knowledge based economy, producing the next generation of engineering talent with needed set of skills is crucial. Hence, is the importance of this investigation conducted in Qatar in light of contextual needs and global advances in engineering innovation and education.

Purpose of the Study

To enable the national shift toward a KBE, Qatar Engineering Education System is expected to advance and graduate engineers who are skilfully prepared for the future practice of 2030. For this to happen, future engineer attributes need to be identified in the context of Qatar and roadmap to achieve them should be developed.

The study aimed to achieve the following objectives:

- Identify attributes required by entry-level graduate engineers in the industry currently and in the future by 2030
- Propose recommendations for the engineering education system to better prepare skilled graduates based on the identified gaps

Global Competency Framework

A comprehensive literature review, reported in details in Abdulwahed et al. (2013), revealed several national investigations of engineering skills needs conducted in countries across the world, such as USA, Australia, UK, Malaysia; etc. Synthesis of skills and competencies from over 200 studies was performedd, and led to categorize the skills founded into a set of 22 items of global skills under 4 main dimensions:

1- **Dimension I - Core Knowledge and Practice**, which includes the following competencies: 1- Science knowledge (Math, Physics & Science Fundamentals); 2- Disciplinary fundamentals, 3- Interdisciplinary fundamentals, 4- Multidisciplinary knowledge, 5- Practical experience, and 6- ICT skills

2- **Dimension II – Cognition, Mental, and Thinking**, which includes the following competencies: 1- Lifelong learning, 2- Problem solving, 3- Decision making, 4- Analytical thinking, 5- Systems thinking, 6- Critical thinking, 7- Creative & Innovation, 8- Design

3- **Dimension III – Professional and Interpersonal**, which includes the following competencies: 1- Professionalism, 2- Ethics and Responsibility, 3- Adaptability, 4- Communications, 5- Teamwork, 6- Foreign languages

4- **Dimension IV – Business and Management**, which includes the following competencies: Management, Leadership, and Entrepreneurship

The developed model aforementioned before was utilized as a base of quantitative investigation using surveys.

Methodology

A comprehensive surveys (included 20+ questions such as demographics, open-ended, and Likert-scale up type questions) have been developed and delivered to the targeted stakeholders groups. Surveys are one of the widely used methodologies in conducting research due to their practicality.

Stakeholders were asked to evaluate a set of 22 engineering skills on two main levels:

- Skills importance currently.
- Skills importance in 2030.

Stakeholders were asked to provide their answers on a Likert scale from 1 to 5 where: 1= "Strongly Disagree", 2= "Disagree", 3= "Neutral", 4= "Agree", and 5= "Strongly Agree".

Future Trends of Skills Demands in in Qatar by 2030

The majority of stakeholders (industry, faculty, students) expected changes in the demand of competencies set of engineering graduates in Qatar in the future by 2030. In particular qualitative feedback from stakeholder highlighted the following set to be of higher demand: more demanded by various stakeholders: ICT skills, management, leadership, entrepreneurship, innovation, problem solving, research, practical skills, communications, teamwork, and decision making.

In quantitative analysis overall, all temporal gaps for all stakeholders (industry, faculty, and students) were in the positive side indicating a higher importance or demand of engineering graduates competencies in the future (2030) than currently. All temporal gaps recorded a statistically significant difference, the gaps were relatively higher for the expert group (industry and faculty) than those reported by students. Industry in general scored higher means than faculty (faculty scored higher than students) on the future (2030) importance of practical skills, and the majority of Dimension 2 (cognition, thinking, and mental), Dimension 3 (professional and interpersonal skills), and Dimension 3 (Business and Management); however no statistically significant difference was detected between faculty and industry.

It worth mentioning that analysis of the top temporal gaps reported by the expert group (industry and faculty) have shown tendency to value the importance KBE oriented competencies higher in the future than other evaluated skills (temporal gaps were top highest for these competencies). Expert group in particular reported highest temporal gaps in highly correlated KBE competencies such as: innovation, entrepreneurship, leadership (and related attributes such as decision making, adaptability, and system thinking), management, interdisciplinary knowledge, multidisciplinary knowledge, design, and problem solving. See Table 1 for further details.

Table 1: Stakeholders Temporal gaps

Temporal gap	
Industry	**Faculty**
Management (0.46)	Decision making (0.65)
Leadership (0.46)	management (0.56)
Entrepreneurship (0.45)	Adaptability (0.56)
Interdisciplinary (0.43)	Entrepreneurship (0.52)
Multidisciplinary (0.36)	Innovation (0.51)
Innovation (0.35)	Problem solving (0.49)
Design/ Decision making (0.34)	System thinking (0.49)

Recommendations, and Conclusions

This study provided empirical analysis on engineering talent competency needs in Qatar. Targeted stakeholders (Industry, academia & students) were asked to evaluate a set of 22 engineering skills on three main levels; Importance currently, and in the future.

Communications, teamwork, management and practical experience were indeed on the very top competencies industry required in engineering graduates. Skills temporal gaps have shown higher tendency to value the importance of skills and competencies to Qatar`s Future. Respondent from all stakeholders expected changes in the demand of competencies set of engineering graduates in Qatar in the future by 2030. The engineering education system will need to provide integrated engineering education curriculum that responds to current needs and future evolutions. As for the current needs, more emphasize on practical experience and professional skills such as communications and teamwork seems essential. This would need to be accompanied by developing modules, courses, programs and activities with more focus on innovation, design, leadership, and technology entrepreneurship to accommodate future emerging needs of the country in light of vision to develop into a knowledge based economy by 2030.

References

Abdulwahed, M., Balid, W., Hasna, M. O., & Pokharel, S., 2013. *Skills of engineers in knowledge based economies: A comprehensive literature review, and model development. In Teaching, Assessment and Learning for Engineering (TALE), 2013 IEEE International Conference on (pp. 759-765). IEEE.*

Bourn, D., & Neal, I. (2008). *The global engineer: incorporating global skills within UK higher education of engineers. Engineer.*

Jamieson,L.,Brophy, S., House, N., Harris, M., Delaurentis, D., Howell, K., Benner, A., Cekic, O., Moseir,N., Okos, M., (2009) *"Purdue's Engineer Of 2020: The Journey," in 2009 Annual Conference & Exposition; New Learning Paradigms II.*

Hundley, S. (2012). AC 2012-4233 : ATTRIBUTES OF A GLOBAL ENGINEER : FIELD-INFORMED PERSPECTIVES , RECOMMENDATIONS , AND IMPLICATIONS. Retrieved from http://www.asee.org/public/conferences/8/papers/4233/view

Knight, D. B. (2012). In Search of the Engineers of 2020: An Outcome-Based Typology of Engineering Undergraduates, AC 2012-3337. In *Proceedings of the 119th Annual Conference of the American Society for Engineering Education,*. San Antonio, TX.,.

Mena I.B., Zappe, S. E. , and Litzinger, T. A. , *(2012), "Preparing the Engineer of 2020: Analysis of Alumni Data," in 2012 ASEE Annual Conference, American Society for Engineering Education.*

NAE. (2004). *The Engineer of 2020: Visions of Engineering in the New Century.*

NAE. (2005). Educating the engineer of 2020: Adapting engineering education to the new century. *IEEE Engineering Management Review*, 1–208. doi:10.1109/EMR.2009.4804343

Spinks, N., Silburn, N., & Birchall, D. (2006). *Educating Engineers for the 21st Century : The Industry View. Engineering*. Greenlands, UK.

Empowering Women in Engineering

Dr. Saud A Ghani, Qatar University

•Associate Professor •Associate Dean for development and industrial relations. •Ph.D., in Mechanical Engineering, School of Mechanical, Materials and Manufacturing Engineering, University of Nottingham (UK), 2002.• •Research: Thermofluids, CFD, Experimental Fluid Mechanics and Flow Visualization, Biomedical Engineering, Airflow Through Human Airways, and CSF flow. Engineering Education

Women and Engineering in Qatar

Foteini Bakochristou, Esmail ElBialy, Waled Mukahal, Saud Ghani

Introduction

Engineering is a STEM subject (Science, Technology, Engineering and Mathematics) offering to young students the equipment, skills and knowledge to evaluate evidence, make sense of information and solve problems **[1]**. Therefore the ability to solve problems creatively has been identified as one of the imperative competencies for graduating students. The need to engage more female students to study STEM subjects is a worldwide concern and was also highlighted by US President Barak Obama **[2]**. By providing knowledge platforms, regardless gender differentiations, helps scientific society to broadness its achievements and consequently improve the industry outcomes. Gender discriminations in educational and occupational field though, have always been under question and investigation. This paper is an attempt to present efforts, motives and incentives that help Qatari female students to excel in their overall engineering studies and career through promoting participation in events, conferences and external competitions.

Educating and Inspiring Female Engineers

Primary education and school policies targeting in promoting females' interest in engineering followed by further innovative university education will assist future Qatari female engineering students to reveal and apply more qualification and abilities. US Department of Education, stated that during 1990-2005 in the USA there has been a continuous improvement at the high school girls' average grade point average (GPA) in math and science, lessons that are preparatory and basic for obtaining a future bachelor's degree in engineering **[3].** Hill et. al (2010)stated that girls are benefited in math and science, due to a bigger mentality development, in a USA report **[4]** was stated that in K-12 girls take as many as high level mathematics and science courses like same aged boys, and in Australia there has been a substantial increase to female students, with high grades in math and science subjects (Gill et al. 2008). Hacker (1981) described the role of math performance as a ''critical filter'' in succeeding in engineering. Robinson and Mcilwee (1991) confirmed women's superiority in math performance and (Jagacinski and LeBold 1981) argued that females can demonstrate equal technical proficiency as men, preventing stereotypes and misconceptions regarding women **[5].** Figure 1 describes the GPA of male and female high school graduates in math and science, for a period of 15 years, in USA; Figure 2, the number of women being awarded engineering degrees, proving that engineering is not a sex segregated field of work nor a male orientated field of education (Youn and Choi 2015). Highlighting STEM women's devotion to their careers, their mathematical abilities can be beneficial to their advanced confidence (Oh and Lewis 2011). Regarding professional performance, (McIlwee and Robinson 1992) stated that both genders display equal desires and wishes when it comes to get involved in strategic projects, applying practical contribution and obtaining administrative knowledge.

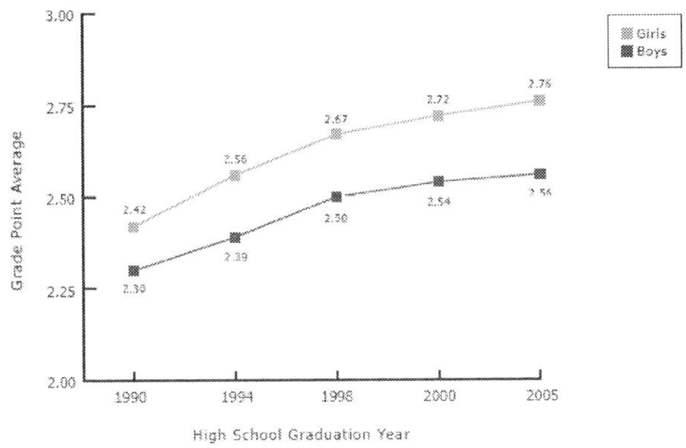

Figure 1. GPA in Math-Science [6]

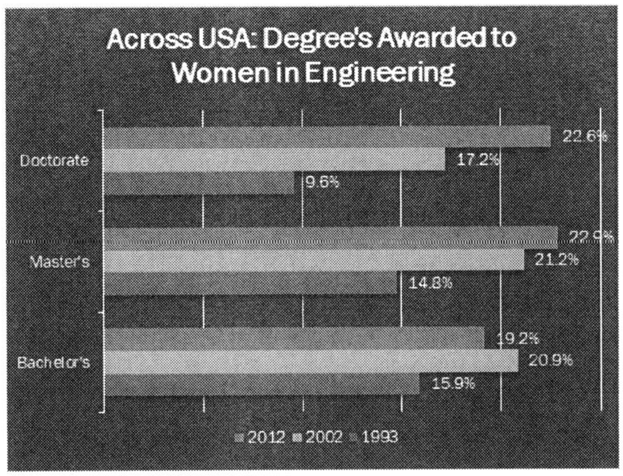

Figure 2. Women awards in Engineering [7]

The importance of family and working environment was also examined; engineers' family and friend background is more supportive than men's, whereas men are considered to find obstacles mainly from family (Inda, Rodríguez, and Peña 2013). Lent et al. (2005) confirmed that females receive more support and less barriers than male. Women can be identified as more systematic, better at listening and taking notes and more selfless (Stonyer 2002). In India (Escueta, Saxena, and Aggarwal 2013), reported, females' advanced confidence, respect and motivation is useful in all kind of working fields; specially women undergraduate engineers are more confident when compared with males of other discipline groups. Khazanee (1996) referred to female engineers, such as less aggressive attitude towards colleagues than males, tendency of listening more and acting not spontaneously, and being more attentive, accurate, and organized. Additionally, successful ability writing in math and science by females, can prove advantageous in all academic fields (Halpern et al. 2007), while introduction of unique and important perspectives and priorities will conclude into positive social outcomes and greater ethical accountability (Eagly and Carli 2003). Females are characterized as high self-confident regarding keeping up with the company's values and ethics, dealing with multiple roles and generally accomplishing different tasks (multirole behavior), can result into less friction experiences in both work and non-work obligations. The importance of promoting more women into leadership (Hoyt and Murphy 2016) and engineering roles is greater than just fulfilling the promise of equal opportunity and making businesses, institutions, and governments more representative. Other assets, include advanced self confidence in processing tasks, in navigating organizational culture/climate and enhanced ability to manage multiple life roles. Finally, academically oriented women are characterized by broad interests, emotional sensitiveness and tolerance of intellectual creativity (McIlwee and Robinson 1992), aspects that assist for a successful engineering career.

As a result, an overall advanced education system, enhanced with chances of participation in conferences and contests can prove to be well suited for females, for enhancing institutional productivity and boosting future engineering practices through articular skills, efficiencies and efficacies. Actions may include R&D tasks in science, math and engineering classes, among Qatari women and girls at all levels, by government, educational, corporate and professional engineering institutions groups. Qatar's current boosting in construction and infrastructure is a valuable asset for motivating females in following engineering paths.

Making the Difference by Changing the Culture

Members of the IEEE and the IEEE Women in Engineering (WIE) have launched a three stage project towards encouraging females for following engineering professions. In the first stage, females should be getting familiar with engineering and scientific achievements from an early age and in high frequency. Secondly, advanced recruitment programs emphasizing on electrical and mechanical engineering careers before enrolment in undergraduate majors can assist females of following such paths and break down of the social barriers.

In Rochester Institute of Technology **[8]**, there is an organized girl engineers' camp where school girls participate in communities and events getting more familiar with STEM field. Women in Engineering Society **[9]**, declared the 23rd of June as the ''National Women in Engineering Day'' for raising awareness and marking female engineering accomplishments and STEMNET has created the ''STEM Ambassador'' **[10]** role. The University of New South Wales (UNSW) **[11]** organizes specially designed for girls visits events (workshops on campus and camps) and scholarships to women applicants. Washington State University **[12],** has its own Society of Women Engineers Officers (SWE) that organizes ''Kids Engineering Day'' for local elementary schools. Other international organizations that promote female engineers are Society of Women Engineers **[13]**, Women's Engineering Society **[9]**, National Girls Collaborative Project **[14],** and IEEE Women in Engineering (WIE) **[15]** including IEEE for girls, Engineer Girl, Women of NASA Projects and Nerd Girl projects. Other examples can be found in Scotland **[16]**, where all engineers involved in the construction of Forth Bridge were strictly females, during World War I-World War II where female enrolling in engineering popular schools was described by ''invasion'' (Bix 2000), and boosting productivity (Douie 1950; Hart 2007).

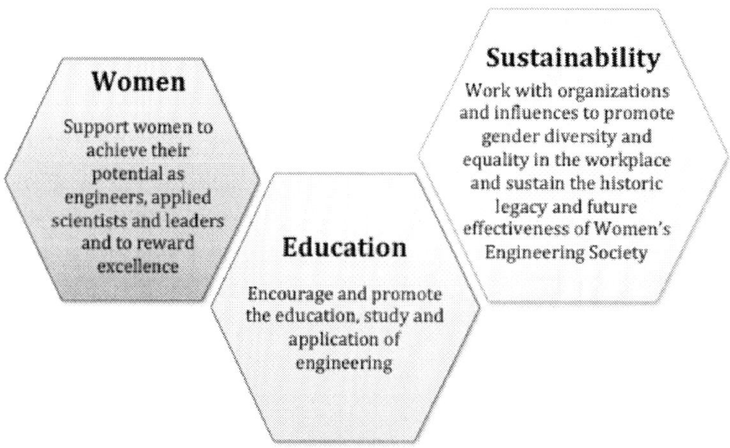

Figure 1. Women Engineering Society's targets [9]

All efforts are considered as paths to proficiency through learn from disapproval and be inspired by others' success. As a result, although negative stereotypes question females' abilities, they manage to endeavor, surpass all difficulties and strengthen their personality through stages of maturity.

College of Engineering (CENG) – Qatar University

Qatar University (QU) is the first national College of Education of the State of Qatar founded in 1973. Hence in 1977, Qatar University was founded with four new colleges, namely Education, Humanities & Social Sciences, Sharia, Law & Islamic Studies, and Science. The College of Engineering was later established in 1980. In its first year the college admitted 57 male and 93 female students. Currently, CENG is a house of six academic departments for undergraduate programs: Architecture and Urban Planning, Computer Engineering and Industrial Engineering, Civil Engineering, Mechanical Engineering, Chemical Engineering, Electrical Engineering and Computer Science. Over the years CENG experienced slow enrolment growth until women were admitted in 2002. CENG displays active participation in research, by establishing collaborations with more than 70 Qatari and international research institutions. [**17**].

Figures 4-9, describe the demographic profile of CENG students by gender.

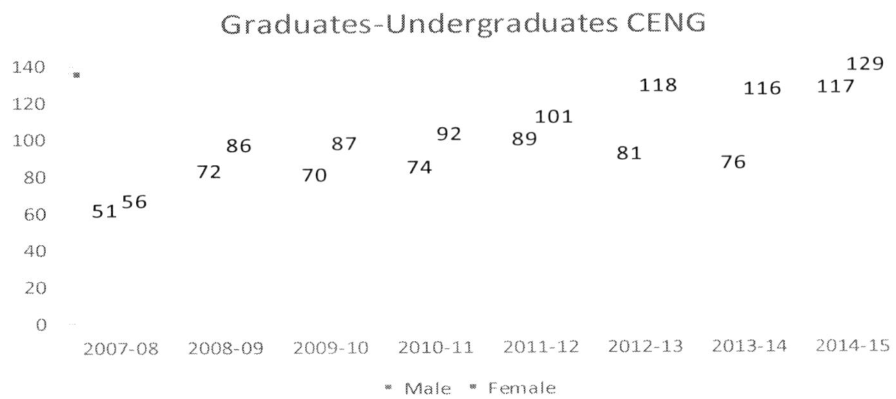

Figure 4.Number of CENG Graduates –Undergraduates Students

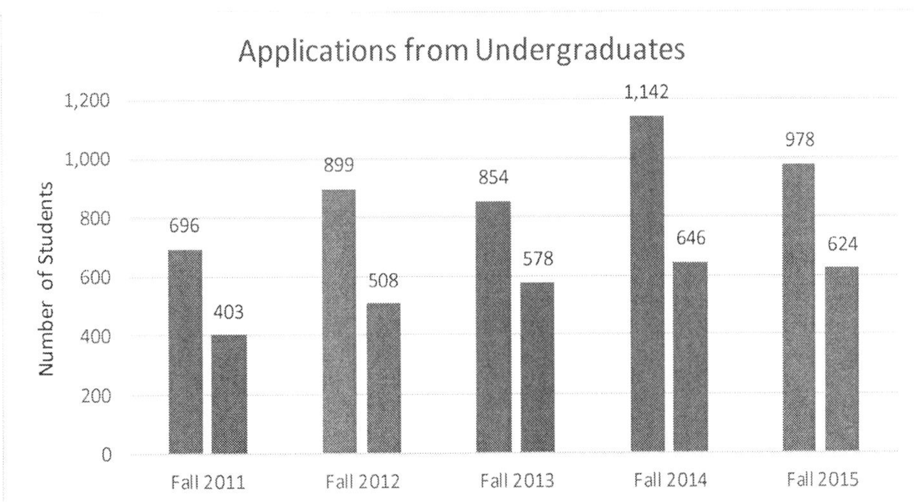

Figure 5. Number of students applying at CENG

Gender classification (Figure 4), reveals that female presence in CENG is continuously rising throughout all 8 academic years (from 2007-2015) by 130% with almost the same rate as males (129%) with 2013-2014 the year were females were found to be 40 more than males. From 2011 until 2014, female applicants in CENG have been steadily growing with a rate of approximately 60%, while the overall increase from 2011 to 2015 was 55% (Figure 5).

Throughout 2011-2014 among registered students, males were continuously more than females, until Fall 2015 when for the first time females were 19 more than males (Figure 6)

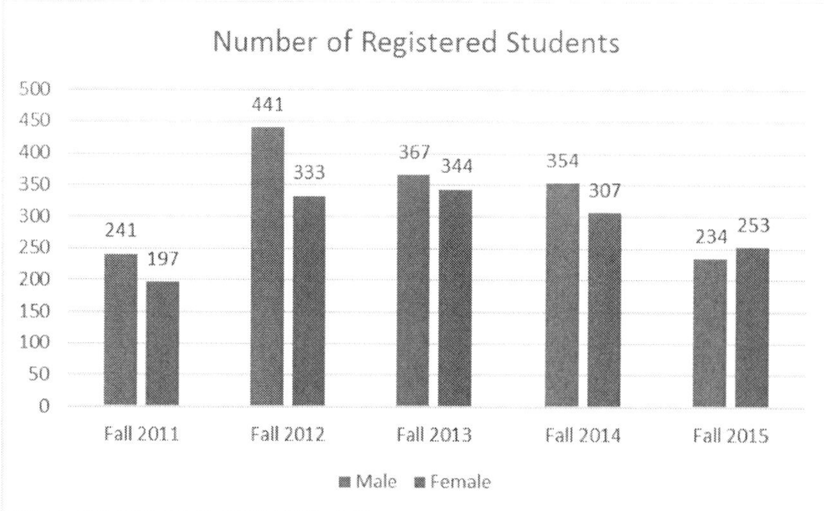

Figure 6. Number of Registered Students

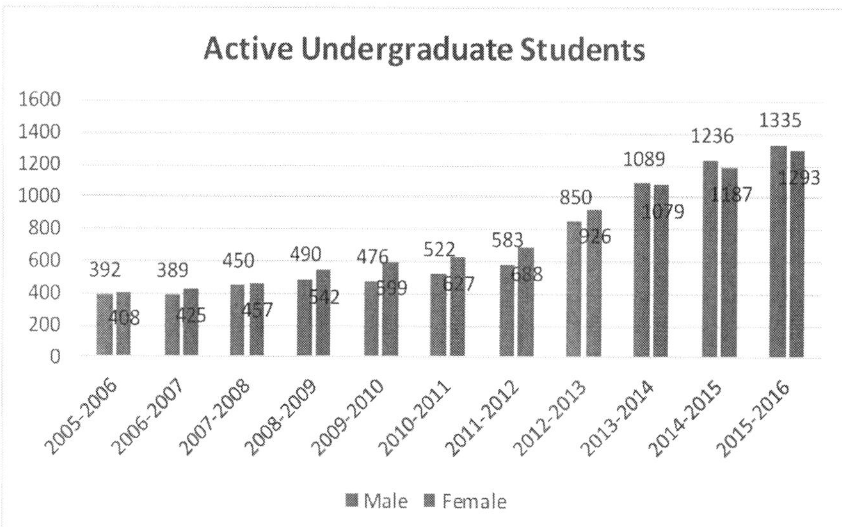

Figure 7.Number of Active undergraduate students

From 2005 until 2013 female engineers appear to be more active than their male peers with the difference in 2010-2012 period to be up to 150 more females. The very next academic year this difference fell (to 76) whereas in 2014 until 2016 males appeared to be more active than females (Figure 7).

Advanced extracurricular participation, higher civic engagement and activism can lead to advanced academic performance (Perez et al. 2010), by inspiring students with higher self-esteem, and motivation to continue learning (Eccles and Barber 1999). Under this light, CENG encouraged students' participation in contests, training and conferences, that referred to locally, regionally and internationally actions, specially designed to promote expertise among engineering students.

Figure 8 illustrates the number of students participating in contests, training and conferences and Figure 9 the obtained GPAs by both male and female students. In 2013-2014 although participating males were doubled than females, the latter obtained overall higher GPAs over two continuous academic years. Experience gained from such events benefited girls to perform more adequately than boys, although lower percentage of participation. In 2014-2015 female participation was more than doubled exceeding that of males, with the GPA performance to be increased as well up to ''B'' grade, proving that female engineering dynamics is continuously grown and developed through such participations. During the present academic year females' event and competition involvement has already surpassed that of males, indicating higher future GPAs.

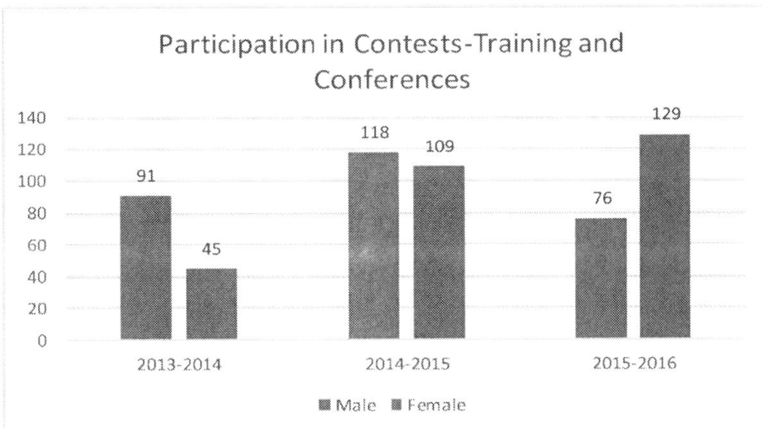

Figure 8. Number of Student Participants in Contests Training and Conferences

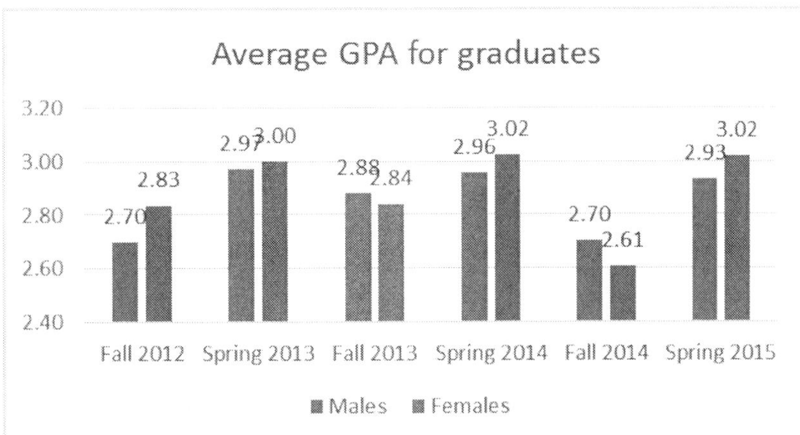

Figure 9. Average GPA by Gender

Conclusions

The importance of the research reveals the advantages and special characteristics that describe women engineers. High self-esteem, confidence, self-reliance, understanding, compassion, advanced sense of ethics, interests in researching, and designing, advanced competitiveness in software operating, and ability in troubleshooting problems, are the most salient and noticeable domains that make female engineers equal or even higher than their male colleagues capable of fulfilling demanding projects. Focusing on the present Qatari industrial and constructing based economy, the nurturing of female engineers is a task of high importance; emergence of 'alpha girls' and female emotional engineering (through emotional business and blue ocean strategies), will lead the industries in the future and help them face the challenges. Predominate misleading perceptions among teenagers that math and science prowess is inherent and therefore engineering is gender oriented are being degrade; engineers are made, not born, and recruiters should inculcate females towards this direction. All in all, the key issue appears to be 'motivation' (Fingleton et al. 2014).

References

Bix, Amy Sue. 2000. "Engineering Education in the United States Has a Gendered." *IEEE, Technology and Society Magazine* 19(1):20–26.

Douie, Vera. 1950. *Daughters of Britain*. edited by Ronald. London.

Eagly, Alice H. and Linda L. Carli. 2003. "The Female Leadership Advantage: An Evaluation of the Evidence." *The Leadership Quarterly* 14(6):807–34.

Eccles, J. S. and B. L. Barber. 1999. "Student Council, Volunteering, Basketball, or Marching Band: What Kind of Extracurricular Involvement Matters?" *Journal of Adolescent Research* 14(2012):10–43.

Escueta, Maya, Tushar Saxena, and Varun Aggarwal. 2013. *Women in Engineering : A Comparative Study of Barriers across Nations.*

Fingleton, Ann, Michael Loughnane, Louise McGuinness, and Killian McKenna. 2014. *UCD Engineering Graduates Association: Towards Gender Balance in Engineering.*

Gill, Judith, Rhonda Sharp, Julie Mills, and Suzanne Franzway. 2008. "I Still Wanna Be an Engineer! Women, Education and the Engineering Profession." *European Journal of Engineering Education* 33(May):391–402.

Hacker, Sally L. 1981. "The Culture of Engineering: Woman, Workplace and Machine." *Women's Studies International Quarterly* 4(3):341–53.

Halpern, Diane F. et al. 2007. "The Science of Sex Differences in Science and Mathematics. [References]." *Psychological Science in the Public Interest* 1–51.

Hart, Robert A. 2007. "Women Doing Men's Work and Women Doing Women's Work: Female Work and Pay in British Wartime Engineering." *Explorations in Economic History* 44(1):114–30.

Hill, Catherine, Christianne Corbet, and Andresse Rose. 2010. *Why So Few Women in Science, Technology, Engineering and Mathematics?* Washington, United States of America.

Hoyt, Crystal L. and Susan E. Murphy. 2016. "Managing to Clear the Air: Stereotype Threat, Women, and Leadership." *The Leadership Quarterly.*

Inda, Mercedes, Carmen Rodríguez, and José Vicente Peña. 2013. "Gender Differences in Applying Social Cognitive Career Theory in Engineering Students." *Journal of Vocational Behavior* 83(3):346–55.

Jagacinski, Carolyn and William LeBold. 1981. "A Comparison of Men and Women Undergraduate and Professional Engineers." *Engineering Education* 72(3):213–20.

Khazanee, By V. L. 1996. "Women in Civil Engineering and Science : It ' S Time for Recognition and Promotion." 122(April):65–68.

Lent, Robert W. et al. 2005. "Social Cognitive Predictors of Academic Interests and Goals in Engineering: Utility for Women and Students at Historically Black Universities." *Journal of Counseling Psychology* 52(1):84–92.

McIlwee, Judith and Gregg Robinson. 1992. *Women in Engineering:Gender, Power and Workplace Culture*. New York, New York, USA: State University of New York.

Oh, Seong Soo and Gregory B. Lewis. 2011. "Stemming Inequality? Employment and Pay of Female and Minority Scientists and Engineers." *The Social Science Journal* 48(2):397–403.

Perez, W., R. Espinoza, K. Ramos, H. Coronado, and R. Cortes. 2010. "Civic Engagement Patterns of Undocumented Mexican Students." *Journal of Hispanic Higher Education* 9(3):245–65.

Robinson, Gregg and Judith Mcilwee. 1991. "Men, Women, and the Culture of Engineering." *Sociological Quarterly* 32(3):403–21.

Stonyer, Heather. 2002. "Making Engineering Students-Making Women: The Discursive Context of Engineering Education." *International Journal of Engineering Education* 18(4):392–99.

Youn, Jong-Tae and Song-Ah Choi. 2015. "Women Included Engineering Education in Korea." *Procedia - Social and Behavioral Sciences* 174:1678–83.

Web References

1. USA Department of education, http://www.ed.gov/stem
2. The White House, Office of Science and Technology Policy, https://www.whitehouse.gov/administration/eop/ostp/women
3. U.S. Department of Education, National Center for Education Statistics, 2007, The Nation's Report Card: America's high school graduates: Results from the 2005 NAEP High School Transcript Study, by C. Shettle et al. (NCES 2007-467) (Washington, DC: Government Printing Office).
4. The state of girls and women in STEM, June 2013 http://ngcproject.org/sites/default/files/documents/ngcpstemstats_web.pdf
5. Atkins 2013: Britain's got talented female engineers, http://www.flipsnack.com/95A89DCF8D6/britains-got-talented-female-engineers.html
6. Why So Few? Women in Science, Technology, Engineering, and Mathematics, The American Association of University Women, www.aauw.org
7. Research Center for Science and Mathematics Education at the Universitat Autonoma de Barcelona
8. Rochester Institute of Technology, Women in Engineering, https://www.rit.edu/kgcoe/women/highlight/encouraging-more-girls-choose-career-engineering
9. http://wes.org.uk/NWED
10. STEMNET, http://www.stemnet.org.uk/ambassadors/what-do-stem-ambassadors-do/
11. University of New South Wales, https://www.unsw.edu.au/about-us
12. https://vcea.wsu.edu/swe/
13. http://societyofwomenengineers.swe.org/
14. https://ngcproject.org
15. https://www.ieee.org/membership_services/membership/women/women_in_engineering.html
16. Edinburgh News, Forth ''Bridge used to encourage female engineers'' by Alistair Grant, 21 Sept 2015 http://www.edinburghnews.scotsman.com/news/transport/forth-bridge-used-to-encourage-female-engineers-1-3893246
17. Qatar University History Booklet, http://www.qu.edu.qa/theuniversity/history.php

Australasian Partnership in a First Year Engineering Course: Deakin University and Wuhan University of Science and Technology

Dr. Sivachandran Chandrasekaran P.E., Deakin University

Dr. Sivachandran Chandrasekaran is a Research Fellow in Engineering Education at Deakin University. He has graduated his BE (CSE) in India and ME, MES (Electronics) from Victoria University and PhD (Engineering Education) from Deakin University respectively. He is active member of Deakin engineering education research Centre (DEERC), School of engineering in the Faculty of science, Engineering and Built Environment at Deakin University. Siva is an active researcher and his research interests include creativity and innovation in learning and teaching, Design based learning, Cloud learning & located learning and engineering education innovation. His education philosophy is founded on the Project Oriented Design Based Learning (PODBL) approach at Deakin University.

Dr. John Matthew Long, Deakin University

Dr. John M. Long completed his undergraduate degree in physics at the University of Michigan (Flint) in 1987, while working as an analytical chemist at AC Spark Plug, General Motors Corporation. In 1995 he completed a PhD in physics at Monash University in Melbourne, Australia. Since then he has worked in the School of Engineering at Deakin University, where he teaches physics, materials, and electronics. His research interests include materials-analysis techniques and engineering education.

Dr. Yanan Wang P.E., School of Engineering, Deakin University, Waurn Ponds, Geelong, Victoria, Australia
Dr. Junior Nomani JN, Deakin University
Prof. Qiang Zhao
Prof. Zhigang Jiang, Wuhan University of Science and Technology
Prof. Roger Dianlei Geng PhD, Wuhan University of Science and Technology

Education: 2008-2015 PhD: Shanghai International Studies University Work: 2010-Present Professor of English and Cultural Studies Director, Office of International Relations/Dean, International School Wuhan University of Science and Technology, China

Prof. Bernard Rolfe, Deakin University

Bernard Rolfe completed a combined Economics and Engineering degree with honours in 1995 from the Australian National University (ANU). During his degree he was employed as a graduate research engineer at the BHP Research Labs, investigating better control systems for hot strip steel mills. After he obtained his degree he worked for several years as a Business Systems consultant with Andersen Consulting (Accenture) before returning to the ANU to pursue a PhD in novel methods of inverse modeling metal forming processes (completed in 2002). This research included an IMechE award winning journal paper. In 2005 Bernard joined Deakin as a Lecturer in Mechanical Engineering. In 2008 Bernard led the Deakin's successful team for "Creating the Model-T for the 21st Century – a Global University Challenge". This vehicle concept outclassed several other top Universities around the world to be awarded joint winner along with the concept from the University of Aachen (RWTH), along with $25000 prize money. In 2015 Bernard received the Vice Chancellor's award for Industry Engagement. Currently Bernard is an Associate Professor (Mechanical) at Deakin University in Australia. This is equivalent to a tenured Professor in the US system. He has been a part of over fifteen successful nationally competitive large research grants, totalling over $15 million in awarded funds. He has published over 150 refereed articles. His current research focus is the forming of light weight structures, including the development of better material models for metal forming.

Australasian Partnership in a First-Year Engineering Course: Deakin University and Wuhan University of Science and Technology

Abstract

This paper presents the results of domestic Chinese undergraduate engineering course taught by international Australasian teaching staff. The project is a part of a teaching collaboration between Deakin University and Wuhan University of Science and Technology. The cohort of students from Wuhan was a freshman undergraduate engineering course in mechanical engineering. The particular subject was a freshman engineering-materials course taught in English. The course covered an introduction to material-science principles and practices. A survey was used for evaluating student perceptions. It is aimed that this study will help academics from Deakin University to better understand student experiences, and to identify the current challenges and barriers faced in student learning. Analysis of the survey has shown that 90% of students agreed that they were motivated to learn and achieve the learning goals through this collaborative program. Around 90% of students found that group-based practical activities were helpful in achieving learning goals. Overall, 90% of students strongly agreed they were satisfied with the method of teaching.

Introduction

Engineering is increasingly seen as a global, international field. As such, one sees an increasing number of international collaborations in teaching undergraduate engineering. There are many examples of international teaching initiatives found within the ASEE conference proceedings. (See for example, papers by Jaselskis, Liu, Parsaei, Fayez, and their colleagues.[1-4]) This paper describes a novel teaching partnership between two university engineering departments in Australia and China.

Wuhan University of Science and Technology[5] (WUST) is a public university providing formal higher education courses and programs in Wuhan, Hubei Province, the People's Republic of China (figure 1). In 2012, Deakin University[6] (Deakin) and WUST agreed to establish a cooperative teaching program in China, the Bachelor of Engineering degree in Mechanical Engineering and Automation, which was subsequently approved by Ministry of Education, China in 2013. The first intake of students from Hubei province was recruited in the same year. The purpose of the cooperation between the two universities is to establish a collaborative relationship for complementing the similar program existing at WUST; and to provide students with education of an international standard through exchange of internationally advanced teaching methods and resources. It is the hope of both universities that this relationship will provide opportunities for educational and cultural exchange between students and staff of the two universities and contribute positively to the relationship and goodwill between Australia and China.

In this program, Deakin sends its teaching staff to WUST each semester to deliver intensive teaching over a two-week period (40 hours) for 13 core courses from sophomore year to the senior year. There are also WUST academic staff who co-deliver each of these courses for 32 hours, which results in 72 teaching hours in total for each course. All the teaching, learning and assessment materials are provided by Deakin. Some minor adjustments are made before they are used to teach and assess students at WUST. The focus of this paper is to discuss the learning-and-teaching approach involved in this collaborative program. It also will discuss some of the challenges faced by the students in the program.

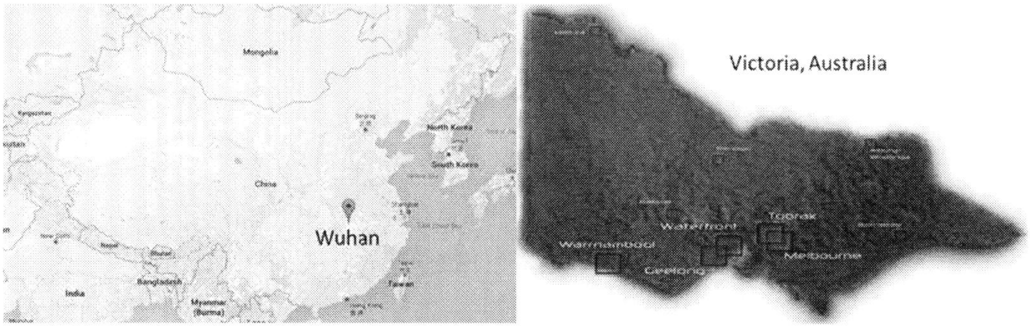

Figure 1: Locations of Wuhan University of Science and Technology (left), and Deakin University in Victoria, Australia (right).

In a multi-disciplinary curriculum setting, learner-centred curricula and pedagogical practices are at the heart of international collaborative-learning programs. The vision of this project is to provide students with a collaborative-learning experience specializing in the theories and applications of advanced technology in mechanical engineering. The graduates in this collaborative program will be equipped with the fundamental knowledge, expertise and learning capacity such that a science-literate and innovative mind can prepare them for work in product and process modelling and also in design as a researcher, manufacturer, developer or administrator.

Teaching and learning with a design-centric curriculum

Each participating institution has at least one representative for each course, a teaching academic, an assessor, and an overall program director. A steering committee consisting of engineering academics and academic deans guides the project in WUST. An internal research review team evaluates the project annually as part of an ongoing longitudinal study. The curriculum approach is based on design-based learning (DBL), a combination of problem-based learning and project-based learning. With different learning styles, students (figure 2) are able to express their skills and talents through working on projects or by simply designing experiments in authentic learning environments. Integrating design and technology tools into science education provides students with dynamic learning opportunities to actively investigate and construct innovative design solutions.[7] A design-based learning environment assists the curriculum to move into the twenty-first century with students being hands-on in their work, in addition to using problem solving skills, engaging in collaborative teamwork, creating innovative designs, learning actively, and engaging with real-world problems.[8, 9]

The collaborative teaching in undergraduate engineering education has a number of features:

- A design-based learning curriculum approach in teaching and learning,
- Student projects around problem/design activities,

- Team-based learning/collaborating learning in classroom seminars,
- Teaching and learning assessment/evaluation processes.

Figure 2: WUST students in Deakin-led classes.

The main goal of teaching and learning through a design-focused curriculum is to increase student learning through problem solving, active/collaborative learning, student-directed learning and analytical thinking. This DBL approach is different from traditional teaching. The School of Engineering at Deakin University has always tried to improve its course delivery method to enrich the student experience and to produce capable job-ready engineering graduates. To this end, it has explored new teaching methods to aid in this process. One such method is design-based learning (DBL). Unlike problem-based learning (PBL) and project-based learning (PjBL), DBL is a self-directed learning approach and opens up learning activity so design skills must be learnt and applied.[10] Students must locate the resources required, and analyse any needs in order to create a design.[11] This method gives students the freedom to apply their design skills as they think best. DBL not only looks at the end product but also at the underlying process in creating that product.[12-14]

Course outline

The course was a sophomore engineering-materials unit. It focused on the introduction of basic concepts of material selection and the fundamental properties of the main material classes: metals, polymers, ceramics. The course also taught different types of material failure and how to select materials to avoid failure. The types of failure topics addressed include yielding, creep, wear, fatigue and fracture. The course text material was based around two popular materials textbooks.[15, 16] Traditional teaching in China tends to be rote learning, more passive than active. This course employed an inquiry-based approach with a combination of integrated class/seminars and laboratory-demonstration classes. The Deakin-led classes consisted in two hours of lecture, followed by a two-hour practical class. The practical classes focussed on six aspects of materials science and engineering:

- Stress and strain in materials
- Material properties
- Strength limits in engineering components

- Crack propagation in failure
- Failure by fatigue
- Materials selection.

Students used these experiences as the starting point for design and research-based activities. In the process of teaching, flexible use of various teaching methods, such as presentation, discussion, competition and so on, were employed to arouse student interest in learning materials and improve learning. For example, figure 3 shows students performing a materials-selection exercise, following the procedures developed by Ashby.[17] What's more, this materials course had a higher requirement on practical work than in more traditional Chinese courses. In addition, several experts were invited to give guest lectures on materials. The assessment of this course was 50% final examination, 10% attendance, and 40% mid-semester examination. The minimum passing grade was 60%. Sixty-three students completed the course with their learning assessed.

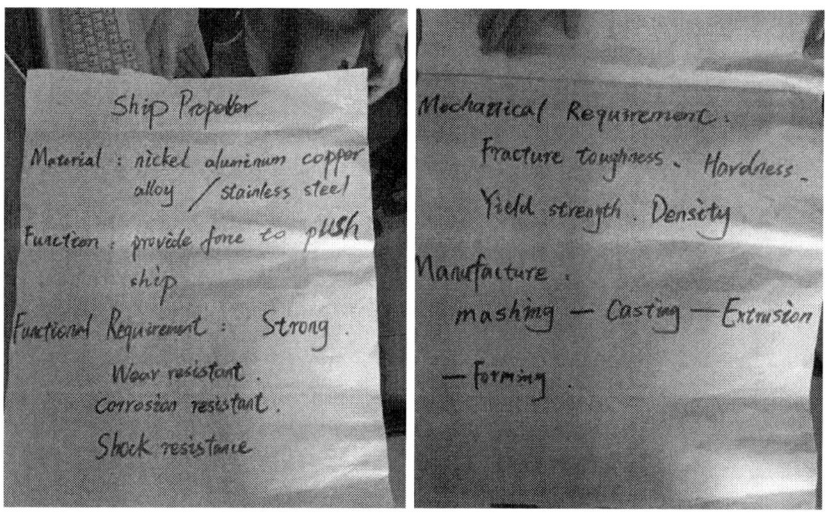

Figure 3: WUST students working on a materials-selection exercise - materials for a ship propeller.

Methodology to determine student perceptions

When considering the significance of student learning outcomes and the teaching requirements of staff in engineering education, this research intends to encapsulate the perspectives of students about a new learning-and-teaching approach. The School of Engineering at Deakin has used DBL as one of its engineering learning principles for further development in teaching and learning. It is required to improve the learning-and-teaching process as a holistic approach from the perspective of students and staff over the entire collaborative teaching program at WUST. A qualitative and quantitative paper-based survey method was used to obtain student perspectives on the course. Qualitative methods are useful for evaluating developing program goals and for involving participants in the evaluation process to gain their insight and perspective.

The survey questions were designed to determine the students' level of experience from their first year of undergraduate engineering. The views of students on collaborative learning using project/design-based learning approach in this research come from all levels of undergraduate engineering. From the quantitative and qualitative analysis performed, the results are analysed and presented from a students' perspective about project/design based learning practice in a particular unit. The survey was given to more than 100 students across multiple courses in the first year of engineering. It was anonymous and non-identifiable. These results are from students' own experiences and the results present various views, which include students' knowledge and expectations. In turn this can inform the collaborative program to enhance the international student-learning experience. In line with ethics process and procedures, a third party conducted the research survey. The questions were prepared to identify the challenges in teaching and learning and in particular to investigate student perspectives on the practice of collaborative teaching. The survey questions used in this research were:

1. Are you motivated to learn and achieve the learning goals in this unit?
2. Are you comfortable to participate and interact in the practical?
3. Does the group based practical activities are most helpful to achieve learning goals?
4. Does the practical activity tasks helped you to obtain skills such as communication skill, teamwork skill, design skill, problem-solving skill?
5. Do you understand the subject and the teaching approach?
6. Overall, are you satisfied with this way of teaching?

Overall assessment grades

Figure 4 summarizes overall assessment grades of students. It shows that over 50% of students achieved greater than 60% of the maximum final-assessment score. A direct comparison between this course and WUST's traditional materials course is not possible because this course and its corresponding traditional course taught in Mandarin had slightly different content and different textbooks. However, from the perspective of similar courses at WUST, this shows an above-average pass rate. Only two students achieved lower than 59% of the maximum possible score and thus failed the course. In a traditional WUST course in materials science, the pass rate is approximately 85%.

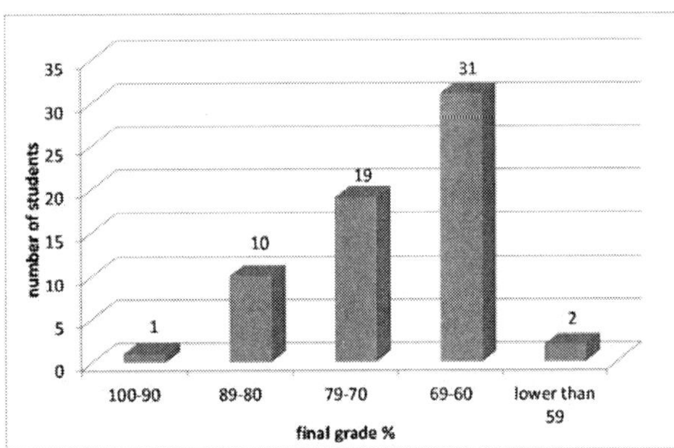

Figure 4: Final overall assessment marks of the students.

Survey results

Of equal importance to academic performance is the student experience. A total of 61 students participated in the survey. In the materials course considered here, the students' perceptions on their learning experience are shown in figures 5-7. For questions 1 to 7, the most common response was agreement. The most questions with which students agreed were questions 1 and 3. A total number of 91% of students agreed that the collaborative teaching had motivated them to achieve the learning goals. 82% of students agreed that the group-based practical activities were most helpful towards achieving the learning goals.

Around 37% of students disagreed that the subject and the teaching approach was not understandable. The student responses to question 5 indicated that the students have a communication issue in the learning environment. It is obvious that communication plays a significant role towards teaching and understanding the subject when it is being taught in a second language. This area should be investigated further to better understand the underlying reasons for the high number of disagree responses.

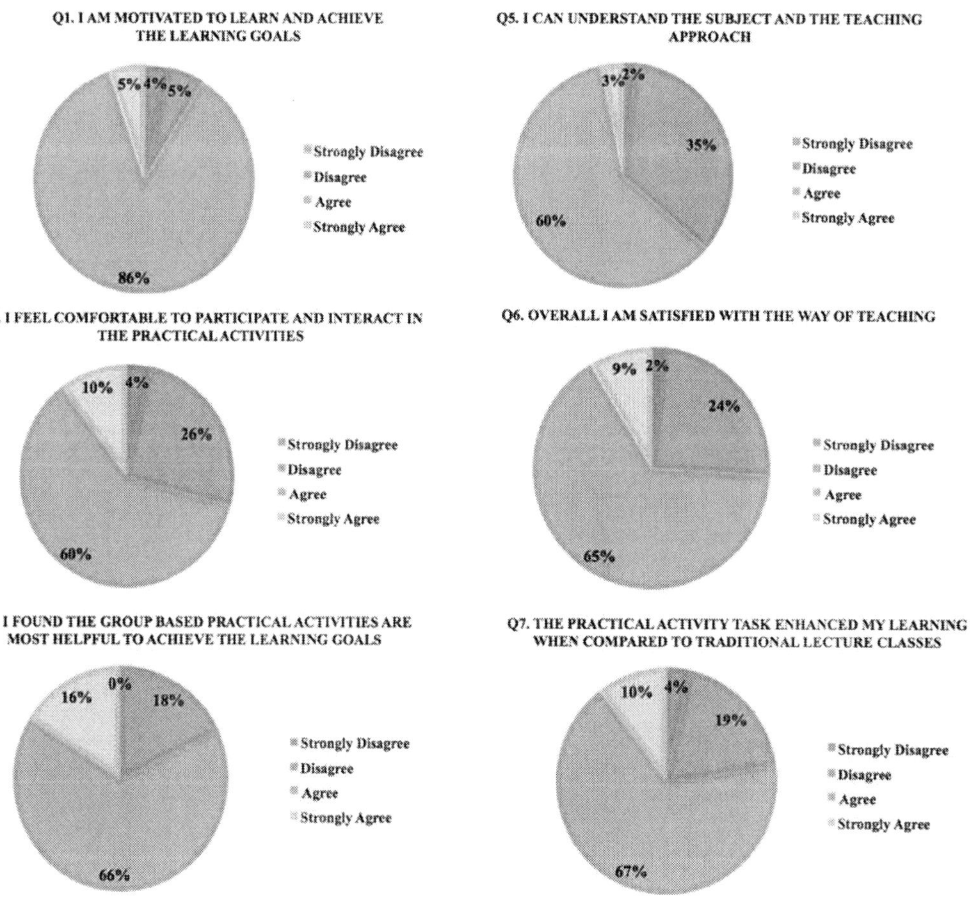

Figure 5: Student feedback responses

The student responses to questions 2, 6 and 7 in the survey were similar. The combined number of agreed responses for these questions ranged between 70-77%. We can say that of the students surveyed, they felt similar about being satisfied with the teaching method and also that the practical activity task had increased their learning compared to traditional learning. Also, the students generally felt comfortable in participating and interacting in the practical activities. That being said, there remains 25-30% of students who were not comfortable with the new learning environment. Future offerings of this course will address this to increase student satisfaction.

When students were asked about what skills they obtained from the practical activities, again most of the responses clustered towards the agree response, shown in figure 5. The most interesting response was the feedback towards design skill, which saw an equal number of responses for agree and disagree at 45% each. However the responses appear be slightly skewed towards agreement, indicated by the greater number of strongly-agree to strongly-disagree responses. Problem-solving was indicated as the higher trend towards agreement, in terms of skills developed from practical activities, followed by teamwork skill and communication.

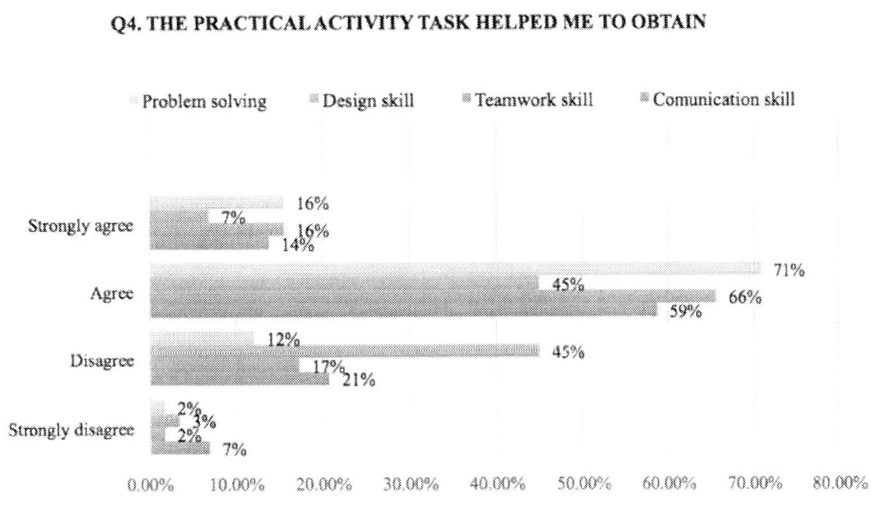

Figure 6: Student feedback response for Question 4.

There is some continuity when comparing the number of agreed responses querying communication skill in question 4 (figure 6), and the number responses to the open-ended question, mentioning communication skill as being what they found most useful in this unit. Figure 7 shows that 56% of open responses mention communication as being most useful. This was followed by 14% of the comments related to teaching being most useful, then practical activity tasks at 9%, concluded by teamwork 7%. Other individual non-related comments made up 14% of the open-question responses.

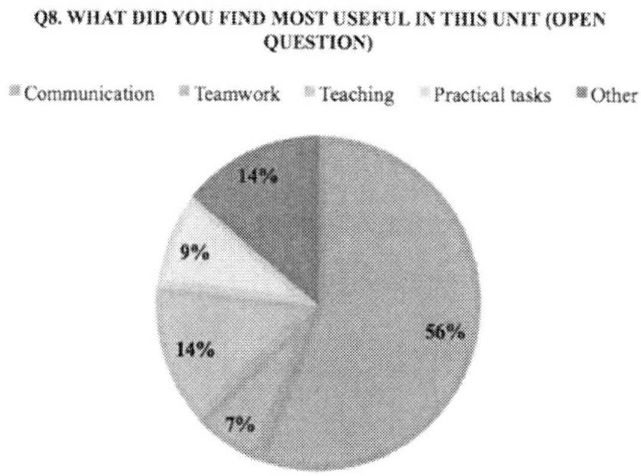

Figure 7. Student responses to the open-ended question

Discussion

The overall assessment shows that most of the students in the class experienced a quality learning experience through collaborative teaching. As indicated in figure 4, around 50% of students achieved higher grades by adopting to a new collaborative teaching. According to the survey results discussed above, over 90% of students mentioned that the new collaborative teaching motivates to learn and around 77% of students felt that the practical activity enhanced their learning compared to traditional learning. Based on analysing the final exam grades, we found that students had a good understanding of theoretical-knowledge-based content. However the students found difficulties in applying this theoretical knowledge to practical problems. For example, on the exam there was a question on how to select a material for a bicycle frame. Ninety percent of the students did not give a correct answer. In order to improve student understanding of the application of theory to real-world problems, in the future more practical case studies will be introduced.

From the survey results, about 82% of students agreed that the group-based practical activities were most helpful towards achieving the learning goals. The project/design-based learning activities have enhanced students learning capability of understanding practical/theoretical prospects in this course. Most of the students mentioned that communication plays an important role towards teaching and understanding of the subject when it is being taught in a second language by the academics from Deakin. On the other hand, students found that understanding the Deakin lecturers' unique Australian accent to be quite a challenge. (Incidentally, one of the authors experienced the same thing with another group of Chinese students who were studying a similar materials course in Australia. One lecturer was Australian, the other American. These students found the American lecturer easier to understand than the Australian one.) Our future work will include even greater cooperation between Chinese and Australian instructors to ensure the best possible learning experience. The course certainly gave the students an opportunity to learn and experience internationalism in an academic environment.

Anecdotally, across the entire Deakin-WUST program, we noticed that at the start, students tended to be quiet and passive. There was little interaction and engagement between the lecturers and students.

(We admit that teaching the classes in English was a contributing factor.) Three semesters later, we noticed that the students became more engaged with the lecturers, more responsive, and more likely to initiate discussion. As this is one of the goals of the program, we see this as a positive result and look forward to backing up this observation with better data.

As far as design-based learning is concerned, in the next few years of the partnership, the teaching team plans to

- Organise DBL workshops for faculty and students,
- Continue the DBL teaching approach through design problems around projects,
- Analyse and improve students learning experiences and difficulties,
- Publish staff and student perceptions for future case studies.

Conclusion

Deakin University in Australia and Wuhan University of Science and Technology in the People's Republic of China have begun a teaching partnership in undergraduate engineering, whereby Deakin courses are taught to WUST students in English. Deakin academics visit Wuhan during the semester to provide intensive classes over a few days. The teaching method centers on design and project-based learning. A trial of this partnership in a sophomore course in materials engineering resulted in good academic results from the WUST students. Surveys of student satisfaction indicated that the students were for the most part satisfied with the course and the novel teaching methods. However, as the course was taught in English and not in the native Chinese, students found communication to be a considerable challenge. Greater cooperation between Australian and Chinese teaching staff will be used to overcome this language barrier.

Acknowledgements

The work described in this paper was partially supported by Hubei Provincial Department of Education of China (2015224). All contributions to this work are gratefully acknowledged.

References

[1] Fayez, H. and Keshawarz, M.S. (2015),"Collaboration of Afghan and American Universities in a New Architectural Engineering Program at Herat University, Afghanistan," in *Proceedings of the 121st ASEE Annual Conference and Exposition* Indianapolos: ASEE.

[2] Jaselskis, E.J., Valdes-Vasquez, R., Kalidindi, S.N., Krute, L.D., Guo, H., Comiskey, D., *et al.* (2015),"Successful Academic Partnership in the Development of an International Construction Practices Course," in *Proceedings of the 2015 ASEE International Forum* Seattle: ASEE.

[3] Liu, A., Morrison, J.R., Dai, Y., and Lu, S.C. (2015), "Design and Evaluation of a Cross-Cultural and Trans-Disciplinary Global Innovation Course," in *Proceedings of the 2015 ASEE International Forum* Seattle: ASEE.
[4] Parsaei, H.R. and Elbashir, N. (2015), "A New Vision for Enginering Education," in *Proceedings of the 121st ASEE Annual Conference and Exposition* Indianapolos: ASEE.
[5] Wuhan University of Science and Technology, http://www.wust.edu.cn/default.html.
[6] Deakin University, http://www.deakin.edu.au.
[7] Doppelt, Y. and Schunn, C.D. (2008), "Identifying Students' Perceptions of the Important Classroom Features Affecting Learning Aspects of a Design-Based Learning Environment," *Learning Environments Research* **11**(3), pp. 195-209.
[8] Dopplet, Y., Christian, M.M.M., Schunn, D., Silk, E., and Krysinski, D. (2008), "Engagement and Achievements: A Case Study of Design-Based Learning in a Science Context," *Journal of Technology Education* **19**(2), pp. 23-39.
[9] Dopplet, Y. (2003), "Implementation and Assessment of Project-Based Learning in a Flexible Environment," *International Journal of Technology and Design Education* **13**(3), pp. 255-272.
[10] Perrenet, J., Aerts, A., and van der Woude, J. (2003), "Design Based Learning in the Curriculum of Computing Science - a Skilful Struggle," in *Proceedings of 2003 International Conference on Engineering Education*, pp. 21-23.
[11] Iwane, N., Ueda, H., and Yoshida, M. (2011), "Design Based Learning by Knowledge Reuse: Towards Its Application to *E*-Learning," in *Proceedings of the IEEE International Conference on Information Reuse and Integration (IRI)*: IEEE, pp. 468-473.
[12] Trigwell, K., Prosser, M., and Waterhouse, F. (1999), "Relations between Teachers' Approaches to Teaching and Students' Approaches to Learning," *Higher Education* **37**(1), pp. 57-70.
[13] Wijnen, W. (1999),"Towards Design-Based Learning," Eindhoven: Educational Service Centre, Eindhoven University of Technology.
[14] Reynolds, B., Mehalik, M.M., Lovell, M.R., and Schunn, C.D. (2009), "Increasing Student Awareness of and Interest in Engineering as a Career Option through Design-Based Learning," *International Journal of Engineering Education* **25**(4), pp. 788-798.
[15] Callister, W.D. and Rethwisch, D.G. (2010), *Materials Science and Engineering: An Introduction*, 8th ed. Wiley.
[16] Ashby, M., Shercliff, H., and Cebon, D. (2014), *Materials: Engineering, Science, Processing and Design*, 3rd ed. Elsevier.
[17] Ashby, M.F. (2011), *Materials Selection in Mechanical Design*, 4th ed. Butterworth-Heinemann.

Supporting STEM Knowledge and Skills in Engineering Education - the PELARS project

Prof. Dorian A. Cojocaru, University of Craiova

Dorian Cojocaru received the Ph.D. in Automation from the Romanian Ministry of Education in 1997. He is active as a full university professor, head of Mechatronics and Robotics Department and PhD co-ordinator. He is acting as expert for Romanian Agency for Quality Assurance in Higher Education and other national and international research and academic bodies. His present areas of research are: computer vision, robotics and mechatronics, and new trends in technical higher education. He has published more than 10 books, more than 200 papers in various scientific journals and international conference proceedings and he coordinated more than 15 research national and international programs.

Prof. Anna Friesel, Technical University-Copenhagen

Anna Friesel is Professor at the Section for Electrical Technology, DTU Diplom - Technical University of Denmark, Campus Ballerup. She is also the president of the EAEEIE - European Association for Education in Electrical and Information Engineering, which is a European non-profit organization, with members from nearly seventy European Universities, most of them teaching in the area of Electrical and Information Engineering (EIE). Anna Friesel is a member of the IEEE Educational Activities Board (EAB) Curricula and Pedagogy Committee (CPC). She collaborates regularly with many technical universities in Europe, Latin America and USA on topics related to improvement of engineering education. Her research interests include mathematical modeling, system dynamics, control theory, and educational methods in automation, robotics, and in engineering in general.

Dr. Daniel Spikol, Malmö University

Spikol is a Senior Lecturer in the Department of Media Technology in the Faculty of Technology. His research interests are in the design and development of technology-enhanced learning that engages learners and support teaching with a focus on engineering and design activities.

Supporting STEM knowledge and skills in engineering education – PELARS project

Abstract

In this paper we present our proposal for improving education with hands-on, project-based and experimental scenarios for engineering students with the use of learning analytics. We accomplish this through teacher and learner engagement, user studies and evaluated trials, performed at UCV (University of Craiova, Romania) and DTU (Technical University of Denmark). The PELARS project (Practice-based Experiential Learning Analytics Research And Support) provides technological tools and ICT-based methods for collecting activity data (moving image-based and embedded sensing) for learning analytics (data-mining and reasoning) of practice-based and experiential STEM. This data is used to create analytics support tools for teachers, learners and administrators, providing frameworks for evidence-based curriculum design and learning systems. The PELARS project creates behavioral recording inputs, proving a new learning analytic that is scalable in application, and bridge qualitative and quantitative methods through reasoning and feedback from input data. The project serves to better understand learners' knowledge in physical activities in laboratory and workshop environments, as well as informal learning scenarios. PELARS traces and helps assess learner progress through technology enhancement, in novel ways building upon current research. The project results in learning analytics tools for practice-based STEM learning that are appropriate for real-world learning environments.

Introduction

The PELARS[1] project is intended as a testing ground and generator of 21st-Century STEM knowledge and skills. The project generates, analyses, uses and provides feedback for analytics derived from hands-on, project-based and experiential learning scenarios. These scenarios are:

1. European High School level STEM classrooms.
2. Postsecondary Science and Engineering practical environments.
3. The more open-form studio environments of the Interaction Design Schools (which form the basis and inspiration for many of the project's orientations).

The project serves to help better understand learners' knowledge in physical activities in laboratory and workshop environments, as well as informal learning scenarios. PELARS traces and helps assess learner progress through technology enhancement, in ways that have been un-attempted and un-scalable until now. The project results in learning analytics tools for practice-based STEM learning that are appropriate for real-world learning environments.

The participants of the PELARS project are twelve European Partners that have all been chosen for their specific expertise and skills in the fields of this project. The Consortium includes seven universities, three small medium enterprises and two non-profit organizations. The partners are from ten different countries (Belgium, Denmark, Germany, Ireland, Italy, Romania, Spain,

Sweden, Switzerland and United Kingdom). This geographical distribution is due to the desire to build a Consortium mainly based on a thematic structure, and aimed at achieving a large interdisciplinary network of excellence. All the named learning objectives are defined across four national settings in the EU.

User Experience in Engineering Education

The objective of Work Package 2 in PELARS-project is to engage, through design ethnography methodologies and on-site experience prototyping, with groups and individuals involved in teaching and learning of STEM subjects in three different contexts[2,3] and one of them is engineering education context[4,5,6]. UCV from Romania and DTU from Denmark are the two engineering higher education institutions where the new PELARS technology has to be tested. The first test has been done at UCV in July 2015 and the second one is scheduled at DTU in April 2016.

Taking into consideration the PELARS objectives we are interested to investigate the different users' opinions (students and teachers) regarding the features offered by PELARS technology and what could be added or modified. An important aspect in our definition is use of learning analytics resulting from the use of these new educational technologies.

In the frame of PELARS we are working to develop new technologies and processes for teaching and learning for design, engineering (as part of STEM) through practical applications[7,8,9,10] .

The technological developments in the last 10-15 years and wide use of the new technologies like tablets, smartphones and information sources on internet, changed everyday life and have the big impact on teaching's methods. It is important to modernize and adapt the educational system to new generation of students, who grown up with and use the named technological tools[11,12,13]. The complexity of new technologies made it also necessary to prepare the students for even more challenging future, where abilities to cooperate and self-study are also included. In this context, we ask the following questions:

- What it is needed to change in the actual educational methods at the two partner universities, UCV and DTU, to perform practical experiments supporting students' motivation?
- How PELARS technology could contribute to support the named abilities/skills?
- What is the impact of the PELARS' technology seen from the final users: students and teachers?
- Will the new technology change the accreditation procedures?

The schedule for trials in PELARS is shown in figure 1.

Creating a comprehensive plan for each of the trials include also ethics considerations. All activities are and will be carried out taking into account FP7 guidelines as well as national guidelines on ethic issues, guaranteed by the Ethics Committee under EU. Participation of users is in line with agreed European and National ethical procedures for user participation and remuneration of test subjects, and subject to sign off by the Ethics and Research committees.

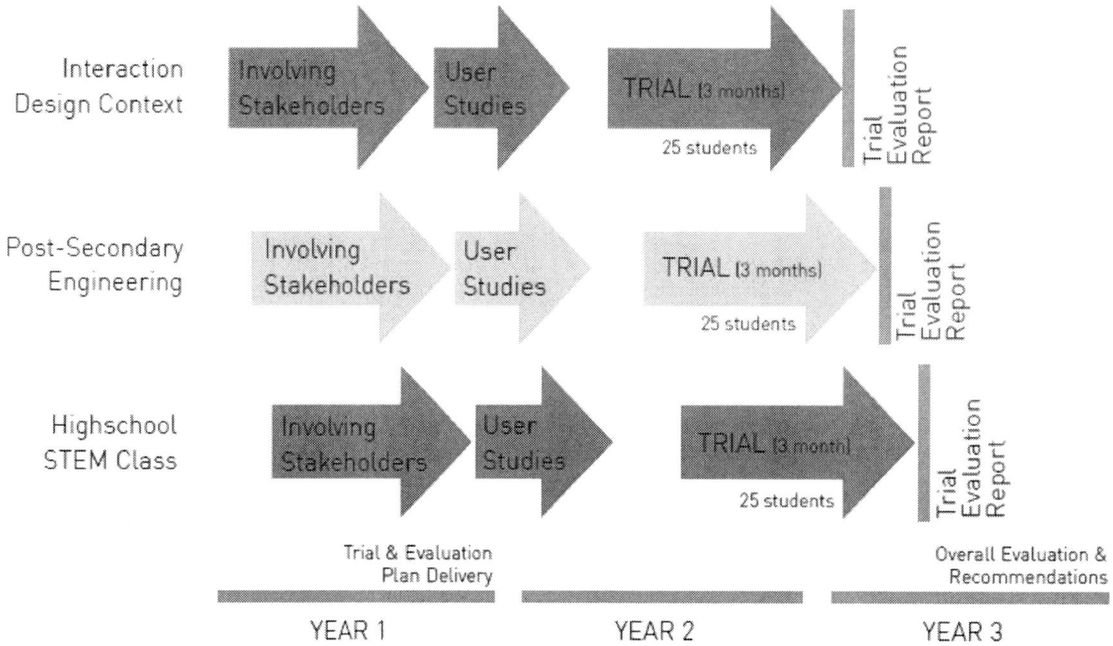

Figure 1. The schedule for PELARS' trials.

Both legal and ethical issues are a priori approved by European Commission and national authorities prior to conducting any trials, and the appropriate materials (project in accordance to the EC's Ethical Guidelines for undertaking ICT research in FP7[14]).

Each year every partner of PELARS project has to provide signed "Ethics declaration", including Section 5, as shown in figure 2.

Section 5 Ethical issues		
What are the ethical issues which may arise in the course of this research, and how will they be addressed? Please consider / address ALL issues that may apply.		
• Potentially vulnerable participants • Safeguarding/child protection • Risks to participants and/or researchers • International research • Sensitive topics • Sampling • Gatekeepers	• Informed consent • Assent • Methods • Confidentiality • Anonymity • Data storage/security • Data transfer/transmission	• Data sharing/encryption • Data documentation • Data management plan • Data protection • Reporting • Dissemination and use of findings

e.g. There are no major ethical issues arising from this phase of the research. The issues that concern us are as follows:
- Voluntary teacher/lecturer participation vis-à-vis participation being imposed by head teachers or management
- Working with young people requires special consideration to their circumstances. In order to address these issues we will take the following steps:
- Participants will be provided with information leaflets and consent forms (see attached samples)
- The project partner is responsible for consent forms signature and storage for 3 years after the project end
- We will ensure that we use participant codes to anonymize the data we collect.
We will only analyse audio, video or photographic data with written consent from teachers, headteachers, students and their parents, and only for the purpose of reporting in academic outputs (for example, in conference presentations). We will publish anonymized reports of the data.
- Students can withdraw anytime
- [...]

Figure 2. Ethics declaration, section 5.

Learning activities/prototyping at UCV

In the summer of 2015 we organized a workshop at UCV, Craiova having the support and direct participation of two partners from Sweden and UK. We had a two-day workshop at UCV with students from two programs: Mechatronics and Robotics (more practical oriented) and Multimedia Engineering Systems (more software oriented and with a better theoretical background) and two group interviews with teachers (mixed subjects). We collected students' feedback on the latest learning activities we had developed at that time. We wanted to hear, in a dialogue with other people than their own teachers, how they consider the actual way to perform laboratory activities (labs) compared to PELARS proposals. We put together engineering students to see how students from different programs are able to work together by proposing solutions for different problems without using a specific theoretical base. The main goal was to obtain and to discuss few proposals generated by students for practical scenarios that can be possibly used to teach with for labs using PELARS technology, methods and equipment. We investigated UCV teachers' perspectives about their current practice of laboratory sessions and their ideas about how to integrate PELARS technologies in their teaching practice. The data collection was done in an informal focus group interview setting and was audio recorded. Workshops are video and audio recorded and interviews are audio recorded for future references.

Student workshops were planned as follows:

- Introduction to PELARS project-Presentation
- Research consent forms
- Introduction to visual programming platform
- Presentation of educational scenarios
- Introduction to brainstorming
- Brainstorming about the learning activities
- Three questions about the learning activities.

The number of participants was 14 for 13th of July, and 15 for the 14th of July. Students were from the programs of Mechatronics & Robotics (third year of study), respectively from Multimedia Engineering Systems (second year of study). Both programs offer a bachelor diploma in engineering after 4 years of study. During the workshop a number of six teachers from UCV (two professors, two associate professors, and two assistant professors), all with PhD in engineering, were involved.

Summarized the comments on PELARS trials at UCV were following:

- The proposed educational technology looks attractive and interesting compared with the methods used at UCV at the moment, which are based on learning a lot of theory without enough creative and cooperative experiments.
- A number of UCV students have a good experience in using Arduino components and building systems using mechanical and electronics parts, and because of that they appreciated the new proposal.
- Students with good practical skills see in the new educational technology a way to counteract with students with better theoretical skills.

Practice-Based and Experiential Learning at DTU

In Denmark there is traditionally close cooperation between engineering educational institutions and the industry, which is the employee for engineering graduates. Due to this tradition DTU make ongoing changes in programs and educational methods, as a result of discussion with our industrial partners. The requirements from industry include also skills like teamwork, project-work and ability to self-study. This is the reason why students at DTU work from the very beginning (first semester) in groups solving practical engineering problems related to the theoretical subjects. Even the basic subjects like mathematics and physics are connected with design-build projects, where students work in teams with the project combining all theoretical subjects. This is quite a different educational approach to teaching theory from the methods used at UCV and this is why our trials at DTU are expected to have different focus from the trials performed at UCV. The DTU trials are scheduled for April 2016 with different groups of DTU's students, involving students from Electronics-, Software- and Electrical Power - programs. At DTU we will focus on several requirements of suitable PELARS toolkits, which should have following properties:

- Flexibility in the adaptation of learning activities to teachers' practice at DTU. PELARS learning activities should define a general curriculum area, but allow teachers to design specific projects within that curriculum area that target their chosen knowledge, skills and competencies;
- A primary focus on skills and competencies, and secondary focus on specific curriculum knowledge;
- A focus on motivating students to engage with STEM;

Relevant learning objectives are:

- Programming: abstracting actions into algorithms, coding an algorithm (knowing the syntax), debugging programs;
- Circuiting: (depending on toolkit:) working with breadboards / soldering / connecting components, debugging circuits;
- Working with sensors, and actuators, mapping analogue values to quantized digital values;
- Planning and conducting projects;
- STEM projects: setting up experiments, measuring data, visualizing data;
- Creativity: idea generation, design, practical skills.

Conclusion

We have identified programs in undergraduate engineering education that will assist the PELARS project to both develop and evaluate learning analytics for practice-based learning. The first practical experiment has been made at UCV and next one is scheduled at DTU in April 2016. The following trials at UCV and DTU will include changes how to best support learning analytics with PELARS due to the conclusions of the first two trials. The two engineering institutions, UCV and

DTU, have similar engineering programs, but the traditional teaching methods are different and the requirements from the national bodies to certificate the educations. Our aim is to develop a framework of learning objectives that are relevant to real-world learning contexts and within which the PELARS consortium can develop and evaluate both the technical possibility and real world relevance of learning analytics for practice-based learning in different countries.

Acknowledgment

The presented results were financially supported by the EU FP7- Technology Enhanced Learning Program, no.: 619738 with the acronym PELARS - Practice-based Experiential Learning Analytics[1].

Bibliography

1. PELARS-project, http://www.pelars.eu/
2. Blikstein, P.: "Multimodal learning analytics"; Proceedings of the Third International Conference on Learning Analytics and Knowledge ((2013, April, pp. 102-106).
3. Worsley, M.: "Multimodal Learning Analytics as a Tool for Bridging Learning Theory and Complex Learning Behaviors"; *Proceedings of the 2014 ACM workshop on Multimodal Learning Analytics Workshop and Grand Challenge* (2014, November, pp. 1-4).
4. Friesel, **A.,** Timcenko, **O.:** "Mathematical Modelling in Electronics and Medialogy – Sharing Experiences for Better Results", ICEE2011 **-** August 2011, Belfast, North Ireland, **UK.**
5. Friesel,A., Avramides, K., Cojocaru, D.: "Identifying how PELARS-project can support the development of new curriculum structures in engineering education", The Experimental International Conference 2015 (**exp.at'15**), June, 2015, University of Azores, Ponta Delgada, São Miguel Island, Azores, Portugal.
6. Krumm, A.E.; Waddington, R.J.; Lonn, S.; Teasley, S.D. :" Increasing Academic Success in Undergraduate Engineering Education using Learning Analytics: A Design-Based Research Project"; Annual Meeting of the American Educational Research Association. Vancouver, BC, Canada, 2012-04 ; . http://hdl.handle.net/2027.42/106032
7. Dragon, T., Mavrikis, M., McLaren, B.M., Harrer, A., Kynigos, C., Wegerif, R., & Yang, Y.: "Metafora: A web-based platform for learning to learn together in science and mathematics". IEEE Transactions on Learning Technologies, 6(3), 197-207; JULY-SEPTEMBER 2013.
8. Minović, M., Milovanović, M. (2013): "Real-time learning analytics in educational games". Proceedings of the First International Conference on Technological Ecosystem for Enhancing Multiculturality - TEEM '13 (pp. 245–251). New York, New York, USA.
9. Cuartielles, D.: "Final STEM Learning Kit with Integrated Learning Analytics". PELARS Project Deliverable D4.2, October 2015. https://docs.google.com/document/d/1Vt-r9PBFGus2AG7SErlGVg8JUkfPLs9n2wX2zfL8rws/edit#heading=h.hcfjfl2xwvzm
10. Martinez-Maldonado, R., Dimitriadis, Y., Martinez-Monés, A., Kay, J., & Yacef, K.: "Capturing and analyzing verbal and physical collaborative learning interactions at an enriched interactive tabletop". International Journal of Computer-Supported Collaborative Learning, 2013, 8(4), pp. 455-485.
11. Martinez-Maldonado, R., Pardo, A., Mirriahi, N., Yacef, K., Kay, J., & Clayphan, A. The LATUX workflow: designing and deploying awareness tools in technology-enabled learning settings. Proceedings of the Fifth International Conference on Learning Analytics And Knowledge (2015, March, pp. 1-10).
12. Mavrikis, M.;, Gutierrez-Santos, S, Geraniou, E, Noss, R & Poulovassilis, A.: "Iterative Context Engineering to Inform the Design of Intelligent Exploratory Learning Environments for the Classroom"; in R Luckin, S Puntambekar, P Goodyear, BL Grabowski, J Underwood & N Winters (eds), Handbook of Design in Educational Technology. Routledge, 2013, pp. 80-92.

13. Vogel, B.; Spikol, D.; Kurti, A.; Milrad,M.: "Integrating Mobile, Web and Sensory Technologies to Support Inquiry-Based Science Learning"; 6th IEEE International Conference on Wireless, Mobile, and Ubiquitous Technologies in Education; April 2010; Kaohsiung, Taiwan; ISBN: 978-0-7695-3992-8
14. The EC's Ethical Guidelines: ftp://ftp.cordis.europa.eu/pub/fp7/docs/guidelines-annex5ict.pdf

ASEE 2016
INTERNATIONAL FORUM
NEW ORLEANS, LOUISIANA / JUNE 25, 2016

Paper ID #17521

Enhancing undergraduate student learning experience in an environmental engineering course through use of technology and industry partnership

Dr. Lokesh P Padhye, The University of Auckland

Dr. Lokesh Padhye received his Masters and PhD from Georgia Institute of Technology, Atlanta. Before joining academics, he was a senior staff engineer with Geosyntec Consultants in Kennesaw, GA. Since 2014, he is working with the University of Auckland as a Senior Lecturer in the Department of Civil and Environmental Engineering.

Increasing undergraduate student learning in an environmental engineering course through use of technology and industry partnership

Abstract

As New Zealand's education system, NCEA, does not mandate chemistry learning in upper two standards of higher secondary education for the admission to engineering, it poses significant challenges for the academic staff involved in environmental engineering courses as environmental chemistry is an integral part of the curriculum. It was evident from past experiences and course evaluations that New Zealand's domestic, undergraduate students, without adequate knowledge of high school chemistry, get easily dispirited by the course content if taught with the traditional approach. Hence, to improve students' understanding of the course content and class participation, we experimented with two strategies for an environmental engineering course: introduction of online quizzes and frequent guest lectures from industry leaders.

The online quiz component was not designed as a replacement for traditional assignments; rather, it was a supplementary tool for enhancing students' learning experience. However, it aided in reducing the length and a total number of assessments. Also, another objective to use online quizzes was to test students on assigned readings in addition to lecture material. Many of these assigned readings were meant for supplementing students' basic chemistry knowledge and to help them in learning advanced environmental chemistry concepts. The instructor generated a multiple choice, online quiz for each of the three modules of the course with the help of an external online testing website, 'ClassMarker'. Each quiz had an average of 50-60 multiple choice questions with five options. Results of the quiz were available upon completion of the quiz so that students were able to review their answers immediately.

In addition to online quizzes, emphasis was placed on making students understand the importance of practical aspects of learning environmental chemistry. Although engineering academics have expertise in teaching and research, they often lack the complete knowledge of current industrial practices. In addition, professionals working in the field can provide first hand evidence of applicability of concepts taught in the class. Hence, six guest lectures from industry leaders were scheduled throughout the course to emphasize importance of learning environmental chemistry concepts for practical applications.

The success was measured by carrying out two surveys after the course completion: first through the University's learning management system about the effectiveness of the course instructor as an indirect measurement and the second was a direct measurement through the external online survey (with SurveyMonkey) with questions designed specifically to address the effectiveness of online quizzes. All respondents of the first survey agreed that the instructor stimulated their interest in the subject and presented material that assisted in their understanding of the subject. Nearly all respondents also agreed that the instructor used educational technologies that supported their learning and encouraged their engagement in the learning process. For the second survey, students felt positive about the opportunity to become more engaged with the course material, to be better prepared for class, and to learn and comprehend the course material than simply memorize it.

Introduction

Course instructors assign textbook and external readings to students for a variety of reasons, most commonly because it is often impossible to present all recommended course material during class time [1]. Completion of assigned readings before a topic is covered in class allows students to better comprehend the class material, and it also increases in-class participation and student interactions [1-3]. Unfortunately, based on my personal experience as well as from the literature, it appears that student compliance with reading assignments in college courses has steadily declined over time [4-6]. There can be a variety of reasons associated with this decline, including increased presence of electronic gadgets in our daily lives [7]. Hence, it is worthwhile to ponder if, instead, can these gadgets and technology be used for the benefit of the student learning?

Teachers need new techniques to increase student reading and class participation for a better understanding of their assigned course material. One such technique involves examining students through frequent quizzes on reading assignments [3]. Having regular in-class quizzes can encourage students to pay closer attention to the assigned reading material as well as the course content, which can improve their understanding not only of the reading material, but also of the lecture material presented in the class [3,8,9]. In addition, frequent quizzes may lead students to study regularly during the semester rather than cramming just before the final exam [3,5,9]. However, it is reported that although scheduled or surprise in-class quizzes can motivate students to complete the assigned reading, students tend to view such quizzes as punishment depending on how those are administered [6,9]. Hence, it is utmost important to administer these quizzes optimally.

A variety of ways to administer in-class quizzes exist and different methods may vary in effectiveness. For example, use of randomly administered quizzes (i.e., neither completely surprise nor scheduled for a particular date) may result in higher levels of reading compliance [10], and selection of subsets of questions from large question banks can reduce the likelihood that students will simply memorize response options without learning the material [11]. Furthermore, the literature suggests that presenting feedback separately after the student answers each question, including both information about accuracy and a source of additional information, can enhance the learning experience for students [1,12]. Pedagogical research findings also suggest that online class quizzes for assigned readings can provide students with a variety of positive learning outcomes [13,14]. For example, the use of online administered quizzes has been shown to motivate students to complete assigned readings, increase participation in class discussion, and improve performance on exams for material covered both on the quizzes and in class [12,15].

However, there is still a disagreement about whether and how frequent online quizzes affect student performance and their overall learning experience. Bonham et al. (2003)[16] and Engelbrecht and Harding (2004)[17] noted that online assessments (as compared to manually marked paper assessments) resulted in no discernable difference in student performances over a range of summative assessments. Likewise, Smith (2007)[18] found that online quiz scores showed higher correlation with final examination marks than laboratory or assignment marks. Grimstad and Grabe (2004)[19] found that students who completed voluntary quizzes significantly improved their exam performances. However, their conclusions were explained by good students being both motivated to take voluntary quizzes and likely to do well on exams. Brothen and Wambach (2001)[8] cautioned that mandatory quizzes only improve exam performance if students employ an efficient strategy of using the quizzes to test their own knowledge of the material, rather than attempting to use the quiz to learn the material.

Furthermore, a set of precautions were suggested by Angus and Watson (2009)[20] for successful administration of periodic online quizzes, including collecting frequent student feedback about e-assessment.

Given the mixed conclusions concerning the impact of quizzes on the class performance and certain limitations of New Zealand's education system, NCEA, we decided to evaluate the effectiveness of the addition of frequent, online quiz component to the assessment of environmental engineering courses in improving the learning experience of students. The study is relevant not only for New Zealand but also for other countries where students admitted to engineering don't necessarily have adequate chemistry background required for environmental engineering courses. Our hypothesis was that introducing frequent online quizzes will assist students in being more interactive in the class by effectively learning chemistry concepts through additional reading material required for quizzes.

I first experimented with the idea of adding online quizzes in my courses in the previous semester. Student verbal feedback was positive for that pilot project. It encouraged me to pursue this project more actively with a motivation to understand if there is indeed a relationship between these two factors. If administered properly, we also hypothesise that online quizzes will be able to save time for instructors in the long run, which is a significant advantage in this age of ever increasing expectations from an academic citizen. In addition, it is expected that with the use of the online quizzes, significantly more class time will be available for active learning activities and other small group activities for instructors. Hence, the project was designed as a service not only to students but also to class instructors.

In addition to online quizzes, we also introduced industry guest lecturing component for helping students learn importance of practical aspects of learning chemistry for environmental engineering careers. Although engineering academics have expertise in teaching and research, they often are not at the forefront of implementation of state of the art engineering solutions for solving real world problems. As demonstrated by D'Este and Perkmann (2011)[21], faculty members benefit from the influx of corporate expertise, and students gain knowledge about high-throughput technology and commercial applications of fundamental principles learnt in the class or through additional readings. Merritt (2001)[22] also highlighted that universities should include industrial research component in their courses and invite industrial experts as guest lecturers who can provide industrial perspective on these issues. Hence, our hypothesis was that students will learn course content more effectively if they can see the practical importance of the class material including additional readings through industry guest lecturing.

Objectives

The motivation for this project was two-fold. The major component was trying to understand the effectiveness of frequent online quizzes in enhancing students' learning experience with more interactive assessment. The second motivating factor was to study if active industry participation in our courses can also benefit student learning and increase their satisfaction and class performance.

Methodology (The Plan of Action)

I introduced the idea of online quizzes in the Semester 1 (March-July 2015) last year. We conducted two online quizzes in the environmental engineering Part 3 (year 3) course (ENVENG 341) of the Semester 1. After receiving a positive verbal feedback from those students, I implemented a more detailed version of the plan in this Semester 2 (July-November) course (ENVENG 342). We designed 4 online quizzes, worth 3% each. Overall student assessment was broken down in the following manner for this course: final examination in theory and problems (40%); midterm examination (20%); assignments (24%); online participation on Piazza (4%); and online quizzes (12%). The following paragraph will focus exclusively on the methodology used in online quizzes.

We generated a multiple choice, online quiz for each of the three modules of the course with the help of the reading material posted through the online learning management system (Cecil). For creating these quizzes, questions were generated from research articles, case studies, and book chapters assigned as recommended readings for this course. The quizzes were written with the third-party software called 'ClassMarker'. Each quiz had an average of 50-60 multiple choice questions with five possible answers each. The 'ClassMarker' has a number of inbuilt functions which allow instructors to design all quiz items in randomized order, along with a randomized test bank of answers. Settings on 'ClassMarker' also can be used to prohibit students from printing material on the screen, thus preventing them from easily printing out the quiz questions. This procedure was used to deter students from cheating by making it difficult to copy and share quiz questions and answers. All students were encouraged to use smartphones to take the quiz. Complaints arising as a result of any technical difficulties were rare and were easily resolved. 'ClassMarker' also allows the quiz settings to be adjusted so that students can navigate through the quiz easily and can change their answers until the quiz is submitted. Results of the quiz were available upon completion of the quiz so that students were able to review their answers immediately.

In line with the above-mentioned objectives, the study sought to investigate the link between frequent online quizzes and student learning by testing the connection between student exposure to the online learning assessment and class performance. As 2015 was my first year of teaching the course, I could not directly compare the class performance of students from previous year. However, we carried out the indirect measurements through student course feedback survey and through external survey ('SurveyMonkey') designed specifically to evaluate effectiveness of online quizzes. In addition, to accomplish the above-mentioned secondary objective of the study, we took the help of student verbal feedback, students' feedback through learning management system survey, students' attendance, and students' enrolment statistics in 2015 and 2016 to evaluate effectiveness of industry guest lectures in increasing student interactions in the class.

Intended Benefits

It is hypothesized that the use of frequent, online quizzes and industry guest lecturing would allow the instructor to engage students in more active learning, group work, and class discussions. It also was expected that the online quizzes coupled with guest lecturing from the industry would be viewed positively by students as an effective means of learning the course material. Students were expected to report that the online quizzes helped them study and gain mastery of the course material throughout the semester rather than simply cramming before a comprehensive exam in order to improve their performance. They were also expected to engage more actively in the class through industry guest lectures as a result of understanding practical importance of material covered in the class. Overall, we expected that students will face fewer difficulties in learning some of the advanced environmental chemistry concepts covered in the class as a result of better understanding of chemistry fundamentals through additional readings and realizing their practical applications.

Measurement of Success

The success of incorporating online quizzes, as an effective means of assessment, and industry guest lecturing was measured in a number of ways:

1. Through Cecil about the effectiveness of the course instructor and the course itself as an indirect measurement,
2. Through external online student survey with questions designed specifically to address the effectiveness of online quizzes as a direct measurement,
3. Through attendance recorded during the industry guest lectures,
4. Through comparison of student enrolment statistics before and after the course, and
5. Through personal feedback obtained from students, either verbally or through written communication with the instructor.

Results and Discussion

Ten out of a possible 25 students (40%) completed the assessment questionnaire on Cecil, based on the effectiveness of instructor for ENVENG 342. Students' responses to all Likert scale questions are presented in Appendix 1 below. Thirteen out of a possible 25 students (52%) completed the assessment questionnaire on 'SurveyMonkey' based on the effectiveness of online quizzes for ENVENG 342. Students' responses to all Likert scale questions are presented in Appendix 2 below.

All respondents of this Part 3 course survey on Cecil agreed that the instructor stimulated their interest in the subject and presented material that assisted in their understanding of the subject. Nearly all respondents also agreed that the lecturer used educational technologies that supported their learning, and the lecturer stimulated their engagement in the learning process. The majority of the respondents of the survey on online quiz effectiveness agreed that online quizzes were helpful to their learning, increased their engagement in the course, increased their interest in the course and made them better learn the course content. Students felt positive about the opportunity to become more engaged with the course material, to be better prepared for class, and to learn and comprehend the course material than simply memorize it. More than half of the respondents indicated that online quizzes may be useful in other courses.

The average attendance for the lectures, including guest lectures, throughout the semester was 90%, and it remained steady from the Week 3 of the Semester (after the official enrolment closure period of the University) till the end of the course. This was particularly significant statistics considering that all lectures were video recorded and were available to students immediately after the class conclusion. The average class attendance for previous year for the same course was 65%. The significant jump in attendance was linked to the students' active participation and interest in the course content.

In addition to attendance, students' enrolment statistics showed significant improvement in 2016 (after the redesigned course) compared to 2015 (before the redesigned course). For 2015, before we implemented the above changes in the curriculum, we had only 25 students enrolled for environmental engineering Part 3 courses (ENVENG 341 and 342). However, after implementing the above measures, enrolment for 2016 has already jumped to 47 (capped by the classroom capacity) for ENVENG 341 and ENVENG 342. Year 3 is a critical year for students of our Department as they decide their sub-specialization for their future careers. There are five sub-specializations available in the Department for getting bachelor's degree. Students can either choose from general civil, transportation, environmental, geotechnical, or construction management. Having almost 100% increase in the student enrolment for environmental engineering courses was interpreted as success for our measures to improve class interactions and student learning for the Part 3 environmental courses.

Apart from these documented surveys and statistics, verbal and visual feedback is a key element in assessing effectiveness of implemented measures. During the duration of these courses, instructor was approached several times by students, personally or through email, to express their satisfaction with our new approaches, online quizzes and industry guest lecturing, implemented in the course. In addition to attendance, the class interaction level was also found to be high as students participated actively in discussions on the class topics. Students also appeared well prepared for each lecture and most of them did not struggle with the background material which was fundamental for most of the environmental chemistry concepts covered in the class. Again, this was attributed to the success of online quizzes in encouraging students to go through the required readings.

Conclusions

The results demonstrate effectiveness of experimenting with online quizzes and industry guest lecturing to increase students learning experience in an environmental engineering course. However, I will add a note that such measures may not be beneficial to all courses. The use of online quizzes is highly recommended for any instructor teaching a course that requires that students to engage in a significant amount of assigned readings as was the case for us due to certain limitations of NZ education system. Many of the environmental engineering courses, however, fall in this realm as the field is highly interdisciplinary and require substantial knowledge of chemistry, microbiology, policy, and toxicology. Extensive knowledge of these fields is not considered mandatory for students' admission to engineering in most of the education systems around the world. In such cases, students will benefit from the use of online quizzes as they become actively engaged in the course material and study with greater frequency throughout the semester. Students will also engage more effectively in the course due to industry guest lecturing as they gain knowledge about practical applications of the class material.

Based on my experience, I will make some additional recommendations for the effective use of online quizzes and industry guest lectures including:

1. Online quiz settings should be designed to provide students with immediate feedback on their quiz grades. Students seem to appreciate such quick feedback.
2. To prevent students from cheating, the order of quiz questions, as well as their multiple choice answers, can be randomized. Settings can also be adjusted to prevent students from printing the quiz directly from the screen.
3. Based on the subjective comments received from the survey, it is also recommended that students may be allowed to review the previous quizzes and their answers before class exams to aid in their course exam preparation.
4. Selection of an industry expert should be done carefully as the person needs to highlight the message that class material learnt has significant practical application.

If administered in the above manner, it appears that frequent online quizzes and industry guest lectures are indeed making a positive impact on enriching the student learning experience. The benefit of increased class engagement of students appears well worth the initial investment of time required to set up the online quizzes and redesign course curriculum to accommodate industry lectures.

Acknowledgments

I would like to thank all supporting academic staff, administrative staff, tutors, and TAs of the University of Auckland involved in successfully conducting above two classes in 2015.

References

1. Ryan, T. E. (2006). Motivating novice students to read their textbooks. Journal of Instructional Psychology, 33, 135–140.

2. Gurung, R. (2003). Pedagogical aids and student performance. Teaching of Psychology, 30, 92–95.

3. Narloch, R., Garbin, C.P., and Turnage K.D. (2006). Benefits of prelecture quizzes. Teaching of Psychology, 33, 109–112.

4. Burchfield, C.M. and Sappington, J. (2000). Compliance with required reading assignments. Teaching of Psychology, 27, 58–60.

5. Clump, M.A., Bauer, H., and Bradley, C. (2004). The extent to which psychology students read textbooks: A multiple class analysis of reading across the psychology curriculum. Journal of Instructional Psychology, 31, 227–229.

6. Sappington, J., Kinsey, K., and Munsayac, K. (2002). Two studies of reading compliance among college students. Teaching of Psychology, 29, 272–274.

7. Gilbert, J. and Fister, B. (2011). Reading, risk, and reality: College students and reading for pleasure. College & Research Libraries, 74(6), 474-495.

8. Brothen, T and Wambach, C. (2001). Effective student use of computerized quizzes. Teaching of Psychology, 28, 292–294.

9. Graham, R.B. (1999). Unannounced quizzes raise test scores selectively for mid-range students. Teaching of Psychology, 26, 271–273.

10. Ruscio, J. (2001). Administering quizzes at random to increase students' reading. Teaching of Psychology, 28, 204–206.

11. Daniel, D.B. and Broida, J. (2004). Using Web-based quizzing to improve exam performance: Lessons learned. Teaching of Psychology, 31, 207–208.

12. Brothen, T and Wambach, C. (2004). The value of time limits on internet quizzes. Teaching of Psychology, 31, 62–64.

13. Hillman, J. (2012). The impact of online quizzes on student engagement and learning, Last Accessed October 20, 2015.

14. Salas-Morera, L., Arauzo-Azofra, A., and Garcia-Hernandez, L. (2012). Analysis of Online Quizzes as a Teaching and Assessment Tool, Journal of Technology and Science Education, 2(1).

15. Johnson, B. C. and Kiviniemi, M. T. (2009). The effect of online chapter quizzes on exam performance in an undergraduate social psychology course. Teaching of Psychology, 36, 33–37.

16. Bonham, S. W., Deardorff, D. L. and Beichner, R. J. (2003). Comparison of student performance using web and paper-based homework in college-level physics. Journal of Research in Science Teaching, 40, 10, 1050–1071.

17. Engelbrecht, J. and Harding, A. (2004). Combing online and paper assessment in a web-based course in undergraduate mathematics. Journal of Computers in Mathematics and Science Teaching, 23, 3, 217–231.

18. Smith, G. (2007). How does student performance on formative assessments relate to learning assessed by exams? Journal of College Science Teaching, 36, 7, 28.

19. Grimstad, K. and Grabe, M. (2004). Are online study questions beneficial? Teaching of Psychology, 31, 143–146.

20. Angus, S. D., and Watson, J. (2009). Does regular online testing enhance student learning in the numerical sciences? Robust evidence from a large data set. British Journal of Educational Technology, 40(2), 255–272.

21. D'Este, P., Perkmann, M. (2011). Why do academics engage with industry? The entrepreneurial university and individual motivations. J Technol Transf 36, 316-339.

22. Merritt, A. (2001). Education: the chemistry between academia and industry. Drug Discovery Today 6, 287.

Appendix 1

ENVENG 342 Evaluation - Padhye

Enrolments: 25 | ENVENG.342 SC 2015 | Evaluation#:44200937

Responses: 10 | Environmental Engineering Design | Print Date: 8/30/2015

LS01	The lecturer was well prepared for the lectures	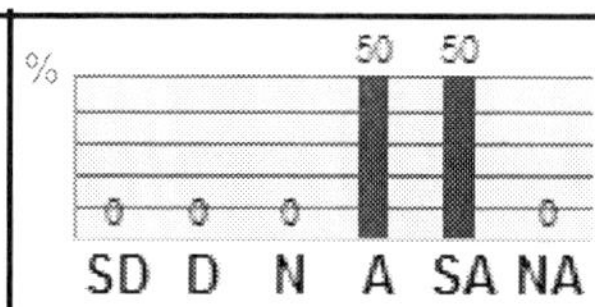	A+SA = 100.0% D+SD = - % NA = 0.0%	Mean: 4.50 StdDev: 0.53
LS02	The objectives of the lectures were clearly explained.	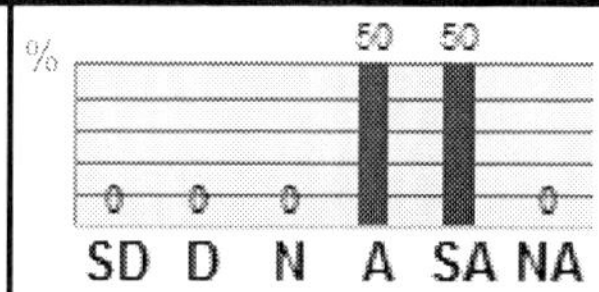	A+SA = 100.0% D+SD = - % NA = 0.0%	Mean: 4.50 StdDev: 0.53
LS03	The lecturer stimulated my interest in the subject.	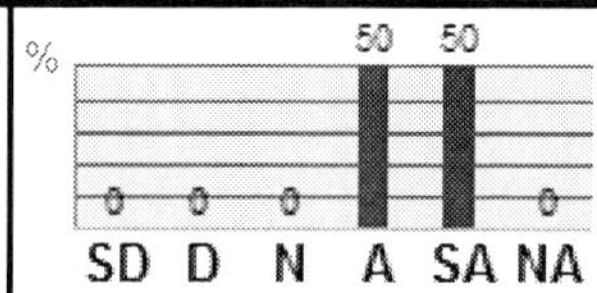	A+SA = 100.0% D+SD = - % NA = 0.0%	Mean: 4.50 StdDev: 0.53
LS04	The way the lecturer presented material assisted my understanding of the subject.	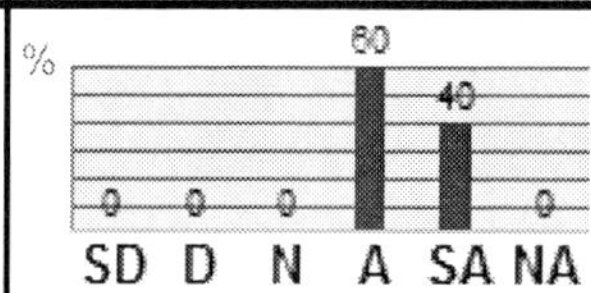	A+SA = 100.0% D+SD = - % NA = 0.0%	Mean: 4.40 StdDev: 0.52
LS05	The lecturer used educational technologies (e.g., e-lectern, CECIL, audio-visual clips) in ways that supported my learning.	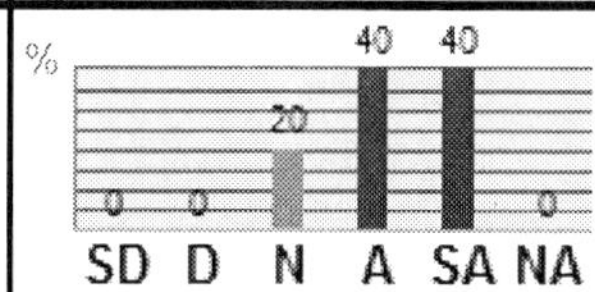	A+SA = 80.0% D+SD = - % NA = 0.0%	Mean: 4.20 StdDev: 0.79
LS06	I found the lecturer approachable.	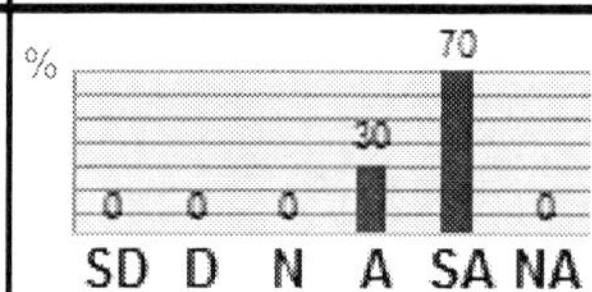	A+SA = 100.0% D+SD = - % NA = 0.0%	Mean: 4.70 StdDev: 0.48
LS07	The lecturer responded to students' questions in a constructive way.	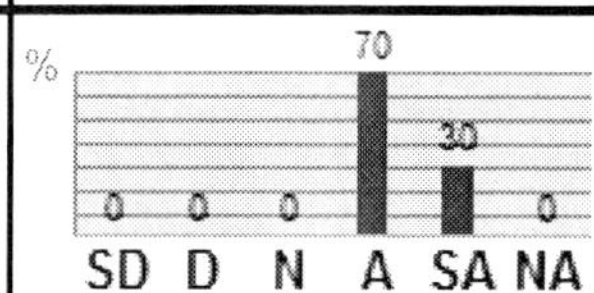	A+SA = 100.0% D+SD = - % NA = 0.0%	Mean: 4.30 StdDev: 0.48
LS08	The lecturer stimulated my engagement in the learning process.	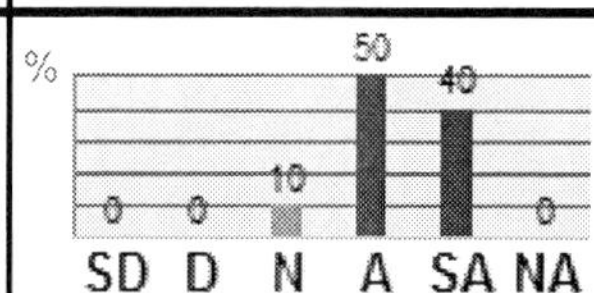	A+SA = 90.0% D+SD = - % NA = 0.0%	Mean: 4.30 StdDev: 0.67
LS09	Overall, the lecturer was an effective teacher.	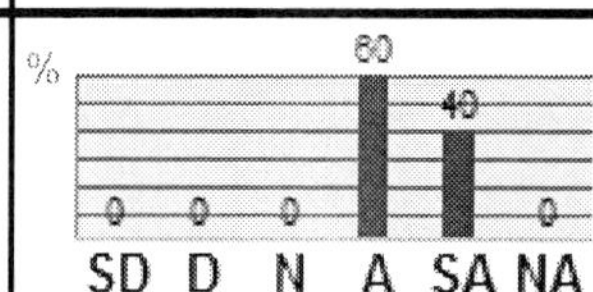	A+SA = 100.0% D+SD = - % NA = 0.0%	Mean: 4.40 StdDev: 0.52

ENVENG 342 Evaluation - Padhye

Enrolments: 25
Responses: 10

ENVENG.342 SC 2015
Environmental Engineering Design

Evaluation#: 44200937
Print Date: 8/30/2015

What was most helpful for your learning?

- -a lot of useful topic was discussed in this paper
  -lecturer is very approachable
- Lecture recordings to go over and make sure I understood the concepts
- Lecture recordings.
- objectives of each topics are clearly written and helps a lot when students are doing revision
- real world examples of concepts
- Thanks for the lecture recordings!

What improvement would you like to see?

- be more selective with class examples and more practice questions with answers provided, so we know we are on the right track when we finish the questions
- may include more exercise to help understand the topic
- Maybe some more practice problems to work through by ourselves :)
- More worked examples done in class or more tutorial sheets provided on cecil to do by ourselves or together in class
- n/a

Appendix 2

Lokesh Padhye

From: Lokesh Padhye
Sent: Tuesday, 22 September 2015 9:28 a.m.
To: 'labb250@aucklanduni.ac.nz'; 'sbur161@aucklanduni.ac.nz'; 'jcha968@aucklanduni.ac.nz'; 'econ458@aucklanduni.ac.nz'; 'ydec845@aucklanduni.ac.nz'; 'cgos005@aucklanduni.ac.nz'; 'kham960@aucklanduni.ac.nz'; 'ghaw668@aucklanduni.ac.nz'; 'ohex319@aucklanduni.ac.nz'; 'ihsu672@aucklanduni.ac.nz'; 'kkud883@aucklanduni.ac.nz'; 'dkur948@aucklanduni.ac.nz'; 'tli553@aucklanduni.ac.nz'; 'jmcc462@aucklanduni.ac.nz'; 'dmil104@aucklanduni.ac.nz'; 'dmou920@aucklanduni.ac.nz'; 'lnew591@aucklanduni.ac.nz'; 'noli859@aucklanduni.ac.nz'; 'korr274@aucklanduni.ac.nz'; 'jqi305@aucklanduni.ac.nz'; 'ssac560@aucklanduni.ac.nz'; 'ssun044@aucklanduni.ac.nz'; 'jtay528@aucklanduni.ac.nz'; 'tvao282@aucklanduni.ac.nz'; 'gyuk716@aucklanduni.ac.nz'
Subject: ENVENG 342 - Online Quizzes - Your Opinion

Dear all,

As we are shifting towards more interactive form of assessment to better engage students and improve learning, it is an appropriate time to understand your opinion about online quizzes. We have tried these for the first time in ENVENG 341 and ENVENG 342 this year and would like to know your opinion about the same.

The survey won't take more than a minute, and needless to say that your responses will be anonymous. Please complete the survey by Thursday night (11:45 pm). I expect everyone to participate as your feedback is important to support our goal of enhancing students' learning experience. Please click the link below to complete the survey.

https://www.surveymonkey.com/r/TZV9GSQ

Thank you,
Lokesh

Lokesh Padhye, Ph.D., P.E., MIPENZ
Senior Lecturer
Department of Civil and Environmental Engineering
University of Auckland
Private Bag 92019
Auckland 1142, New Zealand
https://unidirectory.auckland.ac.nz/profile/l-padhye

Upgrade lokeshpadhye

My Surveys Examples Survey Services Plans & Pricing

+ Create Survey

Upgrade to get meaningful results: View all your responses and get powerful analysis. View Pricing

ENVENG 342 - Online Quizzes

Summary Design Survey Collect Responses Analyze Results

CURRENT VIEW

+ FILTER + COMPARE + SHOW

No rules applied

Rules allow you to FILTER, COMPARE and SHOW results to see trends and patterns. Learn more »

SAVED VIEWS (1)

Original View (No rules applied)

+ Save as...

EXPORTS

SHARED DATA

No shared data

Sharing allows you to share your survey results with others. You can share all data, a saved view, or a single question summary. Learn more »

Share All

RESPONDENTS: 13 of 13

Export All Share All

Question Summaries Data Trends Individual Responses

PAGE 1

Q1

Customize Export

Online quizzes were helpful to my learning of the course material.

Answered: 13 Skipped: 0

	Strongly Disagree	Disagree	Neutral	Agree	Strongly Agree	Total	Weighted Average
(no label)	0.00% 0	0.00% 0	30.77% 4	30.77% 4	38.46% 5	13	4.08

Q2

Customize Export

Online quizzes increased my engagement in the course.

Answered: 13 Skipped: 0

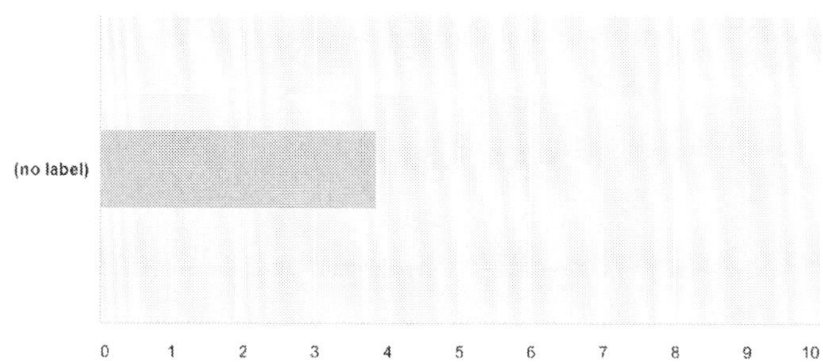

	Strongly Disagree	Disagree	Neutral	Agree	Strongly Agree	Total	Weighted Average
(no label)	0.00% 0	7.69% 1	23.08% 3	46.15% 6	23.08% 3	13	3.85

Q3

Customize Export

Reviewing questions and answers at the end of the online quizzes was helpful.

Answered: 13 Skipped: 0

	Strongly Disagree	Disagree	Neutral	Agree	Strongly Agree	Total	Weighted Average
(no label)	0.00% 0	0.00% 0	23.08% 3	46.15% 6	30.77% 4	13	4.08

Q4

Customize Export

I would recommend adding online quizzes to other courses as well.

Answered: 13 Skipped: 0

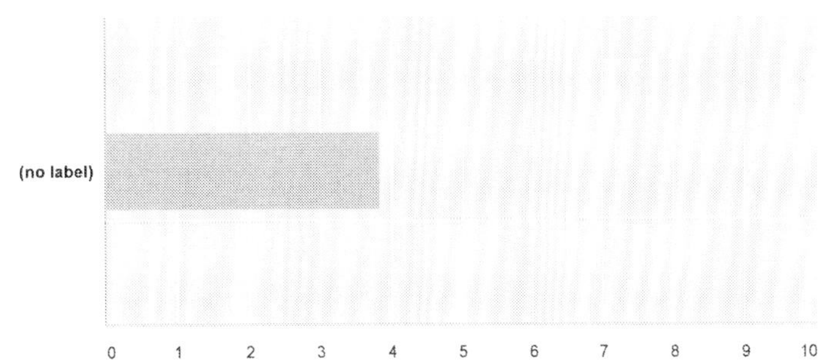

	Strongly Disagree	Disagree	Neutral	Agree	Strongly Agree	Total	Weighted Average
(no label)	0.00% 0	0.00% 0	38.46% 5	38.46% 5	23.08% 3	13	3.85

Q5

Customize Export

Overall, online quizzes increased my interest in the course and made me better learn the course content.

Answered: 13 Skipped: 0

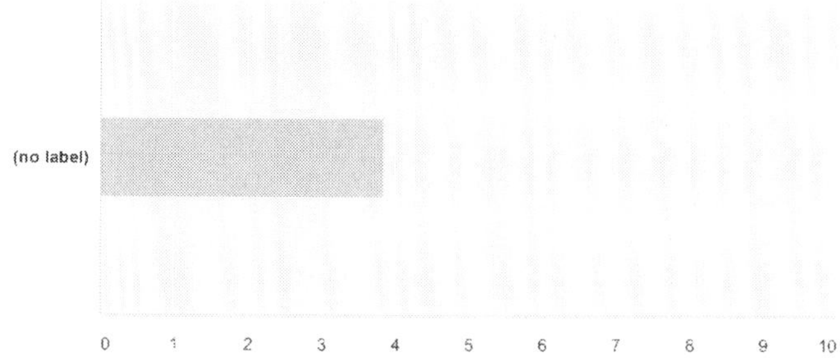

	Strongly Disagree	Disagree	Neutral	Agree	Strongly Agree	Total	Weighted Average
(no label)	0.00% 0	7.69% 1	23.08% 3	46.15% 6	23.08% 3	13	3.85

Q6

Export

Please provide any additional comments you have about online quizzes as a tool for effective class assessment.

Answered: 7 Skipped: 6

Responses (7) Text Analysis My Categories

PRO FEATURE
Use text analysis to search and categorize responses; see frequently-used words and phrases. To use Text Analysis, upgrade to a GOLD or PLATINUM plan.

Upgrade Learn more »

Categorize as... Filter by Category Search responses

Showing 7 responses

One thing that was a bit frustrating was that I couldn't go back onto the site later (i.e. when I got home) to go over any questions I got wrong. Just make sure that people know to print it as a PDF before logging out I suppose.
9/24/2015 10:55 AM View respondent's answers

The quizzes helped motivate me to be more systematic in my study approach, as a result I found studying for the mid-term and exam easier. I believe that the revision of the questions would be more useful if were able to have a copy of our tests. I also think that the quizzes become less effective if they cover more than 3-4 weeks of content. Overall I think that they were a good addition to the course and should be continued.
9/23/2015 8:45 PM View respondent's answers

An improvement would be to include more calculation based questions as they don't get practiced enough in test conditions until the exam/mid term.
9/23/2015 1:51 PM View respondent's answers

Please make the survey questions and answers available on CECIL or ClassMarker. They would be very useful as extra stufy material for mid-term and exam study.
9/22/2015 1:12 PM View respondent's answers

The online quizzes are great for small classes like 324 and 341, but with larger classes it would be too easy to cheat.

The Global Online Laboratory Consortium and its Role in Promoting a Global Cloud of Cyber Physical Laboratories

Prof. Michael E. Auer, CTI Villach, Austria

Dr. (mult.) Michael E. Auer is Professor of Electrical Engineering at the Faculty of Engineering and IT of the Carinthia University of Applied Sciences Villach, Austria and has also a teaching position at the University of Klagenfurt. He is a senior member of IEEE and member of ASEE, IGIP, etc., author or co-author of more than 170 publications and leading member of numerous national and international organizations in the field of Online Technologies. His current research is directed to technology enhanced learning and remote working environments especially in engineering. Michael Auer is Founding-President and CEO of the "International Association of Online Engineering" (IAOE) since 2006, a non-governmental organization that promotes the vision of new engineering working environments worldwide. In 2009 he was appointed as member of the Advisory Board of the European Learning Industry Group (ELIG). Furthermore he is chair of the Advisory Board of the International E-Learning Association (IELA) and member of the Board of Consultants of the International Centre for Innovation in Education (ICIE). In September 2010 he was elected as President of the "International Society of Engineering Education" (IGIP, http://www.igip.org). Furthermore he is one of the founders and Secretary General of the "Global Online Laboratory Consortium" (GOLC). GOLC is the result of an initiative started in 2009 at MIT to coordinate the work on educational online laboratories worldwide.

Prof. Abul K. M. Azad, Northern Illinois University

Abul K. M. Azad is a Professor in the Technology Department of Northern Illinois University. He has a Ph.D. in Control and Systems Engineering and M.Sc. and B.Sc. in Electronics Engineering. His research interests include remote laboratories, mechatronic systems, mobile robotics, and educational research. In these areas, Dr. Azad has over 100 refereed journal and conference papers, edited books, and book chapters. So far, he has attracted around $1.7 million in research and development grants from various national and international funding agencies. He serves on editorial boards of a number of professional journals and is Editor-in-Chief of the International Journal of Online Engineering. He is active with various professional organizations (IEEE, IET, ASEE, and ISA) as well as a member of board of Trustees of CLAWAR Association.

The Global Online Laboratory Consortium and its Role in Promoting a Global Cloud of Cyber Physical Laboratories

1. Introduction

The NMC Horizon Report lists Virtual and Remote Laboratories as one of the most important future technologies in learning.[1] With this in mind researchers and academics have made considerable advances in the design and development of online laboratories. These laboratories are accessible via the Internet 24 hours a day and 7 days a week. These advancements have been possible because of the growing advancement of Internet and networking technologies as well as new developments in e-learning and e-work tools. The forerunners in this area are engineering disciplines and the natural sciences.

Active learning by means of online laboratories is especially valuable for blended learning approaches. Students can access online laboratories from almost any location and the only need is web access. This flexibility is vital for enhancing traditional education as well as for expanding further education and lifelong learning. Online engineering is one of the future directions for advanced teleworking/e-working environments not only in engineering and science (economics and informatics) but also other disciplines that are affecting our society. Especially in engineering education there is an enormous educational value in hands-on laboratory experiences. However, conventional laboratories are becoming increasingly expensive, have complex logistics, limited access time, and cannot be easily shared beyond the physical laboratory's boundaries. Using online laboratories has the potential to significantly reduce obstacles related to cost, time/inefficient use of facilities, inadequate technical support, and limited access. This is especially important for engineering education in developing nations.

Different solutions and technologies exist today to implement online laboratories as well as different communication standards and data exchange protocols. Therefore, each institution/university is likely to adopt its own standards and approaches to performing tasks, such as handling users' accounts and managing experiment data. Because of these challenges, sharing remote laboratories is more difficult. To address these issues as well as to promote and develop the online laboratories through international collaboration, the Global Online Laboratory Consortium (GOLC) has been established. As the first step, a group of enthusiasts and visionaries met in June 2009 at MIT and discussed the objectives and principles of such an association. In June 2010, the GOLC was official founded in Villach (Austria) during its first annual general meeting.

2. What is GOLC?

The mission of the consortium is the creation of sharable, online experimental environments to increase the educational and scientific value of learning that may not be accessible, scalable, or efficient through traditional methods. The GOLC is focused on promoting the development and sharing of, and research into, remotely accessible laboratories for educational use. As the usage of online experiments gains traction in the educational community, there is increasing interest in developing online labs on a common infrastructure. A unified and interoperable architecture is essential to convert the tremendous current interest for online experiments into an economy of labs that can be efficiently shared around the world. The GOLC strategic objectives are [2]

- to encourage and support the creation of new online laboratories and associated curricular materials
- to sponsor the design of an efficient mechanism for sharing, exchanging and trading access to online laboratories by the creation of a global network of shareable experiments
- to support communities of scholars created around online laboratories
- to lead the evolution of an architecture that enables the sharing of online laboratories by unified standards.

3. Activities through GOLC

GOLC is very active in developing collaboration between its members and partner organizations, initiating projects and activities that promote online laboratory area, and organizing and sponsoring professional events (e.g., conferences and workshops). Some of the activities include iLAB Alliance, organizing international conferences and workshops, publishing books, developing reference collection systems, and standardizing activities.

3.1 iLAB Alliance

The iLab Alliance started as an initiative of several partners who agreed to share their online experiments within this network. It is, in this context, a step toward a broader usage and dissemination of online laboratories and has helped to raise issues that must be addressed for the implementation of a more efficient model to govern the practice of sharing labs on a cross institution basis. The software architecture used to maintain the laboratory sessions as well as scheduling service and experiment data storage is the iLab Shared Architecture (ISA) described in the previous section. ISA has an important characteristic, namely its distributed topology that has made it the ideal solution for implementation. ISA has already built-in mechanisms that allow setting up trusted connections between its autonomous network nodes (service brokers) so online laboratories can be seamlessly shared between them. This means that the institutions are free to manage their own online laboratories and their own user accounts and to deliver these labs via their own server. In this way, access for their users to their own laboratories does not depend on the status of other service brokers. On the other hand, to be able to use laboratories from other universities, a user has to authenticate in the main service broker. Each institution member of the network is expected to set up one service broker and deliver at least one experiment via this server. This means that from a pool of laboratories available at one institution, it is up to the institution to decide which laboratories will be available to the other members of the network.

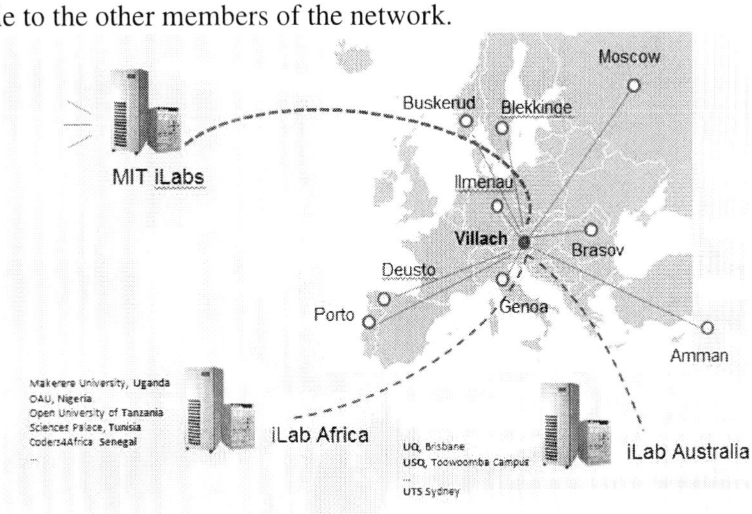

Figure 1: iLAB Alliance network.

Currently a number of entities are connected with this network. These include iLab Europe, MIT iLabs, iLab Africa, and iLab UQ Australia. Figure 1 shows the network layout. In this way GOLC helps to make global learning in engineering a reality. Nowadays more than 100 different laboratories are accessible via the iLab backbone.

3.2 International Conferences

Together with the International Association of Online Engineering (IAOE) GOLC organizes the annual International Conference on Remote Engineering and Virtual Instrumentation (REV). These conferences are a forum for specialists in the fields of Online Laboratories, Internet of Things, Cyber-physical Systems, Collaborative Networks and Grids, Cyber Cloud Technologies, and Service Architectures. An important concern of this conference series is the close collaboration between academic research and industrial partners. The companies supporting the activities include National Instruments, MathWorks, EMONA, Digilent, Keysight, etc. These companies are also members of GOLC. This year's REV2016 conference in Madrid, Spain, was again a very successful one, with more than 120 participants from 29 countries. The next REV conference (REV2017) will be held in New York at Columbia University (US).

An important medium to disseminate the idea of online laboratories is the GOLC *Online Laboratory Award*. The award is presented for outstanding solutions for stable running laboratories in the categories: *Visualized experiments, Simulated experiments, and Remote controlled experiments/laboratories*. Following the REV2015, this award was presented during REV2016 as well. This year's winners came from Greece, Germany, and Czech Republic.

3.3 Organizing Workshops

GOLC organizes workshops in the field of online laboratories to promote and encourage new laboratory development as an emerging area and to disseminate major developments around the world. To ensure an interactive working atmosphere, the participants for these workshops are kept between 20 and 30. The workshops were well attended, and a number of GOLC members took an active part to make these events successful. Typically, we have a mix of onsite and remote presentations during the workshops to show the technology under real conditions.

The first event of this kind was organized during the ASEE Annual Conference in June 2015 (Seattle, US). The second workshop was organized in Pnue, India, on January 10 during the third International Conference on Transformation in Engineering Education (ICTIEE 2016). The next two workshops are planned for during the ASEE Annual Conference in June 2016 (New Orleans, US) and in November 2016, during the World Engineering Education Forum in Seoul, Korea.

3.4 Edited Books

In 2011, members of GOLC published an edited book on online laboratories, titled *Internet Accessible Remote Laboratories: Scalable E-Learning Tools for Engineering and Science Disciplines* [3]. The initiative was to document current developments in the multidisciplinary creation of Internet accessible remote laboratories. This book offers perspectives on teaching with online laboratories, pedagogical designs, system architectures for remote laboratories, future trends, and policy issues regarding the use of remote laboratories. It is a useful resource for educators and researchers who are interested in learning more about the current status of the field as well as various approaches to online laboratory design.

After the success of the first book, we are currently working on a second edited book on online laboratories. The proposal has been reviewed by Springer and we are now in the process of finalizing an agreement. As proposed, the book will have five sections: State of the Art and Future Developments, Mobile Learning with Online Labs, Pedagogy of Online Experimentation, Online Labs: Case Studies from Educational Practice, and Commercial Tools for Online Labs. Call for chapters will be circulated in due course.

3.5 Reference Citation System

Online laboratories are becoming more and more appealing in many educational scenarios due to technology advancements and the rising pedagogical interests. Proper qualitative and quantitative analyses are needed to profile publications and their authors, to highlight trends and most promising research areas, to identify leading authors, and to drive the scientific community toward the right publication targets. Since no tools are currently available to achieve such purposes, a few GOLC member have taken the initiative to develop a data processing pipeline capable of producing a dataset of cleaned and normalized bibliographic references (namely the GRC2014) [4,5]. The initiative has been led by our Italian partners. This will be a biographical reference with an extensive overview of current and historical research and publication trends in the GOLC scenario. A Global Online Laboratory Repository will also soon be available for public use.

3.6 Standardization Activities

As indicated earlier, the potential of online laboratories has generated substantial interest in the areas of research and development. However, the interdisciplinary nature of this research makes the development process rather complex. Various researchers have ventured into the field of remote laboratory development as individuals without having a complete picture of the current status of development by their peers. In the absence of any common framework, each development initiative starts from scratch, working from different assumptions and premises, and making the resulting work sui generis, as there is currently little transferability or integration of the different efforts of persons involved in the development of online laboratories. To address this issue, the GOLC has undertaken an initiative to work on the standardization of remote laboratory development. The initiative was organized via three committees: a technical committee, an education committee, and a membership and communication committee. The technical committee has played a vital role in developing a technical metadata profile and in setting up normative references, terms and definitions. Recently the GOLC has teamed up with the IEEE Standards Committee *P1876* in an effort to drive the initiative forward.

3.7 International Projects

GOLC is also a partner in international research projects. Two recent examples are

- The eScience project funded by the European Union with the aim to establish a network of remote laboratories in the North African countries.
- "Laboratory Pedagogy Action learning based open MOOC" coordinated by the University of Southern Queensland, Australia.

One of the GOLC tasks in the near future is to include online laboratory providers from Asia. In this effort, during the ICTIEE 2016, we have opened an IAOE Indian Chapter.

4. Conclusion

The paper describes the background of the GOLC and its efforts and activities to promote online laboratory discipline. With the seven years since its initiation, the organization has made its mark within the community. The types of activities involve facilitating alliances among the members and organizations, sponsoring and organizing conferences and workshops, driving publications in terms of edited books and technical papers, developing a bibliographic repository system, and collaborating through international projects. The last, but most important, item is to drive a standardization effort for a remote laboratory discipline that will have a long lasting impact for the coming years.

References:

[1] Horizon Report 2016 Higher Education Edition, NMC, nmc.org, ISBN 978-0-9968527-5-3

[2] GOLC web site, http://www.online-engineering.org/GOLC_about.php

[3] A.K.M. Azad, M.E. Auer, and V.J. Harward, *Internet Accessible Remote Laboratories: Scalable E-Learning Tools for Engineering and Science Disciplines*, IGI Global, 2011, DOI: 10.4018/978-1-61350-186-3.

[4] M. Zappatore, A. Longo, and M.A. Bochicchio, The bibliographic reference collection GRC2014 for the Online Laboratory Research community, *12th International Conference on Remote Engineering and Virtual Instrumentation*, Bangkok, Thailand, February 25-27, 2015.

[5] M.A. Bochicchio and A Longo, Profiling the online laboratories research community and its core, 10th *International Conference on Remote Engineering and Virtual Instrumentation*, Sydney, Australia, February 6-8, 2013.

Singapore-U.S. Tactical All-Inclusive Navigation (SUSTAIN) collaborative innovation

Mr. Frederick C Webber, Air Force Research Laboratory

Mr. Webber holds a MS in Electrical Engineering from the Air Force Institute of Technology and a BS in Mechanical Engineering and Computer Science from the Rose-Hulman Institute of Technology. Mr. Webber is currently an autonomy engineer with the United States Air Force Research Laboratory's Warfighter Readiness Research Division, where he is responsible for STEM outreach and for identifying advances in the gaming industry that can improve warfighter training. Mr. Webber is an alternative navigation specialist on the SUSTAIN team.

Lt. KelseyLee Hunter Schafer, Air Force Research Laboratory

KelseyLee H. Schafer, 1st Lieutenant USAF received her bachelor's degree in Mechanical Engineering from University of Cincinnati, (Cincinnati, OH) and serves as a program manager and research engineer in the Air Force Research Laboratory Aerospace Vehicles Directorate. She is currently pursuing her masters degree in Mechanical Engineering at University of Cincinnati.

Dr. Eric T. Vinande, Air Force Research Laboratory
Dr. John P McIntire, US Air Force Research Laboratory

John McIntire is an Engineering Research Psychologist at the US Air Force Research Laboratory, at Wright-Patterson AFB, OH, USA. He received his masters in human factors psychology in 2007 and his PhD in human factors psychology in 2014, both from Wright State University.

Dr. Daniel D. Jensen, U.S. Air Force Academy

Dr. Dan Jensen is a Professor of Engineering Mechanics at the U.S. Air Force Academy where he has been since 1997. He received his B.S. (Mechanical Engineering), M.S. (Applied Mechanics) and Ph.D. (Aerospace Engineering Science) from the University of Colorado at Boulder. He has worked for Texas Instruments, Lockheed Martin, NASA, University of the Pacific, Lawrence Berkeley National Lab and MSC Software Corp. His research includes design of Micro Air Vehicles, development of innovative design methodologies and enhancement of engineering education. Dr Jensen has authored over 100 refereed papers and has been awarded over $4 million of research grants.

Prof. Shaohui Foong, Singapore University of Technology and Design
Mr. Wai Yong Chue, DSO National Laboratories
Mr. Yaohui Li, DSO National Laboratories
Dr. Kevin Ang, Temasek Laboratories @ National University of Singapore

Kevin received his B.Eng. degree in Mechanical and Production Engineering from Nanyang Technological University in 2006. After completing his undergraduate degree, he joined the Singapore Armed Forces as a Maintenance Engineering Officer, fulfilling his scholarship bond. Then upon receiving the DSO Ph.D Research Award in 2010, he pursued and graduated with his Ph.D degree in Electrical and Computer Engineering at the National University of Singapore early 2015. He is now working as a research scientist in the control science group at Temasek Laboratories. His current research interests lie in vision-based obstacle detection, navigation and the development of UAVs in urban environments.

Dr. Richard H. Crawford P.E., University of Texas, Austin

Dr. Richard H. Crawford is a Professor of Mechanical Engineering at The University of Texas at Austin and is the Temple Foundation Endowed Faculty Fellow No. 3. He is also Director of the Design Projects program in Mechanical Engineering. He received his BSME from Louisiana State University in 1982, and his MSME in 1985 and Ph.D. in 1989, both from Purdue University. He teaches mechanical engineering design and geometry modeling for design. Dr. Crawford's research interests span topics in computer-aided mechanical design and design theory and methodology. Dr. Crawford is co-founder of the DTEACh program, a "Design Technology" program for K-12, and is active on the faculty of the UTeachEngineering program that seeks to educate teachers of high school engineering.

ASEE 2016
INTERNATIONAL FORUM
NEW ORLEANS, LOUISIANA / JUNE 25, 2016

Paper ID #17518

Dr. Gabriel Hoongwen Wong, DSO National Laboratories

SUSTAIN International Science and Technology Collaboration: Methods & Best Practices for Non-Co-located Engineering Project Teams

Introduction: The purpose of this paper is to provide important information for others who undertake a non-co-located collaboration. This paper describes the experiences of and lessons learned by the Singapore-U.S. Tactical All-Inclusive Navigation (SUSTAIN) team in their effort to create innovative new navigation technologies for dismounted soldiers. Discussion will 1) highlight the issues that were important to address in SUSTAIN's attempts to effectively collaborate, 2) describe some of the tools used in the effort to facilitate the collaboration, and 3) suggest a set of best practices for future international research collaboration.

Literature Review: Dixon (2007)[1] reports on his experiences teaching project management in an international collaborative setting involving three universities, specifically highlighting the logistical hurdles and coordination challenges he experienced. He notes that "robust communications and communication systems" are vital for international collaborative teams to "survive and thrive." He also observes that the greatest hindrances to communication are often time zone differences, technology infrastructure disparities, and technical proficiency disparities among team members. He noted that the simple initialization of the project floundered due to lack of leadership and proactivity, but eventually was underway after the group established sub-team points-of-contact (POCs), established regular communication methods and times, and developed evaluation rubrics to judge the effectiveness of the collaboration. Dixon also noted that effective international collaboration in his effort hinged on two key technological requirements: international availability and cost. Dixon lists some "lessons learned" in regards to his international team collaboration involving three universities, noting in particular the importance of technological parity between partner, team communication, leadership buy-in and support for the project, cultural influences like time-of-work and interaction styles, and participant (and leadership) levels of interest in the project.

International "field trips" and temporary international "work placement" benefits engineers by providing first-hand knowledge of engineering work in a foreign country, as well as providing cultural insights into their methods of defining and solving problems while challenging conventional notions of problem definitions and solutions[2]. According to Downey et al. (2006)[2], it is "critically important" to gain educational and work experiences with people who were raised and trained in other cultures and countries, to understand dominant science and engineering images and patterns of work. Such exposure also helps develop future leaders and may improve the ability to anticipate, understand, and respect other perspectives, which may also improve conflict understanding and conflict management due to cross-cultural factors. Perhaps most importantly, such international collaborative work seems to provide development of non-technical expertise; i.e., the human, social, and cultural dimensions of engineering and scientific work[2]

The SUSTAIN Mission: The Singapore-U.S. Tactical All-Inclusive Navigation (SUSTAIN) initiative was a twelve month international research and development collaboration between military and academic scientists from both Singapore and the United States culminating in a final technical demonstration to U.S. and Singapore leadership. The concept for SUSTAIN came from a previous project model commonly used in Air Force Research Laboratory (AFRL) called

Commander's Challenge in which a high ranking military customer presents a current real world military problem, and a team of early-career engineers is assembled to rapidly produce an innovative technology solution. However, the SUSTAIN initiative took this project model one step further by including international collaboration with government research scientists from an allied nation. This gave the SUSTAIN initiative a dual purpose: 1) to advance technology in GPS-denied navigation and 2) to build long-lasting international ties, facilitating international collaboration between the two defense research organizations for years to come.

SUSTAIN's overarching mission statement was created based on mutual interests between the U.S. and Singaporean governments. It was agreed that SUSTAIN's mission was to create an innovative process, procedure, and/or technology for dismounted soldier navigation under GPS-degraded and GPS-denied situations. Products demonstrated by SUSTAIN were required to provide precise localization, precise orientation, precise timing, position marking, waypoint management and guidance and be highly portable and easily usable. The technology products presented at the final SUSTAIN technical demonstration were meant to provide innovative, unique solutions to a hard "challenge" problem with high interest from U.S. Department of Defense (DoD) and Singapore Ministry of Defense (MinDef). However, what use cases and specific technology areas were focused on in this effort depended heavily on the individual interests of the two countries. Singapore needed improved GPS–degraded navigation in urban canyon use cases while the U.S. needed improved navigation in triple canopy jungle environments for its special forces. Both countries had previous and ongoing research in different specialized navigation technologies, which influenced their choices of technologies to develop through SUSTAIN. U.S. and Singapore leadership compromised on the scope of the SUSTAIN mission. This agreement set the stage for the next year of collaboration.

Collaboration Challenges: Collaboration efforts in SUSTAIN suffered from problems related to distance and time separating team members. Team members were geographically separated by over 15,000 kilometers and a 13-hour time difference. Co-located teams working during the same time have the ability to communicate verbally, see facial expressions, and exchange ideas in real-time. These luxuries were not available to the SUSTAIN team on a daily basis. Also, because Singapore team members were chosen from different organizations and universities, Singaporean team members could not be co-located on a daily basis. Team members in Singapore were located at three different offices: the National University of Singapore (NUS), the Singapore University of Technology and Design (SUTD), and Defence Science Organization National Laboratories (DSO). It proved difficult to coordinate team members' schedules because each member had additional professional responsibilities outside of SUSTAIN which occasionally took priority.

All teams collaborating remotely face challenges in communication, cohesion, and trust. However, distance and time issues increased these challenges, making international communication especially slow. Early in the project, this caused the SUSTAIN team to struggle with team decision making, problem identification, and problem solving. Eventually, the communication breakdown led to program management obstacles such as difficulties assigning tasks or simply deciding as a team what tasks were relevant and feasible to do. Sharing facilities and resources was also difficult, making collaborative testing of prototypes a rare occurrence.

Early efforts were made at a kick off meeting in Hawaii to bring the entire team and its leadership together for face to face collaboration and team building activities. However, as the project continued, distance and time also became barriers to team interpersonal relations, morale and cohesion, and trust. Motivating team members to complete tasks became difficult because often, the team wasn't able to sit together to agree on which tasks should be done and on what schedule. Such problems led to decreasing morale among team members. Trust was a major component in the ability of the team to function cohesively. Without explicit and consistent communication, team members were often wondered who was working on what task, and if members were actually accomplishing the work promised. As most members did not have prior relationships, there was also a technical trust issue: did teammates have the required expertise?

Other obstacles stemmed from political barriers to international collaboration. SUSTAIN was unique in this because it involved collaboration between the defense organizations of both countries. Team members from both the U.S. and Singapore each came in with their own national secrets and certain proprietary technologies that were not legally sharable with the other country. This meant that certain high level technologies and data were left unavailable to the SUSTAIN team. Because all team members came from military and government backgrounds, it was well understood and accepted that this was the case, and trust between team members did not suffer significantly as a result. Differences in national political focus had a significant effect on team composition, time allocated to project, and project goals. The U.S. and Singapore governments were each interested in different use cases for the navigation technology SUSTAIN was tasked to produce. This widened project scope combined with struggling international communications contributed to a very loosely defined set of goals for SUSTAIN at the beginning of the effort, which caused further program management difficulties later.

Work place structures, resources, and work habits also played important roles in affecting program management, execution, and communication. Multi-tasking, was a common work place habit between both Singapore and U.S. sides of the team. The organizations team members were sourced from often expected team members to continue maintaining their previous workloads along with the SUSTAIN workload. Team members were exposed to additional work commitment from their individual organization without consideration for the SUSTAIN collaboration. For example, two of the Singaporean team members were full-time professors at their universities. This led to team members needing to juggle between making progress for the SUSTAIN project or contributing to other work, which was at times more urgent. U.S. team members faced the same dilemma, however not to as much an extent, most likely because the American members of the team were provided a separate office space in which they could be co-located to focus solely on the SUSTAIN effort. These split motivations caused some rift in international collaboration and trust between team members.

Not only were the team members dispersed across countries, but so were their resources. Team members located at the universities in Singapore were able to delegate tasks to their students to generate and do initial prototyping of ideas for the project, and members from DSO had access to existing hardware resources to expedite the prototyping process. Sending prototypes or hardware across international borders was neither cost nor time efficient. Therefore, U.S. and Singapore sides of the team sometimes had to duplicate prototyping efforts in order to collaborate.

Differences in team members' skillsets and backgrounds resulting from work place structures also played a role in the collaboration. The AFRL focuses on program management and acquisition skills. U.S. team members, gravitated to organizing the SUSTAIN project based on the acquisition principles they were taught. Singapore team members came from research organizations and universities, where work place structures were more academic. They were more apt to focus on prototyping and experimenting than on program management. Individual culture challenges did not affect the team as significantly. All team members from Singapore spoke fluent English and had received a portion of their education in the United States. These team members, though from a foreign country, were already well adjusted to American culture and language. Also, the country of Singapore boasts an impressively diverse population both ethnically and culturally, which resulted in the Singaporean members of the SUSTAIN team being well accustomed to working with others of different nationalities. Members from the U.S. were generally less accustomed to international collaboration. However, one U.S. team member had been a Vietnamese immigrant and was already familiar with Asian culture. And one U.S. team member had already collaborated with members from Singapore on previous projects. These levels of familiarity, which were already in place at the beginning of the SUSTAIN effort paved the way for easy development of relations between team members.

Collaboration Tools and Methods Used: The first step in beginning the SUSTAIN project was building the team. Each country selected team members separately. Member selection on the U.S. side consisted of an advertisement for volunteers from junior force military and government employees in the AFRL and an interview process. Candidates were interviewed and accepted based upon their technical experience, program management skills, creativity, and personality. Specific experience in the field of GPS and navigation was not required. Five team members were selected: three government scientists and two Air Force officers. Singapore leadership selected team members on a non-volunteer basis, emphasizing expertise and research experience in GPS navigation, all four with military backgrounds. Both nations had senior leaders within their defense organizations who championed the project and access to military and technical subject matter experts (SMEs). The result was a diverse team with members from both military and civilian backgrounds and many different areas of expertise: GNSS navigation, electrical and mechanical engineering, psychology, computer programming, program management, etc.

At the start of this collaboration, the SUSTAIN team was made up of nine total strangers put together in a team. All team members and leadership attended a face to face kick off meeting to officially start the project. Tools based on concepts from industrial organizational psychology were used to analyze the make-up of the team and formally educate members about how to effectively work together at a face to face kick off meeting. Team members and leaders participated in a formal analysis of individual personalities and creative habits by taking a Meyers-Briggs personality test and a 6-Hats personality test. The results of these tests were collected, presented to the team, and discussed at length with advice provided on the how to understand and work together with members of different personalities. Initial team bonding and icebreaking exercises were used at the meeting to help members come together as a team. These activities included events such as mountain climbing and dining at local restaurants so team members could interact in a more relaxed environment. Spending time together outside work helped team members to understand the various personalities and areas of expertise in the group and to build trust and familiarity with each other. However, team cohesion activities could only be planned over short periods of time as members could not spend an extended amount of time away from their home organizations. In substitution for face to face time, chat groups (text and video) provided a good alternative for maintaining relationships between team members.

At the four month point in the project, the U.S. members of the team visited Singapore to work closely with their Singaporean counterparts for two weeks. Likewise, at the nine month point in the project, the Singaporean team members visited the U.S. for two to three weeks. It was during these key times that the most effective collaboration was achieved. During these weeks, normal business hours were spent in team meetings, in group efforts to tests and evaluate prototypes, and in touring the facilities each government organization had to offer. Face to face team meetings with all members in attendance provided a platform for vital team decision making. Group testing allowed for all team members to not only feel involved in the technical aspects of the project but also to learn from each other and gain expertise in fields they'd not studied in before. Exchanging laboratory tours allowed each country to showcase its scientific accomplishments, consider possible technical areas for future collaboration, and understand what tools were available for use by the SUSTAIN team. The team spent evening hours hosting each other at dinner and participating in team recreational activities to build cohesion.

The team needed ways to coordinate tasking, communicate ideas, and manage knowledge. The team also needed to accomplish coordinated software development. Because this project was entirely "unclassified" (the lowest level of government information security requirements), both nations permitted use of private cloud services. Due to the small budget and short twelve month time allotted for the project, it was important to find low cost and easy to use communication tools. No one tool could be found that perfectly integrated all necessary forms of communication. So, several separate software communication tools were chosen.

Efforts were made to perform program management through software tools widely used in business today. To coordinate tasking, the Atlassian JIRA software service was purchased. JIRA allows project managers to convert a work breakdown structure into individual tasks. Ultimately, JIRA was set aside because the amount of time required to operate JIRA made the tool less convenient. Also, many tasks in this basic research environment project were too small to warrant capturing in JIRA. To fulfil knowledge management and document and data sharing needs, Atlassian Confluence was chosen. Confluence is a wiki-format tool that permitted attaching external files, simultaneous group document editing, and file-change merging, and searching articles. One of the reasons a feature-rich wiki was successful was because both nations had experience using the tool and immediately saw benefit. Members' initial familiarity with the tool made consensus on its use easily attainable. The team hosted computer code written for the project in Bitbucket, a cloud-based repository from Atlassian.

In addition to written forms of communication, collaboration required frequent conversational communication. Brief daily team discussions were captured in an internet relay chat (IRC). The team used Slack for this service when computers were available but used the group messenger What's App for same-day logistical conversations. Both services are no-cost to send messages internationally. The team hosted a weekly video teleconference (VTC) to check in. At the time of this paper, Google Hangouts provided the best ability to allow personnel to call in from multiple locations with good video and audio quality at no monetary cost. While this method alone was insufficient to accomplish mission objectives, it did provide face to face time and an opportunity for team members to report on their individual activities and plan for future tasks. The U.S. was 13 hours behind Singapore, making both evening and morning video calls viable.

Due to the SUSTAIN project's bilateral military research aspects, the SUSTAIN project had unique security needs. The project required additional legal authorization. The project scope was restricted to basic research, which the U.S government generally considers as "unclassified". Military leadership from both nations signed an international agreement, authorizing personnel to collaborate. This required approval of the U.S. Foreign Disclosure office as well as editing portions of information so that important data, documents, and concepts were shareable without revealing sensitive military operational or technical details. When consulting experts from academia, industry, and both governments, it was important for SUSTAIN members to inform SMEs that the nature of SUSTAIN involved international partners, and that any information SMEs shared must either be: 1) shareable with foreign partners, or 2) clearly defined or noted as non-shareable. Establishing this policy up front and early allowed all parties to agree on how to manage information security.

One key technology area in which communication and collaboration proved challenging was collaborative computer software and hardware development in conjunction with the prototype technologies SUSTAIN produced. The team's software and computer code experts were located in separate countries and came with varying levels of software development experience. Due to the distance between members, it was impractical to integrate software and hardware components routinely or to do in-person identification of development problems. Team members collaborated on how to control versions of software and hardware as well as what commercially available tools and code libraries to use in order to manage system configurations. Software was shared on a joint repository. For simplicity, team members used the same integrated development environment (IDE) for code development. Builds were automated as much as possible using software package managers. The NuGet repository provided libraries in both C++ and .NET(C#). Similar repositories were available in Java and C#. The team decided what hardware would be duplicated at both U.S and Singapore work sites and what would be prototyped at only one site.

The SUSTAIN project officially began at the kick off meeting, and initial attempts were made to set in place traditional project management methods. Team project leaders were chosen. Roles of other team members were defined. A tentative schedule for the project was made using gantt charts. Collaboration tools were chosen, and a strategy for communication was developed. The scope of the project was left wide to leave room for innovative thought. However, it was assumed that the scope would later be narrowed once early background research and ideation efforts were accomplished. Most of these original plans were subject to change as the project went forward due to unforeseen communication, scheduling, and technological complications.

The nine members of the core SUSTAIN team chose from among themselves one project leader from Singapore and one project leader from the U.S. The responsibilities of these project leaders were to lead weekly video chat meetings, serve as an interface between the team and government leaders and subject matter experts, oversee the work of other team members, and conduct overall project management. One team member was tasked with managing budget and inventory. A sub team was created for computer software development, and other sub teams were formed as needed for smaller tasks within the project. These early attempts to assign team members to basic offices faltered as individuals' schedule conflicts surfaced and communication lapsed in the first months. Therefore, the team's organizational structure had to become more fluid. Tasks were assigned on a volunteer basis. Team members chose tasks they could complete based on their individual schedules and expertise. As time went on, new navigation solutions were thought of, and team members joined sub teams for prototypes at their leisure. Often, smaller prototyping efforts that failed to garner support from at least two or three team members did not continue to make progress. In other cases, multiple team members were able to unite around a prototyping effort, and progress on that prototype proceeded rapidly. If a member wished to champion a prototype, he or she had to first lobby for interest in that prototype with the rest of the team, receive feedback, and gain consensus on whether or not the proposed prototype was viable, and appealing to potential users. This interest-based method of prototype selection developed slowly, resulting from the absence of definitive technical direction from government leadership.

A critical step in the research and prototyping process was to narrow and completely define the scope of the SUSTAIN project. Early in the project, background research both inside and outside the field of navigation was encouraged in order to promote creative thought. However, the need for a well-defined problem space and purpose quickly became apparent as team focus lapsed without specific direction. Thus, considerable collaboration time was spent defining scope of problem while identifying problems of individual interest and mutual interest to both nations.

Best Practices for International Collaboration: The SUSTAIN team identified many best practices for international collaboration that other teams should consider. Future teams would benefit from a longer face to face kick off meeting to begin their project. Key tasks accomplished at this meeting should be as follows. 1) Arrange time for team bonding activities and introductions to all cultures and nationalities present in the group. 2) Train team members in using online communication tools, project management skills, and other specific skills necessary for the effort. 3) Establish a communication strategy which includes consistent, regulated communication. It can be concluded that the most effective collaboration will be achieved if all team members agree on what communication tools to use and a strategy of how to use those tools in a uniform way. 4) Create the initial budget plan and schedule for the effort, which should include major milestones and performance measurements as well as scheduled face to face meetings and reviews by leadership. 5) Leadership and upper management should fully define the scope of the effort. If the effort is meant to be an innovative technology effort like SUSTAIN, leadership should clearly define a customer base for the technology to be produced.

Best communication practices as international collaborative projects continue are as follows. 1) Use online chat tools which allow hyperlinks, attachments, and images to be sent as well as text. These should be used to enhance daily conversation between team members. 2) Create knowledge and data repositories for team use. Knowledge management software in the form of a wiki has potential to greatly enhance a team's collaboration ability. For data and document sharing and daily conversation, these written communications were found far superior to email methods. 3) When co-location isn't possible, team members should make use of online video chat tools for regular communications. Weekly video chat meetings with all team members in attendance are highly encouraged. These weekly meetings provide a way for team members to report on the progress of individual tasks, assigned new tasks, and achieve group consensus on team decisions.

Other advisable collaboration practices include maximizing face to face collaboration time, efficiency of workflow, and leadership involvement. If team members are located in separate countries, arrange for all members to exchange visits to each other's countries. This helps team members to gain an appreciation for the different cultures on the team and maximizes face to face collaboration time. Leadership should make the best use of resources provided by their organizations to make these visits cost efficient. Avoid duplicating work due to lack of co-location. If a team member is an expert in a certain area of knowledge or the performance of a certain task, that team member should be tasked to lead that particular portion of the effort, and if necessary, a sub-team should be formed around that person. This will permit the rest of the team to move other tasks forward, promoting efficient workflow. Leadership should ensure the team is provided with its own workspace (with separate sites in different countries if necessary) outfitted with all essential office tools, where team members can be co-located as much as possible, better communicate, and avoid distractions arising from outside workloads. It is recommended to have a member of leadership present at each work site on a regular basis to review team progress and provide accountability, motivation, and guidance.

Conclusion: The success of the SUSTAIN international collaboration depended heavily on new technology in online communication, the expertise of the SUSTAIN team members in both scientific research and program management, and continuous support from high ranking leadership. Not only was the SUTAIN project an effort to bring innovative technology to its customers but the project was also an exploration of new methods and best practices to implement when collaborating internationally on scientific research and engineering efforts. One of the goals of the SUSTAIN project was to pave the way for future international science and technology collaborations between the U.S. and other nations. Teams wishing to undertake collaborations similar to that of SUSTAIN should consider the benefits of global engineering competency to be gained as well as the difficulties of bringing together a multinational team.

Possible language and cultural barriers must be addressed before the project begins with training for team members in order to ensure that the collaboration is feasible. Distance and time between team members must be overcome using methods that are agreeable and usable by all parties. Teams should take advantage of face to face interactions wherever possible. Consensus must be achieved on communication tools used and how to use them to create a uniform communication strategy. Especially if the time frame of a project is short, leadership must give early guidance to form a project management strategy and must remain available to team members throughout the project. Detailed planning, leadership involvement, and cohesive, motivated team members will ultimately decide the success of international collaboration.

Disclaimer: This document reflects the official position of neither the United States Air Force nor the Singapore Ministry of Defense. Opinions are exclusively that of the authors. Mention of products does not constitute endorsement of either the United States or Singaporean government.

1. Dixon, G. (2007). "Teaching project management with international collaboration." Proceedings of the 114th ASEE Conference, Honolulu, HI, USA, June 24-27.
2. Downey, G. L., Lucena, J. C., Moskal, B. M., Parkhurst, R., Bigley, T., Hays, C., Jesiek, B. K., Kelly, L., Miller, J., Ruff, S., Lehr, S. L., & Nichols-Belo, A. (2006). "The globally-competent engineer: Working effectively with people who define problems differently." *Journal of Engineering Education*, April 2006, pp. 107-122.
3. Otto and Wood, *Product Design: Techniques in Reverse Engineering and New Product Development,* 2000.

India and Japan Joint Project-Based Learning -What was Learned from the Design Thinking Workshop-

Ms. Ayano OHSAKI P.E., Okayama University

Ms Ayano OHSAKI is an assistant professor at Okayama University. She also works as an assistant professor at the Innovation Center for Engineering Education, Tottori University since 2012. She is in charge of development new engineering education program. The objectives of the program are improvement of creativity, collaboration skills and problem solving skills. Students learn communication skills, project management skills, analysis, etc. by working on design assignments and projects in this program. More than 400 students are studying in this program. She is writing a textbook and developing an assessment system for this program.

She is pursuing her doctoral research in Computer Supported Collaborative Leaning (CSCL) and the Engineering Design. Prior to this, she also has over 5 years of experiences as a Production Engineer. She designed a jig, production processes and production systems for on-vehicle unit systems.

India and Japan Joint Project Based Learning
-What is the Learnt from Design Thinking Workshop-

This study proposes a new instructional method with design thinking for international engineering education program. I had international PBL (project-based leaning) courses with Indian students twice times. Those PBL aims to learn as follows: (1) Intercultural and diversity mind, (2) Project Management, (3) Design Thinking, (4) Engineering knowledge and skills. To make suggestion for a new instructional method, I analyzed the daily reports and final reports whom Japanese participants of two courses write as Design Based Research (DBR). DBR is the one of the research method for learning. According to BARAB and SQUIR (2004)[1], Design Based Research is "a series of approaches, with the intent of producing new theories, artifacts, and practices that account for and potentially impact learning and teaching in naturalistic settings". Reeves(2006) made a process flowchart of DBR as fig.1 [2].

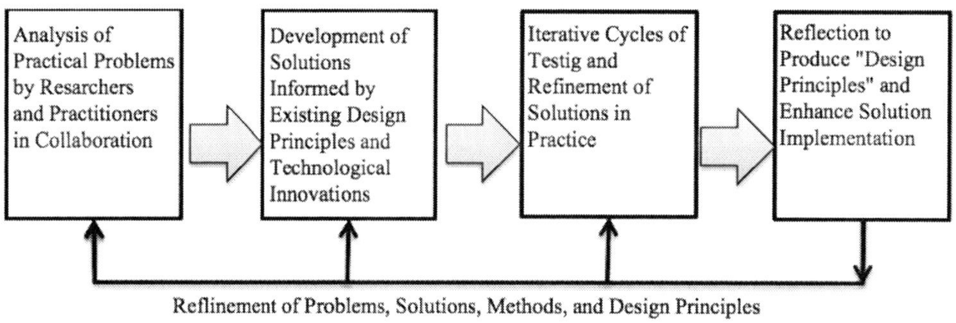

Fig.1 process of DBR [2]

Higher education of engineering needs to develop global engineers with both knowledge and problem-solving skills. Especially, Japanese engineering education has focused on engineering design. The Japan Accreditation Board for Engineering Education defines engineering design as an "open-ended and creative problem-solving process based on the knowledge of science, technology, and social sciences and humanities"[3]. Engineering design is interdisciplinary and diverse problem solving related to engineering. Recently, many Japanese universities have engaged in international PBL [4][5][6]. However, there has been little research on PBL's effects in the international environment. In consequence, we should discuss about the between educational effects of PBL and learning environment. We must define the learning objective in this program. I set "Design Thinking"[7] as main objective, because Design Thinking has paid attention for the methodology of innovation. Many Universities has special curriculum for Design Thinking such as Stanford University,

USA [8], University of Tokyo, Japan [9] etc.

PBL is a construction of Project-based learning or Problem Based Leaning. According to Yuasa at el (2011)[12], it is focused on learning new knowledge and skills in Problem-based learning, and it is focused on using their knowledge and skills in Project Based Leaning. But a teacher has a role of a facilitator for learns in both of PBL. Furthermore, Barron at el (1998)[13] suggested "that there are strong advantage to pairing problem-based leaning and project-based learning", because problem-based learning is scaffold activities in project-based learning. Consequently, there is no distinction between problem-based learning and project-based learning in this paper. I set activities to learn new knowledge and skills and using students' knowledge and skills through this PBL.

Students learn from their experiences and community in PBL. Therefore, the teacher has to design learning environment based on the theory of learning. Scardamalia (2002) defined twelve design principles to clarify about knowledge building and the technology that supports it as table1 [14].

I have started this international workshop program since 2015. I designed learning environment with design principles for knowledge building in both of programs (table 1) (table 2)(table 3).

In 2015, three Japanese students and six Indian students took part in this program. The theme of 2015's program is "Application software with Augmented Reality (AR) for Visitors". Students developed an application software that people can use on android smartphone. Indian students and Japanese students developed new application software for each other. Students ware required knowledge of JAVA, knowledge of AR, graphic design skill and programing skill as engineering knowledge and skills. All of Japanese students came from engineering department, but major was different. One was chemical engineering, another was mechanical engineering and the other was information technology. Three Japanese students had another PBL before visiting India in 2015, They made prototypes of other software. But Japanese participants ware not in the same group. Two students ware in the same group with another student, one student had a project alone. Japanese students had a meeting for an hour and a half after class of the other project weekly. Students and a teacher discussed about Indian culture and prepared visiting India in that meeting.

In case of 2016, three Japanese students and four Indian students took part in this program. The theme of 2016's program is "Health food robotics". Students developed a robot for helping healthy life. Indian students made an application software of the robot, and Japanese students made a body of the robot. Students ware required knowledge of arduino [15], knowledge of circuit design, programing skill and techniques of processing and making something by hand as engineering knowledge and skills. One Japanese student came from agriculture department, however he had experiment of making other robot. The major of other two Japanese students was applied science. They came from engineering department. The program of 2016 started officially from middle of December, 2015..Students did not have much time to prepare for real collaboration term. On the other hand, I revised that program from the experience of 2015. The first revision was setting a starting report. It aimed that students concentrated their learning objective. In addition, I made mixed team with Indian and Japanese to share their knowledge.

Table1. Design principle and learning environment of experiments

	Design principle	2015 program	2016 program
1	Real ideas, and authentic problems	To solve visitor problems	To solve a participant's problem
2	Improvable ideas	Mindset of radical collaboration	
3	Idea diversity	All of participants can ideate	
4	Rise above	Having reflection time	Making reflection report
5	Epistemic agency	Setting a role of teacher as facilitator	
6	Community knowledge, collective responsibility	Distributed responsibility for the project	
7	Democratizing knowledge	Mindset of radical collaboration	
8	Symmetric knowledge advancement	Participants have difference role	Participants have difference knowledge
9	Pervasive knowledge building	Mindset of radical collaboration	
10	Constructive uses of authoritative sources	Mindset of radical collaboration	
11	Knowledge building discourse	Participants define the goal of project	

12	Embedded and transformative assessment	Evaluation by participants

Table2. Details of two years program

	2015 program	2016 program
Project theme	Application Software with Augmented Reality for Visitors	Health Food Robotics
Participants	3 Japanese, 6 Indians	3 Japanese, 4 Indians
Preparation lessons	37.5 hours	10.5 hours
Online meeting with Indian student	Once 35 days before visiting India	Once 5 days before visiting India
Real collaboration term	8 days	8 days
Daily report's theme	What are your today's activities and what did you learn from the activities?	Please explain about your today's activities and what did you learn from the activities.
Starting Report's theme	—	What is your opinion for International project?
Final report's theme	What is your learnt	What is your learnt * Using more than 400 characters

Table3. Schedule of real collaboration term

	2015 program	2016 program
First day	Meeting	Meeting
Second day	One day lecture	Half day lecture
Third day	Half day lecture	One day lecture
Fourth day	One day experiment	One day experiment
Fifth day	One day experiment	One day experiment
Sixth day	One day experiment	One day experiment
Seventh day	One day experiment	One day experiment
Last day	Final presentation	Final presentation

In this section, the results of comparing the two experiences are shown. At first, I counted the description about learning objectives. Table 4 is the percentage of description about learning objective in final report. All of students write the learning objective, which is taken as 100%. Secondly, I counted the number of characters of students' report to analyze students' learnt (Table 5) for analysis the relationship between activities and learnt. It was difficult for us to analyze learning of students from a number of reports' characters, because it had a considerable amount of variance. At last, I categorized the students' report and counted the number of times each word was used in them for quality analysis. I divided the students' reports into three categories: learning objective, relationship with the team and emotion. Moreover, I distributed these descriptions into eight subcategories: intercultural and diversity mind, project management, design thinking, engineering knowledge and skills, contribution, be supported, motivation and incompetent. . Each subcategory's meaning and sample comments are shown in table 6. All of students write the word every day, which is taken as 100% in table 7.

Table4. The percentage of description about learning objective in final report

Leaning objective	2015 program	2016 program
Intercultural & diversity mind	100%	100%
Project Management	67%	67%
Design Thinking	67%	67%
Engineering knowledge & skills	67%	100%

Table5. The amount of characters in students' report

	2015 program		2016 program	
	Average	Variance	Average	Variance
First day	1091.3	568.2	521.7	150.6
Second day	1732.3	724.2	799.0	126.5
Third day	1002.3	211.9	731.3	211.5
Fourth day	779.3	341.0	633.7	75.8
Fifth day	505.0	349.8	864.7	170.7
Sixth day	605.7	373.1	745.7	135.3
Seventh day	540.3	195.7	638.0	272.5
Last day	1202.7	521.6	705.0	187.3

Final Report	653	534.6	407.0	22.1

Table6. The category for daily report analysis

Category	Sub category	Sample
Leaning objectives	Intercultural & diversity mind	I watched Indian style cooking.
	Project Management	I thought that sharing information was important.
	Design Thinking	From the point of view of radical collaboration,
	Engineering knowledge & skills	They had high level programing skill.
Relationship with the team	Contribution	I could suggest about layout of buttons.
	Be supported	He helped me to install new software.
Emotion	Motivation	I want to work with them in the future.
	Incompetent	I was discouraged.

Table7. The percentage of category in daily report

Sub Category	2015 program	2016 program
Intercultural & diversity mind	92%	67%
Project Management	75%	71%
Design Thinking	42%	25%
Engineering knowledge & skills	33%	63%
Contribution	4%	21%
Be supported	21%	46%
Motivation	42%	83%
Incompetent	63%	38%

Students learned about learning objectives through international PBL were confirmed from final report. Additionally, the number of students who wrote about all of learning objective had increase from the program on 2015 to 2016.

The cause of considerable amount of variance in distinguishing the difference of the writing styles among students is as follows: (1) a student wrote in her one of her reports such as

“9.00am breakfast. Today’s menu: bread, curry, strawberry yogurt”. (2) Another student wrote in his one of his report such as “I could study a lot of things from the point of view of intercultural project”. I did not set a format and minimum characters for the daily report, because I was interested in their curiosities. Consequently, the numbers of character of students’ report change depending on students' motivation to deliver a report..

All of students gave a detailed description of their activities in daily reports. It is assumed that students learned lots of things from their experiment. Moreover, we can confirm that students focused on their project more than the other learning objectives. In addition, students’ curiosity moved to “engineering knowledge and skills” and relationship with team by setting theme. All of students had to use engineering knowledge and skills in 2016 program. In case of 2015 program, one student did not use engineering knowledge and skills. He wrote about engineering knowledge only once.

Besides, descriptions of relationship with the team had increased from 2015 to 2016. One of the factors behind this was the 8th design principle of knowledge building. Japanese students did not have as much knowledge and skill as Indian students. They should ask about Engineering knowledge and skills to develop new product together.

On the other hand, descriptions of incompetent and Intercultural & diversity mind ware decreased from 2015 to 2016. There is a possibility that students could not make efforts to build rapport with the teacher. For example, participating students of 2015 used many young people's words and colloquial expressions when they wrote their report. Participating students of 2016 could not have other PBL before visiting India. Hence students could not write their thinking and emotion honestly.

Students learned leaning objectives of the PBL as follows: (1) Intercultural and diversity mind, (2) Project Management, (3) Design Thinking, (4) Engineering knowledge and skills through international PBL. In addition, My work suggests three design principles for International PBL as follows: (1) real problem to solve by the result of the project, (2) Opportunity for Sharing engineering knowledge and skills between participants, (3) Preparation team to build rapport with the teacher.

In this research, I could not analyze the result of learning by static, because participants were very few students. Hence, I analyzed students’ learning and activities using by framework of

DBR. We should analyze more detail of students' activities and descriptions from now on.

Ms. Ayano OHSAKI is an assistant professor at Okayama University. She also works as an assistant professor at the Innovation Center for Engineering Education, Tottori University since 2012. She is in charge of development new engineering education program. The objectives of the program are improvement of creativity, collaboration skills and problem solving skills. Students learn communication skills, project management skills, analysis, etc. by working on design assignments and projects in this program. More than 400 students are studying in this program. She is writing a textbook and developing an assessment system for this program.
She is pursuing her doctoral research in Computer Supported Collaborative Leaning (CSCL) and the Engineering Design. Prior to this, she also has over 5 years of experiences as a Production Engineer. She designed a jig, production processes and production systems for on-vehicle unit systems.

[1] Barab, S., & Squire, K. (2004). Design-Based Research: Putting a Stake in the Ground, *Journal of the Learning Sciences*, 13(1), 1-14

[2] Reeves, T. (2006). Design research from a technology perspective. *Educational design research,* 1(3), 52-66.

[3] Quality Assurance for education and Engineering education (1990). The Japan Accreditation Board for Engineering Education. http://www.jabee.org/public_doc/download/?docid=27

[4] Cross Border PBL (n.d.). Yamaguchi University. http://mot.yamaguchi-u.ac.jp/AIC/pbl.html

[5] Kumamoto, K., Zhou, H., & Yoshimura, T. (2015). International PBL in Osaka Institute of Technology. *IEICE Communications Society GLOBAL NEWSLETTER*, 39(2), 9-10

[6] The report of international capstone design competition (2014). Kumamoto University. http://cedec.kumamoto-u.ac.jp/project/kokusai_mono.html

[7] Brown, T. (2008). Design thinking. *Harvard business review*, 86(6)

[8] d.school (n.d.). Stanford University. http://dschool.stanford.edu/

[9] i.school (n.d.). University of Tokyo. http://ischool.t.u-tokyo.ac.jp/

[10] Yuasa, K., Oshima, J., & Oshima, R. (2011). Characteristics and Effectiveness of PBLs : Problem-Based Learning & Project-Based Learning, *The research of information, Shizuoka University*, 16, pp.15-22.

[11] Barron, B., Schwartz, D., Vye, N., Moore, A., Petrosino, A., Zech, L., & Bransford, J. (1998) Doing With Understanding: Lessons From Research on Problem- and Project-Based Learning, *Journal of the Learning Sciences*, 7(3-4), 271-311,

[12] Scardamalia, M. (2002). Collective cognitive responsibility for the advancement of knowledge. *Liberal education in a knowledge society*, 97, 67-98.

[13] arduino. (n.d.). https://www.arduino.cc/

TEACHING ORAL COMMUNICATION AT A RUSSIAN UNIVERSITY: HELPING ENGLISH LANGUAGE LEARNERS PRESENT THEIR ENGINEERING DESIGNS

Ms. Jennifer Craig, Massachusetts Institute of Technology

Jennifer Lynn Craig is a lecturer in the Department of Comparative Media Studies/Writing at the Massachusetts Institute of Technology where she teaches writing, oral communication, and collaborative skills to engineering students in the Department of Aeronautics and Astronautics. She has consulted in the graduate program of the Department of Mechanical Engineering as well as in industry. Ms. Craig has also consulted and taught in universities in Russia, Singapore, Chile, Costa Rica, and Mexico. She is the author of "Integrating Writing Strategies in EFL/ESL University Contexts: A Writing-across-the-Curriculum Approach." She is a co-author of "Learning to Communication in Science and Engineering: Case Studies from MIT", a book that was the 2012 winner of the CCCC's Advancement of Knowledge award.

TEACHING ORAL COMMUNICATION AT A RUSSIAN UNIVERSITY: HELPING ENGLISH LANGUAGE LEARNERS PRESENT THEIR ENGINEERING DESIGNS

Jennifer Craig[1]
Department of Comparative Media Studies/Writing
Massachusetts Institute of Technology

ABSTRACT:
A writing-across-the-curriculum approach was used to teach and refine oral communication skills of English language learners (ELL) at Skoltech, a Russian university. The objective was to develop disciplinary communication skills in English so that students could present their engineering designs during a rapid prototyping project. A pre/post survey assessed changes in self-efficacy as a measure of success in the instruction about, practice and performance of oral presentations. The post-test survey showed a statistically significant increase in self-efficacy for a majority of the students. Survey data combined with faculty observation indicates that the communication pedagogy combined with practice was effective in increasing self-efficacy and in facilitating and refining oral communication skills for the majority of the Skoltech students. However, the presence of a less active population within this group illustrates some of the challenges in teaching and strengthening disciplinary oral communication skills in a group of ELL with varying levels of English proficiency. Suggestions are made for teaching communication skills to low proficiency ELL in disciplinary settings.

KEYWORDS:
writing-across-the-curriculum (WAC), writing-in-the-discipline (WID), oral communication, ELL, English language learners, self-efficacy

INTRODUCTION

This study assessed the effectiveness of the communication pedagogy, Writing Across the Curriculum (WAC), when used with a cohort of English language learners (ELL) in a Russian university. This paper describes the background for this work, the theoretical framework for the work, the methodology used, and the results. This paper concludes by describing 'lessons learned' and offers suggestions for applying what has been learned to other contexts.

BACKGROUND

The work described in this paper took place during fall, 2013 at Skoltech University in Moscow, Russia as a cohort of 55 students participated in a rapid prototyping design project. In 9 days, small teams (4-5 students) worked with a communication lecturer[2] and a design team[3] from the Massachusetts Institute of Technology (MIT). The project was to design and rapid-prototype a mobile device that would motivate end users to monitor and modify their energy use effectively and efficiently.

Established in 2011 by a group of nine Russian institutions and organizations and in collaboration with the Massachusetts Institute of Technology, Skoltech is a graduate

---

[1] Massachusetts Institute of Technology, 77 Massachusetts Avenue, Cambridge, MA, USA, 02139; jcraig@mit.edu
[2] Jennifer Craig
[3] Federico Casalegno, Catherine Winfield, Cagri Zaman, Jason Lipshin, Laura Poluzzi, Stephanie Lee

university focused on science, engineering and entrepreneurship (Skoltech University, 2014) However, Skoltech's emphasis is not only on high quality education and research but also on innovation and entrepreneurship in these fields. At the time of this project, there were 55 students, but by the end of 2019, the enrollment will increase to 1200 students, most of who may be English language learners (ELL). Thus the underlying research question was how well a pedagogy developed for native speakers would work with ELL's and moreover, how well would it work in a fast-paced, technical project with limited time for instruction and little time for rehearsal.

THEORETICAL FRAMEWORK:

Two theoretical frameworks were useful to this work. Previous work on WAC/WID was the basis for the pedagogy used in this project. Previous work in self-efficacy and group dynamics with ELL was the basis for the survey questions and also helped interpret some of the lack of participation of a small group of students.

Writing-across-the-curriculum pedagogy (WAC) has expanded to encompass not only writing but also oral and graphical communication. In WAC pedagogy, instruction and practice is situated in various parts of a curriculum so that students practice often and with different kinds of material. When WAC is associated with writing-in-the-disciplines (WID), this means that the communication instruction and practice takes place in a disciplinary context, mentored by disciplinary experts with the collaboration of communication instructors. Extensive information about WAC/WID pedagogy can be found on the WAC Clearinghouse web site ("WAC Clearinghouse," n.d.).

WAC/WID pedagogy is characterized by moderate amounts of targeted instruction, frequent practice, feedback from the instructor involved, and meaningful revision by students. In WAC/WID pedagogy, communication is always connected with critical thinking and substantive content. In this way, students learn the standards of excellence in their disciplines and begin to emerge as young professionals who can communicate as knowledgeable 'insiders' in their professions (Bazerman et al., 2005; McCleod & Soven, 1992; Poe, Lerner, & Craig, 2010; Russell, 1991; Zawacki & Rogers, 2011). With its communication-intensive curriculum, the Massachusetts Institute of Technology has used this pedagogy in most departments since 2001, and the overall result has been stronger student outcomes in communication skills (MIT Communication Requirement, 2014)

Although WAC/WID pedagogy is widely practiced and researched in the United States and in Western Europe, it is a pedagogy developed with native English speakers, and its use with English language learners working in the target language has not been widely explored. Thus more precise data on the effectiveness of this method was sought through a survey based on Bandura's classic work in self-efficacy, the beliefs that a person holds about his/her capability in a specific domain(Bandura, 1994). While self-efficacy is not the only predictor of intention, self-efficacy is highly correlated not only with intention (or motivation) but also persistence toward a specific task (Maddux, 2002; Mau, 2003). Moreover, high self-efficacy appears to allow people to withstand unsuccessful experiences since a person with high self-efficacy attributes that lack of success to lack of effort, not lack of skill or overwhelming environmental circumstances (Bandura, 1971) Since success in a variety of tasks correlates with persistent repetition, then assessing self-efficacy can give a good estimate of the strength of a skill and the likelihood that a person will repeat and refine it.

Yet in a fast-paced, collaborative project in which all of the instruction and the demonstration of knowledge is done in English—a second or other language for all but one Skoltech student— intention and persistence were probably influenced not only by a person's self-efficacy and internal control, but also by external factors, actual individual level of language acquisition, and group dynamics (Ajzen, 1991; Chang, 2010; Dornyei & Malderez, 1999;

Dornyei, 1994) As will be described, many of the Skoltech students were still refining their understanding and use of English, thus presenting a very different dynamic than a room full of native English speakers.

Moreover, the practice of oral communication in the target language is more public than writing. A speaker may then be vulnerable not only to his/her internal assessment of skill displayed but also to the possible competition or comparison by peers. In language learning classes, students are usually separated into homogenous groups so the differences in skill usually are not too disparate. However, in a class or project that uses a target language and in which participants are not sorted by language skill level, such a heterogeneous group is likely to exhibit a range of productive and receptive language skills. Then the unevenness or gaps in language abilities may create obstacles to student participation in the technical work. As Peirce points out, the meaningful use of a second or other language can only be refined in the actual act of communication, and each act of communication either with a teacher or a group of peers involves a display of linguistic control as the speaker/writer can understand and then produce the target language (Peirce, 1995). Peirce goes on to suggest that control over a second or other language is linked to power and influence within a specific social network or domain. In addition, obstacles or imbalances in power and influence may also affect group cohesion and collaboration as well as the motivation of the student with lower language level proficiency(Cox, 2011). Understandably, frustration and vulnerability can reduce a student's sense of self-efficacy in that domain, and low self-efficacy correlates with lack of persistence, reticence, and withdrawal (Idrus, Salleh, Ali, & Hassan, 2013). Bandura notes that people who fall into this cycle are also likely to blame themselves for their lack of ability, often experiencing stress and depression (Bandura, 1994).

THE ORAL COMMUNICATION TASK:
As the eight teams worked on their projects, each team gave five collaborative oral presentations during the nine days. The first presentation was a 5-minute informal talk on the first day's brainstorming. The teams continued with a short presentation of each team's ethnographic research and a day or so later, of their progress, and culminated at the end of the project with a 15-minute formal presentation and demonstration to the design team and their peers. As a final event, each team participated in a pitch competition before a general audience in downtown Moscow where one speaker from each team had three minutes and three slides to promote his/her team's innovative design. All of these presentations were given in English.

The focus of the nine days was on design and innovation, but Skoltech leaders asked for an oral communication element, knowing that communication is a key skill for scientists, engineers and designers, especially those working in entrepreneurial environments. Thus the objective of the oral communication pedagogy was to teach students how to compose and deliver various modes of presentations effectively even as they worked on a fast-paced design project. These outcomes were reflected in the pre/post survey questions and guided by the CDIO standards (CDIO, 2014). CDIO is an international engineering education framework developed at MIT and adopted by Skoltech that focuses not only on the technical knowledge an engineer must have but also on the professional, communication, and interpersonal skills so essential to an engineer's success.

RATIONALE and PEDAGOGY:

Oral presentation is often taught as if it is a single genre and as if one lecture can encompass all the complexities of that genre. In fact, oral presentation is a range of smaller genres with some striking and also subtle differences. Thinking of oral presentation not only as performative but also as communicative, we can see that the spectrum is wide. When the topic is framed this way, students begin to grasp the way that oral communication functions not only at the end point of a project but also how communication allows work to proceed. At one end of the spectrum, there may be the traditional, formal oral presentation or design review. However, the spectrum also may include progress reports, briefings in meetings, design 'crits', desk reviews, bench reviews, to name only a few. Some of these oral communication acts may be minutes long while others may last for several hours. Rhetorical situations and purposes may vary during the life of a project and with shifts in audience. Moreover, engineers and scientists often compose and present as a team, not as a single speakers, so this reality involves a more collaborative and at times heavily negotiated team approach. Clearly, one lecture cannot prepare students to excel along this spectrum.

To be successful, students must be taught to analyze each separate situation. At first, this focus may seem overly elaborate, but students pick up this skill quickly when prompted to think critically about the audience: the listener's objectives in listening, the roles, responsibilities and technical background of those listeners, and the multiple nature of any audience. Then students must be taught to think about their own purpose in speaking. Of course, in engineering and science, rigorous technical content is necessary and that content must be well organized and specific. Based on this analysis (often done in a matter of minutes), a speaker can then determine the scope, style, and organization of what s/he wants to say.

Experienced communicators do this analysis and synthesis without much difficulty, moving efficiently to the appropriate oral communication act. But students are usually novices at this kind of thinking and speaking. Moreover, engineering students may be unused to the standards of excellence in their technical discipline and thus fall back onto older strategies, often adopted from the humanities. Because they are novices, they need to learn actively under the guidance of instructors who help shape student skills.

The design project took place during the first application period at Skoltech, a time in which students were asked to apply the technical skills they had learned during their recent course work. With almost no break, students were plunged into the demanding project. Consequently there was little time to instruct students. Moreover, distance was a factor since the communication instructor did not arrive in Moscow until the first day of the application period.

First, the cohort was surveyed via Survey Monkey to assess their sense of self-efficacy.[4] The response rate was 78% as students answered twelve questions about their confidence in their abilities to perform certain skills in oral communication. The list of survey questions can be found in Table 3 in the appendix.

When the pre-test survey had closed, the instructor posted instructional material posted on MIT's academic site. Students were welcomed to the learning module and told that these would be the concepts discussed in the course on the first day. The concept of the 'flipped' classroom was the basis of this decision (Schell, 2012). The first goal was to have them review this material before they arrived on the first day so that they could begin to apply those concepts to their work on the design project.

Second, since they were ELL, it was important for them to have enough time to read and to absorb information. ELL students often need textual support since the act of listening comprehension in the target language can be taxing for them.

---

[4] MIT Committee on the Use of Human Subject Approval,#1310005921

The third goal was to create a resource to which they could return during and after the course. Again, young professionals----especially ELL---benefit when they have resource material that they can use independently and review as needed.

In the Skoltech classroom, the communication instructor used active learning techniques rather than long lectures. When a student has numerous opportunities—small, large--- to apply concepts, integrating what s/he has learned into what s/he already knows, the student's abilities are displayed, the instructor can support the nascent skills, and students apply their sklls rather than stay at the conceptual level (Bean, 2011).

As in all WAC/WID pedagogy, feedback on communication organization, style, and content was linked to the tasks that the design team asked Skoltech teams to perform. As teams began brainstorming, the communication instructor moved through the large classroom, making contact with each team. If students were working in a productive way, there was contact but little interruption. However, if a team seemed to struggle as they tried to pull their ideas together for communication, the instructor was more active: questioning, clarifying the expectations of the design team, helping limit the scope of material and perhaps suggesting patterns of organization.

Second, as teams stood up to brief their ideas or to present, the communication instructor took notes and emailed these comments to each team. If a team had noticeable problems, they received support and feedback in a face-to-face contact. The pace of the design project prohibited much rehearsal, but the goal was to improve the next oral communication act.

During the nine days, the communication instructor gave short lectures: pointing out common difficulties that teams were having, discussing audience analysis, giving feedback on graphics, and talking about the more effective modes of persuasion, as well as facilitating discussion and answering questions about communication. However, much of the work was done through interactions with individual teams and then through specific feedback as they presented. What this meant was that each team got feedback tailored to their particular challenges and they got the feedback 'just-in-time' as they worked. The design team also gave verbal feedback not only on the design but also on the ways in which the ideas were communicated, so student teams were consistently directed to not only on the substance of their communication but also to the way in which they were composing that oral communication and the way in which they delivered it.

After the last communication event---the pitch competition in downtown Moscow---the post-test survey was administered via Survey Monkey™. Students were asked about their sense of self-efficacy as a result of the oral communication pedagogy. Post-test respondents also answered about frequency of practice of oral communication acts, not only in the nine-day project but also in the preceding academic year. The response rate was 81%

RESULTS AND DISCUSSION

In all measures, a statistically significant increase in self-efficacy was documented after the nine days of instruction, practice, and feedback. Tables 4 and 5 in the Appendix list the self-efficacy measures and show that the sharpest increases came in items 3, 5, 6 and 8. In those items, respondents described increased confidence in their abilities to know what strategy is best suited to a specific context and audience (Item 3; mean = 75; S.D. = 16)). Respondents also described enhanced confidence in their abilities to choose the appropriate combination of media and an appropriate style (Item 5; mean = 75; S.D.=21) (Item 6; mean=77; S.D. 19) Lastly, respondents described an increase in confidence in preparing for oral communication acts (Item 8; mean= 75; S.D. =18).

At the other end of the spectrum, students reported the least increase in their confidence on Item 1, their ability to identify objectives in speaking about their work, (mean = 76;S.D.= 18). As Table 4 shows, there still was a statistically significant increase in this measure, but not the strong increases shown in Items 3, 5, 6, and 8. Part of this lower measure of self-efficacy may have come from the highly directive approach of the design team in which students were not invited to think about their objectives but instead had a specific activity prescribed. Students then were instructed to complete that activity. A slower pace and more autonomy might have given students time to think critically about their objectives, always a key skill for oral communicators.

In regard to frequency of practice, 45% of the post-test respondents reported that they presented in English in a variety of circumstances at least two or three times in the nine days. And nearly a quarter of the respondents presented four or five times. However, as Table 1 also shows, about one third of the group presented once or not at all, a marginal rate of participation with so many opportunities and such small teams.

Table 1: number of presentations given during the 9-day project period

0-1 presentations	2-3 presentations	4-5 presentations
33%	45%	22%

Varying proficiency in spoken English within the cohort may explain the lower frequency of presentation for 33% of the respondents. For example, 57% of students reported they had been speaking English in their technical disciplines for fewer than 3 years. While that length of time might produce conversational fluency, it might not be sufficient to build the skills, disciplinary vocabulary and critical thinking strength in a second language necessary for students to feel confident in an oral presentation.

Despite the relatively short time in speaking English, 61% of respondents reported that they had given more than 6 oral presentations in English in the preceding year at Skoltech (Table 2). Thirty nine per cent of respondents said they had given 6 or fewer presentations in English in the last year. Clearly, the majority of the students surveyed were active in oral communication both in the nine-day project and in the preceding year.

Table 2: Number of oral presentations given in the preceding year of study.

More than 6 oral presentations in English in the preceding year	Fewer than 6 oral presentations in English in the preceding year
61%	39%

To summarize briefly, the communication-intensive WAC/WID pedagogy was used with a population of ELL working in the target language but in an international setting. There were statistically significant increases in self- efficacy on all measures, and the majority of ELL students practiced their English oral communication skills frequently during the nine-day project. Thus, the data reported here answers part of the question asked earlier: yes, WAC/WID pedagogy combined with practice seems to strengthen the oral communication skills of ELL students even in a fast-paced design project. Based on Bandura's work, it is reasonable to think that the increases reported in this study predict intention, which is likely to lead to persistent effort and hence the progressive refinement and transferability of the skill being practiced (Bandura, 1977; Bandura, 1994).

.

However, just as informative is the smaller population that both Tables 1 and 2 show was less active during the nine-day project as well as in the preceding year. No other data was collected on this smaller group and close observation of student behavior was difficult in the classroom that was informal, busy, and fluid as most design classrooms are. However, it was noticed that on multiple occasions, some students disappeared when there was a briefing or presentation opportunity or they yielded the floor to more fluent (in English) students or tended to stand in the back of the large room or even in the hallway, looking in. When asked about where they had been or why they were not part of a particular presentation, those students shyly explained they *'had not felt well'* or had had *'to leave early'*. However, on two occasions, students said of their peers who were presenting, *"_____ is good at English. It's better for the team if she presents."* or *"_____ knows how to say it better."* Even if the reticent students were clearly engaged in teamwork where some of the discussion was often in Russian, these same intelligent, active, and voluble students drew back when there was an opportunity to present in English, formally or informally.

Subjective observation and several informal conversations cannot offer conclusive data on the reasons for the reticence of students in giving oral presentations in English. However, the phenomenon of the reticent ELL student in the midst of busy collaborative discussion has been observed, and it has been conjectured that this behavior may be linked to a lower level of productive---specifically verbal--language proficiency(Cox, 2011; Craig, 2014; Idrus et al., 2013). Such reticent behavior is understandable. Many ELL are accustomed to speaking in language acquisition classes and in those settings with circumscribed topics, their language abilities may seem adequate to the normative language level of that group. However, when there is a shift to a language use setting—the real world---, a student may find himself unable to perform fluently. Without a significant and readily retrievable vocabulary (perhaps in a specific technical discipline) and a ready command of grammar, the lower proficiency student cannot generate enough language to make meaning in that specific context or perhaps he cannot generate it quickly enough. Moreover, his receptive skills may play a part since oral presentation and interaction usually include listening and then responding quickly. Lastly, oral communication is nearly always a public act so the low proficiency student understandably may feel vulnerable. Thus his intention to practice is limited not only by low-self-efficacy, but also by his/her actual skill level, the way in which s/he compares to the norms of those around him/her, and his/her guess about what is feasible to accomplish successfully in that specific context. Here, Ajzen's theory of planned behavior helps us understand a student's sense of efficacy as only one factor in a complex behavioral scene (Ajzen, 1991)

Bandura claims that learners build a sense of efficacy in several ways(Bandura, 1994). Three of these ways are more relevant for the Skoltech study. First, a learner gains self-efficacy by a mastery experience. Second, a learner gains efficacy vicariously as he watches others perform a task. Third, a learner also can bolster his sense of efficacy through persuasion or encouragement from another. Applying this structure to the high proficiency ELL Skoltech students, we see that the majority of them seem to have had one or more mastery experiences. All of the students had the opportunity to learn vicariously from a range of generally well-done range of presentations. And all of the students received support and encouragement.

However, for a lower proficiency ELL student in the Skoltech study, mastery experiences may have been (or seemed to be) out of reach in the fast-paced project. Since the cohort had not been sorted by language level, opportunities to practice oral communication may have felt like potential failure experiences and public ones, at that. The lower proficiency students certainly could have learned vicariously, but in a heterogeneous classroom where some students are very advanced, vicarious learning may quickly swing into unfavorable comparisons and a lower sense of self-efficacy which then may precede less intention, less persistence and more self-blaming. Of course, support and encouragement from peers and from the communication instructor and design team may have persuaded the lower proficiency students of their abilities but only temporarily. Such persuasive effects would be quickly erased by any less than successful experience or even the potential loss of face and stature within the group.

Moreover, the psychological effects of low self-efficacy ELL within a heterogeneous group then can affect group dynamics (Chang, 2010; Clement et al., 1994; Leki, 2001) (Craig, unpublished data). Again, data in the Skoltech study can only suggest but not support this conjecture. However in a small cohort of well-acquainted students, it seems likely that the reticence of small sub-group that may feel discouraged, defensive, and unable to meet the norms of the larger group would have some effect on the group dynamic. Their affect and general stress could not go unnoticed by their friends and their teammates. As Peirce argues, communication exists in a network of social relations that incorporate unspoken agreements about power, authority, and hierarchy (Peirce, 1995). When a group of students does not or cannot conform to the group norm, the difference in their self-image and their self-efficacy can begin to influence their identity and their role within the group and to affect their overall experience.

LESSONS LEARNED; SUGGESTIONS FOR APPLICATION

This brief study seems to indicate that using WAC/WID pedagogy helps strengthen the oral communication skills of high proficiency ELL in their discipline through moderate instruction, repeated practice, and feedback. However, working on a fast-paced design project and also practicing oral communication in English does not seem to produce progress in low proficiency ELL. In fact, subjective observation indicates that there may be a disadvantage to self-esteem in the low proficiency ELL and group dynamics in general.
WAC/WID pedagogy can be used effectively with low proficiency ELL since the principles of WAC/WID apply even with restricted language skills, but this success is more likely in groups that are sorted by language level. (Craig, 2012)(Craig, unpublished data)

However, cohorts in engineering classrooms are rarely if ever sorted by English language levels. Moreover the proportion of ELL within a course may vary from semester to semester, and the types of language issues that they present may change, too. Thus, even when engineering faculty design multiple opportunities for students to practice oral communication, those professors may be challenged by the needs of the lower proficiency ELL in their classrooms.

Engineering faculty who are interested in strengthening English language skills among all ELL students in their courses may incorporate a modified WAC/WID approach in their courses:

- Design a sequence of assignments that allows the ELL to use critical thinking, group discussion, and/or writing to organize his/her English language skills and build vocabulary. Oral communication acts that follow such a sequence are more likely to be successful for the ELL.
- Design 'low stakes' assignments in which an ELL can practice oral communication skills without being penalized for low proficiency but instead can focus on building skills.
- If communication in English is a desired course objective, adopt a pace in the course that allows time for this cognitive task.
- Avoid the temptation of sending the ELL to an external 'English course.' This referral may be well intentioned, but it effectively removes the ELL from the vocabulary and the disciplinary discourse s/he is trying to master.
- If working with a cohort with widely differing English language skills, bringing an English language tutor into the course may be efficient and effective for the low proficiency ELL while allowing the professor to focus on engineering content.

SUMMARY:

The data from the Skoltech study suggests that the WAC/WID pedagogy used here was effective in improving the self-efficacy of the majority of this cohort of ELL as they practiced along a range of individual and collaborative oral communication skills in an intense, focused, prescriptive design project. Data does show statistically significant increases on all measures of self-efficacy. Moreover, observation of the active classroom, and the high quality of oral presentations also showed students' application of these skills. However, it is less clear that this pedagogy was effective with the lower proficiency ELL who were less confident speakers of English and who seemed to evade the numerous opportunities to make oral presentations. However, there are modifications that may allow faculty to see the benefits of WAC/WID pedagogy while still working with both high and low proficiency ELL in engineering courses and design projects..

SOURCES CITED

Ajzen, I. (1991). The Theory of Planned Behavior. *Organizational Behavior and Human Decision Processes, 50*, 179–211.

Bandura, A. (1971). Self-efficacy: Toward a Unifying Theory of Behavioral Change. *Psychological Review, 84*(2), 191–215.

Bandura, A. (1977). Self-efficacy: toward a unifying theory of behavioral change. *Psychological Review, 84*(2), 191–215. Retrieved from http://www.ncbi.nlm.nih.gov/pubmed/847061

Bandura, A. (1994). Self-efficacy. In V. S. Ramachaudran (Ed.), *Encyclopedia of Human Behavior* (pp. 71–81). New York: Academic Press.

Bazerman, C., Little, J., Bethel, L., Chavkin, T., Fouquette, D., & Garufis, J. (2005). *Reference Guide to Writing Across the Curriculum.* (C. Bazerman, J. Little, L. Bethel, T. Chavkin, D. Fouquette, & J. Garufis, Eds.). West Lafayette IN: Parlor Press.

Bean, J. (2011). *Engaging Ideas: A Professor's Guide to Integrating Writing, Critical Thinking, and Active Learning in the Classroom* (2nd ed.). Jossey-Bass.

Chang, L. Y.-H. (2010). Group Processes and EFL Learners' Motivation: A Study of Group Dynamics in EFL Classrooms. *TESOL Quarterly, 44*(1), 129–154.

Clement, R., Dornyei, Z., & Noels, K. A. (1994). Motivation, Self-Confidence, and Group Cohesion in the Foreign Language Classroom. *Language Learning, 44*(3), 417–448.

Cox, M. (2011). WAC: Closing Doors or Opening Doors for Second Language Writers? *Across the Disciplines, 8 (4).* Retrieved from http://wac.colostate.edu/atd/ell/cox.cfm

Craig, J. (2012). *Integrating Writing Strategies in EFL/ESL University Contexts: A Writing-across-the-Curriculum Approach.* New York: Routledge Taylor Francis.

Craig, J. (2014). Teaching Writing in a Globally Networked Learning Environment: Diverse Students at a Distance. In T. Zawacki & M. Cox (Eds.), *WAC and Second Language Writers: Research towards Linguistically and Culturally Inclusive Programs and Practices* (pp. 369–386). West Lafayette IN: Parlor Press.

Dornyei, Z. (1994). Motivation and Motivating in the Foreign Language Classroom. *The Modern Language Journal, 78*(3), 273–284.

Dornyei, Z., & Malderez, A. (1999). The Role of Group Dynamics in Foreign Language Learning and Teaching. In J. Arnold (Ed.), *Affective Language Learning* (pp. 155–169). Cambridge UK:: Cambridge University Press.

Idrus, H., Salleh, R., Ali, R. M. M., & Hassan, W. N. W. (2013). Self-efficacy in English language oral communication skills of technical trainees. In *IEEE International Conference on Teaching, Assessment and Learning for Engneering* (pp. 800–805).

Leki, I. (2001). " A Narrow Thinking System ": Nonnative-English-Speaking Students in Group Projects across the Curriculum. *TESOL Quarterly*, *35*(1), 39–67.

Maddux, J. E. (2002). Self-Efficacy: The Power of Believing You Can. In C. R. Snyder & S. J. Lopez (Eds.), *Handbook of Positive Psychology* (pp. 277–287). New York: Oxford University Press.

Mau, W.-C. (2003). Factors That Influence Persistence in Science and Engineering Career Aspirations. *The Career Development Quarterly*, *51*(3), 234–243.

McCleod, S. H., & Soven, M. (1992). *Writing Across the Curriculum: A Guide to Developing Programs.* (S. H. McCleod & M. Soven, Eds.). Newbury Park CA: Sage Publishing.

N.a. (n.d.). Skoltech Institute of Science and Technology. Retrieved from http://www.skoltech.ru

N.a. (2014). MIT Communication Requirement, http://web.mit.edu/commreq/

Peirce, B. N. (1995). Social Identity, Investment and Language Learning. *TESOL Quarterly*, *29*(1), 9–31.

Poe, M., Lerner, N., & Craig, J. (2010). *Learning to Communicate in Science and Engineering: Case Studies from MIT.* Cambridge, MA: MIT Press.

Russell, D. (1991). *Writing in the Academic Disciplines: A Curricular History* (1st ed.). Carbondale IL: Southern Illinois University Press.

Schell, J. (2012). Student-Centered University Learning: Turning Traditional Education Models Upside Down. *ReVista: Harvard Review of Latin America*, *Fall*, 20–23.

WAC Clearinghouse. (n.d.). Retrieved from http://wac.colostate.edu/

Zawacki, T., & Rogers, P. M. (2011). *Writing Across the Curriculum: A Critical Sourcebook.* (T. Zawacki & P. M. Rogers, Eds.). Boston MA: Bedford St Martin's.

BIOGRAPHICAL INFORMATION:

Jennifer Craig, M.S., M.A, is a Lecturer II in the Department of Comparative Media Studies/Writing at the Massachusetts Institute of Technology. She teaches in the Department of Aeronautics and Astronautics. She is the author of *Integrating Writing Strategies in EFL/ESL University Contexts: A Writing-across-the-Curriculum Approach* (2012*)* and co-author of *Learning to Communicate in Science and Engineering: Case Studies from MIT* (2010*)*, Her work has been included in collections about teaching and tutoring English language learners. She has taught at and consulted with universities in Chile, Brazil, Russia, Singapore, Costa Rica, and Mexico.

ACKNOWLEDGEMENTS: At Skoltech in Cambridge, MA, I appreciate the support of Brian Anthony, Diane Soderholm, Jason Torres, Lissa Natkin, Gerri Powers, and Samantha Young. Special thanks to Bill Lucas and Yashu Kaufman for their help with data collection. At Skoltech in Moscow, I appreciate the support of Mats Hanson, Bram Caplan, Ekaterina Malyuzhinets, Olga Tenisheva, Madina Tyuryaeva, and Vitaly Buldakov.

APPENDIX

Table 3: Pre/post averages for the oral communication module during the Application Period, Skoltech University, Fall, 2013

How confident are you in your current ability to do each of these tasks without any further preparation?	**Pre-test average** **n=43**	**Post-test average** **n=45**
Identify my objectives in speaking about my work?	8.23	8.59
Understand the objectives and expectations of the audience in listening?	7.67	8.58
Know what oral communication strategy is best suited to that context and audience?	6.44	8.50
Understand the appropriate level of formality expected in an oral communication event?	7.14	8.74
Choose the appropriate combination of media for a specific oral communication event?	6.40	8.56
Select the appropriate style for the oral communication?	6.49	8.74
Select and organize the right content for a specific oral communication event?	7.19	8.52
Prepare and participate in an interactive briefing?	6.14	8.61
Prepare and practice formal oral communication?	6.64	8.45
Use appropriate delivery style (stance, gestures, eye contact, voice quality) in oral communication?	6.55	8.27
Demonstrate answering questions effectively during and after an oral communication event?	7.00	8.33
Work as part of a team in preparing and delivering oral communication?	7.19	8.73

Table 4. Paired sample statistics in oral communication survey

		Mean	N	Std. Deviation	Std. Error Mean
Pair 1	Identify my objectives in speaking about my work?	69.1%	33	22.131	3.852
		76.1%	33	18.865	3.284
Pair 2	Understand the objectives and	65.6%	32	20.936	3.701

	expectations of the audience in listening?	75.9%	32	18.984	3.356
Pair 3	Know what oral communication strategy is best suited to that context and audience?	53.3%	33	23.274	4.051
		75.2%	33	16.978	2.956
Pair 4	Understand the appropriate level of formality expected in an oral communication event?	58.2%	33	20.533	3.574
		77.6%	33	18.376	3.199
Pair 5	Choose the appropriate combination of media for a specific oral communication event?	50.9%	33	26.500	4.613
		75.8%	33	21.799	3.795
Pair 6	Select the appropriate style for the oral communication?	53.0%	33	23.517	4.094
		77.6%	33	19.208	3.344
Pair 7	Select and organize the right content for a specific oral communication event?	59.4%	32	20.625	3.646
		74.4%	32	17.586	3.109
Pair 8	Prepare and participate in an interactive briefing?	55.0%	32	23.962	4.236
		75.3%	32	18.489	3.268
Pair 9	Prepare and practice formal oral communication?	57.1%	32	19.344	3.420
		74.1%	32	21.532	3.806
Pair 10	Use appropriate delivery style (stance, gestures, eye contact, voice quality) in oral communication?	53.!%	32	23.086	4.081
		73.1%	32	20.703	3.660
Pair 11	Demonstrate answering questions effectively during and after an oral communication event?	57.0%	32	21.317	3.768
		73.1%	32	23.478	4.150
Pair 12	Work as part of a team in preparing and delivering oral communication?	62.0%	32	24.814	4.387
		76.0%	32	21.013	3.715

Table 5. Paired differences in pre/post test surveys in oral communication survey

Paired Differences		Difference	Sig 2-tail	Std. Error Mean	t	df	Sig. 2-tail
Pair 1	Identify my objectives in speaking about my work?	6.9%	24.249	4.253	-1.639	32	ns
Pair 2	Understand the objectives and expectations of the audience in listening?	10.1%	25.334	4.479	-2.303	31	.05
Pair 3	Know what oral communication strategy is best	22.0%	28.445	-4.406	4.051	32	.001

	suited to that context and audience?						
Pair 4	Understand the appropriate level of formality expected in an oral communication event?	19.4%	23.041	4.011	-4.835	32	.001
Pair 5	Choose the appropriate combination of media for a specific oral communication event?	25.0%	27.964	4.868	-5.104	32	.001
Pair 6	Select the appropriate style for the oral communication?	24.6%	28.514	4.964	-4.945	32	.001
Pair 7	Select and organize the right content for a specific oral communication event?	15.0%	24.626	5.353	-3.446	31	.005
Pair 8	Prepare and participate in an interactive briefing?	20.3%	27.880	4.929	-4.121	31	.001
Pair 9	Prepare and practice formal oral communication?	16.5%	29.139	5.151	-3.215	31	.001
Pair 10	Use appropriate delivery style (stance, gestures, eye contact, voice quality) in oral communication?	19.7%	29.236	5.168	-3.809	31	.001
Pair 11	Demonstrate answering questions effectively during and after an oral communication event?	16.3%	25.495	4.507	-3.606	31	.001
Pair 12	Work as part of a team in preparing and delivering oral communication?	15.0%	28.398	5.010	-2.988	31	.004

ASEE 2016
INTERNATIONAL FORUM
NEW ORLEANS, LOUISIANA / JUNE 25, 2016
Paper ID #17502

Experience of teaching an introductory Biomedical Engineering Course for Undergraduate ECE Students in Pakistan

Mr. Ijlal Haider, The University of Lahore

Ijlal Haider is currently employed at The University of Lahore, Pakistan in Department of Electrical Engineering. He is actively working on Cardiac and Neural Signal Analysis and Interdisciplinary Engineering Education.

Mr. Farhan Ahmad, UOL

I am an engineer and fond of doing research in my field as well as bio medical field.

Mrs. Nishwa Fayyaz, Virtual University, Pakistan

Nishwa Fayyaz is a graduate student in Bioinformatics at Virtual University, Pakistan. She is also involved in research at Medical Informatics, Control & Signals (MEDICS) Research Lab and Science, Technology & Engineering Education Research (STEER) Network at The University of Lahore as part time research staff.

Experience of teaching an introductory Biomedical Engineering Course for Undergraduate ECE Students in Pakistan

1 Introduction

In early 20th century, Electrical and Mechanical Engineering fields were well established and Medical Sciences were experiencing cutting edge research. However, as research in Medical Sciences progressed, it came across rehabilitating body structures and diagnosing and treating through control and communication signals inside the body. Right at this moment, it started to knock doors of Engineering for the expertise they had in dealing with structures and signals. Since Medical Sciences required assistance from various domains of Engineering, researchers felt need for a new specialized field aiming at bridging the gap between different disciplines. Later, with the advent of modern computing techniques, new avenues of research were opened in Biomedical Engineering which is a merger of traditional engineering and computing disciplines, life science and medicine. At this moment Biomedical Engineering is dealing with as diverse fields as Medical Tribology and Bioinformatics.

This field, however, emerged very late in developing countries. For example in Pakistan first Biomedical Engineering program was offered in early 1990's. The major cause was lack of resources and awareness which eventually resulted in absence of defined career path neither in academia nor in industry. The situation is changing now with a positive rate. However, there still is only a small fraction of students opting for this field in their undergraduate studies. This has posed a challenge to academia to introduce students of different background to this multidisciplinary field. A new introductory course for undergraduate ECE students was needed especially. Several courses are offered at various institutes for non-biomedical students; however a single course covering the breadth of this field without going into depth yet giving essential knowledge to the students was still sought.

In this paper, we have presented our experience of designing and teaching a new introductory course for undergraduate students of Department of Electrical Engineering at The University of Lahore, Pakistan. First, objectives and design basis of the course will be discussed. This will be followed by course details, syllabus, learning and assessment methodology. A survey was conducted among students who enrolled this course from Fall 2012 to Fall 2014 to get their feedback on usefulness of this course. Results of the survey are discussed at the end.

2 Objectives and basis of design of the course

Table 1 Prominent areas of Biomedical Engineering pertaining to ECE

Area of Study	Sample Topics
Biomedical Instrumentation	Signal acquisition (EMG, ECG), Measurements
Biomedical Signal Processing/ Analysis	Automated diagnosis of diseases from the signals
Medical Physics	Fundamentals of MRI
Medical Imaging	Medical image acquisition and automated analysis
Bioinformatics	Decoding the molecular chains and patterns
Medical Information Systems/Tele-health	Remote patient monitoring

The main objective of this course was to acquaint students of Electrical Engineering background with Biomedical Engineering. Biomedical Engineering can be subdivided into various areas pertaining to Electrical & Computer Engineering as shown in Table 1.The course was expected to introduce briefly about these areas and strong emphasis on the area of research related to their particular background.

An important factor was the consideration that a graduate with interest in Biomedical Engineering can go for either academia or in industry as both of the areas are well groomed now. The design and development of Biomedical Systems in both directions requires considering several design aspects along with concentrated technical knowledge. To explain this we give several examples to students such as pacemaker batteries cannot be recharged/ replaced every year due to surgical complexity associated with it so it needs to be energy efficient. Figure 2 shows importance of some of the design considerations in sample implantable biomedical systems.

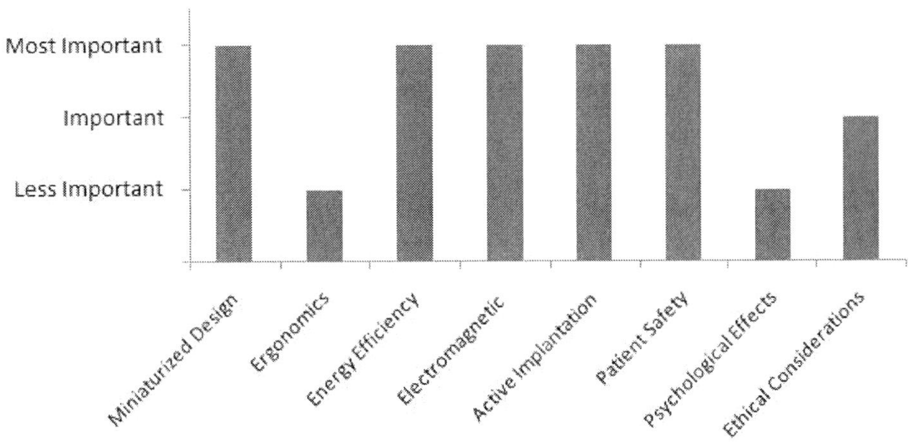

Figure 1. Importance of various design aspects in implantable Medical Engineering design

The course covers several topics related to three focal areas Introduction to Biomedical Engineering, its Application and Designing as per following proportion:
I: Introduction (15%)
II: Application (50%)
o Diagnostic:
o Therapy and Rehabilitation:
III: Design (35%)

3 Review of literature on the course design

Since teaching of Biomedical Engineering is quite new topic and can be offered at different levels of education, a wide spectrum of topics can be studied under this area. Several authors have presented their ideas and experience in different research articles.

Gibson et al. has developed a teaching pack of Medical Physics for UK secondary schools and discussed advantages of teaching Medical Physics[1]. Robinson[2] has presented their experience of offering an introductory Biomedical Engineering course to students from various backgrounds.

This course was offered as Science, Technology & Society (STS) elective for engineering and non-engineering students. They examined the technological basis of innovation, ethics and economics of medical technology. Ramon has discussed teaching medical instrumentation using MATLAB. They cover several topics including EEG, ECG, Biosensors and Electrodes, Blood flow and MRI[3].Toft-Nielsen et al. has explained the importance of teaching virtual instrumentation to students of Biomedical Engineering. They have designed a one credit course "LabVIEW Application in Biomedical Engineering" in which students are taught skills in virtual instrumentation for developing biomedical device prototypes[4].

Multidisciplinary studies require students to take more courses then other traditional courses. This has caused to reduce load of laboratory modules. Beebe has proposed teaching hands on biomedical engineering to the students of biomedical engineering instead of traditional teaching. This will give students an insight into lab work without performing any lab experiments[5]. Ton-Tai Pan has presented design of a myoelectric controlled prosthesis to reinforce mechatronic education offered to Electrical Engineering and Bio-industrial Mechatronic Engineering students at KWIT and NTU in Taiwan[6]. Sodager has presented his experience of a new course on Implantable Biomedical Microsystems offered to graduate students[7].

LaPlaca et al. proposed and investigated the use of Problem Base Learning for teaching biomedical engineering. PBL is well suited to educating students of multidisciplinary fields such as Biomedical Engineering. They found it effective in graduate programs and based on this they were embarking on introducing PBL for undergraduate courses[8].
Several other articles were used to conduct study to complete this research such as Rashid's article describing importance of teaching Memristors[9]. Fielder has discussed fundamental issues in designing and teaching of engineering curriculum[10].

4 Course Information

This course is offered with code and title ECE4552: Medical Electronics in Electrical Engineering Department. The course has 48 hours of instruction spanning over 16 weeks. This is a senior undergraduate course open to senior undergraduate students and masters students as an elective course. Prerequisite courses are Electronic Devices & Circuits, Signal Processing, Instrumentation and High School Biology (one week remedial lectures are planned to overcome this deficiency, in case).

4.1 Course Description

The short introductory description provided to students is as follows:
"*This course provides an introduction to the emerging field of Medical Electronic i.e. application of Electronic System Designing for Medical Applications. These application ranges from diagnosis such as electrophysiological measurement and medical imaging to treatment & rehabilitation techniques such as FES and Prosthesis. This course involves very few concepts in Clinical Diagnosis procedures and Medical Instrumentation. Major emphasis is given to the design of electronic systems for medical application which involves designing for resilience to noise and interference amplifier and filter, electromagnetic compatibility, energy efficiency. Role*

of signal processing and pattern recognition are highlighted in this course for applications in Cardiac and Neural signals as well.

The course will start with Introduction to Medical Electronics. This will be followed by its application in Diagnosis, Treatment & Rehabilitation and appropriate discussion of Medical Electronic design fundamentals."

4.2 Study Material

Ample literature is available on this field in the form of Books, Magazines and research articles. Major part of introduction was covered from Medical Physics by J.R. Cameron. For other advanced topics articles from various books, magazines and application notes were referred.

4.3 Course Syllabus

As the course is designed to cover a wide range of topics, we have divided the contents in several modules. Most of the modules are covered in 6 hours spanned over two weeks. Medical Imaging is covered in approximately 9 hours. Major topics covered in the course are shown in Table 2.

Table 2. Major topics of the course and time allocation

Topic(s)	Time given
Scope of Medical Electronics, Diagnosis, Therapy and Rehabilitation, Introductory concepts on basic body function, Bigger Picture	3 hours
Neuromuscular system, Action Potentials, Nerve Conduction Velocity	6 hours
Electrophysiological measurements EMG, EEG, Intro non-traditional measurements EOG, EGaG	6 hours
Neural Stimulation, Evoked Potentials, Prosthesis and Exoskeletons	6 hours
Cardiovascular system, ECG, Parameter Detection	6 hours
Pacemakers, Defibrillators	6 hours
Blood Pressure and Blood Flow Measurements, Plethysmography	6 hours
Medical Imaging (image acquisition & basic processing techniques)	9 hours

4.4 Learning Methodologies

Several other topics were covered by giving short articles to students in their in-class assessments. These articles included Memristors for memory building circuits, digestible telehealth monitoring systems, cognition and perception etc. Two research assignments were given to students in each half of the course. In first research assignment, students were expected to read a given research article, understand it and present it in the class while in other, they had to prepare a short research proposal on a topic from a list of topics given and submit it to the instructor.

4.5 Course Assessments

The assessment was divided throughout the semester in form of Exams (one semester mid and one semester final), in-class tests, assignments and long research assignments (presentation for one and report for other). The revised Bloom Taxonomy was evaluated to design assessment and teaching[11]. Sessional assessments were designed to be open book to ease the students in remembering the medicine terminology used in the course.

5 Student Survey

The effectiveness of the proposed course structure has been evaluated by feedback of students who attended the course. Quantitative analysis was done in this research by conducting the survey. Students from Fall 2012 to Winter 2015 were asked about the importance of medical electronics course and their experience.
The survey questionnaire was constructed on the same pattern as used by Sodagar[7]. The questionnaire was distributed to the students.

5.1 Survey Results

The questionnaire was distributed online and in print. Survey was participated by approximately 40% of the students. Results of the survey are as follows:

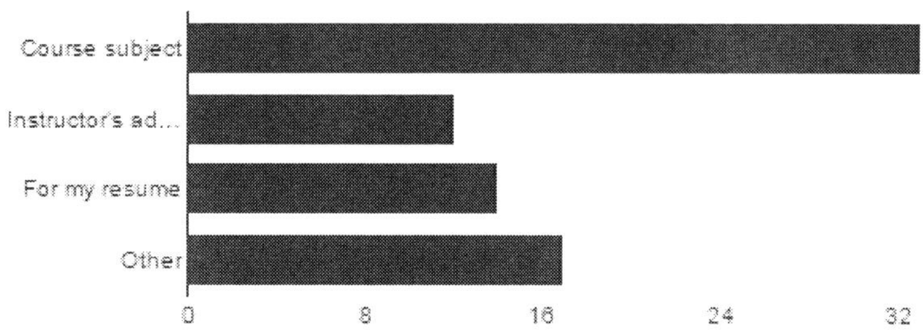

Course subject	33	49.3%
Instructor's advice	12	17.9%
For my resume	14	20.9%
Other	17	25.4%

Figure 2. Why did you take the Medical Electronics course?

Approximately 60% of the students took this to use Engineering for help.

Engineering design	19	28.4%
Engineering for helping patients and people with disabilities	42	62.7%
Knowledge of Health Sciences	29	43.3%
Other	2	3%

Figure 3. What did you find the most interesting about Medical Electronics course?

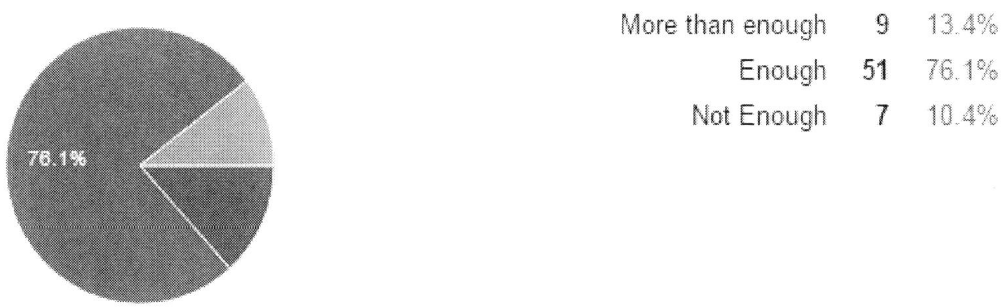

Figure 4. Did the course provide you enough material?

Figure 5. Was this the first time you have taken a multi-disciplinary course?

There were some comments such as “I'm pretty much satisfied with my Job as a starter, because this field is still to be explored in our country, there is huge space for the research and development in this field. I strongly recommend that rather than offering just a single course there must be a complete degree in Bio Medical Engineering/Technology in BS and MS as well.” Whereas majority of the students emphasized to include practical sessions and lab work in the course.

6 Conclusion

It was a pleasant experience overall to see positive response from the students. Despite very static academic environment in developing countries, students take interest in multidisciplinary courses designed for direct application in society. The course material and teaching strategy was revised continuously based on student’s feedback. A one credit practical work will be added in the course in future.

REFERENCES

[1] A. Gibson, E. Cook, A. Newing, “Teaching Medical Physics”, Physics Education, vol. 41, no. 4, pp. 301-306, 2006.

[2] C. J. Robinson, “First course on biomedical engineering for desperate students”, ASEE Annual Conference, Vancouver, Canada, 2011.

[3] C. Ramon, “Teaching Medical Instrumentation at the University of Washington”, MATLAB Digest, 2007.

[4] J. Nielsen, J. Boharquez, Q. Shen, “Teaching Virtual Instrumentation to Biomedical Engineering Students”, ASEE Annual Conference, StarkVille, MS, USA, 2012.

[5] D. J. Beebe, "Teaching Hands On Biomedical Instrumentation", ASEE Annual Conference, Washington, DC, 1996.
[6] T. Pan, P. Fan, H. Chiang, R. Chang and J. Jiang, "Mechatronic Experiments Course Design: A Myoelectric Controlled Partial-Hand Prosthesis Project", IEEE Transaction on Education, vol. 47, no. 3, pp. 348-355, 2004.
[7] A. Sodagar, "Implantable Biomedical Microsystems: A New Graduate Course in Biomedical Circuits and Systems", IEEE Transaction on Education, vol. 57, no. 1, pp. 48-53, 2014.
[8] M. LaPlaca, W. Newstetter, A. Yoganathan, "Problem-based learning in biomedical engineering curricula", ASEE Annual Conference, Albuquerque, NM, 2001.
[9] M. Rashid, "Electronics - what should we teach and why?" ICECE, Dhaka, 2010, pp. 66-69.
[10] R. Felder and R. Brent, "The ABCs of engineering education: ABET, Bloom's taxonomy, cooperative learning, and so on", ASEE Annual Conference, Salt Lake City, UT, 2004.
[11] S. Goel and N. Sharda, "What do engineers want? Examining engineering education through Bloom's Taxonomy," 15th Annual Conference for AAEE, Sep 2004, Queensland, Australia

Using Service Oriented Remote Laboratories in Engineering Courses

Prof. Hamadou Saliah-Hassane ing., TELUQ -University of Quebec

Professor Saliah-Hassane is a senior researcher at L@d (Laboratory at Distance | TELUQ - University of Quebec), and member of the Ordre des ingenieurs du Quebec (OIQ); of the IEEE Education Society, Communication Society and Computer Society); He teaches informatics and computer networks at TELUQ l of University of Quebec. Professor Saliah-Hassane has a PhD in Computer Aided Analysis and Design from the Electrical and Computer Engineering at McGill University in Montreal and a Master of Applied Science degree from Ecole Polytechnique de Montreal. Professor Saliah-Hassane research interest is, among other, Online Engineering Education and Distibuted Systems.

Dr. Mamane Moustapha Dodo Amadou ing., École de technologie Supérieure

Mamane Moustapha, Dodo Amadou received the M.S. degree from the Kharkiv Polytechnic Institute, Ukraine, in 1994, and the Ph.D. degree from the same institute in 2001. Since 2003, he is a lecturer for undergraduate students in the department of Electrical Engineering of École de technologie Supérieure, Quebec, Canada. His research interests are in power systems modeling, simulation, and control, as well as in virtual and remote laboratories.

Dr. Maarouf Saad, Ecole de Technologie Superieure

Maarouf Saad received a bachelor and a master degrees in electrical engineering from Ecole Polytechnique of Montreal respectively in 1982 and 1984. In 1988, he received a Ph.D. from McGill University in electrical engineering. He joined Ecole de Technologie Supérieure in 1987 where he is teaching control theory and robotics courses. His research is mainly in nonlinear control and optimization applied to robotics and flight control system.

Dr. Willie K. Ofosu, Pennsylvania State University, Wilkes-Barre Campus

Dr. Willie K. Ofosu is an Associate Professor of Electrical Engineering Technology at Penn State, Wilkes-Barre, where he teaches telecommunications, wireless systems, networking, optoelectronics and analog and digital electronics. He is a member of IEEE, IET (England), and a Chartered Engineer (CEng) of England. His research interests include RF components and antennas, and Powerline Communication. He is an advocate of diversity in the educational environment. Dr. Ofosu received his Ph.D. from the Electronic Systems Engineering Department at University of Essex in England.

Using Service Oriented Remote Laboratories in Engineering Courses

Abstract

This paper suggests a new approach to perform the laboratories sessions in engineering courses using remote laboratory. Our approach is that for each laboratory session, there are sets of user interfaces designed as services available for the learners. Each service allows applying some inputs (if any) to the physical equipment and returns some selected output signals from the equipment. These outputs signals can be measured and/or saved. In order to obtain the required results formulated by the teacher in his/her learning scenario, the remote learner has to select the services that match the best, and be able to perform all necessary calculations and analysis We refer to this approach as constructional, since, to achieve a given task, the user has to search and choose the appropriate user interfaces designed as services among many others available. The teacher is also able to assess whether the learner has understood the theoretical concepts previously studied in class. The proposed approach is used in a Control System Laboratory session using a DC motor remotely controlled and in Electric circuit analysis course.

Keywords

Online Laboratory, Remote Laboratory, Control System, Service Oriented, Constructional-Approach.

Introduction

During their studies in engineering courses, students carry out simulations and/or real experiments in order to understand or verify some theoretical concepts. The simulations with software tools have a unique benefit in that they can be used by an unlimited number of the students without geographical or temporal limitations. Online laboratories based on simulations are called Virtual Laboratories [1-5]. On the other hand, hands-on/physical experiments require equipment. With the traditional approach of performing experiment directly with equipment, the need for equipment grows proportionally to the increase of the number of students, often resulting in considerable expenses. Following hands-on approach, the only way for universities to share the same equipment is if the students of a other universities are moving physically to a given university to perform their laboratory sessions. We all now know that one way to overcome this drawback is the use of remote laboratories [6-14]. Remote laboratories allow students to access and manipulate real equipment located anywhere in the world through the Internet via a dedicated interfaces, thanks to the development of Information and Communication Technologies. In most proposed remote laboratory work, the users only remotely interact with the real device to change some parameters, perform some calculations and analyze the obtained results.

The main contribution of this paper is to propose an approach which allows the students not only to change some parameters of the dedicated interface of the remote laboratory, but also apply critical thinking to reconfigure the system to optimize the performance using the available set of service oriented resources for each experiment.

The paper consists of the following sections. Section 2 gives the description of the proposed approach. Section 3 illustrates the application of the proposed approach using a DC Motor control experiment, and circuit analysis, and finally conclusions are drawn in section 4.

Proposed approach

With the traditional approach, generally the experiment session results in systematically performing predetermined steps and then analyzing the obtained results. The proposed approach is different, allowing students to select by themselves the appropriate services for a given experiment in order to perform the required tasks set in the teacher's learning goals. The flowchart of the proposed approach is shown in Figure 1.

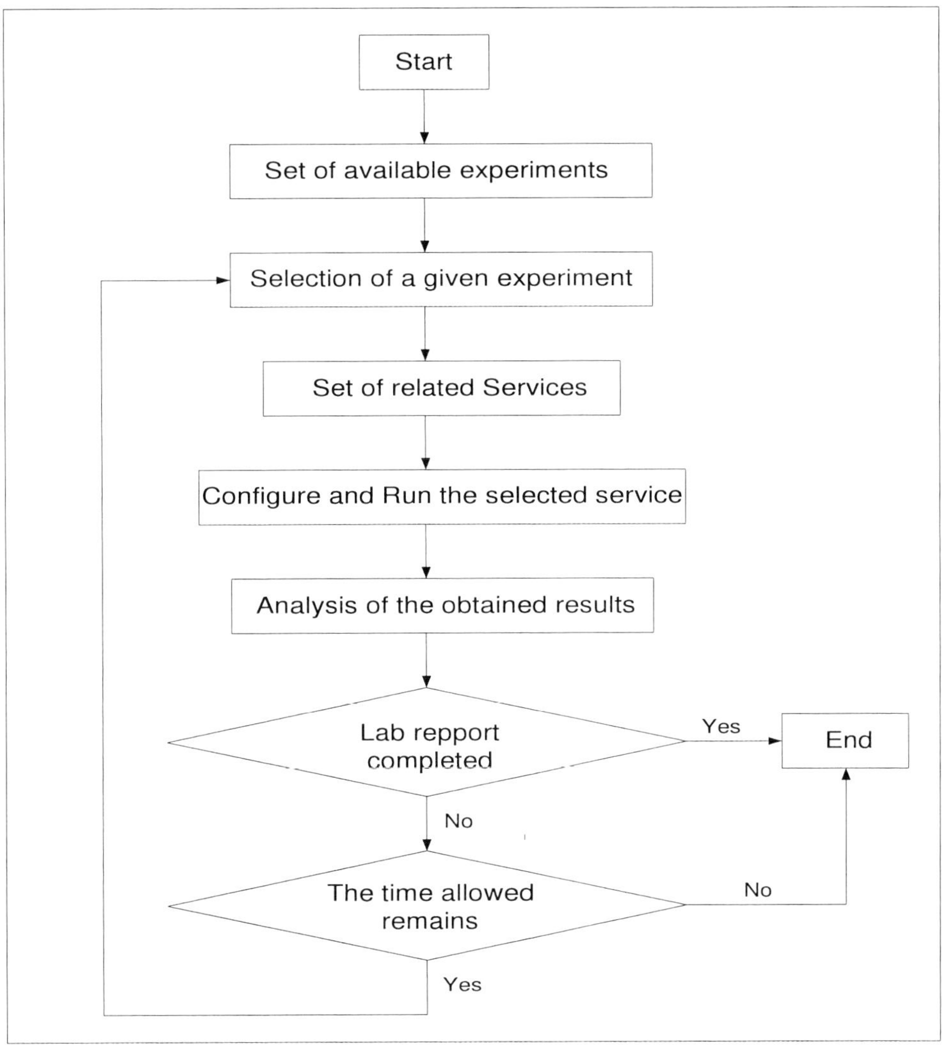

Fig. 1. Flow chart of the proposed approach

As one can see from Figure 1, there are the following main steps:

1) **Set of available experiments:** All currently available experiments are displayed.
2) **Selection of a given experiment**: At this step the user selects a given experiment among the available experiments. Usually this selection is imposed by the teacher.
3) **Services related to the selected experiment:** For each experiment a set of services is available.
4) **Selection of a given service**: Ordering selection of some or all services to be executed will allow achievement of all goals fixed by the teacher. Before executing the next service, it is sometimes necessary to calculate some parameters from the results obtained from the previous services.
5) **Configure and run the selected service:** Only one service can be executed at a time. Generally, in this case the service configuration is limited to choosing appropriate inputs values in the interface of the service. After the execution of the service the related outputs parameters will be available. If necessary the service can be executed again and again.
6) **Analysis of the obtained results:** Before putting the results in the lab report, it is sometimes necessary to analyze the adequateness of the obtained results.

We used LabVIEW™ 2013 Service Oriented feature to implement all the programs controlling the real processes.

The generic interface of each service consists of:

- The remote panel that is the front panel of the VI (a LabVIEW Software component) viewed in a web page [15]. The remote panel has the same functionality as the front panel.
- A live video to watch the real process when the service is running. This is to imitate the traditional experiment session. For example for the DC Motor Control experiment which will be used as case study in section 3 of this paper the user can watch the rotation of the motor during the execution of the service.
- A link to go back to all services of a given experiment. This link is useful, since generally to achieve required goals of an experiment session it necessary to use several services.

Experimenting with the proposed approach

The proposed approach is applied to an experiment called "DC Motor Control Experiment". The setup used for this experiment is a QNET™ DC Motor Control Trainer acquired from Quanser™ shown in figure 2.

Figure 2. QNET™ DC Motor Control Trainer.

This type of the experiment is proposed to the undergraduate students in control systems course in an engineering school in Montreal, Quebec, Canada.

For this experiment the following services are available:

Service 1: Open Loop Response. With this service the user first selects the voltage to apply to the motor. The amplitude of the applied voltage is limited to a given interval. After selecting the applied voltage the user pushes the Start bouton and the motor's speed variation is displayed on the graph. The motor runs for 2 seconds and stops.

Service 2: Closed Loop Response with P Controller. With this service the user first selects the desired speed reference, calculates the gain value (Gp) of the proportional controller depending ~~of~~ on the desired performance and pushes the *Start* bouton. The motor's speed variation will be displayed on the graph. The motor runs for 3 seconds and stops.

Service 3: Closed Loop Response with PI Controller. This service is the same as Service 2 but the user needs to calculate and then set the gains Gp and GI of the proportional integral controller.

Service 4: Closed Loop Response with PID Controller. This service is the same as Service 2 but the user needs to calculate and then set the gains Gp, GI, Gd of the proportional integral derivative controller.

To understand how these services are used, assume that the teacher provides the following learning scenario.

Learning scenario and Lab Report (From the teacher)

Use the DC Motor Control Experiment to determine the value of the gain Gp that allows reaching a steady state error (ess) equal to 15% when the reference speed is a step signal of amplitude 150 rad/s. The setup has the following block diagrams:

The block diagram of the DC Motor in open loop is:

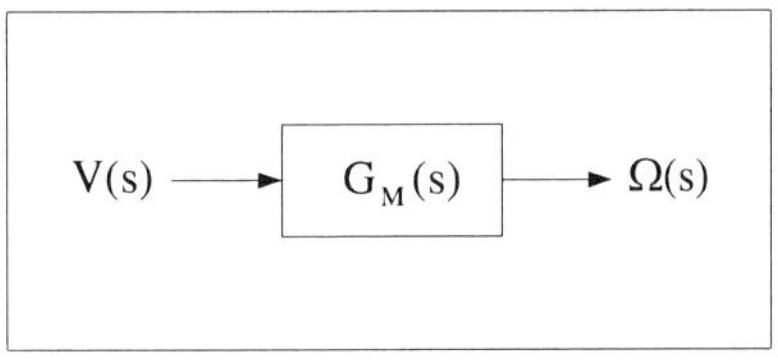

Figure 3. Block diagram of the DC Motor in open loop.

V(s) and Ω(s) are the Laplace transforms of the applied voltage to the motor and speed respectively, and GM(s) is the transfer function of the motor.

The block diagram of the DC Motor in closed loop is:

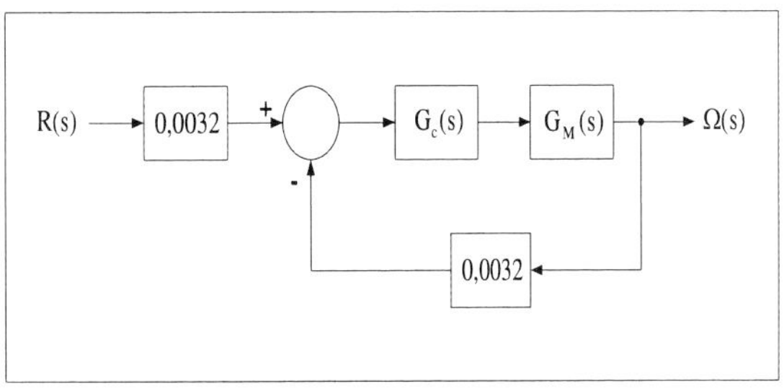

Figure 4. Block diagram of the DC Motor in closed loop.

R(s) and Ω(s) are the Laplace transform of the reference speed (desired speed) or and actual speed respectively; $G_c(s)$ and $G_M(s)$ are the transfer functions of the used controller and motor respectively.

Indicate in the lab report all services and computation procedures used to achieve the set goal.

Provided Lab Report (From the student)

Determination of Gp

The steady-state error e_{ss} for a step input is determined by [16]:

$$e_{ss} = \frac{1}{1+K_p} \quad (1)$$

Where K_p is the position constant error.

According to the requirement we need e_{ss} = 15% (0.15). From equation (1) the required value of K_p is 5,666.

Now determine the expression of ~~the~~ K_p from the block diagram of the system in the closed loop. The controller is a proportional one with gain G_p, $G_c(s) = G_p$. In this case the block diagram of Figure 4 can be reduced to the following block diagram with unity feedback:

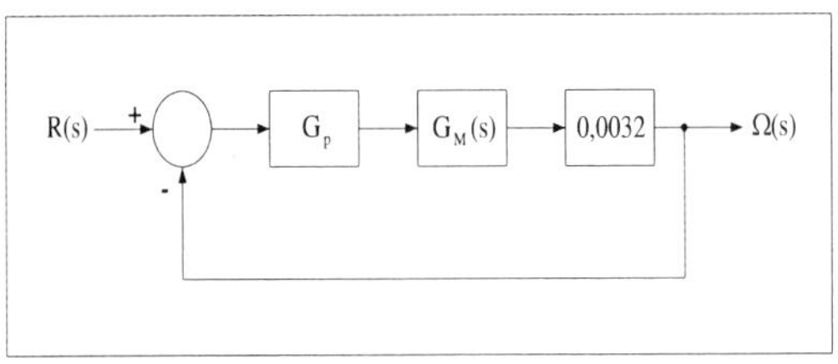

Figure 5. Block diagram of the DC Motor with unity feedback

From Figure 5:

$$K_p = 0{,}0032 G_p \lim_{s\to 0} G_M(s) \quad (2)$$

For the sake of simplicity let:

$$K_{ss} = \lim_{s \to 0} G_M(s) \quad (3)$$

Taking into account (3) and the required value of K_p:

$$G_p = \frac{5,666}{0,0032K_{ss}} \quad (4)$$

K_{ss} is called a steady state gain of the system. It is defined as the ratio of the output (speed in this case) value in steady state to the input value (applied voltage to the motor in this case). The procedure to determine K_{ss} is illustrated in the subsection below.

Determination of K_{ss}

Based on the definition of K_{ss}, the Open Loop Response service is the only service which allows reaching this gain. The screenshot after configuring (Setting the applied voltage in this case to 4,5 Volts) and running the service is shown in Figure 6.

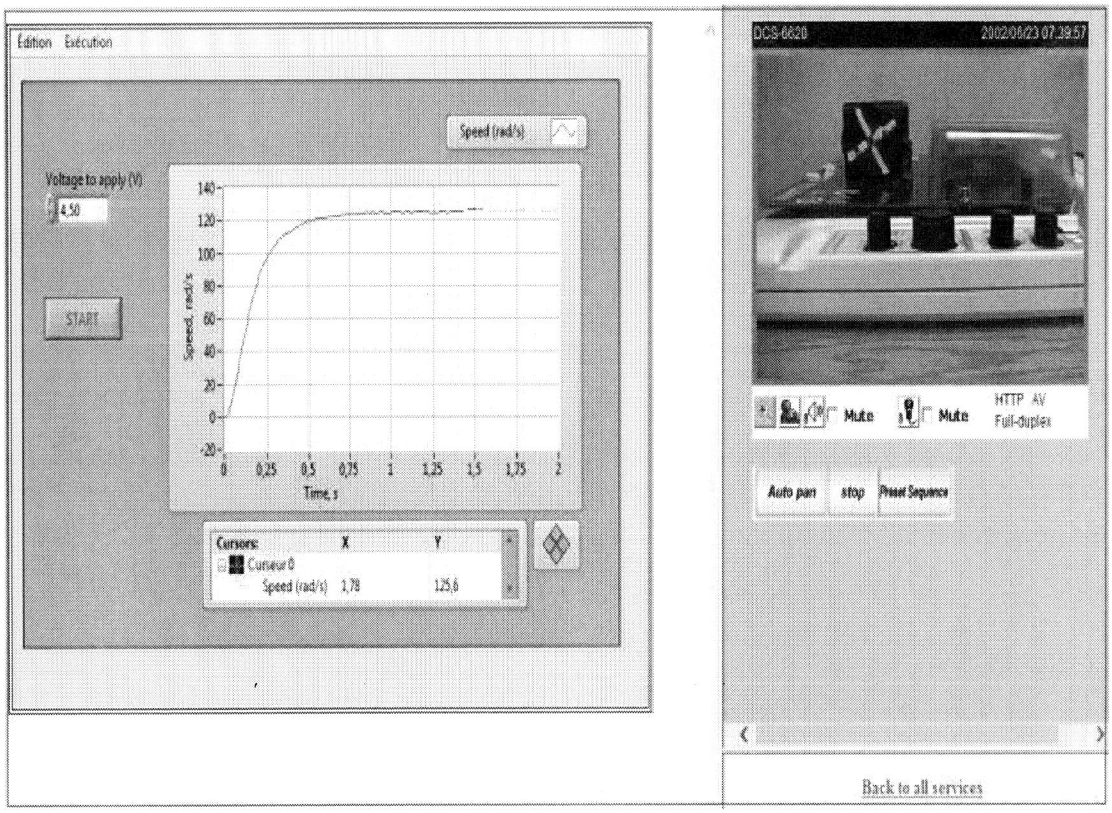

Figure 6. Interface of the Open Loop Response Service

The steady state gain is determined by the following:

$$K_{ss} = \frac{\omega_{ss}}{V} \qquad (5)$$

Where ω_{ss} is the value of the motor's speed in steady state and V is the amplitude of the applied voltage to motor. From figure 6, placing the cursor 0 in steady state, we can read that ω_{ss} = 125,6 rad/s. Taking into account that the amplitude of the applied voltage is 4,5 V (see figure 6), we have: K_{SS} = 27,911 rad/s.V.

Validation of the obtained result

Taking into account that K_{SS} = 27,911, $G_p = 63{,}438$.

In order to validate the determined value of G_p we configure (Setting the reference speed to 150 rad/s and Gp = 63,438) and run the service Closed loop with P Controller. The screenshot of the service interface is shown in figure 7.

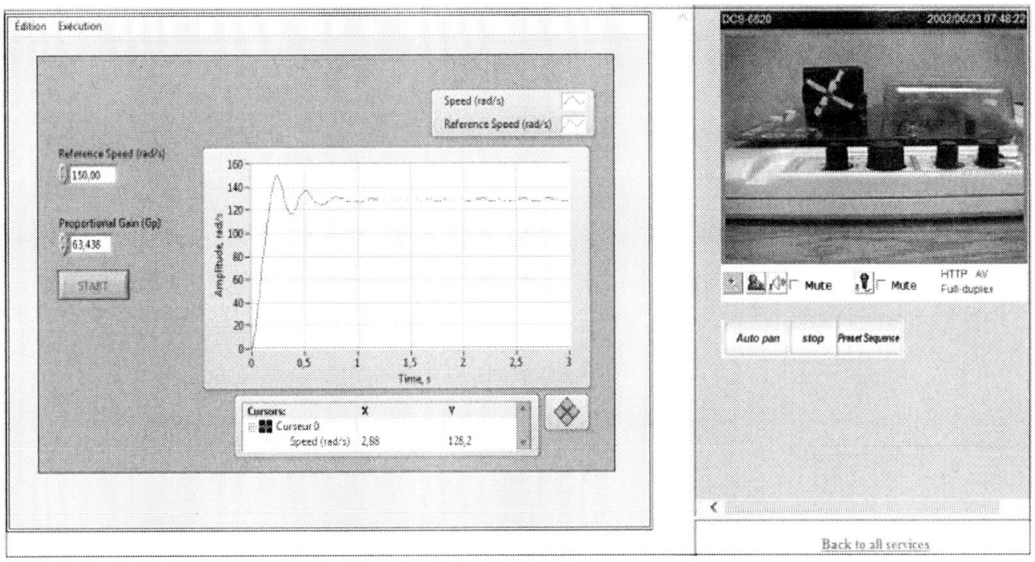

Figure 7. Interface of the closed loop with P Controller service.

From Figure 7, placing the cursor 0 in steady state, we can read that ω_{ss} = 128,2 rad/s.

The real steady state error obtained is:

$$e_{ss}^{r} = \frac{150-128{,}2}{150}100 = 14{,}533\%$$

Note that the obtained real steady state error (14,533%) is nearly the desired steady state error (15%)

The proposed approach is also applied to Kirchhoff's Laws experiments for undergraduate students in electrical circuit course.

For this experiment the following services are available:

Service 1: Measurements of Voltages. With this service for a given circuit the interested nodes are shown on the circuit. In order to measure voltage across given nodes, the user has to:

a) Select the appropriate nodes (number and polarity for each node);
b) Select the amplitude of the voltage source to be applied to the circuit;
c) Read the correspondent value on a numerical indicator.

Service 2: Measurements of currents. With this service, for the nodes where the circuit can be opened are shown. In order to measure a current flowing across a given branch the user has to:

a) Select the amplitude of the voltage source to be applied to the circuit;
b) Select the appropriate nodes (number and polarity for each node);
c) Read the correspondent value on a numerical indicator.

Learning scenario and Lab Report (From the teacher)

Given the circuit shown in Figure 8:

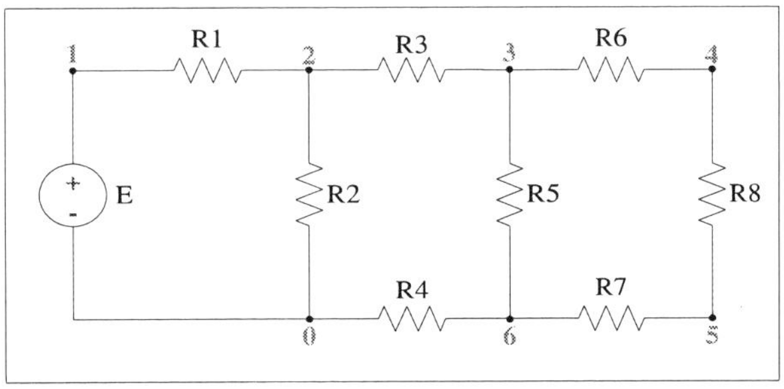

Figure 8. Analyzed circuit.

Use Kirchhoff's Laws Experiment to verify the Kirchhoff voltage law for the loop formed by resistors R2, R3, R4 and R5 given that the value of the amplitude the generator E = 10 Volts.

In your Lab report do the following:

a) Select the appropriate service(s).

The interface of the appropriate service is shown below:

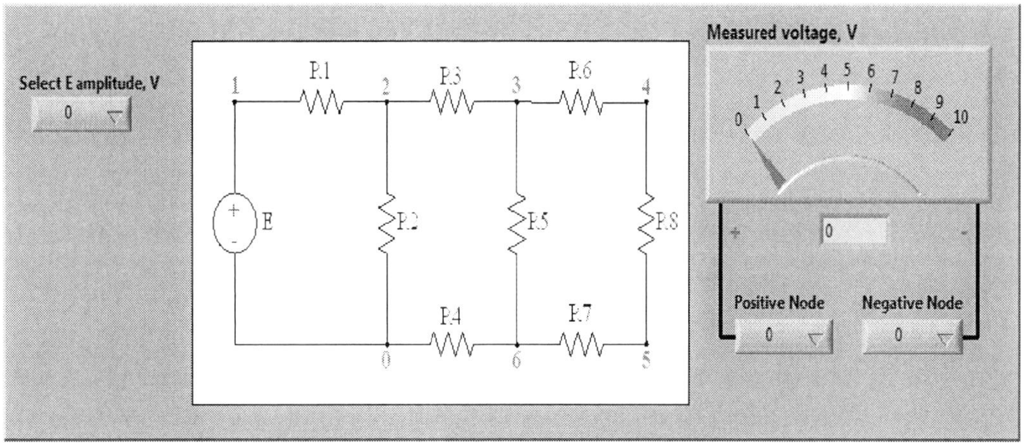

Figure 9. Interface of the Measurements of Voltages Service

b) Write the Kirchhoff's voltage law and verify this law with the measured voltages.

Expected Lab Report to be filled by the student

a) Selection of the service

Kirchhoff's voltage law is based on summing voltages around loops, so in this case the Service 1 is the appropriate service as it allows to measure voltages across all elements of the circuit.

b) Applying Kirchhoff's voltage law (KVL)
The selected signs for the nodes are shown in the table below

Node	**0**	**2**	**3**	**6**
Sign	-	+	-	+

The measured voltages are shown in the table below

Voltages	U_{R2}	U_{R3}	U_{R4}	U_{R5}
Values, V	1,501	1,03	2,007	1,503

In the table above U_{Ri} is the voltage across the resistor R_i

In this the KVL equation clockwise around the loop using passive sign convention is:

$$U_{R3} - U_{R5} + U_{R4} - U_{R2} = 0 \quad (6)$$

Taking into account the measured values of the voltages, we obtain:

1,03 - 1,503 + 2,007 -1 ,501 = 0,033.

Note that the obtained sum is close to zero.

Conclusion

With the proposed approach, the teacher can choose either to ask students to perform the predetermined steps of a given experiment or give them the opportunity to choose for themselves all the steps required in order to achieve the goals of the experiment. The latter approach is the best way to assess the student's learning process. Applying the approach we were able to find out if the learner truly understood the theoretical concepts learned in Control System course. Moreover, we found that expanding several services of each experiment allows more flexibility when many institutions remotely share the same devices. Furthermore, the teacher is then able to formulate the goals of an experiment according to the learning outcomes he/she expects.

In this paper the application of the proposed approach to experiment in Control systems course has been illustrated. We provide an example of how we are using the same approach for Electric circuit analysis. The approach can also be used in the context of some other courses. The usefulness of the proposed approach depends on the ability of the teacher to provide adequate set of services for each experiment and the detailed description of the functionality of each service. In a future work we will specifically address, in the framework of this approach, the issue of the collaborative use of a remote laboratory with not only one single service at a time but in cases where multiples services are needed to perform a laboratory learning task. We will also orient our effort toward the compliance of our approach with the ongoing IEEE-SA P1876™ Working Group achievements [17].

References

[1] M. M. Dodo Amadou, M. Saad, J. P. Kenne, and V. Nerguizian, "Virtual and Remote Laboratories," in Proc. 1st IEEE ICELIE, Dec. 2006, pp. 173–176.

[2] P. Permezel, O. Marcoux, H. H. Saliah, M. Saad and C. Nerguizian, "Generic Simulator Model for Training and Learning in Virtual Laboratory Environments," in E-Learn, World Conference on E-Learning in Corporate, Government, Healthcare and Higher Education, Montreal, Canada, October 15-19, 2002.

[3] H. Kliffi, J. Desbiens, M. Cheriet, M. Saad and H. H. Saliah, "Building a Virtual Laboratory over a Duplicated Object Systems," in 2nd International Workshop on Tele-Education in ngineering Using Virtual Laboratories, August 8-9, 2002, Sherbrooke, Canada.

[4] K.M. Lee, W. Daley, and T. McKlin, "An interactive learning tool for dynamic systems and control," in Proc. Int. Mechanical Engineering Congr. Expo., Anaheim, CA, 1998, pp. 71–76.

[5] C. Schmid, "The virtual lab VCLAB for education on the Web," in Proc. American Control Conf., Philadelphia, PA, 1998, pp. 1314–1318.

[6] E. Sancristobal Ruiz et al., "Virtual and remote industrial laboratory: Integration in learning management systems," IEEE Ind. Electron. Mag.,vol. 8, no. 4, pp. 45–58, Dec. 2014.

[7] T. de Jong, M. C. Linn, and Z. C. Zacharia, "Physical and virtual laboratories in science and engineering education," Science, vol. 340, no. 6130, pp. 305–308, 2013.

[8] Agel Valera, José Luis Diez, Marina Vallés and Pedro Albertos, "Virtual and Remote Control Laboratory Development, " IEEE Control Systems Magazine, February 2005, vol. 25, no. 1, pp. 35-39.

[9] M. Casini, D. Prattichizzo and A. Vicino, "The Automatic Control Telelab," IEEE Control Systems Magazine, June 2004, vol. 24, no. 3, pp.36-44.

[10] M. M. Dodo Amadou, H. Saliah Hassane and M. Saad, "Laboratories in Engineering Courses via Internet," in iCEER 2004 Conference, Ostrava, 2004.

[11] W. E. Dixon, D. M. Dawson, B. T. Costic, and M. S. de Queiroz, "A MATLAB-based control systems laboratory experience for undergraduate students: Toward standardization and shared resources,"IEEE Transactions on Education, vol. 45, no. 3, August 2002, pp.218-226.

[12] C. C. Ko, Ben M. Chen, Jianping Chen, Yuan Zhuang, and Kay Chen Tan, "Development of a Web –Based Laboratory for Control Experiments on Coupled Tank Apparatus, "IEEE Transactions on education, vol.44, no. 1, February 2001, pp.76-86.

[13] H. H. Saliah, M. Saad, L. Villardier, B. Assogba, C. Kedowide and T. Wong, "Resource Management Strategies for Remote Virtual Laboratory Experimentation," in Proceedings of the 2000 Frontier in Education Conference; Building on a Century of Progress in Engineering Education, Kansas City, October 18-21, 2000.

[14] D. Gillet, T. de Jong, S. Sotirou, and C. Salzmann, "Personalised learning spaces and federated online labs for STEM education at school" in Proc. IEEE Global Eng. Edu. Conf. (EDUCON), Mar. 2013, pp. 769–773.

[15] Remote Panels in LabVIEW -Distributed Application Development,

http://www.ni.com/tutorial/4791/en/

[16] Norman S. Nise "Control Systems Engineerings", Seventh Edition, Wiley.

[17] IEEE-SA P1876™ Standard for Networked Smart Learning Objects for Online Laboratories: http://standards.ieee.org/email/2012_09_cfp_P1876wg_web.html

AUTHOR INDEX

AUTHOR INDEX